International Series in Operations Research & Management Science

Volume 278

Series Editor
Camille C. Price
Department of Computer Science, Stephen F. Austin State University,
Nacogdoches, TX, USA

Associate Editor
Joe Zhu
Foisie Business School, Worcester Polytechnic Institute, Worcester, MA, USA

Founding Editor
Frederick S. Hillier
Stanford University, Stanford, CA, USA

More information about this series at http://www.springer.com/series/6161

Allen Holder • Joseph Eichholz

An Introduction to Computational Science

 Springer

Allen Holder
Department of Mathematics
Rose-Hulman Institute of Technology
Terre Haute, IN, USA

Joseph Eichholz
Department of Mathematics
Rose-Hulman Institute of Technology
Terre Haute, IN, USA

Additional material to this book can be downloaded from http://extras.springer.com.

ISSN 0884-8289 ISSN 2214-7934 (electronic)
International Series in Operations Research & Management Science
ISBN 978-3-030-15677-0 ISBN 978-3-030-15679-4 (eBook)
https://doi.org/10.1007/978-3-030-15679-4

© Springer Nature Switzerland AG 2019

This work is subject to copyright. All rights are reserved by the Publisher, whether the whole or part of the material is concerned, specifically the rights of translation, reprinting, reuse of illustrations, recitation, broadcasting, reproduction on microfilms or in any other physical way, and transmission or information storage and retrieval, electronic adaptation, computer software, or by similar or dissimilar methodology now known or hereafter developed.

The use of general descriptive names, registered names, trademarks, service marks, etc. in this publication does not imply, even in the absence of a specific statement, that such names are exempt from the relevant protective laws and regulations and therefore free for general use.

The publisher, the authors, and the editors are safe to assume that the advice and information in this book are believed to be true and accurate at the date of publication. Neither the publisher nor the authors or the editors give a warranty, express or implied, with respect to the material contained herein or for any errors or omissions that may have been made. The publisher remains neutral with regard to jurisdictional claims in published maps and institutional affiliations.

This Springer imprint is published by the registered company Springer Nature Switzerland AG.
The registered company address is: Gewerbestrasse 11, 6330 Cham, Switzerland

Dedicated to the educators who have so nurtured our lives.

Joseph Eichholz: Weimin Han
Allen Holder: Harvey J. Greenberg
and
Jeffrey L. Stuart.

Foreword

The interplay among mathematics, computation, and application in science and engineering, including modern machine learning (formerly referred to as artificial intelligence), is growing stronger all the time. As such, an education in computational science has become extremely important for students pursing educations in STEM-related fields. This book, along with undergraduate courses based on it, provides a modern and cogent introduction to this rapidly evolving subject.

Students pursuing undergraduate educations in STEM-related fields are exposed to calculus, differential equations, and linear algebra. They are also exposed to coursework in computer science in which they learn about complexity and some algorithmic theory. These courses commonly require students to implement algorithms in popular languages such as C, Java, Python, MATLAB, R, or Julia. However, many students experience unfortunate disconnects between what is learned in math classes and what is learned in computer science classes. This book, which is designed to support one or two undergraduate terms of study, very nicely fills this gap.

The book introduces a wide spectrum of important topics in mathematics, spanning data science, statistics, optimization, differential equations, and randomness. These topics are presented in a precise, dare I say rigorous, mathematical fashion. However, what makes this book special is that it intertwines mathematical precision and detail with an education in how to implement algorithms on a computer. The default mathematical language is MATLAB/Octave, but the interface to other programming languages is also discussed in some detail.

The book is filled with interesting examples that illustrate and motivate the ideas being developed. Let me mention a few that I found particularly interesting. One homework exercise asks students to access "local" temperature data from the National Oceanic and Atmospheric Association's website and to use that data as input to a regression model to see if average daily temperatures are on the rise at that location. Another exercise in the chapter on modern parallel computing requires students to generate the famous fractal called the Mandelbrot set. Yet another exercise in the material on linear regression models asks students to reformulate a nonlinear regression model that computes an estimate of the center and radius of a set of random data that lies around a circle. The reformulation is linear, which allows students to solve the problem with their own code from earlier exercises.

The computational part of the book assumes the reader has access to either MATLAB or Octave (or both). MATLAB is perhaps the most widely used scripting language for the type of scientific computing addressed in the book. Octave is freely downloadable and is very similar to MATLAB. The book includes an extensive collection of exercises, most of which involve writing code in either of these two languages. The exercises are very interesting, and a reader/student who invests the effort to solve a significant fraction of these exercises will become well educated in computational science.

Robert J. Vanderbei

Preface

The intent of this text is to provide an introduction to the growing interdisciplinary field of Computational Science. The impacts of computational studies throughout science and engineering are immense, and the future points to an ever increasing reliance on computational ability. However, while the demand for computational expertise is high, the specific topics and curricula that comprise the arena of computational science are not well defined. This lack of clarity is partially due to the fact that different scientific problems lend themselves to different numerical methods and to diverse mathematical models, making it difficult to narrow the broad variety of computational material. Our aim is to introduce the "go to" models and numerical methods that would typically be attempted before advancing to more sophisticated constructs if necessary. Our goal is not to pivot into a deep mathematical discussion on any particular topic but is instead to motivate a working savvy that practically applies to numerous problems. This means we want students to practice with numerical methods and models so that they know what to expect. Our governing pedagogy is that students should understand the rudiments of answers to: How expensive is a calculation, how trustworthy is a calculation, and how might we model a problem to apply a desired numerical method?

We mathematically justify the results of the text, but we do so without undue rigor, which is appropriate in an undergraduate introduction. This often means we assert mathematical facts and then interpret them. The authors share the melancholy of a matter-of-fact presentation with many of our mathematical colleagues, but we hope that students will continue to pursue advanced coursework to complete the mosaic of mathematical justification. Our humble approach here is to frame the future educational discourse in a way that provides a well-honed skill for those who might consider working in computational science.

The intended audience is the undergraduate who has completed her or his introductory coursework in mathematics and computer science. Our general aim is a student who has completed calculus along with either a traditional course in ordinary differential equations or linear algebra. An introductory course in a modern computer language is also assumed. We introduce topics so that any student with a firm command of calculus and programming should be able to approach the material. Most of our numerical work is completed in MATLAB, or its free counterpart Octave, as these computational platforms are standard in engineering and science. Other computational resources are introduced to broaden awareness. We also introduce parallel computing, which can then be used to supplement other concepts.

The text is written in two parts. The first introduces essential numerical methods and canvases computational elements from calculus, linear algebra, differential equations, statistics, and optimization. Part I is designed to provide a succinct, one-term inauguration into the primary routines on which a further study of computational science rests. The material is organized so that the transition to computational science from coursework in calculus, differential equations, and linear algebra is natural. Beyond the mathematical and computational content of Part I, students will gain proficiency with elemental programming constructs and visualization, which are presented in their MATLAB syntax. Part II addresses modeling, and the goal is to have students build computational models, compute solutions, and report their findings. The models purposely intersect numerous areas of science and engineering to demonstrate the pervasive role played by computational science. This part is also written to fill a one-term course that builds from the computational background of Part I. While the authors teach most of the material over two (10 week) terms, we have attempted to modularize the presentation to facilitate single-term courses that might combine elements from both parts.

Terre Haute, IN, USA   Allen Holder
Terre Haute, IN, USA   Joseph Eichholz
Summer 2018

Acknowledgments

Nearly all of this text has been vetted by Rose-Hulman students, and these careful readings have identified several areas of improvement and pinpointed numerous typographical errors. The authors appreciate everyone who has toiled through earlier drafts, and we hope that you have gained as much from us as we have gained from you.

The authors have received aid from several colleagues, including many who have suffered long conversations as we have honed content. We thank Eivind Almaas, Mike DeVasher, Fred Haan, Leanne Holder, Jeff Leader, John McSweeney, Omid Nohadani, Adam Nolte, and Eric Reyes. Of these, Jeff Leader, Eric Reyes, and John McSweeney deserve special note. Dr. Leader initiated the curriculum that motivated this text, and he proofread several of our earlier versions. Dr. Reyes' statistical assistance has been invaluable, and he was surely tired of our statistical discourse. Dr. McSweeney counseled our stochastic presentation. Allen Holder also thanks Matthias Ehrgott for sabbatical support. Both authors are thankful for Robert Vanderbei's authorship of the Foreword.

The professional effort to author this text has been a proverbial labor of love, and while the authors have enjoyed the activity, the task has impacted our broader lives. We especially want to acknowledge the continued support of our families—Amanda Kozak, Leanne Holder, Ridge Holder, and Rowyn Holder. Please know that your love and support are at the forefront of our appreciation.

We lastly thank Camille Price, Neil Levine, and the team at Springer. Their persistent nudging and encouragement has been much appreciated.

Contents

Part I Computational Methods

1 **Solving Single Variable Equations** 3
   1.1 Bisection 4
   1.2 Linear Interpolation 7
   1.3 The Method of Secants 8
   1.4 Newton's Method 13
      1.4.1 Improving Efficiency with Polynomials 15
      1.4.2 Convergence of Newton's Method 17
      1.4.3 MATLAB Functions 20
   1.5 Exercises 21

2 **Solving Systems of Equations** 33
   2.1 Systems of Linear Equations 34
      2.1.1 Upper- and Lower-Triangular Linear Systems 34
      2.1.2 General $m \times n$ Linear Systems 36
   2.2 Special Structure: Positive Definite Systems 44
      2.2.1 Cholesky Factorization 47
      2.2.2 The Method of Conjugate Directions 49
   2.3 Newton's Method for Systems of Equations 52
   2.4 MATLAB Functions 55
   2.5 Exercises 58

3 **Approximation** 67
   3.1 Linear Models and the Method of Least Squares 67
      3.1.1 Lagrange Polynomials: An Exact Fit 72
   3.2 Linear Regression: A Statistical Perspective 74
      3.2.1 Random Variables 75
      3.2.2 Stochastic Analysis and Regression 83

		3.3	Cubic Splines 98
		3.4	Principal Component Analysis 103
			3.4.1 Principal Component Analysis and the Singular Value Decomposition 111
		3.5	Exercises 112
4	**Optimization** 123
	4.1	Unconstrained Optimization 123
		4.1.1	The Search Direction 127
		4.1.2	The Line Search 131
		4.1.3	Example Algorithms 132
	4.2	Constrained Optimization 136
		4.2.1	Linear and Quadratic Programming 151
	4.3	Global Optimization and Heuristics 164
		4.3.1	Simulated Annealing 165
		4.3.2	Genetic Algorithms 171
	4.4	Exercises 177
5	**Ordinary Differential Equations** 191
	5.1	Euler Methods 192
	5.2	Runge-Kutta Methods 198
	5.3	Quantifying Error 202
	5.4	Stiff Ordinary Differential Equations 207
	5.5	Adaptive Methods 211
	5.6	Exercises 219
6	**Stochastic Methods and Simulation** 237
	6.1	Simulation 238
	6.2	Numerical Integration 239
		6.2.1	Simpson's Rule 241
		6.2.2	Monte Carlo Integration 245
		6.2.3	Bootstrapping 249
		6.2.4	Deterministic or Stochastic Approximation 251
	6.3	Random Models 254
		6.3.1	Simulation and Stochastic Differential Equations 255
		6.3.2	Simulation and Stochastic Optimization Models 259
	6.4	Exercises 262
7	**Computing Considerations** 269
	7.1	Language Choice 270
	7.2	C/C++ Extensions 271
	7.3	Parallel Computing 276
		7.3.1	Taking Advantage of Built-In Commands 277
		7.3.2	Parallel Computing in MATLAB and Python 278
		7.3.3	Parallel Computing in Python 284

Contents   xv

    7.3.4 Pipelining 287
    7.3.5 Ahmdal's Law 291
    7.3.6 GPU Computing 292
  7.4 Exercises 296

Part II Computational Modeling

8 Modeling with Matrices 309
  8.1 Signal Processing and the Discrete Fourier Transform 310
    8.1.1 Linear Time Invariant Filters 311
    8.1.2 The Discrete Fourier Transform 315
    8.1.3 The Fast Fourier Transform 320
    8.1.4 Filtering Signals.................................... 324
    8.1.5 Exercises 325
  8.2 Radiotherapy 326
    8.2.1 A Radiobiological Model to Calculate Dose 326
    8.2.2 Treatment Design 332
    8.2.3 Exercises 338
  8.3 Aeronautic Lift... 339
    8.3.1 Air Flow 340
    8.3.2 Flow Around a Wing 343
    8.3.3 Numerical Examples 348
    8.3.4 Exercises 353

9 Modeling with Ordinary Differential Equations 355
  9.1 Couette Flows 355
    9.1.1 Exercises 362
  9.2 Pharmacokinetics: Insulin Injections 363
    9.2.1 Exercises 370
  9.3 Chemical Reactions....................................... 371
    9.3.1 Exercises 375

10 Modeling with Delay Differential Equations................ 377
  10.1 Is a Delay Model Necessary or Appropriate? 378
  10.2 Epidemiology Models 379
  10.3 The El-Niño–La-Niña Oscillation 383
  10.4 Exercises 385

11 Partial Differential Equations 389
  11.1 The Heat Equation 389
  11.2 Explicit Solutions by Finite Differences.................... 392
  11.3 The Wave Equation....................................... 397
  11.4 Exercises 401

12 Modeling with Optimization and Simulation 403
12.1 Stock Pricing and Portfolio Selection 403
   12.1.1 Stock Pricing 404
   12.1.2 Portfolio Selection 408
   12.1.3 Exercises 415
12.2 Magnetic Phase Transitions 418
   12.2.1 The Gibbs Distribution of Statistical Mechanics 419
   12.2.2 Simulation and the Ising Model 421
   12.2.3 Exercises 428

13 Regression Modeling 431
13.1 Stepwise Regression 435
13.2 Qualitative Inputs and Indicator Variables 437
13.3 Exercises 441

A Matrix Algebra and Calculus 447
A.1 Matrix Algebra Motivated with Polynomial Approximation .. 447
A.2 Properties of Matrix–Matrix Multiplication 453
A.3 Solving Systems, Eigenvalues, and Differential Equations 455
   A.3.1 The Nature of Solutions to Linear Systems 456
   A.3.2 Eigenvalues and Eigenvectors 460
A.4 Some Additional Calculus 463

Index 465

Part I
Computational Methods

Chapter 1
Solving Single Variable Equations

> No more fiction for us: we calculate; but that we may calculate, we had to make fiction first. – Friedrich Nietzsche

The problem of solving the equation $f(x) = 0$ is among the most storied in all of mathematics, and it is with this problem that we initiate our study of computational science. We assume functions have their natural domains in the real numbers. For instance, a function like $\sqrt{x}$ exists over the collection of nonnegative reals. The right-hand side being zero is not generally restrictive since solving either $f(x) = k$ or $g(x) = h(x)$ can be re-expressed as $f(x) - k = 0$ or $g(x) - h(x) = 0$. Hence looking for roots provides a general method to solve equations.

The equation $f(x) = 0$ may have a variety of solutions or none at all. For example, the equation $x^2 + 1 = 0$ has no solution over the reals, the equation $x^3 + 1 = 0$ has a single solution, and the equation $\sin(x) = 0$ has an infinite number of solutions. Such differences complicate searching for a root, or multiple roots, especially without forehand knowledge of f. Most algorithms attempt to find a single solution, and this is the case we consider. If more than one solution is desired, then the algorithm can be re-applied in an attempt to locate others.

Four different algorithms are presented, those being the time-tested methods of bisection, secants, interpolation, and Newton. Each has advantages and disadvantages, and collectively they are broadly applicable. At least one typically suffices for the computational need at hand. Students are encouraged to study and implement fail-safe programs for each, as they are often useful.

1.1 Bisection

The theoretical underpinnings of bisection lie in one of the most important theorems of Calculus, that being the Intermediate Value Theorem. This theorem ensures that a continuous function attains all intermediate values over an interval.

Theorem 1 (Intermediate Value Theorem). *Let $f(x)$ be continuous on the interval $[a, b]$, and let k be between $f(a)$ and $f(b)$. Then, there is a c between a and b such that $f(c) = k$.*

The method of bisection is premised on the assumption that f is continuous and that $f(a)$ and $f(b)$ differ in sign. We make the tacit assumption that $a < b$, from which the Intermediate Value Theorem then guarantees a solution to $f(x) = 0$ over the interval $[a, b]$. To illustrate, the function $f(x) = x^3 - x$ satisfies

$$f(1.5) = 1.875 > 0 \text{ and } f(-2) = -6 < 0.$$

So f is guaranteed to have at least one root over the interval $[-2, 1.5]$. Indeed, f has three roots over this interval, although the Intermediate Value Theorem only guarantees one.

Bisection proceeds by halving the interval. The midpoint of $[-2, 1.5]$ is -0.25, and since $f(-0.25) = 0.2344 > 0$, we again have from the Intermediate Value Theorem that f has a root over the interval $[-2, -0.25]$. Two observations at this point are (1) this new interval removes two of the three roots, and (2) we are guaranteed to be within $(2 - (-1.5))/2 = 3.5/2$ of a solution to $f(x) = 0$. The algorithm continues by halving the new interval and removing the half that no longer guarantees a root. Bisection is called a bracketing method since a solution is always "bracketed" between two other values. The first five iterations for this example are listed in Table 1.1. The iterates show that the interval collapses onto the solution $x = -1$. These iterations are depicted in Fig. 1.1, and general pseudocode is listed in Algorithm 1.

One of the benefits of bisection is that it has a guarantee of how close it is to a root. The algorithm's estimate of the root for the example above is within $(1.5 + 2)/2^5 = 0.1094$ of the real solution after the fifth iteration. Indeed, after k iterations the algorithm is within $(1.5 + 2)/2^k$ of the root, a fact that allows us to calculate how many iterations are needed to ensure a desired accuracy. So if we want to be within 10^{-8} of a solution, i.e. 8 digits of accuracy, then finding the first integer value of k such that $3.5/2^k \leq 10^{-8}$ shows that bisection will need

$$k = \left\lceil \frac{\ln(3.5 \times 10^8)}{\ln(2)} \right\rceil = 29$$

iterations for this example.

1.1 Bisection

Iteration	Interval	Best value of $|f|$
0	$[-2.000000,\ 1.500000]$	1.875000
1	$[-2.000000,\ -0.250000]$	0.234375
2	$[-1.125000,\ -0.250000]$	0.234375
3	$[-1.125000,\ -0.687500]$	0.298828
4	$[-1.125000,\ -0.906250]$	0.161957
5	$[-1.015625,\ -0.906250]$	0.031986

Table 1.1 Five iterations of bisection solving $f(x) = x^3 - x = 0$ over $[-2, 1.5]$. The best value of f in the third column is the smallest value of $|f(x)|$ at the end points of the iteration's interval.

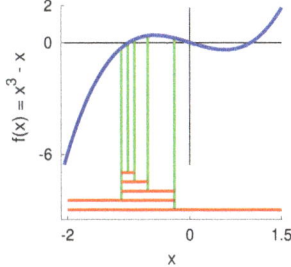

Fig. 1.1 An illustration of bisection solving $f(x) = x^3 - x = 0$ over $[-2, 1.5]$. The intervals of each iteration are shown in red at the bottom of the figure, and vertical green lines are midpoint evaluations.

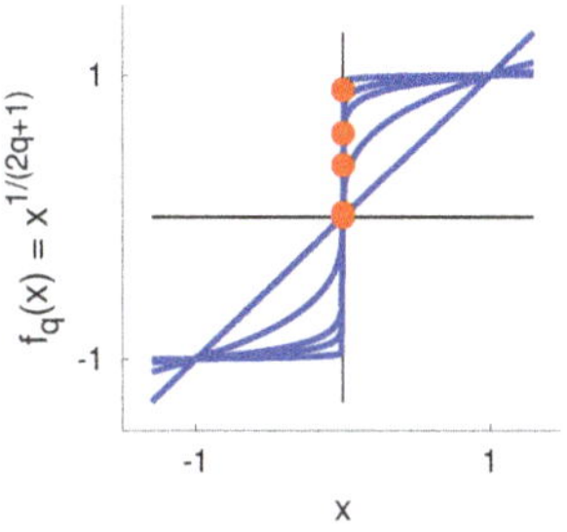

Fig. 1.2 The function $f_q(x) = x^{1/(2q+1)}$ for $q = 0, 1, 5, 10$, and 50. The red dots show the value of $f_q(x)$ at the approximate solution to $f_q(x) = 0$ after 15 iterations of bisection initiated with $[0.5, 1.3]$.

While the distance to a root can be controlled by the number of iterations, this fact does not ensure that the function itself is near zero. For example, the function $f_q(x) = x^{1/(2q+1)}$ is increasingly steep at $x = 0$ as the integer q increases, see Fig. 1.2. Indeed, for any x between -1 and 1 other than zero we have

$$\lim_{q \to \infty} \left| x^{1/(2q+1)} \right| = 1.$$

So any approximate solution to $f_q(x) = 0$ can have a function value near 1 for large q. In other words, x being near a solution does not ensure that $f(x)$ is near 0. If we start with the interval $[-0.5, 1.3]$, then 15 iterations guarantee a solution within a tolerance of 10^{-4} and end with the interval $[-1.2 \times 10^{-5},\ 4.3 \times 10^{-5}]$ independent of q. However, for $q = 1, 5, 10$, and

Algorithm 1 Pseudocode for the method of bisection
```
k = 1
while unmet termination criteria do
    x̂ = (a + b)/2
    if sign(f(x̂)) == sign(f(a)) then
        a = x̂
    else if sign(f(x̂)) == sign(f(b)) then
        b = x̂
    else if sign(f(x̂)) == 0 then
        set termination status to true
    else
        set status to failure and terminate
    end if
    k = k + 1
end while
return  best estimate of root
```

50 the values of f_q at the midpoint of this interval are, respectively, 0.0247, 0.3643, 0.5892, and 0.8959, see Fig. 1.2.

Alternatively, the value of $f(x)$ can be close to zero even though x remains some distance from a solution. For example, functions of the form

$$g_q(x) = \frac{x}{1 + (10^q\, x)^2}$$

have the property that $g_q(x) \to 0$ as $q \to \infty$ for all nonzero x. So any real number can appear to be a solution for sufficiently large q. Suppose the termination criterion is $|g_q(x)| \leq 10^{-4}$. If bisection is initialized with the interval $[-0.9, 0.8]$, then the terminal interval is $[-1.95 \times 10^{-4}, 1.2 \times 10^{-5}]$ for $q = 2$. The value of g_q at the midpoint of this interval is reasonable at 1.2×10^{-5}. However, for $q = 3$ the value at the initial left end point is $g_q(-0.9) = -1 \times 10^{-6}$, and bisection terminates immediately with an estimated solution of $x = -0.9$, which is some distance from the desired $x = 0$.

A dilemma of writing code for general use is deciding the termination criteria. Two common convergence conditions are

$$|f(x)| \leq \varepsilon, \text{ for the best computed value of } x, \text{ and}$$
$$|x - x^*| \leq \delta, \text{ where } x^* \text{ is a solution to } f(x) = 0,$$

where ε and δ are user supplied convergence tolerances. Satisfying both is preferred to satisfying one, but requiring both could be computationally burdensome and possibly unrealistic.

Care is warranted when implementing pseudocode like that of Algorithm 1. The termination criteria should include tests for the original end points, for the value of the function at the newly computed midpoint, and for the width of the interval. Moreover, we might add tests to

- ensure that the original interval is well defined, i.e. $a < b$,
- see that the signs of $f(a)$ and $f(b)$ differ, and
- validate that the function value is real, which is not always obvious in MATLAB and Octave, e.g. (-1)^(1/3) isn't real, but nthroot(-1,3) is.

Returning a status variable in addition to the computed solution lets a user know if she or he should trust the outcome. In many cases a status variable encodes one of several termination statuses. Moreover, it is generally good advice to undertake the prudent step of verifying the assumed properties of the input data and of the computed result, even though doing so lengthens code development and can seem to pollute a clean, succinct, and logical progression. However, including measures to vet computational entities will lead to trustworthy algorithms that can be used for diverse tasks.

1.2 Linear Interpolation

The method of linear interpolation is similar to that of bisection, but in this case the algorithm uses the function's values at the endpoints and not just their signs. As with bisection, the method of linear interpolation uses the Intermediate Value Theorem to guarantee the existence of a root within an interval. Unlike the method of bisection, the method of linear interpolation approximates the function to estimate the next iterate. If the function is itself linear, or more generally affine, then linear interpolation will converge in one step, an outcome not guaranteed with bisection.

Assume that f is continuous on the interval $[a, b]$ and that $f(a)$ and $f(b)$ differ in sign. The line through the points $(a, f(a))$ and $(b, f(b))$ is called the interpolant and is

$$y - f(a) = \frac{f(a) - f(b)}{a - b}(x - a).$$

Solving for x with $y = 0$ gives the solution $\hat{x}$ so that

$$\hat{x} = a - f(a)\left(\frac{a - b}{f(a) - f(b)}\right). \tag{1.1}$$

If $f(\hat{x})$ agrees in sign with $f(a)$, then the interval $[\hat{x}, b]$ contains a root and the process repeats by replacing a with $\hat{x}$. If $f(\hat{x})$ agrees in sign with $f(b)$, then the interval $[a, \hat{x}]$ contains a root and the process repeats by replacing

b with $\hat{x}$. If $|f(\hat{x})|$ is sufficiently small, then the process terminates since $\hat{x}$ is a computed root. The pseudocode for linear interpolation is in Algorithm 2.

Algorithm 2 Pseudocode for the method of linear interpolation
---
$k = 1$
while unmet termination criteria **do**
    $\hat{x} = a - f(a)(a-b)/(f(a) - f(b))$
    if $\text{sign}(f(\hat{x})) == 0$ **then**
        set termination status to true
    else if $\text{sign}(f(\hat{x})) == \text{sign}(f(a))$ **then**
        $a = \hat{x}$
    else if $\text{sign}(f(\hat{x})) == \text{sign}(f(b))$ **then**
        $b = \hat{x}$
    else
        set status to failure and terminate
    end if
    $k = k + 1$
end while
return best estimate of root

The first five iterations of linear interpolation solving $f(x) = x^3 - x = 0$ over the interval $[-2, 1.5]$ are listed in Table 1.2, and the first few iterations are illustrated in Fig. 1.3. Linear interpolation converges to $x = 1$ as it solves $f(x) = x^3 - x = 0$ initiated with $[-2, 1.5]$, whereas the method of bisection converges to $x = -1$. So while the methods are algorithmically similar, the change in the update produces different solutions. The convergence is "one-sided" in this example because the right end point remains unchanged. Such convergence is not atypical, and it can lead to poor performance. As a second example, iterations solving $f(x) = \cos(x) - x = 0$ over $[-0.5, 4.0]$ are listed in Table 1.2 and are shown in Fig. 1.4. In this case convergence is fairly rapid due to the suitable linear approximation of f by the interpolant.

Interpolation does not provide a bound on the width of the k-th iterate's interval. In fact, there is no guarantee that the width of the intervals will converge to zero as the algorithm proceeds. Therefore, the only reasonable convergence criterion is $|f(x)| < \varepsilon$. As with bisection, care and prudence are encouraged during implementation.

1.3 The Method of Secants

The method of secants uses the same linear approximation as the method of linear interpolation, but it removes the mathematical certainty of capturing a root within an interval. One advantage is that the function need not

1.3 The Method of Secants

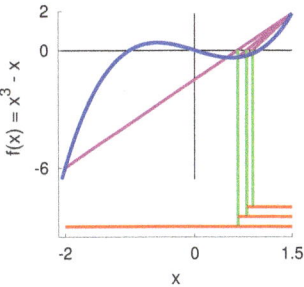
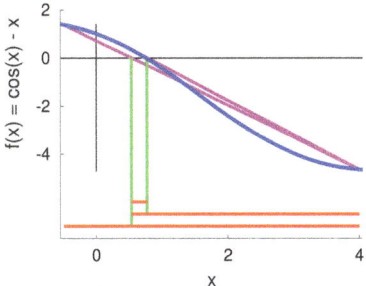

Fig. 1.3 An illustration of the first few iterations of linear interpolation solving $f(x) = x^3 - x = 0$ over $[-2, 1.5]$. The red lines at the bottom show how the intervals update according to the roots of the linear interpolants (shown in magenta).

Fig. 1.4 An illustration of the first few iterations of linear interpolation solving $f(x) = \cos(x) - x = 0$ over $[-0.5, 4]$. The red lines at the bottom show how the intervals update according to the roots of the linear interpolants (shown in magenta).

	$f(x) = x^3 - x$		$f(x) = \cos(x) - x$	
Iteration	Interval	Best $|f|$	Interval	Best $|f|$
0	[−2.0000, 1.5000]	1.8750	[−0.5000, 4.0000]	1.3776
1	[0.6667, 1.5000]	0.3704	[0.5278, 4.0000]	0.0336
2	[0.8041, 1.5000]	0.2842	[0.5278, 0.7617]	0.0336
3	[0.8957, 1.5000]	0.1771	[0.7379, 0.7617]	0.0019
4	[0.9479, 1.5000]	0.0963	[0.7391, 0.7617]	0.0000
5	[0.9748, 1.5000]	0.0484	[0.7391, 0.7617]	0.0000

Table 1.2 The first five iterations of linear interpolation solving $x^3 - x = 0$ over $[-2, 1.5]$ and $\cos(x) - x = 0$ over $[-0.5, 4.0]$.

change sign over the original interval. A disadvantage is that the new iterate might not provide an improvement without the certainty guaranteed by the Intermediate Value Theorem.

The method of secants is a transition from the method of linear interpolation toward Newton's method, which is developed in Sect. 1.4. The mathematics of the method of secants moves us theoretically from the Intermediate Value Theorem to the Mean Value Theorem.

Theorem 2 (Mean Value Theorem). *Assume f is continuous on $[a, b]$ and differentiable on (a, b). Then, for any x in $[a, b]$ there is c in (a, x) such that*

$$f(x) = f(a) + f'(c)(x - a).$$

This statement of the Mean Value Theorem is different than what is typically offered in calculus, but note that if $x = b$, then we have the common observation that for some c in (a, b),

$$f'(c) = \frac{f(b) - f(a)}{b - a}.$$

The statement in Theorem 2 highlights that the Mean Value Theorem is a direct application of Taylor's Theorem, a result discussed more completely later.

The method of secants approximates $f'(c)$ with the ratio

$$f'(c) \approx \frac{f(b) - f(a)}{b - a},$$

which suggests that

$$f(x) \approx f(a) + \frac{f(b) - f(a)}{b - a}(x - a).$$

The method of secants iteratively replaces the equation $f(x) = 0$ with the approximate equation

$$0 = f(a) + \frac{f(b) - f(a)}{b - a}(x - a),$$

which gives a solution of

$$\hat{x} = a - f(a)\left(\frac{b - a}{f(b) - f(a)}\right).$$

This update is identical to (1.1), and hence the iteration to calculate the new potential root is the same as the method of linear interpolation. What changes is that we no longer check the sign of f at the updated value. Pseudocode for the method of secants is in Algorithm 3.

The iterates from the method of secants can stray, and unlike the methods of bisection and interpolation, there is no guarantee of a diminishing interval as the algorithm progresses. The value of $f(x)$ can thus worsen as the algorithm continues, and for this reason it is sensible to track the best calculated solution. That is, we should track the iterate x_best that has the nearest function value to zero among those calculated. Sometimes migrating outside the original interval is innocuous or even beneficial. Consider, for instance, the first three iterations of solving $f(x) = x^3 - x = 0$ over the interval $[-2, 1.5]$ in Table 1.3 for the methods of secants and linear interpolation. The algorithms are the same as long as the interval of linear interpolation agrees with the last two iterates of secants, which is the case for the first two updates. The third update of secants in iteration 2 is not contained in the interval $[a, b] = [0.6667, 0.8041]$ because the function no longer changes sign at the

1.3 The Method of Secants

Algorithm 3 Pseudocode for the method of secants

$k = 1$
while unmet termination criteria **do**
    if $f(b) == f(a)$ **then**
        set status to failure and terminate
    else
        $\hat{x} = a - f(a)(a-b)/(f(a) - f(b))$
        $a = b$
        $b = \hat{x}$
        $k = k + 1$
    end if
end while
return best estimate of root

	Linear interpolation		Secants		
Iteration	Interval	Update ($\hat{x}$)	a	b	Update ($\hat{x}$)
0	$[-2.0000, 1.5000]$	0.6667	-2.000	1.5000	0.6667
1	$[\ 0.6667, 1.5000]$	0.8041	1.5000	0.6667	0.8041
2	$[\ 0.8041, 1.5000]$	0.8957	0.6667	0.8041	1.2572
3	$[\ 0.8957, 1.5000]$	0.9479	0.8041	1.2572	0.9311

Table 1.3 The first three iterations of linear interpolation and the method of secants solving $f(x) = x^3 - x = 0$ over $[-2, 1.5]$.

end points, and in this case, the algorithm strays outside the interval to estimates the root as 1.2572. Notice that if $f(a)$ and $f(b)$ had been closer in value, then the line through $(a, f(a))$ and $(b, f(b))$ would have been flatter. The resulting update would have been a long way from the earlier iterates. However, the algorithm for this example instead converges to the root $x = 1$, so no harm is done. The first ten iterations are in Table 1.4, and the first few iterations are depicted in Fig. 1.5.

Widely varying iterates are fairly common with the method of secants, especially during the initial iterations. If the secants of the most recent iterations favorably point to a solution, then the algorithm will likely converge. An example is solving $f(x) = \cos(x) - x = 0$ initiated with $a = 10$ and $b = 20$, for which linear interpolation would have been impossible because $f(a)$ and $f(b)$ agree in sign. The first two iterates provide improvements, but the third does not. Even so, the algorithm continues and converges with $|f(x)| < 10^{-4}$ in seven iterations, see Table 1.5. Indeed, the method converges quickly once the algorithm sufficiently approximates its terminal solution. This behavior is

Iteration	a	b	$\hat{x}$	x_{best}	$f(x_{best})$
0	−2.0000	1.5000	0.6667	0.6667	−0.3704
1	1.5000	0.6667	0.8041	0.8041	−0.2842
2	0.6667	0.8041	1.2572	0.8041	−0.2842
3	0.8041	1.2572	0.9311	0.9311	−0.1239
4	1.2572	0.9311	0.9784	0.9784	−0.0418
5	0.9311	0.9784	1.0025	1.0025	0.0050
6	0.9784	1.0025	0.9999	0.9999	−0.0002
7	1.0025	0.9999	1.0000	1.0000	−0.0000

Table 1.4 Iterations of the method of secants solving $f(x) = x^3 - x = 0$ initiated with $a = -2.0$ and $b = 1.5$.

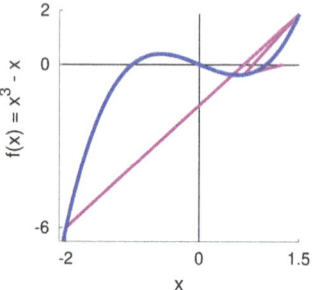

Fig. 1.5 The first three iterations of the secant method solving $f(x) = x^3 - x = 0$ initiated with $a = -2.0$ and $b = 1.5$. The third secant line produces an iterate outside the interval of the previous two.

typical and concomitant with the favorable convergence of Newton's method, which is being mimicked (see Sect. 1.4.2).

The splaying of unfavorable iterates does not always lead to convergence, and in some cases the algorithm diverges. An example is solving $f(x) = x^2/(1 + x^2) = 0$ initiated with $a = -2$ and $b = 1$. In this case the magnitude of the iterates continues to increase, and our best solution remains $x = 1$ with a function value of $f(1) = 0.5$. Iterations are listed in Table 1.5.

The method of secants often converges if the initial points are sufficiently close to the root. However, what constitutes close enough varies from function to function and is usually difficult to determine, particularly without prior knowledge of the function and its solution. It is typical to employ a "try and see" strategy with the method of secants. We start with two reasonable guesses of the root and simply see if the method converges.

1.4 Newton's Method

	Solving $\cos(x) - x = 0$ Initiated with $a = 10$ and $b = 20$			Solving $x^2/(1+x^2) = 0$ Initiated with $a = -2$ and $b = 1$		
Iteration	Iterate ($\hat{x}$)	x_{best}	$f(x_{\text{best}})$	Iterate ($\hat{x}$)	x_{best}	$f(x_{\text{best}})$
1	−2.3835	−2.3835	1.6574	6.0000	1.0	0.5
2	−0.6377	−0.6377	1.4412	−4.2857	1.0	0.5
3	11.0002	−0.6377	1.4412	−400.71	1.0	0.5
4	0.7109	0.7109	0.0469	7277.9	1.0	0.5
5	0.7546	0.7546	−0.0260	-1.2×10^9	1.0	0.5
6	0.7390	0.7390	0.0002	6.6×10^{16}	1.0	0.5
7	0.7391	0.7391	0.0000			

Table 1.5 The third iterate of the method of secants solving $\cos(x) - x = 0$ lacks improvement, but the algorithm continues to converge. The iterates attempting to solve $x^2/(1+x^2) = 0$ diverge.

1.4 Newton's Method

Newton's method is a numerical stalwart and is commonly the method of choice. Unlike the method of secants, which estimates f', Newton's method uses the derivative itself to estimate the function locally. Assuming f is continuously differentiable, we have from the Mean Value Theorem (Theorem 2 in the previous section) that for any x there is a c between x and a such that

$$f(x) = f(a) + f'(c)(x - a).$$

So, if x is close to a, then f is approximated by the linear function

$$f(x) \approx f(a) + f'(a)(x - a).$$

The substitution of $f'(a)$ for $f'(c)$ points to the analytic assumption that f' be continuous.

Newton's method iteratively replaces the equation $f(x) = 0$ with the approximate equation

$$f(a) + f'(a)(x - a) = 0,$$

which has the solution

$$\hat{x} = a - \frac{f(a)}{f'(a)}$$

as long as $f'(a) \neq 0$. The iterate $\hat{x}$ is the new estimate of the root, and unless the process reaches a termination criterion, $\hat{x}$ replaces a and the process continues. Pseudocode for Newton's method is in Algorithm 4. The first six iterates solving $x^3 - x = 0$ and $\cos(x) - x = 0$ are listed in Table 1.6. Several of the first iterations are illustrated in Figs. 1.6 and 1.7.

Algorithm 4 Pseudocode for Newton's method

while unmet termination criteria **do**
    if $f'(a) \neq 0$ **then**
        $a = a - f(a)/f'(a)$
    else
        set status to failure and terminate
    end if
end while

	Solving $x^3 - x = 0$ Initiated with $a = -0.48$		Solving $\cos(x) - x = 0$ Initiated with $a = -1$	
Iteration	Iterate (a)	$f(a)$	Iterate (a)	$f(a)$
0	-0.4800	0.3694	1.0000	1.5403
1	0.7163	-0.3488	8.7162	-9.4755
2	1.3632	1.1702	2.9761	-3.9624
3	1.1075	0.2508	-0.4258	1.3365
4	1.0139	0.0283	1.8512	-2.1279
5	1.0003	0.0006	0.7660	-0.0454
6	1.0000	0.0000	0.7392	-0.0002

Table 1.6 The first six iterates of Newton's method solving $x^3 - x = 0$ initialized with $x = -0.48$ and $\cos(x) - x = 0$ initiated with $x = -1$.

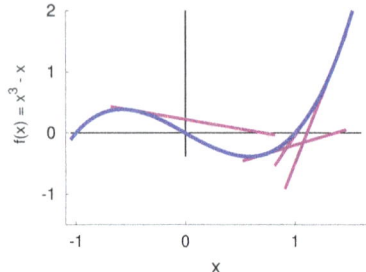

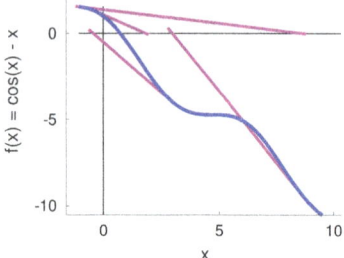

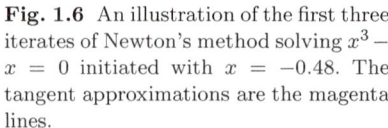

Fig. 1.6 An illustration of the first three iterates of Newton's method solving $x^3 - x = 0$ initiated with $x = -0.48$. The tangent approximations are the magenta lines.

Fig. 1.7 An illustration of the first three iterates of Newton's method solving $\cos(x) - x = 0$ initiated with $x = -1.0$. The tangent approximations are the magenta lines.

1.4 Newton's Method

The calculation of $\hat{x}$ assumes that $f'(a)$ is not zero, and indeed, this is a computational concern. If r is a root and $f'(r) = 0$, then the assumed continuity of f' ensures that $|f'(a)|$ is arbitrarily small as the process converges to r. Division by small values challenges numerical computation since it can result in less accurate values or exceed the maximum possible value on the computing platform. A reasonable assumption for both computational and theoretical developments is that $f'(r) \neq 0$, and a reasonable termination criterion is to halt the process if $|f'(a)|$ is sufficiently small.

A notational convenience is to let $\Delta x = \hat{x} - a$, which succinctly expresses the update as
$$\hat{x} = a + \Delta x, \text{ where } f'(a)\,\Delta x = -f(a).$$
The change between iterations, Δx, is called the Newton step. The expression $f'(a)\Delta x = -f(a)$ removes the algebraic concern of dividing by zero, and it nicely extends to the higher dimensional problems of the next chapter. Of course, the solution is unique if and only if $f'(a) \neq 0$.

Newton's method shares a loss of guaranteed convergence with the method of secants, and iterates can similarly stray during the search for a solution. However, it is true that if the starting point is close enough to the root, then Newton's method will converge. A common approach is to "hope for the best" when using Newton's method, and for this reason the exit statuses of both the method of secants and Newton's method are particularly critical.

1.4.1 Improving Efficiency with Polynomials

The most computationally expensive calculation of solving $f(x) = 0$ with an iterative method is commonly the evaluation of the function itself. A function's evaluation in many real-world problems depends on substantial calculations that can take hours, days, weeks, or months. This reality suggests that we pay particular attention to the cases for which the computational burden can be reduced, and we demonstrate how to hasten the evaluation of polynomials. Restricting our attention to polynomials may seem quaint against the backdrop of possible functions that we might consider, but polynomials are regularly used to approximate more complicated functions. Hence, an efficient calculation method finds wide application.

Evaluating polynomials and their derivatives can be accomplished more efficiently than is commonly first perceived. We consider the quadratic $f(x) = 3x^2 + 2x + 1$ as a small example. Evaluating $f(x)$ as expressed requires 3 multiplications and 2 additions: one multiplication for x^2, another for $3x^2$, and one for $2x$. However, we may rewrite the polynomial as

$$f(x) = 3x^2 + 2x + 1 = (3x + 2)x + 1, \tag{1.2}$$

and this last expression only requires 2 multiplications and 2 additions. Multiplications are more computationally burdensome than additions, and the factored form shows that we can reduce this burden.

Calculating the factored form is often tabulated in a synthetic division table. As an example, the table to evaluate $f(-2)$ is

$$-2 \;\begin{array}{|ccc} 3 & 2 & 1 \\ & -6 & 8 \\ \hline & 3 & -4 & 9 \end{array} \tag{1.3}$$

The process is to "drop" the leading 3, which is then multiplied by -2 to form $3x$ at $x = -2$. The resulting -6 is then aligned with and added to the 2 to form $3x + 2$. Multiplications and additions repeat until $(3x + 2)x + 1$ is evaluated at $x = -2$, with the resulting value being 9.

The linearity of the derivative provides another advantage of the factored form of the polynomial. Differentiating all expressions in (1.2) shows

$$f'(x) = 6x + 2 = (3x + 2) + 3x. \tag{1.4}$$

The expression on the right appears to have an added multiplication over the one on the left, but this is untrue if we re-use the already calculated value of -4 for $3x + 2$ at $x = -2$. This shows that $f'(-2) = (-4) + 3(-2) = -10$, which again can be nicely tabulated using the first two computed values from the bottom row of the synthetic division table in (1.3),

$$-2 \;\begin{array}{|cc} 3 & -4 \\ & -6 \\ \hline & 3 & -10 \end{array} \;.$$

There is nothing special about quadratics in these calculations, and in general, the factored form reduces the number of multiplications from $n(n-1)/2$ to n for a polynomial of degree n. This is a substantial reduction for large n. Since Newton's method requires both $f(x)$ and $f'(x)$, the speedup on polynomials is significant for large n.

The monikers associated with the method above are synthetic division and Hörner's method. Recall that if $p(x)$ is a polynomial of degree n, then division by the degree one polynomial $x - c$ can be expressed as

$$p(x) = q_c(x)(x - c) + r_c, \tag{1.5}$$

where $q_c(x)$ is the $n - 1$ degree polynomial called the quotient, and the constant r_c is called the remainder. From the observation that

$$p(c) = q_c(c)(c - c) + r_c = r_c,$$

1.4 Newton's Method

we conclude that $p(c)$ is the remainder of $p(x)$ upon division by $x - c$. The value of r_c is the right-most entry in the bottom row of the table, e.g. $f(-2) = r_{-2} = 9$ as indicated in the synthetic division table in (1.3). The tabular method further calculates the coefficients of $q_c(x)$ in the first $n - 1$ bottom entries, and $q_{-2}(x)$ is $3x - 4$ for our example. The resulting factorization is

$$3x^2 + 2x + 1 = (3x - 4)(x + 2) + 9.$$

We have from the factored form of Eq. (1.5) that

$$p'(x) = q'_c(x)(x - c) + q_c(x),$$

and therefore

$$p'(c) = q_c(c).$$

We can thus calculate $p'(c)$ by evaluating $q_c(c)$, which for the example means that we can evaluate the derivative of $3x^2 + 2x + 1$ at $x = -2$ by simply evaluating $3x - 4$ at $x = -2$. As noted above, $q_c(x)$ was already calculated during the synthetic division process calculating $p(c)$. It is therefore already available and may be evaluated using a separate synthetic division process.

1.4.2 Convergence of Newton's Method

Newton's method is often the gold standard due to its favorable convergence properties, which we analyze by updating our previous notation so that each iterate is denoted by x_k. This notational change mandates that

$$x_{k+1} = x_k - \frac{f(x_k)}{f'(x_k)},$$

where x_0 is the initial estimate of the solution. If $f(x) = x^3 - x$ with $x_0 = 5$, then $x_k \to 1$ as $k \to \infty$. The error at iteration k is $|x_k - 1|$, which is plotted in Fig. 1.8. The rapid decrease to zero is characteristic of the method and can be explained mathematically.

The mathematical intent of explaining the rate at which a sequence converges is typically defined by order. Specifically, the sequence x_k converges to x^* as $k \to \infty$ with order p if

$$\lim_{k \to \infty} \frac{|x_{k+1} - x^*|}{|x_k - x^*|^p} < \infty. \tag{1.6}$$

A sequence converges linearly if $p = 1$ and quadratically if $p = 2$. We comment that if a sequence converges with order 2, then it also converges with order p so long as $1 \le p \le 2$.

The concept of order intuitively explains how much faster the sequence is approaching its limit as it progresses. For example, if for some k we have $|x_k - x^*| < 1/10$ and $p = 2$, then roughly $|x_{k+1} - x^*| < M/100$, where M is some suitably large value independent of k. Figure 1.8 graphs the ratio $|x_{k+1} - x^*|/|x_k - x^*|^p$ as Newton's method solves $x^3 - x = 0$ with $x_0 = 5$. The three cases of $p = 1.9$, $p = 2.0$, $p = 2.1$ are shown. For $p = 2$ the ratio flattens to a constant, which provides an estimate for M. For $p = 2.1$ the ratio climbs, which suggests that the value of p is too large since M cannot be decided independent of k. For $p = 0.9$ the ratio decreases, which suggests that a larger p is possible.

Consider $x_k = 1/k^2$ as a second example, which converges to $x^* = 0$. Then,

$$\lim_{k \to \infty} \frac{|x_{k+1} - x^*|}{|x_k - x^*|^p} = \lim_{k \to \infty} \frac{1/(k+1)^2}{(1/k^2)^p} = \lim_{k \to \infty} \left(\frac{k^p}{k+1} \right)^2.$$

This limit diverges toward infinity for any $p > 1$, but for $p = 1$ we have

$$\lim_{k \to \infty} \left(\frac{k}{k+1} \right)^2 = 1.$$

Hence the order at which $1/k^2$ converges to zero is $p = 1$. Errors and ratios for this sequence are graphed in Fig. 1.9. In this case we have that the ratio of sequential errors flattens for $p = 1$ but climbs for the larger value of $p = 1.1$. The ratio falls for $p = 0.9$. The graphs agree with our calculated result that the maximum order is $p = 1$.

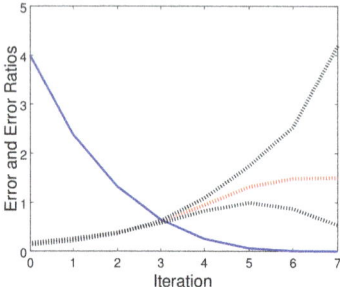
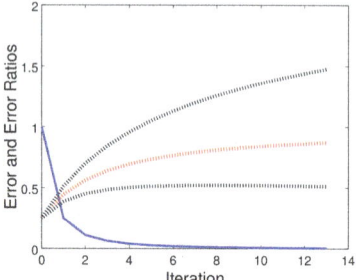

Fig. 1.8 The blue curve is the error $|x_k - 1|$ for the k-th iteration of Newton's method solving $x^3 - x = 0$ with $x_0 = 5$. The ratio $|x_{k+1} - 1|/|x_k - 1|^p$ is displayed for $p = 2$ (red), $p = 1.9$ (black bottom), and $p = 2.1$ (black top).

Fig. 1.9 The blue curve is the error $|x_k - 0|$ for the sequence $x_k = 1/k^2$. The ratio x_{k+1}/x_k^p is shown for $p = 1$ (red), $p = 0.9$ (black bottom), and $p = 1.1$ (black top).

1.4 Newton's Method

Analyzing the convergence of Newton's method hinges on two standard results from calculus, which we briefly review. The first is a classic result from Karl Weierstrass stating that a continuous function on a closed and bounded interval attains its maximum and minimum.

Theorem 3. *If f is a continuous function on $[a, b]$, then*

$$f(\hat{x}) = \max\{f(x) : a \leq x \leq b\} \text{ and } f(\tilde{x}) = \min\{f(x) : a \leq x \leq b\},$$

for some $\hat{x}$ and $\tilde{x}$ in $[a, b]$.

Although the applications of this result are many, our immediate need is the modest consequence that a continuous function is bounded on a closed and bounded interval. Hence, if f is continuous on $[a, b]$, then there is a β such that $|f(x)| < \beta$ for any x in $[a, b]$.

The second result from calculus is Taylor's theorem, which provides a convenient way to approximate a function with a polynomial.

Theorem 4 (Taylor's Theorem). *Let f have n derivatives over the interval (a, b). Then for any w and y in (a, b) there is a z between w and y such that*

$$f(w) = \sum_{k=0}^{n-1} \frac{1}{k!} f^{(k)}(y)(w - y)^k + \frac{f^{(n)}(z)}{n!}(w - y)^n.$$

Taylor's theorem reduces to the Mean Value Theorem if $n = 1$. However, the relationship between these two results is more than this reduction, as the proof of Taylor's theorem rests on repeated applications of the Mean Value Theorem.

If we further assume that $f^{(n)}$ is continuous, then we have from Theorem 3 that $f^{(n)}$ is bounded over the closed interval whose end points are w and y. Specifically, for $n = 2$ this guarantees the existence of β such that

$$\left| \frac{f''(x)}{2} \right| \leq \beta \text{ for any } x \text{ between } w \text{ and } y.$$

Let x^* and x_k be sufficiently close to each other in an interval over which the second derivative is continuous. From Taylor's theorem we have

$$f(x^*) = f(x_k) + f'(x_k)(x^* - x_k) + \frac{1}{2} f''(c)(x^* - x^k)^2,$$

where c is between x^* and x_k. Subsequently, we have from Theorem 3 that

$$|f(x^*) - f(x_k) - f'(x_k)(x^* - x_k)| \leq \frac{1}{2}|f''(c)||x^* - x_k|^2 \leq \beta|x^* - x_k|^2.$$

Since the existence of f'' implies that f' is continuous, we have that $f'(x) \neq 0$ as long as $f'(x^*) \neq 0$, provided that x is sufficiently close to x^*. Hence,

$1/f'(x)$ is continuous near x^*, and from Theorem 3 there is a γ such that $|1/f'(x_k)| \leq \gamma$, assuming again that x_k is sufficiently close to x^*.

Combining the bounds on $|f''(x)|$ and $|1/f'(x)|$, we find that if $f(x^*) = 0$ and x_k is sufficiently close to x^*, then the Newton iterations satisfy

$$|x_{k+1} - x^*| = \left| x_k - x^* - \frac{f(x_k)}{f'(x_k)} \right|$$
$$= \left| \frac{1}{f'(x_k)} \right| |f(x^*) - f(x_k) - f'(x_k)(x^* - x_k)|$$
$$\leq \gamma \beta |x^* - x_k|^2.$$

The ratio of the first and last statements shows that the order of convergence of Newton's method is at least 2, which establishes the following theorem.

Theorem 5. *Let f be a twice differentiable function with f'' continuous, $f(x^*) = 0$, and $f'(x^*) \neq 0$. If the Newton iterates x_k converge to x^*, then as x_k approaches x^* the order of convergence is at least 2.*

Quadratic convergence is desirable since it roughly means that the number of accurate significant digits doubles per iteration as the method converges. If the error is $1/10$ at iteration k, then the error at iteration $k+1$ is approximately $1/100$ and at iteration $k+2$ about $1/10,000$. We note that there are problem classes for which the order can be increased, providing extremely rapid convergence.

1.4.3 MATLAB Functions

MATLAB and Octave both provide the function `fzero` for finding a root of a scalar nonlinear function. This function may be called with a variety of options, see the documentation for a complete list. The expression,

```
[x,fx,status]=fzero(f,x0)
```

returns an approximate root, x, the function value, fx, and a status flag, status.

The arguments to `fzero` are a function, `f`, and an initial bracket, i.e. an initial interval, or starting point x0. If x0 is a vector of length two, i.e. x0 =[a,b], then the initial bracket $[a,b]$ is used in the search. If x0 is a point, then `fzero` attempts to bracket a root in the neighborhood of x0 prior to initiating its search.

For example, the code

```
f=@(x) cos(x)-x;
[x,fx,status]=fzero(f,[0,1]);
```

finds a root of $f(x) = \cos(x) - x$ in the interval $[0,1]$. The code

1.5 Exercises

```
f=@(x) cos(x)-x;
[x,fx,status]=fzero(f,1);
```

finds a root of $f(x) = \cos(x) - x$ near the initial point $x_0 = 1$.

fzero is a bracketing based method, and the initial interval must satisfy $\text{sign}(f(a)) \neq \text{sign}(f(b))$. If an initial point is given rather than an initial interval, then MATLAB and Octave attempt to find an appropriate interval about x0. The algorithm aborts if no such interval is found. For example, the function $f(x) = x^2$ has a root at $x = 0$, but the code

```
f=@(x) x^2;
[x,fx,status]=fzero(f,1.2);
```

fails because it is impossible to appropriately bracket the root.

MATLAB and Octave both provide the function fsolve as an alternative to fzero. fsolve is not a bracketing method, and thus it can find a root of a function even if no bracket is possible. However, this increase in ability comes at a cost, and just as we saw with Newton's method, fsolve may fail to find a root even if one exists. Despite this potential issue, fsolve is regularly reliable in practice. fsolve is also applicable to systems of nonlinear equations, a feature not shared by fzero or any bracketing method.

The code

```
f=@(x) x^2;
[x,fx,status]=fsolve(f,1.2);
```

successfully finds the root of x^2, where fzero failed.

The special structure of polynomials allows us to construct code that can find all roots. The function

```
r=roots(c)
```

returns all the roots, both real and complex, of the polynomial $p(x)$ whose coefficients are stored in the vector c. For example, the code

```
r=roots([1 0 -2]);
```

returns both roots of the polynomial $p(x) = x^2 - 2$.

1.5 Exercises

1. Suppose the method of bisection is initialized with the interval $[-10^4, 10^4]$. How many iterations are needed to guarantee that the k-th interval is no wider than 10^{-8}?

2. Let
$$g_q(x) = \begin{cases} (x/|x|)e^{-1/x^{2q}}, & x \neq 0 \\ 0, & x = 0. \end{cases}$$

Show that if x is a nonzero value in $(-1, 1)$, then $g_q(x) \to 0$ as $q \to \infty$. Explain why the methods of bisection and linear interpolation might have an advantage over the method of secants and Newton's method as q increases.

3. Show that the secant method can diverge even though the error, i.e. the value of $|f(x_k)|$, in solving

$$f(x) = \frac{x}{1+x^2} = 0$$

diminishes at each iteration.

4. Show that the sequence defined by the recursive statement $d_{k+1} = d_k/2$, with $d_1 = a$ converges with order 1 for any constant a. What does this say about the bisection method, especially if compared to Newton's method?

5. Write a function with declaration

        [xbest,fxbest,nitr,status]=Bisection(f,a,b,
            epsilon,delta,maxitr,loud)

that attempts to calculate a root of the function f with the bisection method.

The input arguments are:

f   - The function whose root is being approximated. Bisection should accept f either as a string containing a file name or as a function handle.
a   - An endpoint of the search interval.
b   - A second endpoint of the search interval. An error should occur if $[a, b]$ does not form a bracket of a root.
epsilon - A convergence criterion on $|f(x_{best})|$.
delta   - A convergence criterion on $|x^* - x_{best}|$, where x^* is an exact root of f. The code should terminate when both $|x^* - x_{best}| < \delta$ is guaranteed and $|f(x_{best})| < \varepsilon$, or when $f(x_{best}) = 0$.
maxitr   - The maximum number of iterations that should be calculated.
loud   - An optional variable that defaults to 0. If loud is nonzero, then the code should produce output in the format below.

The output arguments are:

xbest   - The best computed approximation of a root of f, i.e. the value of x with the smallest observed value of $|f|$.
fxbest - The value of f at x_{best}.

nitr   - The number of iterations used to compute xbest, where
            the initial bracket counts as iteration 0.
status   - A status variable encoded as follows:

status 0: Success, meaning that either both convergence criteria
          are satisfied or that an exact root is calculated. Use
          sign(fxbest) == 0 as the test for the latter case.
status 1: Failure, the iteration limit was reached.
status 2: Some failure other than the iteration limit occurred.
          For example, $[a, b]$ does not form a bracket of a root of
          f, $\varepsilon \leq 0$, or $\delta \leq 0$.

For example, running

```
[x,fx,nitr,stat]=Bisection(f,0,5,1e-3,1e-4,50,1)
```

will attempt to find a root of f in the interval $[0, 5]$ with the bisection method. The search will stop once $|f(x_{best})| < 10^{-3}$ and $|x^* - x_{best}| \leq 10^{-4}$ or once 50 iterations are reached. Output is desired, and it should be displayed like the example below (iteration information could be on the same line, and we have skipped several iterations for brevity):

```
Itr:0 a:0.00 b:5.00  |b-a|:5.00e+00
      |f(a)|:5.00e+00 |f(b)|: 2.00e+01  xstar: 0.00
Itr:1 a:0.00 b:2.50  |b-a|:2.50e+00
      |f(a)|:5.00e+00 |f(b)|: 1.25e+00  xstar: 2.50
Itr:2 a:1.25 b:2.50  |b-a|:1.25e+00
      |f(a)|:3.44e+00 |f(b)|: 1.25e+00  xstar: 2.50
Itr:3 a:1.88 b:2.50  |b-a|:6.25e-01
      |f(a)|:1.48e+00 |f(b)|: 1.25e+00  xstar: 2.50
Itr:4 a:2.19 b:2.50  |b-a|:3.12e-01
      |f(a)|:2.15e-01 |f(b)|: 1.25e+00  xstar: 2.19
          .
          .
          .
Itr:16 a:2.24 b:2.24  |b-a|:7.63e-05
       |f(a)|:2.01e-04 |f(b)|: 1.40e-04  xstar: 2.24
```

6. Repeat Exercise 5 for the method of linear interpolation. The function declaration should be

```
[xbest,fxbest,nitr,status]=Interpolation(f,a,b,
    epsilon,maxitr,loud)
```

The input arguments are:

f   - The function whose root is to be approximated, either as
         a function handle or as a string containing a file name.
a   - One endpoint of a bracket of a root of f.
b   - A second endpoint of a bracket of a root of f. An error
         should occur if $[a, b]$ does not form a bracket of a root.

epsilon   - The convergence criterion on $|f(x_{best})|$.
maxitr   - The maximum number of iterations to perform.
loud   - An optional variable that defaults to 0. If loud is nonzero, then output in the format indicated below should be produced.

The output arguments are:

xbest   - The best computed approximation of a root of f.
fxbest - The value of f at xbest.
nitr   - The number of iterations required to satisfy the convergence criterion, counting the initial bracket as iteration 0.
status - A status variable, encoded as follows:

  status 0: Success, an xbest such that $|\text{fxbest}| < \varepsilon$ has been found in less than maxitr iterations.
  status 1: Failure, the iteration limit was reached.
  status 2: Failure for some other reason, such as $\varepsilon \leq 0$.

If output is desired, then it should be in the same format as Exercise 5.

7. Repeat Exercise 6 for the secant method. The function declaration should be

    [xbest, fxbest, nitr, status] = Secant(f, x0, x1,
        epsilon, maxitr, loud)

The input arguments are:

f   - The function whose root is to be approximated, either as a function handle or as a string containing a file name.
x0   - One guess at a root of f.
x1   - A second guess at a root of f.
epsilon - The convergence criterion on $|f(x_{best})|$.
maxitr   - The maximum number of iterations to perform.
loud   - An optional variable that defaults to 0. If loud is nonzero, then output in the format indicated below should be produced.

The output arguments are:

xbest   - The computed approximation of a root of f.
fxbest - The value of f at xbest.

1.5 Exercises

nitr   -   The number of iterations required to satisfy the convergence criterion, counting x_0 as iteration 0 and x_1 as iteration 1.

status   -   A status variable, encoded as follows:

status 0: Success, an xbest such that $|fxbest| < \varepsilon$ has been found in less than maxitr iterations.

status 1: Failure, the iteration limit was reached.

status 2: Failure for some other reason, such as $\varepsilon \leq 0$.

If output is desired, then it should appear as:

```
Itr:0  x:0.00  |f(x)|:5.00e+00
Itr:1  x:5.00  |f(x)|:2.00e+01
Itr:2  x:1.00  |f(x)|:4.00e+00
Itr:3  x:1.67  |f(x)|:2.22e+00
Itr:4  x:2.50  |f(x)|:1.25e+00
Itr:5  x:2.20  |f(x)|:1.60e-01
Itr:6  x:2.23  |f(x)|:9.05e-03
Itr:7  x:2.24  |f(x)|:7.37e-05
```

8. Write a function with declaration

```
[xbest,fxbest,nitr,status]=Newton(f,fp,x0,
    epsilon,maxitr,loud)
```

that uses Newton's method to attempt to find a root of f.

The input arguments are:

f   -   The function whose root is to be approximated, either as a function handle or as a string containing a file name.

fp   -   The derivative of f, f'.

x0   -   An initial guess at a root of f.

epsilon   -   The convergence criterion on $|f(x_{best})|$.

maxitr   -   The maximum number of iterations to perform.

loud   -   An optional argument, with default value 0. If loud is nonzero, then output should be formatted like that of Exercise 7.

The function should return the same information as Exercise 7. The output format is also identical to that of Exercise 7. Note that the case in which $f'(x_k) = 0$ is of special concern, and status should return as 1 in this case.

9. The value of $\sqrt[n]{d}$, for $d > 0$, is a solution to $x^n - d = 0$. Calculate $\sqrt[n]{d}$ for all primes between 2 and 97 as n ranges from 2 to 5. You may find the command isprime useful. Each calculation should be undertaken in four ways,

   - the method of bisection initialized over the interval $[0, d]$,
   - the method of linear interpolation over the interval $[0, d]$,
   - the method of secants initialized with $a = 0$ and $b = d$, and
   - Newton's method initialized with $x = d/2$.

   Limit the number of iterations to 1000, and assume all tolerances are 10^{-10}. Display a table for each d containing the number of iterations required by each method. For example, the table could look like

d	n	Bisect.	Lin. Int.	Secants	Newton
2	2	35	15	8	4
	3	36	30	11	5
	⋮	⋮	⋮	⋮	⋮
	5	37	95	*	5
3	2	35	20	8	4
	⋮	⋮	⋮	⋮	⋮

   Denote an algorithm's failure with a *. Which algorithms seem especially well suited for the task of evaluating roots?

10. Write a function with declaration

    `[px,q]=Horner(c,x0)`

    that evaluates a polynomial whose coefficients are stored in a vector c at a point x_0 using Hörner's method.

    The input arguments are:

    c   - A vector that contains the coefficients of p in order from highest to lowest degree. For example, $p(x) = 4x^3 + 2x + 9$ is stored as [4 0 2 9].
    x0 - The point at which p is to be evaluated.

    The output arguments are:

    px - The computed value of $p(x_0)$.
    q   - A vector that stores the coefficients of the quotient of $p(x)$ upon division by $(x - x_0)$, stored from highest degree to lowest degree.

1.5 Exercises

Note that this exercise is almost a direct re-implementation of the built-in function `polyval`.

11. Create a function with declaration

    ```
    [xbest,pxbest,nitr,q,status]=NewtonPoly(c,x0,
        epsilon,maxitr,loud)
    ```

    that uses Newton's method to solve $p(x) = 0$, where $p(x)$ is the polynomial whose coefficients are stored in `c` in order from highest to lowest degree. The function should use Hörner's method to evaluate $p(x)$ and $p'(x)$.

    The input arguments are:

    c   - A vector that stores the coefficients of p in order from highest to lowest degree.
    x0   - The initial guess at the root.
    epsilon - The convergence tolerance on $|p(x_{best})|$.
    maxitr   - The maximum number of iterations to perform.
    loud   - An optional argument, with default value 0. If `loud` is nonzero, then output should be in the same format as Exercise 8.

    The output arguments are:

    xbest   - The approximation to a root of p.
    pxbest - The value of $p(x_{best})$.
    nitr   - The number of iterations required to obtain x_{best}, with the initial guess counted as iteration 0.
    q   - The quotient of p upon division by $(x - x_{best})$, stored as a vector.
    status - A status variable, encoded as follows:

    status 0: Success, a value of x_{best} has been found such that $|p(x_{best})| < \varepsilon$ in less than `maxitr` iterations.
    status 1: Failure, the iteration limit was reached or an iteration x_k was found such that $p'(x_k) = 0$.
    status 2: Failure for some other reason.

    In the event that output is desired, it should be in the same format as Exercise 8. Your code should use the `Horner` function from Exercise 10 to evaluate both $p(x)$ and $p'(x)$.

12. One solution to $\cos(x) = -1$ is $x = \pi$. So solving $\cos(x) + 1 = 0$ in the neighborhood of 3 calculates π. Solve this equation with bisection, linear interpolation, secants, and Newton's method to calculate π to a desired accuracy. Comment on each algorithm's ability to accurately calculate π. Use reasonable initial conditions such $a = 3$ and $b = 3.5$ for the methods of bisection, linear interpolation, and secants; and $x_0 = 3$ for Newton's method.

13. Replace $\cos(x)$ in Exercise 12 with its approximate truncated Taylor expansion

$$\cos(x) \approx \sum_{n=0}^{N} \frac{(-1)^n}{(2n)!} x^{2n},$$

and use the efficient polynomial adaptation of Newton's method to approximate π by solving

$$1 + \sum_{n=0}^{N} \frac{(-1)^n}{(2n)!} x^{2n} = 0.$$

Explore the computational relationship between N and the possible accuracy of our estimate of π.

14. Write a function with declaration

```
[r,status]=AllRoots(c,x0,epsilon,maxitr)
```

that attempts to find all the roots, real or complex, of the polynomial whose coefficients are in c.

The input arguments are:

c   - A vector containing the coefficients of p in order from highest to lowest degree.
x0   - A starting point for Newton's method.
epsilon   - A convergence tolerance for Newton's method.
maxitr   - A maximum number of iterations to perform on each application of Newton's method.

The output arguments are:

r   - A vector containing all n roots of p, where n is the degree of p. Repeated roots should be listed according to their multiplicity.

1.5 Exercises

status - A status variable encoded as follows:

status 0: Success.

status 1: Failure due to Newton's method hitting the iteration limit or due to a zero derivative. r should contain all the roots of p that have been successfully identified in this case.

status 2: Failure for some other reason.

Your algorithm should work by repeatedly calling `NewtonPoly` from Exercise 11 with starting point x0, convergence tolerance epsilon, and iteration limit maxitr. Note that in order to identify complex roots, the starting point must be a complex number. Your code should print a warning message, but not return an error, if x0 is a real number.

Your code should use a deflation method. A deflation method works by observing that if x^* is a root of $p(x)$, then

$$p(x) = (x - x^*)q(x),$$

and so all other roots of $p(x)$ must also be roots of $q(x)$. Using this observation we can successfully detect repeated roots of $p(x)$ as well as avoid convergence to the same root on multiple runs of Newton's method. Pseudocode for the technique is in Algorithm 5.

Algorithm 5 Pseudocode for the deflation method
---
q=c
for k from 1 to n **do**
  Find a root r_k and quotient q_k of the polynomial with coefficients in q using the synthetic division method
  q=q_k
end for
return list of roots

---

15. Use one of the root finding methods to calculate an initial a such that Newton's method applied to

$$f(x) = \begin{cases} \dfrac{x^2}{1+x^2}, & x \geq 0 \\ \dfrac{-x^2}{1+x^2}, & x < 0 \end{cases}$$

will cycle between a and $-a$.

16. Is it possible for Newton's method to generate a single iterate between every consecutive pair of solutions to $\sin(x) = 0$?
17. The function ProjectileLanding.m, available at http://www.springer.com, has declaration

    d=ProjectileLanding(theta,v,m,alpha,w).

    This function returns the distance a projectile with mass m will travel down a flat range if fired with initial angle θ and an initial velocity v if its drag coefficient is α and in the presence of wind with velocity w.

    Create a function Velocity.m with declaration

    [v]=Velocity(theta,m,alpha,w,d)

    that returns the initial velocity needed to fire a projectile with mass m at an initial angle θ if the drag coefficient is α and there is a wind with velocity w in order to hit a target d meters downrange. Your code should rely on ProjectileLanding and the function fzero in MATLAB.
18. A 1×1 ft^2 metal plate is kept in a room with an ambient temperature $-5\,°$F. The metal plate has a known thermal conductivity, and the plate must be heated by a torch prior to one of the manufacturing stages. The temperature of the torch is $2000\,°$F, and it's centered at position $(0.6, 0.7)$ on the plate. The plate must have a temperature of $10\,°$F at position $(0.5, 0.5)$ for the next manufacturing stage to proceed. The question is, how long must we apply the torch in order for the plate to have the correct temperature? We intend for the next manufacturing stage to start in less than 4 s after we begin applying the torch, and experience suggests that 4 s exceeds the necessary heating time.

    The function HeatSimulation, available at http://www.springer.com, has declaration

    U=HeatSimulation(HeatSourceCenter,HeatSourceTmp,
        HeatSourceCutoffTime,SimulationEndTime,
        MeasurementPosition,ShowAnimation)

    This code is a general tool that simulates the distribution of heat in the metal plate as it is heated by the torch. The returned value of U is the temperature of the plate at a certain point at the end of the simulation.

    The arguments of HeatSimulation are:

    HeatSourceCenter   - The location on the plate at which the torch is centered.

1.5 Exercises

HeatSourceTmp	-	The temperature of the torch ($°F$).
HeatSourceCutoffTime	-	The amount of time the torch will run.
SimulationEndTime	-	The amount of time that will be simulated by the program.
MeasurementPosition	-	The position on the plate whose temperature at the end of the simulation will be returned.
ShowAnimation	-	A flag indicating whether or not an animation of the heat distribution should be shown.

Combine HeatSimulation and fzero to create a function with declaration

```
[t,status]=RequiredHeatingTime(MeasurementTime,
    MeasurementPosition,HeatingPosition,
    TargetTemperature)
```

that returns the heating time necessary in order to make sure position measurement_position has exactly temperature TargetTemperature at time TargetTime if it is heated with a 2000 °F torch applied to position heating_position. Your code should not produce animations as it runs. Return status 1 if it is impossible to hit the desired temperature by the desired time. Otherwise, return status 0. Watch out! If you ask for an impossible result, for example heating a particular point to 100,000 °F in no more than 2 s, then it will take a very long time for fzero to determine that this is impossible.

Chapter 2
Solving Systems of Equations

> But all of my efforts served only to make me better acquainted with the difficulty, which in itself was something. – Henri Poincaré

The problem of finding a solution that simultaneously solves several equations does not lend itself to the sequential application of the computational methods for single equations. As an example, consider finding an x that solves both
$$x^2 - x = 0 \text{ and } x^2 + x = 0. \tag{2.1}$$
The first equation has solutions $x = 0$ and $x = 1$, and the second has solutions $x = 0$ and $x = -1$. If solved separately, then it would be possible to calculate the $x = 0$ solution for the first and the $x = -1$ solution for the second. We might erroneously conclude that there is no common answer, and hence, it is imperative that both equations be considered simultaneously.

We use the following notation to consider several functions simultaneously,
$$f : \mathbb{R}^n \to \mathbb{R}^m : x \mapsto (f_1(x), f_2(x), \ldots, f_m(x))^T.$$

This notation shows that f is a multi-dimensional function of the vector $x = (x_1, x_2, \ldots, x_n)^T$. The function f produces m values that are the components of the vector $(f_1(x), \ldots, f_m(x))^T$, where the real-valued functions f_i are called the component functions of f. For example, if
$$f : \mathbb{R}^1 \to \mathbb{R}^2 : x \mapsto (x^2 - x, x^2 + x)^T,$$
then the equation $f(x) = 0$ is the same as system (2.1). Similarly, the equation
$$f(x_1, x_2) = \begin{pmatrix} x_1^2 + x_2^2 - 1 \\ x_1 + 2x_2 - 2 \end{pmatrix} = 0$$

is a reformulation of the system

$$x^2 + y^2 = 1$$
$$x + 2y = 2.$$

The change in notation from x and y to x_1 and x_2 is common, and in general, we use notation convenient for the presentation.

2.1 Systems of Linear Equations

Linear systems are the bedrock of computational science, and a lucid command of how they are solved is important. Indeed, even nonlinear systems are most commonly estimated linearly as

$$f(x) \approx f(z) + D_x f(z)(x - z),$$

where $D_x f$ is the general extension of df/dx and is developed in Sect. 2.3. The equation $f(x) = 0$ is then replaced iteratively with the approximate linear equation $f(z) + D_x f(z)(x - z) = 0$, illustrating that even nonlinear systems rely on solving related linear systems.

We assume throughout this section that

$$f : \mathbb{R}^n \to \mathbb{R}^m$$

is the affine map

$$f(x) = Ax - b,$$

where A is an $m \times n$ matrix and b is an m-vector. Thus solving $f(x) = 0$ is the same as solving the linear system $Ax = b$. The term "affine" means that f is the sum of a linear function, which is Ax, and a constant (vector), which is b. We begin by considering the special linear cases in which a matrix is either upper- or lower-triangular.

2.1.1 Upper- and Lower-Triangular Linear Systems

Let L be an $n \times n$ lower triangular matrix,

$$L = \begin{bmatrix} l_{11} & 0 & \cdots & & 0 \\ l_{21} & l_{22} & 0 & \cdots & 0 \\ \vdots & & \ddots & & \\ l_{n1} & & \cdots & & l_{nn} \end{bmatrix},$$

and consider the linear system of equations

2.1 Systems of Linear Equations

$$Lx = b. \tag{2.2}$$

If we look at the component form of (2.2), then we see that the system of equations is

$$\begin{aligned} l_{11}\,x_1 &= b_1 \\ l_{21}\,x_1 + l_{22}\,x_2 &= b_2 \\ &\;\;\vdots \\ l_{n1}\,x_1 + l_{n2}\,x_2 + \cdots + l_{nn}\,x_n &= b_n. \end{aligned} \tag{2.3}$$

The solution is guaranteed if the diagonal elements of L are nonzero, and in this case we have

$$\left. \begin{aligned} x_1 &= b_1/l_{11} \\ x_2 &= (b_2 - l_{21}x_1)/l_{22} \\ x_3 &= (b_3 - l_{31}x_1 - l_{32}x_2)/l_{33} \\ &\;\;\vdots \\ x_n &= (b_n - l_{n1}x_1 - \cdots - l_{n,n-1}x_{n-1})/l_{nn}. \end{aligned} \right\} \tag{2.4}$$

Consider the system for which

$$L = \begin{bmatrix} 1 & 0 & 0 \\ 1 & 1 & 0 \\ 2 & 2 & 1 \end{bmatrix} \quad \text{and} \quad b = \begin{pmatrix} -2 \\ 2 \\ 3 \end{pmatrix}.$$

Proceeding from top to bottom, we have

$$Lx = b \iff \begin{bmatrix} 1 & 0 & 0 \\ 1 & 1 & 0 \\ 2 & 2 & 1 \end{bmatrix} \begin{pmatrix} x_1 \\ x_2 \\ x_3 \end{pmatrix} = \begin{pmatrix} -2 \\ 2 \\ 3 \end{pmatrix} \begin{array}{l} \Rightarrow x_1 = -2 \\ \Rightarrow x_2 = 2 - x_1 = 4 \\ \Rightarrow x_3 = 3 - 2x_1 - 2x_2 = -1. \end{array}$$

Similarly, if U is an $n \times n$ upper triangular matrix with no zeros on its main diagonal, then the solution to

$$Ux = b$$

is

$$\left. \begin{aligned} x_n &= b_n/u_{nn} \\ x_{n-1} &= (b_{n-1} - u_{n-1,n}x_n)/u_{n-1,n-1} \\ x_{n-2} &= (b_{n-2} - u_{n-2,n-1}x_{n-1} - u_{n-2,n}x_n)/u_{n-2,n-2} \\ &\;\;\vdots \\ x_1 &= (b_1 - u_{12}x_2 - u_{13}x_3 - \cdots - u_{1n}x_n)/u_{11}. \end{aligned} \right\} \tag{2.5}$$

So, if
$$U = \begin{bmatrix} 1 & -2 & 1 \\ 0 & 3 & -1 \\ 0 & 0 & 1 \end{bmatrix} \quad \text{and} \quad b = \begin{pmatrix} -2 \\ 4 \\ -1 \end{pmatrix},$$
then we solve $Ux = b$ by proceeding from bottom to top,
$$Ux = b \iff \begin{bmatrix} 1 & -2 & 1 \\ 0 & 3 & -1 \\ 0 & 0 & 1 \end{bmatrix} \begin{pmatrix} x_1 \\ x_2 \\ x_3 \end{pmatrix} = \begin{pmatrix} -2 \\ 4 \\ -1 \end{pmatrix} \begin{matrix} \Rightarrow x_1 = -2 + 2x_2 - x_3 = 1 \\ \Rightarrow x_2 = (4 + x_3)/3 = 1 \\ \Rightarrow x_3 = -1. \end{matrix}$$

A straightforward calculation of the number of multiplications and divisions in (2.4) and (2.5) shows that it is possible to solve an upper- or lower-triangular linear system with approximately $n^2/2$ multiplications. This computational efficiency is important since solving a general, square linear system requires about $n^3/3$ multiplications, which is far greater for large n.

2.1.2 General m × n Linear Systems

We now extend our success with upper- and lower-triangular systems to general linear systems. Computing a solution to a linear system can be accomplished by combining the three row operations:

- multiply any row by a nonzero scalar,
- swap a pair of rows, and
- add a multiple of one row to another.

None of these operations alters the set of solutions, and hence, these can be applied to the system $Ax = b$ without losing a solution or introducing a faux solution.

Our strategy is to use row operations to find a lower-triangular matrix L, with ones on the main diagonal, and an upper-triangular matrix U so that $A = LU$. Expressing A as a product of two matrices is called factoring, and the pair L and U form a factorization of A. We can re-express the linear system of equations once L and U are determined using the associative property of matrix multiplication. The result is
$$Ax = LUx = L(Ux) = b.$$

This last equality shows that $Ax = b$ can be solved sequentially by

$$\begin{aligned} &\textbf{Step 1: Solving } Ly = b, \text{ and} \\ &\textbf{Step 2: Solving } Ux = y. \end{aligned} \qquad (2.6)$$

2.1 Systems of Linear Equations

We can use the algorithms from Sect. 2.1.1 to solve each step, resulting in an efficient method to solve a general linear system. Calculating a solution from the factorization requires a forward (Step 1) and a backward (Step 2) substitution.

The calculation of L and U follows the standard row reduction process of solving a linear system, and this is how we begin our development. Consider the linear system $Ax = b$ given by

$$\begin{bmatrix} 1 & -2 & 1 \\ 1 & 1 & 0 \\ 2 & 2 & 1 \end{bmatrix} x = \begin{bmatrix} -2 \\ 2 \\ 3 \end{bmatrix}. \tag{2.7}$$

The augmented matrix for system (2.7) is

$$\left[\begin{array}{ccc|c} 1 & -2 & 1 & -2 \\ 1 & 1 & 0 & 2 \\ 2 & 2 & 1 & 3 \end{array}\right].$$

We use notation like $R_1 + 2R_2 \to R_1$ to show that row one is being replaced with 2 copies of row two added to row one. Reducing the augmented matrix is accomplished with the following row operations, and the system is solved with a backward substitution,

$$\left[\begin{array}{ccc|c} 1 & -2 & 1 & -2 \\ 1 & 1 & 0 & 2 \\ 2 & 2 & 1 & 3 \end{array}\right]$$

$$-R_1 + R_2 \to R_2 \quad \left[\begin{array}{ccc|c} 1 & -2 & 1 & -2 \\ 0 & 3 & -1 & 4 \\ 2 & 2 & 1 & 3 \end{array}\right]$$

$$-2R_1 + R_3 \to R_3 \quad \left[\begin{array}{ccc|c} 1 & -2 & 1 & -2 \\ 0 & 3 & -1 & 4 \\ 0 & 6 & -1 & 7 \end{array}\right]$$

$$-2R_2 + R_3 \to R_3 \quad \left[\begin{array}{ccc|c} 1 & -2 & 1 & -2 \\ 0 & 3 & -1 & 4 \\ 0 & 0 & 1 & -1 \end{array}\right] \begin{array}{l} \Rightarrow x_1 = (-2 + 2x_2 - x_3) = 1 \\ \Rightarrow x_2 = (4 + x_3)/3 = 1 \\ \Rightarrow x_3 = -1. \end{array}$$

The first nonzero element of each row in the reduced matrix is called a pivot, and these values are used in the reduction process to "zero out" the elements below them.

The coefficient matrix of the augmented system's row echelon form is upper triangular, and this coefficient matrix is U in our factorization. The result of the row operations being U suggests that multiplying by L should somehow encode the row operations that transform A into U. This is indeed true and is simply expressed by embedding row operations in matrix multiplication.

An elementary matrix is the result of performing a single row operation on the identity. The first row operation's elementary matrix is

$$-R_1 + R_2 \to R_2 \iff \begin{bmatrix} 1 & 0 & 0 \\ -1 & 1 & 0 \\ 0 & 0 & 1 \end{bmatrix}.$$

If we multiply the coefficient matrix on the left by this elementary matrix, then the result is the performance of the row operation,

$$\begin{bmatrix} 1 & 0 & 0 \\ -1 & 1 & 0 \\ 0 & 0 & 1 \end{bmatrix} \begin{bmatrix} 1 & -2 & 1 \\ 1 & 1 & 0 \\ 2 & 2 & 1 \end{bmatrix} = \begin{bmatrix} 1 & -2 & 1 \\ 0 & 3 & -1 \\ 2 & 2 & 1 \end{bmatrix}.$$

Repeating with the other two row operations shows

$$\begin{bmatrix} 1 & 0 & 0 \\ 0 & 1 & 0 \\ 0 & -2 & 1 \end{bmatrix} \begin{bmatrix} 1 & 0 & 0 \\ 0 & 1 & 0 \\ -2 & 0 & 1 \end{bmatrix} \begin{bmatrix} 1 & 0 & 0 \\ -1 & 1 & 0 \\ 0 & 0 & 1 \end{bmatrix} \begin{bmatrix} 1 & -2 & 1 \\ 1 & 1 & 0 \\ 2 & 2 & 1 \end{bmatrix} = \begin{bmatrix} 1 & -2 & 1 \\ 0 & 3 & -1 \\ 0 & 0 & 1 \end{bmatrix}.$$

Let the three elementary matrices be E_1, E_2, and E_3 so that we can write the last equality as

$$E_3 E_2 E_1 A = U.$$

Each elementary matrix is invertible, from which we have

$$A = \left(E_1^{-1} E_2^{-1} E_3^{-1} \right) U = LU. \tag{2.8}$$

Hence, L is the product of the inverses of the elementary matrices, and as expected, it reverses the row operations to transform U back into A.

While Eq. (2.8) defines L, it hides some of its structure and form. The inverse of the row operation $-2R_1 + R_3 \to R_3$ adds 2 copies of row one back to row three. In other words, the inverse of $-2R_1 + R_3 \to R_3$ is $2R_1 + R_3 \to R_3$. In terms of elementary matrices we have

$$\begin{bmatrix} 1 & 0 & 0 \\ 0 & 1 & 0 \\ -2 & 0 & 1 \end{bmatrix}^{-1} = \begin{bmatrix} 1 & 0 & 0 \\ 0 & 1 & 0 \\ 2 & 0 & 1 \end{bmatrix}.$$

In general, the inverse of an elementary matrix is another elementary matrix (of the same form). The product of the inverses forming L is

$$E_1^{-1} E_2^{-1} E_3^{-1} = \begin{bmatrix} 1 & 0 & 0 \\ 1 & 1 & 0 \\ 0 & 0 & 1 \end{bmatrix} \begin{bmatrix} 1 & 0 & 0 \\ 0 & 1 & 0 \\ 2 & 0 & 1 \end{bmatrix} \begin{bmatrix} 1 & 0 & 0 \\ 0 & 1 & 0 \\ 0 & 2 & 1 \end{bmatrix} = \begin{bmatrix} 1 & 0 & 0 \\ 1 & 1 & 0 \\ 2 & 2 & 1 \end{bmatrix} = L.$$

2.1 Systems of Linear Equations

The elements below the main diagonal of L are individually the non-identity components of the elementary matrices, which is always true with this type of row operation.

We now demonstrate that it is possible to factor A into L and U by overwriting the values of A itself, which reduces the use of memory. The process is illustrated for the example, with the boxed items being the elements of L as they are calculated,

$$\begin{bmatrix} 1 & -2 & 1 \\ 1 & 1 & 0 \\ 2 & 2 & 1 \end{bmatrix}$$

$$-R_1 + R_2 \to R_2 \quad \begin{bmatrix} 1 & -2 & 1 \\ \boxed{1} & 3 & -1 \\ 2 & 2 & 1 \end{bmatrix}$$

$$-2R_1 + R_3 \to R_3 \quad \begin{bmatrix} 1 & -2 & 1 \\ \boxed{1} & 3 & -1 \\ \boxed{2} & 6 & -1 \end{bmatrix}$$

$$-2R_2 + R_3 \to R_3 \quad \begin{bmatrix} 1 & -2 & 1 \\ \boxed{1} & 3 & -1 \\ \boxed{2} & \boxed{2} & 1 \end{bmatrix} \Rightarrow \begin{bmatrix} 1 & -2 & 1 \\ 1 & 3 & -1 \\ 2 & 2 & 1 \end{bmatrix}$$

$$\Rightarrow L = \begin{bmatrix} 1 & 0 & 0 \\ 1 & 1 & 0 \\ 2 & 2 & 1 \end{bmatrix} \text{ and } U = \begin{bmatrix} 1 & -2 & 1 \\ 0 & 3 & -1 \\ 0 & 0 & 1 \end{bmatrix}.$$

Multiplying LU confirms that $A = LU$.

Although the previous example illustrates the procedure of calculating L and U, it does not exhibit the care required to generalize the algorithm so that it aptly factors matrices of arbitrary size in a fashion that is numerically accurate. Since any matrix can be reduced to an upper triangular matrix with the three operations, it is immediate that any matrix can be factored as LU if we include a permutation of the rows. An improvement in numerical accuracy comes from the fact that we want to avoid division by relatively small numbers, which we accomplish with appropriate row swaps to divide by the largest possible values. Row swaps lead to a factorization that depends on a permutation of the rows of A, and the technique is called partial pivoting.

To illustrate a more general case, consider the non-square matrix

$$A = \begin{bmatrix} 60 & 40 & -85 & 13 \\ 20 & 15 & 55 & -4 \\ 80 & 40 & 20 & 4 \end{bmatrix}.$$

The biggest element of column 1 is in the 3rd row, and the first permutation represents a swap between the first and third rows. We don't actually swap the rows since doing so would necessitate added memory and extra computational effort. We instead continue as though we had swapped the rows by referencing row indices relative to a permutation vector p, which is more efficient to update. If we start with the original row indices listed in $p = (1, 2, 3)^T$, then swapping rows 1 and 3 is recorded by swapping p_1 and p_3, resulting in $p = (3, 2, 1)^T$.

The reduction process would normally proceed by making the elements of the column below the pivot zero. However, with the pivot being in the third row, we instead reduce the matrix so that the first two elements of column 1 are zero. This reduction is shown below, with the boxed elements being the components of L from the row operations,

$$p = \begin{pmatrix} 3 \\ 2 \\ 1 \end{pmatrix} \quad \begin{matrix} -(1/4)R_{p_1} + R_{p_2} \to R_{p_2} \\ -(3/4)R_{p_1} + R_{p_3} \to R_{p_3} \end{matrix} \quad \begin{bmatrix} \boxed{3/4} & 10 & -100 & 10 \\ \boxed{1/4} & 5 & 50 & -5 \\ 80 & 40 & 20 & 4 \end{bmatrix}.$$

The second column's largest element excluding the first row indexed by p, i.e. excluding row 3, is $A_{12} = 10$. Hence we want to think of row 1 as the second row, which means p_2 needs to become 1. The updated permutation is $p = (3, 1, 2)^T$, which means that we are using $p_1 = 3$ as the "first" row, $p_2 = 1$ as the "second" row, and $p_3 = 2$ as the "third" row. The row operation using $A_{12} = 10$ to eliminate $A_{22} = 5$ is

$$p = \begin{pmatrix} 3 \\ 1 \\ 2 \end{pmatrix} \quad -(1/2)R_{p_2} + R_{p_3} \to R_{p_3} \quad \begin{bmatrix} \boxed{3/4} & 10 & -100 & 10 \\ \boxed{1/4} & \boxed{1/2} & 100 & -10 \\ 80 & 40 & 20 & 4 \end{bmatrix}.$$

If we re-order the rows as indicated by the permutation p, then we have

$$\begin{matrix} \text{row } p_1 = 3 \to \\ \text{row } p_2 = 1 \to \\ \text{row } p_3 = 2 \to \end{matrix} \begin{bmatrix} 80 & 40 & 20 & 4 \\ 3/4 & 10 & -100 & 10 \\ 1/4 & 1/2 & 100 & -10 \end{bmatrix}.$$

The resulting L and U are reported in terms of the permutation, and hence,

2.1 Systems of Linear Equations

$$L = \begin{bmatrix} 1 & 0 & 0 \\ 3/4 & 1 & 0 \\ 1/4 & 1/2 & 1 \end{bmatrix} \quad \text{and} \quad U = \begin{bmatrix} 80 & 40 & 20 & 4 \\ 0 & 10 & -100 & 10 \\ 0 & 0 & 100 & -10 \end{bmatrix}.$$

In this case we have

$$LU = \begin{bmatrix} 1 & 0 & 0 \\ 3/4 & 1 & 0 \\ 1/4 & 1/2 & 1 \end{bmatrix} \begin{bmatrix} 80 & 40 & 20 & 4 \\ 0 & 10 & -100 & 10 \\ 0 & 0 & 100 & -10 \end{bmatrix}$$

$$= \begin{bmatrix} 80 & 40 & 20 & 4 \\ 60 & 40 & -85 & 13 \\ 20 & 15 & 55 & -4 \end{bmatrix} = \text{A(p,:)}.$$

The notation A(p,:) is how MATLAB and Octave would re-order the rows of A relative to the permutation vector.

Unfortunately, the LU factorization is not necessarily unique, a fact demonstrated by changing A_{12} from 40 to 30 and A_{22} from 15 to 10. The new matrix is

$$A = \begin{bmatrix} 60 & 30 & -85 & 13 \\ 20 & 10 & 55 & -4 \\ 80 & 40 & 20 & 4 \end{bmatrix}.$$

The first two row operations remain the same, and the result is

$$p = \begin{pmatrix} 3 \\ 2 \\ 1 \end{pmatrix} \quad \begin{array}{c} -(1/4)R_{p_1} + R_{p_2} \to R_{p_2} \\ -(3/4)R_{p_1} + R_{p_3} \to R_{p_3} \end{array} \quad \begin{bmatrix} \boxed{3/4} & 0 & -100 & 10 \\ \boxed{1/4} & 0 & 50 & -5 \\ 80 & 40 & 20 & 4 \end{bmatrix}. \quad (2.9)$$

The second column is already zero except for the row first indexed by p, i.e. row 3, and the process can halt. Re-ordering the rows with respect to the permutation vector gives

$$\begin{array}{c} \text{row } p_1 = 3 \to \\ \text{row } p_2 = 2 \to \\ \text{row } p_3 = 1 \to \end{array} \begin{bmatrix} 80 & 40 & 20 & 4 \\ 1/4 & 0 & 50 & -5 \\ 3/4 & 0 & -100 & 10 \end{bmatrix}.$$

The resulting factorization is

$$LU = \begin{bmatrix} 1 & 0 & 0 \\ 1/4 & 1 & 0 \\ 3/4 & 0 & 1 \end{bmatrix} \begin{bmatrix} 80 & 40 & 20 & 4 \\ 0 & 0 & 50 & -5 \\ 0 & 0 & -100 & 10 \end{bmatrix}$$

$$= \begin{bmatrix} 80 & 40 & 20 & 4 \\ 20 & 10 & 55 & -4 \\ 60 & 30 & -85 & 13 \end{bmatrix} = \text{A(p,:)}.$$

The last two rows of U suggest that we perform an additional row operation to eliminate either the 50 or the -100. Indeed, such a row operation would be expected if we were working toward one of the echelon forms. Since the largest magnitude element is -100, our scheme to divide by the largest magnitude value mandates that we use -100 to eliminate the 50. Working from (2.9), we have

$$p = \begin{pmatrix} 3 \\ 1 \\ 2 \end{pmatrix} \quad (1/2)R_{p_2} + R_{p_3} \to R_{p_3} \quad \begin{bmatrix} \boxed{3/4} & 0 & -100 & 10 \\ \boxed{1/4} & \boxed{-1/2} & 0 & 0 \\ 80 & 40 & 20 & 4 \end{bmatrix}. \quad (2.10)$$

The new element of L is $-1/2$, and its position takes some care to identify. First, the row being updated is $p_3 = 2$, and hence, the new element of L must reside in the 2nd row of our unpermuted matrix. The element being used to do the elimination is in the 2nd row as indexed by p, i.e. -100 is in the second row after permuting the rows since $p_2 = 1$. This means that the new element of L is in the $(2,2)$-position. We again permute the rows to identify L and U as lower- and upper-triangular matrices,

$$\begin{array}{c} \text{row } p_1 = 3 \to \\ \text{row } p_2 = 1 \to \\ \text{row } p_3 = 2 \to \end{array} \begin{bmatrix} 80 & 40 & 20 & 4 \\ \boxed{3/4} & 0 & -100 & 10 \\ 1/4 & \boxed{-1/2} & 0 & 0 \end{bmatrix}.$$

The resulting factorization is

$$LU = \begin{bmatrix} 1 & 0 & 0 \\ 3/4 & 1 & 0 \\ 1/4 & -1/2 & 1 \end{bmatrix} \begin{bmatrix} 80 & 40 & 20 & 4 \\ 0 & 0 & -100 & 10 \\ 0 & 0 & 0 & 0 \end{bmatrix} = \begin{bmatrix} 80 & 40 & 20 & 4 \\ 60 & 30 & -85 & 13 \\ 20 & 10 & 55 & -4 \end{bmatrix},$$

which is again $\mathtt{A(p,:)}$, but with a different L, U, and p.

With an L, U, and p we can solve $Ax = b$ with the same two-step procedure listed in (2.6). However, in this case the system has either no solution or an infinite number of solutions. This is because the rank of A is 2. The rank being less than the number of rows means that the system could be inconsistent. The rank being less than the number of columns means that the system has free variables and is thus guaranteed to have an infinite number of solutions should it be consistent. If $b = (-72, 51, 24)^T$, then using the last LU factorization of A allows us to solve $Ly = b(p)$ as

$$\begin{bmatrix} 1 & 0 & 0 \\ 3/4 & 1 & 0 \\ 1/4 & -1/2 & 1 \end{bmatrix} \begin{pmatrix} y_1 \\ y_2 \\ y_3 \end{pmatrix} = \begin{pmatrix} 24 \\ -72 \\ 51 \end{pmatrix} \begin{array}{l} \Rightarrow y_1 = 24 \\ \Rightarrow y_2 = -72 - (3/4)y_1 = -90 \\ \Rightarrow y_3 = 51 - (1/4)y_1 + (1/2)y_2 = 0. \end{array}$$

Notice that the elements of b are permuted according to p to align the elements of b with their corresponding equations. Step 2 is

2.1 Systems of Linear Equations

$$\begin{bmatrix} 80 & 40 & 20 & 4 \\ 0 & 0 & -100 & 10 \\ 0 & 0 & 0 & 0 \end{bmatrix} \begin{pmatrix} x_1 \\ x_2 \\ x_3 \\ x_4 \end{pmatrix} = \begin{pmatrix} 24 \\ -90 \\ 0 \end{pmatrix} \begin{array}{l} \Rightarrow x_1 = (6 - 40x_2 - 6x_4)/80 \\ \Rightarrow x_3 = -(-90 - 10x_4)/100 \\ \Rightarrow \text{consistency.} \end{array}$$

The expression for x_1 follows by replacing x_3 with $(-90 - 10x_4)/100$ so that

$$\begin{aligned} 80x_1 &= 24 - 40x_2 - 20x_3 - 4x_4 \\ &= 24 - 40x_2 + (20/100)(-90 - 10x_4) - 4x_4 \\ &= 6 - 40x_2 - 6x_4. \end{aligned}$$

Although a solution exists for any x_2 and x_4, a reasonable answer to compute has these values set to zero, which results in the solution

$$x = \begin{pmatrix} 0.0750 \\ 0 \\ 0.9000 \\ 0 \end{pmatrix}.$$

There is no need to actually permute the rows to calculate a solution for a specific right-hand side. Instead, the matrix in (2.10) could be used directly with the permutation vector p. If dealing with a special, possibly sparse, matrix format, then leaving the matrix intact and using the permutation to index a solution's calculation is typically preferred.

The command lu in MATLAB and Octave alters L and U depending on what is requested. If a permutation is requested, then L and U are upper and lower triangular, and the product LU is A after the rows are permuted. If a permutation is not requested, then LU is A, but L is not necessarily lower triangular and is instead a permuted matrix of the lower triangular matrix that could be had by appropriately ordering the rows of A. For example, the command

```
[L, U] = lu([1, 2; 3, 4]);
```

returns

$$L = \begin{bmatrix} 0.3333 & 1.0000 \\ 1.0000 & 0.0000 \end{bmatrix} \text{ and } U = \begin{bmatrix} 3.0000 & 4.0000 \\ 0.0000 & 0.6667 \end{bmatrix},$$

whereas

```
[L, U, P] = lu([1, 2; 3, 4]);
```

returns

$$L = \begin{bmatrix} 1.0000 & 0.0000 \\ 0.3333 & 1.0000 \end{bmatrix}, U = \begin{bmatrix} 3.0000 & 4.0000 \\ 0.0000 & 0.6667 \end{bmatrix}, \text{ and } P = \begin{bmatrix} 0 & 1 \\ 1 & 0 \end{bmatrix}.$$

The matrix P represents the permutation $p = (2,1)^T$, as it is the 2×2 identity with the first and second rows swapped. For any 2-element vector we have

$$Pv = \begin{bmatrix} 0 & 1 \\ 1 & 0 \end{bmatrix} \begin{pmatrix} v_1 \\ v_2 \end{pmatrix} = \begin{pmatrix} v_2 \\ v_1 \end{pmatrix},$$

which correctly swaps the first and second components. Subsequently, the product PA, where A has two rows, swaps the first and second rows of A. In general, a permutation matrix is the result of swapping the rows of the identity according to a permutation, and the action of the permutation is accomplished by matrix multiplication.

Factoring A into LU and then solving $Ax = b$ requires approximately the same number of calculations as row reducing the augmented system $[A|b]$, which is about $n^3/3$. However, the LU factorization stores the reduction process and can be efficiently re-used for different right-hand sides. In this scheme, the first solve of $Ax = b$ requires approximately $n^3/3$ operations, but every subsequent solve of $Ax = b$, where b has been updated, only requires about n^2 operations. This significant time savings is critical in some applications. Solving from an LU factorization is also numerically preferred in some cases to performing Gauss-Jordan elimination on $[A|b]$, i.e. performing row operations so that $[A|b]$ is in reduced row echelon form.

2.2 Special Structure: Positive Definite Systems

Numerous computational applications lead to systems with special structures, and solution methods can be tailored to take advantage of such structures. One of the most prominent of special cases is to have a positive definite system, and this is the case we consider here.

Assume we want to solve the system $Ax = b$, where A is an $n \times n$ positive definite matrix, which is defined below.

Definition 1. The symmetric $n \times n$ matrix A is positive definite, denoted $A \succ 0$, if for any nonzero n-vector x we have

$$x^T A x > 0.$$

A couple of examples of positive definite matrices are

$$A = \begin{bmatrix} 2 & 0 & 0 \\ 0 & 1 & 0 \\ 0 & 0 & 4 \end{bmatrix} \text{ and } Q = \begin{bmatrix} 4 & 2 & 1 \\ 2 & 4 & 1 \\ 1 & 1 & 6 \end{bmatrix}. \tag{2.11}$$

The first matrix is diagonal with positive diagonal elements. Hence,

$$x^T A x = 2x_1^2 + x_2^2 + 4x_3^2 \geq 0,$$

2.2 Special Structure: Positive Definite Systems

from which we immediately have that $x^T A x > 0$ as long as x is not zero. This example easily extends to argue that any square diagonal matrix with positive diagonal elements is positive definite, and in many ways, the class of positive definite matrices generalizes these matrices. We postpone verifying that $Q \succ 0$ until we have alternative ways to classify positive definite matrices. A fact used in many of the forthcoming chapters is that $A^T A \succ 0$ as long as the columns of A are linearly independent, see Exercise 4.

Positive definite matrices induce a geometry on the space of n-vectors that deserves at least a brief note to every student of computational science. Recall that the dot product and the vector norm are related by

$$\|v\| = \sqrt{v \cdot v} = \sqrt{v^T v} = \sqrt{v^T I v}.$$

Notice that the square root is real because $v^T I v$ provides a nonnegative value for all v. Moreover, $v^T I v = 0$ if and only if v is zero, which means that the only vector whose length is zero is $v = 0$. The definition of positive definite suggests that we might be able to replace I with any positive definite matrix, which is indeed true. Notationally, we let the norm induced by $Q \succ 0$ be

$$\|v\|_Q = \sqrt{v^T Q v}.$$

If no subscript is used, then the typical norm with $Q = I$ is assumed.

Consider the positive definite matrices

$$Q' = \begin{bmatrix} 0.8 & 0 \\ 0 & 3 \end{bmatrix} \quad \text{and} \quad Q'' = \begin{bmatrix} 4 & 1 \\ 1 & 2 \end{bmatrix} \tag{2.12}$$

to illustrate the effect of altering, or skewing, our standard concept of length. Then

$$\|v\|_{Q'} = \sqrt{0.8 v_1^2 + 3 v_2^2} \quad \text{and} \quad \|v\|_{Q''} = \sqrt{4 v_1^2 + 2 v_2^2 + 2 v_1 v_2}.$$

One way to depict how these norms alter the concept of length is to plot all vectors of length one, i.e. all solutions to $\|v\|_Q = 1$. Graphs for Q' and Q'' against the standard $Q = I$ are shown in Figs. 2.1 and 2.2.

Positive definite matrices more generally induce an inner product, a norm, and a metric. We forego formal definitions, but the idea is that if $Q \succ 0$, then the dot (inner) product

$$v \cdot w = \langle v, w \rangle = v^T w \quad \text{generalizes to} \quad \langle v, w \rangle_Q = v^T Q w,$$

the norm

$$\|v\| = \sqrt{v^T v} \quad \text{generalizes to} \quad \|v\|_Q = \sqrt{v^T Q v},$$

and the metric

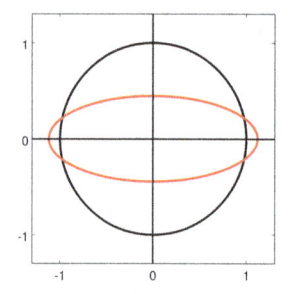

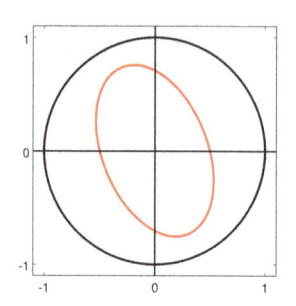

Fig. 2.1 The black circle is the standard unit circle of all vectors satisfying $\|v\| = 1$. The red ellipse depicts all solutions to $\|v\|_{Q'} = 1$ with Q' from (2.12).

Fig. 2.2 The black circle is the standard unit circle of all vectors satisfying $\|v\| = 1$. The red ellipse depicts all solutions to $\|v\|_{Q''} = 1$ with Q'' from (2.12).

$$\|v - w\| = \sqrt{(v-w)^T(v-w)} \text{ generalizes to}$$

$$\|v - w\|_Q = \sqrt{(v-w)^T Q(v-w)}.$$

In the first case we extend the concept of orthogonality from

the vectors v and w are orthogonal if and only if $v^T w = 0$ to *the vectors v and w are Q-conjugate if and only if $v^T Q w = 0$.*

In this extension we have that I-conjugacy is the same as orthogonality. Importantly, Q-conjugate vectors form an independent set, a fact that we use to develop one of our forthcoming computational methods.

There are several ways to classify positive definite matrices, and there are multiple consequences that can be drawn from the positive definite property. We list several of these theoretical consequences in the following theorem.

Theorem 6. *The following are equivalent:*

- $Q \succ 0$,
- *the eigenvalues of Q are all positive and $Q = Q^T$,*
- *there is a unique lower triangular matrix L such that $Q = LL^T$,*
- *there are a matrix U and a diagonal matrix D such that*

  1) $Q = UDU^T$,
  2) $U^T = U^{-1}$, and
  3) $D \succ 0$.

Moreover, if $Q \succ 0$, then

- $\langle v, w \rangle_Q = v^T Q w$ *is an inner product,*
- $\|v\|_Q = \sqrt{v^T Q v}$ *is a norm,*

2.2 Special Structure: Positive Definite Systems

- $\|v - w\|_Q$ is a metric,
- Q is nonsingular, and
- there is a unique solution to $\min_v v^T Q v + b^T v$.

The two factorizations $Q = LL^T$ and $Q = UDU^T$ are of particular note. The first is called the Cholesky factorization, and it relates to our earlier LU factorization. Indeed, we can calculate L by adapting the LU algorithm, which we do momentarily. The second factorization shows that Q is (unitarily) similar to a positive diagonal matrix, which means that we can alter the basis of the space of n-vectors so that angle and distance are preserved and so that Q behaves like the positive diagonal matrix D. So all positive definite matrices act like positive diagonal matrices from the perspective of this factorization.

2.2.1 Cholesky Factorization

Let Q be the matrix in (2.11). Then the first row operation associated with factoring Q into LU is $(-1/2)R_1 + R_2 \to R_2$. Pre-multiplying by the corresponding elementary matrix gives

$$\begin{bmatrix} 1 & 0 & 0 \\ -1/2 & 1 & 0 \\ 0 & 0 & 1 \end{bmatrix} \begin{bmatrix} 4 & 2 & 1 \\ 2 & 4 & 1 \\ 1 & 1 & 6 \end{bmatrix} = \begin{bmatrix} 4 & 2 & 1 \\ 0 & 3 & 1/2 \\ 1 & 1 & 6 \end{bmatrix}.$$

The symmetry of Q means that we can eliminate the 2 in the first row, second column by performing the same row operation but on the columns instead of the rows, i.e. $(-1/2)C_1 + C_2 \to C_2$. This is accomplished by post-multiplying by the transpose of the elementary matrix,

$$\begin{bmatrix} 1 & 0 & 0 \\ -1/2 & 1 & 0 \\ 0 & 0 & 1 \end{bmatrix} \begin{bmatrix} 4 & 2 & 1 \\ 2 & 4 & 1 \\ 1 & 1 & 6 \end{bmatrix} \begin{bmatrix} 1 & -1/2 & 0 \\ 0 & 1 & 0 \\ 0 & 0 & 1 \end{bmatrix} = \begin{bmatrix} 4 & 0 & 1 \\ 0 & 3 & 1/2 \\ 1 & 1/2 & 6 \end{bmatrix}.$$

Repeating the same process for the next row and column operations results in

$$\begin{bmatrix} 1 & 0 & 0 \\ 0 & 1 & 0 \\ -1/4 & 0 & 1 \end{bmatrix} \begin{bmatrix} 4 & 0 & 1 \\ 0 & 3 & 1/2 \\ 1 & 1/2 & 6 \end{bmatrix} \begin{bmatrix} 1 & 0 & -1/4 \\ 0 & 1 & 0 \\ 0 & 0 & 1 \end{bmatrix} = \begin{bmatrix} 4 & 0 & 0 \\ 0 & 3 & 1/2 \\ 0 & 1/2 & 23/4 \end{bmatrix}.$$

The final row and column operations show

$$\begin{bmatrix} 1 & 0 & 0 \\ 0 & 1 & 0 \\ 0 & -1/6 & 1 \end{bmatrix} \begin{bmatrix} 4 & 0 & 0 \\ 0 & 3 & 1/2 \\ 0 & 1/2 & 23/4 \end{bmatrix} \begin{bmatrix} 1 & 0 & 0 \\ 0 & 1 & -1/6 \\ 0 & 0 & 1 \end{bmatrix} = \begin{bmatrix} 4 & 0 & 0 \\ 0 & 3 & 0 \\ 0 & 0 & 17/3 \end{bmatrix}.$$

Allowing the three consecutive elementary matrices to be E_1, E_2, and E_3, we have

$$E_3 E_2 E_1 A E_1^T E_2^T E_3^T = D,$$

where D is the final positive diagonal matrix. From the fact that the inverses of the elementary matrices only reverse the sign of the non-identity elements, we find that

$$Q = E_1^{-1} E_2^{-1} E_3^{-1} D (E_3^T)^{-1} (E_2^T)^{-1} (E_1^T)^{-1}$$

$$= \begin{bmatrix} 1 & 0 & 0 \\ 1/2 & 1 & 0 \\ 1/4 & 1/6 & 1 \end{bmatrix} \begin{bmatrix} 4 & 0 & 0 \\ 0 & 3 & 0 \\ 0 & 0 & 17/3 \end{bmatrix} \begin{bmatrix} 1 & 1/2 & 1/4 \\ 0 & 1 & 1/6 \\ 0 & 0 & 1 \end{bmatrix}$$

$$= \left(\begin{bmatrix} 1 & 0 & 0 \\ 1/2 & 1 & 0 \\ 1/4 & 1/6 & 1 \end{bmatrix} \begin{bmatrix} \sqrt{4} & 0 & 0 \\ 0 & \sqrt{3} & 0 \\ 0 & 0 & \sqrt{17/3} \end{bmatrix} \right) \left(\begin{bmatrix} \sqrt{4} & 0 & 0 \\ 0 & \sqrt{3} & 0 \\ 0 & 0 & \sqrt{17/3} \end{bmatrix} \begin{bmatrix} 1 & 1/2 & 1/4 \\ 0 & 1 & 1/6 \\ 0 & 0 & 1 \end{bmatrix} \right)$$

$$= \begin{bmatrix} 2 & 0 & 0 \\ 1 & \sqrt{3} & 0 \\ 1/2 & \sqrt{3}/6 & \sqrt{17/3} \end{bmatrix} \begin{bmatrix} 2 & 1 & 1/2 \\ 0 & \sqrt{3} & \sqrt{3}/6 \\ 0 & 0 & \sqrt{17/3} \end{bmatrix}$$

$$= LL^T.$$

The altered LU algorithm that performs row and column operations to factor Q into LL^T is called the outer product Cholesky algorithm. The final step requires the ability to calculate $\sqrt{D}$, which in the example works because the diagonal elements are all positive. Although not immediately obvious from the definition of positive definite, it is true that the diagonal elements are positive as long as $Q \succ 0$.

While the outer product algorithm is comfortably similar to LU, another algorithm, called the inner product Cholesky algorithm, is generally preferred. This method has a simple elegance that follows directly from matrix multiplication. As an example, the Cholesky factorization for Q in (2.11) satisfies

$$LL^T = \begin{bmatrix} L_{11} & 0 & 0 \\ L_{21} & L_{22} & 0 \\ L_{31} & L_{32} & L_{33} \end{bmatrix} \begin{bmatrix} L_{11} & L_{21} & L_{31} \\ 0 & L_{22} & L_{32} \\ 0 & 0 & L_{33} \end{bmatrix} = \begin{bmatrix} 4 & 2 & 1 \\ 2 & 4 & 1 \\ 1 & 1 & 6 \end{bmatrix} = Q.$$

Thus the first row of Q is the result of the following inner products,

2.2 Special Structure: Positive Definite Systems

$$4 = (L_{11}, 0, 0)(L_{11}, 0, 0)^T = L_{11}^2 \qquad \Rightarrow L_{11} = 2$$
$$2 = (L_{11}, 0, 0)(L_{21}, L_{22}, 0)^T = L_{11}L_{21} \quad \Rightarrow L_{21} = 1$$
$$1 = (L_{11}, 0, 0)(L_{31}, L_{32}, L_{33})^T = L_{11}L_{31} \Rightarrow L_{31} = 1/2.$$

Proceeding to the second row of Q we find

$$2 = (L_{21}, L_{22}, 0)(L_{11}, 0, 0)^T = 2 \cdot 1$$
$$4 = (L_{21}, L_{22}, 0)(L_{21}, L_{22}, 0)^T = 1 + L_{2,2}^2 \qquad \Rightarrow L_{2,2} = \sqrt{3}$$
$$1 = (L_{21}, L_{22}, 0)(L_{31}, L_{32}, L_{33})^T = 1/2 + L_{22}L_{32} \Rightarrow L_{32} = \sqrt{3}/6.$$

The first equation provides no information outside what has already been calculated and can be skipped. The third row of Q gives

$$1 = (L_{31}, L_{32}, L_{33})(L_{11}, 0, 0)^T = 2 \cdot 1/2$$
$$1 = (L_{31}, L_{32}, L_{33})(L_{21}, L_{22}, 0)^T = 1 \cdot 1/2 + (\sqrt{3}/6)(\sqrt{3})$$
$$6 = (L_{31}, L_{32}, L_{33})(L_{31}, L_{32}, L_{33})^T = 1/3 + L_{33}^2 \qquad \Rightarrow L_{33} = \sqrt{17/3}.$$

The first two equations don't provide new information, but the third decides the last element of L. Removing all the calculations that don't decide an element of L, we see that the inner product algorithm requires one equation to be solved for each element of L (ignoring the upper triangular elements of zero), and that each of these equations requires at most a single division. As with the outer product algorithm, the method works as long as the equations defining the diagonal elements allow the square roots to be evaluated as real values. This property is guaranteed if and only if Q is positive definite, and in fact, the inner product algorithm is commonly the computational test to decide if a matrix is positive definite.

There are three significant computational advantages to performing a Cholesky factorization. First, rounding error can be shown to be insignificant as the Cholesky factorization is calculated. Thus there is no need for complicated pivoting strategies. Second, the number of calculations required for the inner Cholesky decomposition is roughly $n^3/6$, compared against the $n^3/3$ for the LU factorization. Finally, we only need to store the lower triangular matrix L as opposed to both L and U, which reduces storage by a factor of two. Moreover, don't forget that we can use the two-step backsolve procedure to find a solution to $Qx = b$ once a positive definite matrix Q is factored. That is, we can first solve $Ly = b$ and then $L^T x = y$.

2.2.2 The Method of Conjugate Directions

Unlike our previous methods for solving linear systems, all of which compute solutions from a factorization, the method of conjugate directions is an iterative procedure. This means that we initialize the process with an initial guess and then update it to progress toward a solution. Consider the $n \times n$

positive definite matrix Q, and assume we want to solve $Qx = b$. Let x_0 be an initial guess at the solution. The iterative procedure updates this initial guess by:

Step 1: $x_1 = x_0 + \lambda_0 \, d_0$
Step 2: $x_2 = x_1 + \lambda_1 \, d_1 = x_0 + \lambda_0 \, d_0 + \lambda_1 \, d_1$
$\vdots$
Step n: $x_n = x_{n-1} + \lambda_{n-1} \, d_{n-1} = x_0 + \lambda_0 \, d_0 + \lambda_1 \, d_1 + \ldots + \lambda_{n-1} \, d_{n-1}.$

The first step seeks to improve the initial guess of x_0 along the direction d_0, and the scalar λ_0 is selected to improve x_0 as much as possible. The second step similarly attempts to improve x_1 by moving along d_1 a step size of λ_1. If the search directions d_i form a collection of independent vectors, then at step n we know that we can search the entire space by adjusting the λ scalars, i.e. any n-vector is a linear combination of the d_i's. So, if x^* is the unique solution to $Qx = b$, then we are guaranteed the existence of scalars λ_i such that

$$x^* - x_0 = \lambda_0 \, d_0 + \lambda_1 \, d_1 + \ldots + \lambda_{n-1} \, d_{n-1}. \tag{2.13}$$

This observation raises the question of how to select the search directions and the resulting scalars so that the solution is known after n-steps.

The matrix Q being positive definite is particularly important since a collection of Q-conjugate vectors forms a linearly independent set. The method of conjugate directions uses this fact and iteratively selects the direction vectors so that they form a linearly independent set of Q-conjugate directions. There are several such updates, but the traditional search directions are defined recursively as

$$d_{k+1} = -(Q\, x_{k+1} - b) + \left(\frac{d_k^T Q (Q\, x_{k+1} - b)}{d_k^T Q \, d_k} \right) d_k \tag{2.14}$$

$$= -(Q\, x_{k+1} - b) + \left(\frac{d_k^T Q (Q\, x_{k+1} - b)}{\|d_k\|_Q} \right) d_k.$$

As evidence that this update generates Q-conjugate direction vectors, notice that the consecutive iterates d_k and d_{k+1} satisfy

$$d_k^T Q \, d_{k+1} = d_k^T Q \left(-(Q\, x_{k+1} - b) + \left(\frac{d_k^T Q (Q\, x_{k+1} - b)}{d_k^T Q \, d_k} \right) d_k \right)$$

$$= -d_k^T Q (Q\, x_{k+1} - b) + \left(\frac{d_k^T Q (Q\, x_{k+1} - b)}{d_k^T Q \, d_k} \right) d_k^T Q \, d_k$$

$$= -d_k^T Q (Q\, x_{k+1} - b) + d_k^T Q (Q\, x_{k+1} - b)$$

$$= 0.$$

2.2 Special Structure: Positive Definite Systems

A straightforward, albeit tedious, extension of this calculation shows that $d_k^T Q\, d_t = 0$ as long as $k \neq t$.

We now turn to the question of deciding the λ_k scalars to achieve our goal of solving the system $Qx = b$. From (2.13) we have for any direction d_k that

$$d_k^T Q(x^* - x_0) = \lambda_0\, d_k^T Q\, d_0 + \ldots + \lambda_k\, d_k^T Q\, d_k + \ldots + \lambda_{n-1}\, d_k^T Q\, d_{n-1}$$
$$= 0 + 0 + \ldots + \lambda_k\, d_k^T Q\, d_k + \ldots + 0$$
$$= \lambda_k\, d_k^T Q\, d_k,$$

where the zero terms in the third expression are due to the Q-conjugacy of the search directions. Hence, using the fact that $Qx^* = b$, we have

$$\lambda_k = \frac{d_k^T Q(x^* - x_0)}{d_k^T Q\, d_k} = \frac{d_k^T(b - Q\, x_0)}{d_k^T Q\, d_k}.$$

The factor $b - Qx_0$ is called a residual, and it measures the componentwise incorrectness of the original guess. The λ_k scalars are typically re-expressed in terms of the current residual $b - Qx_k$ instead of the original residual $b - Qx_0$. This is possible since the iterative process forces

$$x_k - x_0 = \lambda_0\, d_0 + \lambda_1\, d_1 + \ldots + \lambda_{k-1}\, d_{k-1},$$

and hence,

$$d_k^T Q(x_k - x_0) = \lambda_0\, d_k^T Q\, d_0 + \lambda_1\, d_k^T Q\, d_1 + \ldots + \lambda_{k-1}\, d_k^T Q\, d_{k-1} = 0.$$

We conclude with a little algebra that

$$\lambda_k = \frac{1}{d_k^T Q\, d_k}\left(d_k^T Q(x^* - x_0)\right)$$
$$= \frac{1}{d_k^T Q\, d_k}\left(d_k^T Q(x^* - x_k + x_k - x_0)\right)$$
$$= \frac{1}{d_k^T Q\, d_k}\left(d_k^T Q(x^* - x_k) + d_k^T Q(x_k - x_0)\right)$$
$$= \frac{1}{d_k^T Q\, d_k}\left(d_k^T Q(x^* - x_k)\right)$$
$$= -\frac{d_k^T(Q\, x_k - b)}{d_k^T Q\, d_k}.$$

Combining the linear independence of the search directions with these λ_k values leads to the following theorem, which guarantees a solution to $Qx = b$ after n iterations.

Theorem 7. *Suppose $Q \succ 0$ is an $n \times n$ matrix. Then starting from any non-solution x_0, the following n-step iterative process terminates with the unique solution to $Qx = b$,*

1. Set $k = 0$ and $d_k = -(Qx_0 - b)$.
2. Set

$$x_{k+1} = x_k - \left(\frac{d_k^T(Qx_k - b)}{d_k^T Q d_k}\right) d_k$$

and

$$d_{k+1} = -(Qx_{k+1} - b) + \left(\frac{d_k^T Q(Qx_{k+1} - b)}{d_k^T Q d_k}\right) d_k.$$

3. If $k \leq n$, then set $k = k+1$ and repeat step 2; otherwise, x_n solves $Qx = b$.

Theorem 7 provides a mathematical certificate of correctly identifying the unique solution to the $n \times n$ positive definite system $Qx = b$ after n iterations. However, this certificate requires calculations free from computational error. A particular source of concern is the division by $d_k^T Q d_k = \|d_k\|_Q$, as errors tend to magnify if this quantity is near zero. The value of $\|d_k\|_Q$ is zero if and only if $d_k = 0$, and the calculation in (2.14) shows that d_k is zero if the residual is zero. In other words, the search directions are zero once we have a solution. This analysis extends to approximate solutions, and hence, the residuals become small as the algorithm approaches a solution. Computational errors can subsequently increase as the algorithm approaches termination.

The method of conjugate directions can proceed until a sufficiently good solution is established instead of assuming error free arithmetic and ending after n iterations. Altering the termination criterion to $\|Qx - b\| \leq \varepsilon$ has a twofold advantage. First, it is common to have a reasonable solution in less than n iterations. So ending once $\|Qx - b\| \leq \varepsilon$ can reduce the computational burden by decreasing the number of iterations. Second, should the algorithm fail to meet the termination criterion after n iterations, say due to rounding errors, then it can continue in the pursuit of an adequate solution. This can be perceived as the algorithm starting anew with the result from the first n iterations being the initial guess.

As with most iterative processes, it is efficient computationally to hold only what is required to make an iteration. All that is needed in the case of conjugate directions is the current iterate x and the current direction d, which can be replaced at each step. Pseudocode for the method of conjugate directions is in Algorithm 6.

2.3 Newton's Method for Systems of Equations

Nonlinear systems lack much of the theoretical and computational analysis afforded linear systems, and they regularly present difficult and important challenges in the study of computational science. Indeed, many physical phe-

2.3 Newton's Method for Systems of Equations

Algorithm 6 Pseudocode for the method of conjugate directions

set $d = -(Qx - b)$
while termination criteria is unmet **do**
$$x = x - \left(\frac{(Qx-b)^T d}{d^T Q d}\right) d$$
if $\|Qx - b\| \leq \varepsilon$ **then**
    set termination status to true
else
$$d = -(Qx - b) + \left(\frac{d^T Q(Qx-b)}{d^T Q d}\right) d$$
end if
end while

nomena are inadequately represented linearly and require nonlinear solution methods.

A natural and reasonable first attempt at solving a nonlinear system is to harness the depth of theoretical and computational ability provided by imposing linearity. The process resulting from such an imposition is a multidimensional version of Newton's method for systems, and this fundamental algorithm remains the gold standard for nonlinear equations.

Let $f : \mathbb{R}^n \to \mathbb{R}^n$ be such that the component functions $f_1, f_2, \ldots, f_n$ are twice smooth, i.e. that
$$\frac{\partial f_i}{\partial x_j} \quad \text{and} \quad \frac{\partial^2 f_i}{\partial x_j \, \partial x_k}$$
exist and are continuous for any choice of i, j, and k. In this case the multidimensional extension of the derivative is called the Jacobian, which is an $n \times n$ matrix denoted by

$$D_x f = \begin{bmatrix} \partial f_1/\partial x_1 & \partial f_1/\partial x_2 & \cdots & \partial f_1/\partial x_n \\ \partial f_2/\partial x_1 & \partial f_2/\partial x_2 & \cdots & \partial f_2/\partial x_n \\ \vdots & \vdots & & \vdots \\ \partial f_n/\partial x_1 & \partial f_n/\partial x_2 & \cdots & \partial f_n/\partial x_n \end{bmatrix}.$$

A brief comment on notation is warranted. Other notations for the Jacobian are J, J_f, ∇f, and F. The J and J_f notations are short for "Jacobian" and "Jacobian of f." The ∇ operator most commonly indicates a gradient, i.e. a vector of first partials, but it is also used to denote the Jacobian matrix of first partials. The F notation is less common. Any student of computational science should be notationally nimble and able to perceive notational differences.

The linear approximation of f at vector v is

$$f(x) \approx f(v) + D_x f(v)(x - v),$$

and as in the single variable case, Newton's method replaces $f(x) = 0$ with

$$f(v) + D_x f(v)(x - v) = 0.$$

If we let $\Delta v = x - v$ be the Newton step, then we solve the linear system

$$D_x f(v) \Delta v = -f(v)$$

for Δv and replace v with $v + \Delta v$. The common termination criteria are $\|f(v)\| \leq \varepsilon$ and/or $\|\Delta v\| \leq \delta$. The parameters ε and δ are tolerances indicating, in the first case, that the current iteration sufficiently satisfies $f(v) = 0$, and in the second case, that the update from one iteration to the next has stalled.

The square system $D_x f(v) \Delta v = -f(v)$ has a unique solution if and only if $D_x f(v)$ is invertible. This requirement is the extension of $f'(x) \neq 0$ in the single variable situation, and a standard assumption is that $D_x f(v)$ is invertible sufficiently close to an actual solution, i.e. $D_x f(x^*)$ is nonsingular if $f(x^*) = 0$.

As an example, consider the system

$$f(x) = \begin{pmatrix} x_1^2 + x_2^2 + x_3^2 - 1 \\ x_1^2 + x_2^2 - x_3 \\ x_1 + x_2 + x_3 \end{pmatrix} = 0. \tag{2.15}$$

The Jacobian is

$$D_x f(x) = \begin{bmatrix} 2x_1 & 2x_2 & 2x_3 \\ 2x_1 & 2x_2 & -1 \\ 1 & 1 & 1 \end{bmatrix}.$$

Notice that we are unable to initiate Newton's method with either $v = 0$ or $v = (1, 1, 1)^T$ because the Jacobian is singular in both cases. Assume initially that $v = (0.1, 0, -0.1)^T$. Then

$$f(v) = \begin{pmatrix} -0.9800 \\ 0.1100 \\ 0.0000 \end{pmatrix} \text{ and } D_x f(v) = \begin{bmatrix} 0.2000 & 0.0000 & -0.2000 \\ 0.2000 & 0.0000 & -1.0000 \\ 1.0000 & 1.0000 & 1.0000 \end{bmatrix}.$$

The Newton step and update are

$$v + \Delta v = \begin{pmatrix} 0.1 \\ 0.0 \\ -0.1 \end{pmatrix} + \begin{pmatrix} 6.2625 \\ -7.6250 \\ 1.3625 \end{pmatrix} = \begin{pmatrix} 6.3625 \\ -7.6250 \\ 1.2625 \end{pmatrix}.$$

2.4 MATLAB Functions

This update results in

$$f(v) = \begin{pmatrix} 99.216 \\ 97.360 \\ -2.2 \times 10^{-16} \end{pmatrix} \text{ and } \|f(v)\| = 139.01.$$

The process continues and requires eight iterations to calculate a solution satisfying $\|f(v)\| \leq \varepsilon = 10^{-6}$. These iterates are listed in Table 2.1. The pseudocode for Newton's method is shown in Algorithm 7.

Iteration	v	$\|f(v)\|$
0	$(0.10000, 0.00000, -0.10000)^T$	0.9862
1	$(6.36250, -7.62500, 1.26250)^T$	139.01
2	$(3.15053, -3.88639, 0.73586)^T$	34.554
3	$(1.47835, -2.10200, 0.62365)^T$	8.4663
4	$(0.64570, -1.26375, 0.61805)^T$	1.9742
5	$(0.28017, -0.89821, 0.61803)^T$	0.37793
6	$(0.16678, -0.78481, 0.61803)^T$	0.036367
7	$(0.15327, -0.77130, 0.61803)^T$	5.2×10^{-4}
8	$(0.15307, -0.77111, 0.61803)^T$	1.1×10^{-7}

Table 2.1 Newton iterates for solving system (2.15).

Algorithm 7 Pseudocode for Newton's method applied to systems

  while unmet termination criteria **do**
    if $D_x f(v)$ is nonsingular **then**
      solve $D_x f(v) \Delta v = -f(v)$
      $v = v + \Delta v$
    else
      set status to failure and terminate
    end if
  end while

2.4 MATLAB Functions

MATLAB and Octave both include a number of tools for solving linear systems. In both languages the backslash operator \ and forwardslash operator / are robust and powerful tools for solving linear systems. In order to solve the system of equations $Ax = b$ in MATLAB and Octave we use the backslash operator, sometimes called "left division." As an example, the code

```
A = [1,2; 3,4];
b = [5;4];
A\b
```

results in the solution $(-6.0, 5.5)^T$. Notice that using "right division" in place of "left division" in this example results in a dimension mismatch. The "right division" operator, /, assumes that you are trying to solve the system of equations $xA = b$, and in this case the dimensions of A and b are incompatible.

If A is $m \times n$ with $m > n$, then $Ax = b$ may not have a solution. In this case A\b returns the least squares solution to the equation, a topic developed in Sect. 3.1. If $m < n$, then there may be infinitely many solutions to the system. In this case A\b returns a single solution to $Ax = b$, but there is no guarantee of which solution it might be.

Both MATLAB and Octave provide the command inv for computing the inverse of a square matrix. The code below computes A^{-1} and then left-multiplies b by A^{-1} to solve $Ax = b$.

```
A = [1,2; 3,4];
b = [5;4];
inv(A)*b
```

The difference between using inv and \ is worth discussion. The first method is actually calculating the matrix A^{-1}, which takes about $2n^3/3$ calculations. It then performs a matrix–vector multiply to find $x = A^{-1}b$, which takes an additional n^2 calculations. The second method defaults to calculating an LU decomposition of A and then performing forward and backward substitutions. A different algorithm may be used if a special structure of A can be exploited to improve efficacy. Indeed, the backslash command is not a sole algorithm but is instead an expert system that first analyzes the matrix to decide how it should be solved. The calculation of the LU factorization takes approximately $n^3/3$ calculations, and backward and forward substitution again require about n^2 calculations. So the calculation requirements for the two methods are similar, with the LU factorization having a slight advantage regarding the number of calculations. However, for reasons of computational accuracy, **solving the system of equations using an LU factorization is almost always preferred to multiplying by A^{-1}.**

Systems with the same coefficient matrix often need to be solved repeatedly, i.e. solving the k systems

$$Ax = b_i, \ i = 1..k,$$

is common. It is efficient to calculate the LU decomposition of A once and re-use the factors to solve subsequent systems. The function lu returns the LU decomposition of A. There are many different ways to call lu. Two basic uses are

```
[L,U]=lu(A)
```

2.4 MATLAB Functions

and

```
[L,U,P]=lu(A).
```

The first returns an upper triangular U and a permuted lower triangular matrix L such that $LU = A$. The second returns a lower triangular matrix L, an upper triangular matrix U, and a permutation matrix P such that $LU = PA$. A third option is

```
[L,U,p] = lu(A,'vector'),
```

which returns L, U, and a permutation vector p.

MATLAB and Octave recognize the special structure of triangular matrices if the \ operator is used, and both perform forward or backward substitution in this case. An example of storing the LU factorization of a matrix in order to facilitate the repeated solution of $Ax = b$ follows.

```
A=rand(10,10);
[L,U]=lu(A);
for i = 1: 10000
    b=rand(10,1);
    y=L\b;
    x=U\y;
end
```

MATLAB and Octave provide the chol command for calculating the Cholesky factorization of a positive definite matrix A. The command U=chol(A) returns an upper triangular matrix U such that $U^T U = A$.

MATLAB and Octave provide a preconditioned conjugate gradient method (the topic of preconditioning is not discussed in this text; however, it is a common extension of the conjugate gradient method). The code

```
A=[84 76 116;76 109 113;116 113 165];
b=[1;2;3];
pcg(A,b)
```

solves the equation $Ax = b$ using the preconditioned conjugate gradient method. Octave requires additional options to pcg, but the idea is similar. Both MATLAB and Octave use a conjugate gradient method if \ is used and A is positive definite, so there is little use in using pcg directly.

Solving systems of nonlinear equations is accomplished with fsolve, e.g. with the following.

```
f=@(x) [(x(1)-1)^2+x(2)^2-2; (x(1)+1)^2+x(2)^2-3];
[x,fx,status]=fsolve(f,[0,1]);
```

On exit x contains an approximate solution, fx contains $f(x)$, and status is 1 if the algorithm completes successfully. There are a variety of options to fsolve; the reader should consult the documentation for further information.

2.5 Exercises

1. Show that the inverse of an elementary matrix is another elementary matrix. Give 3×3 examples.
2. Show that the product of upper (lower) triangular matrices is upper (lower) triangular.
3. Let $Q \succ 0$ be an $n \times n$ matrix and let $v_1, v_2, \ldots v_n$ be a collection of Q-conjugate n-vectors. Show that the collection is linearly independent.
4. Suppose A is an $m \times n$ matrix such that the rank$(A) = n$. Show that $A^T A \succ 0$.
5. Show that d_n from Theorem 7 is zero. Hence, if the iterative process were allowed to continue past n steps, then all iterates would be identical after step n.
6. Write a function with declaration

    ```
    [x,status] = LinearSolver(A,b)
    ```

    that attempts to solve $Ax = b$ by row reducing the augmented matrix $[A|b]$. The algorithm should incorporate row pivots and should use the technique of backsolving once $[A|b]$ is reduced to an upper triangular form. A should not be assumed to be square, and the dimensions of b should not be assumed to be consistent with those of A.

    The input arguments are:

    A - The coefficient matrix A of the linear system.

    b - The right-hand side vector b of the linear system. The code should return an error status if b is not a column vector or has dimensions that are inconsistent with A. See the status definition.

    The output arguments are:

    x   - A solution to $Ax = b$ if one exists.

    status - A status variable encoded as follows:

    status 0: $Ax = b$ is consistent with a unique solution.

    status 1: $Ax = b$ is consistent with an infinite number of solutions. In this case a solution to the linear system should be returned in x.

    status 2: $Ax = b$ is inconsistent.

    status 3: The code did not succeed for some other reason, such as inconsistent dimensions of A and b.

2.5 Exercises

7. Write a function with declaration

    `[L,U,status]=LUfact_NoPivot(A)`

    that factors a matrix A into the product LU. The algorithm should not perform pivoting.

    The input argument is:

    A - The matrix A to be factored; A is not assumed to be square.

    The output arguments are:

    L   - A square lower triangular matrix L, with ones on the main diagonal.
    U   - An upper triangular matrix U, with the property that $LU = A$ if the code succeeds.
    status - A status variable encoded as follows:

    status 0: The code has succeeded.

    status 1: The code failed for some reason; the value of L and U should not be trusted.

8. Write a function with declaration

    `[L,U,p]=LUfact(A)`.

    The function should factor the matrix A into the product LU up to a permutation. The algorithm should perform row swaps so that the pivot is always the largest magnitude element of the appropriate column.

    The input argument is:

    A - The matrix A to be factored; A is not assumed to be square.

    The output arguments are:

    L - A square lower triangular matrix with ones along the main diagonal.
    U - An upper triangular matrix.
    p - A permutation vector. The equality `LU==A(p,:)` should be very nearly true in every position.

9. Adapt Problem 8 to have a declaration of

    [Afac,p] = LUfactOneMat(A).

   The function should perform an LU factorization of A and return both L and U in one matrix rather than in separate matrices. In particular, strictly lower triangular entries of L should be stored in the strictly lower triangular portion of Afac(p,:) and the upper triangular entries of U should be stored in the upper triangular portion of Afac(p,:).

   The input argument is:

   A - The matrix A to factor; A is not be assumed to be square.

   The output arguments are:

   Afac - A matrix with the same dimensions as A. Upon permutation by p, the strictly lower triangular portion of Afac should contain the strictly lower triangular portion of L. The upper triangular portion of Afac should contain the upper triangular portion of U.

   p - A permutation vector encoding the row swaps necessary during partial pivoting.

10. Write a function with declaration

    [x,status] = Backsolve(L,U,p,b)

    that solves $Ax = b$ after A has been factored into LU up to permutation by p. The algorithm is the two-step process in (2.6).

    The input arguments are:

    L - A square lower triangular square matrix L with ones on the main diagonal.
    U - An upper triangular matrix U. L and U are not assumed to have consistent dimensions.
    p - A permutation vector p. It is assumed that $LU = A(p,:)$.
    b - The right-hand side vector in $Ax = b$, b is not assumed to have consistent dimensions with A.

    The output arguments are:

    x - A solution x to $Ax = b$ if the problem is consistent.

status - A status variable encoded as follows:

  status 0: $Ax = b$ is consistent with a unique solution.

  status 1: $Ax = b$ is consistent with an infinite number of solutions, in which case a solution to $Ax = b$ is returned in x.

  status 2: $Ax = b$ is inconsistent.

  status 3: The function failed for some other reason, such as input arguments with inconsistent dimensions.

11. Adapt Problem 10 to have declaration

    `[x,status] = Backsolve(Afac,p,b)`.

    The function should solve $Ax = b$ as in Problem 10, but the elements of L and U should be referenced from the matrix Afac without reordering the rows of Afac and without creating L and U.

12. Write a function with declaration

    `[L,status] = CholeskyOuter(A)`

    that factors a positive definite matrix A into LL^T, where L is lower triangular. The calculation method is to be the outer product algorithm.

    The input argument is:

    A - A matrix A to be factored. A is not assumed to be positive definite or square.

    The output arguments are:

    L - A lower triangular matrix L satisfying $LL^T = A$ upon success.

    status - A status variable encoded as follows:

      status 0: The function succeeded.

      status 1: The function failed because A is square but is not positive definite.

      status 2: The function failed for some other reason, such as A being non-square.

13. Write a function with declaration

    `[L,status] = CholeskyInner(A)`

that factors a positive definite matrix A into LL^T, where L is lower triangular. The calculation method is to be the inner product algorithm. The function should have the same input and output arguments as Problem 12.

14. Write a function with declaration

    ```
    [x,itr,status] = ConjugateGradient(A,b,x0,
        maxiter,epsilon)
    ```

    that solves $Ax = b$ with the conjugate gradient method.

    The input arguments are:

    A   - The matrix A, which should be tested with chol to ensure that it is positive definite.
    b   - The right-hand side vector b.
    x0   - An initial guess at the solution to $Ax = b$.
    maxiter - The maximum number of iterations to perform.
    epsilon - The convergence tolerance, the code should stop iterating once $\|Ax_n - b\| < \varepsilon$.

    The output arguments are:

    x   - The computed approximate solution to $Ax = b$.
    itr   - The number of iterations performed by the code.
    status - A status variable, encoded as follows:

        status 0: The algorithm completed successfully.
        status 1: The algorithm reached the maximum number of iterations without finding an approximate solution.
        status 2: The algorithm encountered some other error.

15. A random $n \times n$ positive definite matrix Q with eigenvalues between 1 and 10 can be generated with

    ```
    Q=20*rand(n)-10;
    [U,D]=eig(Q'+Q);
    Q=U*diag(9*rand(n,1)+1)*U';
    ```

    Design a computational experiment to assess the average number of required iterations to solve $Qx = b$ with the method of conjugate directions, where b is a random $n \times 1$ vector whose components are between 0 and 10 and Q is randomly generated as noted above. Use your code from Exercise 14, with epsilon $= 10^{-8}$ and x0 being the vector of ones. The

2.5 Exercises

experiment should test n from 100 to 1000. Create graphs to illustrate your conclusions.

16. Write a function with declaration

    ```
    [x,nrmfx,numitr,status]=NewtonSys(f,df,x0,
        epsilon,maxiter)
    ```

    that attempts to solve the nonlinear system $f(x) = 0$ with Newton's method, where $f : \mathbb{R}^n \to \mathbb{R}^n$.

    The input variables are:

f	- A function handle or filename that defines $f(x)$. Assume that $f(x)$ is a column vector.
df	- A function handle or filename that defines $D_x f(v)$. Assume that $D_x f(v)$ is always a matrix with dimensions consistent with $f(v)$.
x0	- An initial guess at a solution to $f(x) = 0$. Check to ensure that x0 is a column vector.
epsilon	- The termination tolerance. Quit if $\|f(x_n)\| < \varepsilon$.
maxiter	- An iteration limit.

    The output variables are:

x	- The computed solution x.
nrmfx	- The value of $\|f(x)\|$ at the computed solution.
numitr	- The number of iterations used to compute the solution.
status	- The status of the search, encoded as follows:

    status 0: The code was successful.

    status 1: The code reached the iteration limit before finding an acceptable solution.

    status 2: Some other problem occurred, for instance x0 was not a column vector.

17. Many important problems, such as creating cubic splines (Sect. 3.3), solving ordinary differential equations (Chap. 5), and solving partial differential equations (Chap. 11), require solving linear systems in which the coefficient matrix has a special structure. We can often leverage this structure into an efficient solution algorithm.

    One such structure is that of a tridiagonal matrix, which has zero entries other than on the main-, super-, and sub-diagonals. For instance, a 6×6 tridiagonal matrix has the form

$$A = \begin{bmatrix} a_{11} & a_{12} & 0 & 0 & 0 & 0 \\ a_{21} & a_{22} & a_{23} & 0 & 0 & 0 \\ 0 & a_{32} & a_{33} & a_{34} & 0 & 0 \\ 0 & 0 & a_{43} & a_{44} & a_{45} & 0 \\ 0 & 0 & 0 & a_{54} & a_{55} & a_{56} \\ 0 & 0 & 0 & 0 & a_{65} & a_{66} \end{bmatrix}.$$

We can develop a specialized LU factorization algorithm for tridiagonal matrices by noticing that we only need to eliminate the leading entry in row $i+1$ when doing elimination on column i. The pseudocode for this specialized algorithm is in Algorithm 8. This special LU algorithm requires about $3n$ operations to complete, which means that the number of operations is linear in the size of the matrix.

Inspection of the lower and upper triangular factors of a tridiagonal matrix reveals that L has the form

$$L = \begin{bmatrix} 1 & 0 & \cdots & & & 0 \\ l_{21} & 1 & 0 & \cdots & & 0 \\ 0 & l_{32} & 1 & 0 & \cdots & 0 \\ \vdots & & \ddots & \ddots & & \\ 0 & & & \cdots & l_{n,n-1} & 1 \end{bmatrix}. \qquad (2.16)$$

We can exploit this structure to implement a fast forward substitution algorithm such as Algorithm 9, which uses about n calculations to complete.

Create a function with declaration

[x,status]=TDMS(A,b)

that solves the system of equations $Ax = b$ if A is tridiagonal. Extend the idea of fast forward substitution to implement a fast backward substitution, and combine these substitution routines with the tridiagonal LU factorization to complete the task of solving $Ax = b$ in about $5n$ calculations. Note that this algorithm may not always complete successfully even if A is invertible, as it does not implement any pivoting.

2.5 Exercises

Algorithm 8 Pseudocode for tridiagonal LU factorization

Set $U = A$
Set $L = I$
for i from 1 to $n-1$ **do**
　Set $L_{i+1,i} = U_{i+1,i}/U_{i,i}$
　Set $U_{i+1,:} = U_{i+1,:} - L_{i+1,i}U_{i,:}$
end for

Algorithm 9 Pseudocode for forward substitution when L has special structure

Set $x_1 = b_1$
for i from 2 to n **do**
　Set $x_i = b_i - l_{i,i-1}x_{i-1}$
end for

18. Use the `tic` and `toc` commands to compare the run time of your `LinearSolver` routine from Exercise 6 versus the backslash command \ in MATLAB. Use random $n \times n$ matrices A and random $n \times 1$ vectors b where $n = 10^k$, with $k = 1, 2, 3, 4$ as the basis of your comparison. You should find that \ is extremely efficient.

19. The file `RandSspd.m`, available at http://www.springer.com, runs as

    `[sA,dA]=RandSspd(n).`

    The returns are a sparse positive definite matrix `sA` and a dense perturbation of the same matrix `dA`. Use `tic` and `toc` to compare the time required to solve $(sA)\,x = b$ with your own conjugate gradient code versus solving $(dA)\,x = b$ with \. Use dimensions $n = 2^k$, with $k = 5 \ldots 10$.

20. The function `RandTrid`, available at http://www.springer.com, returns a random tridiagonal matrix T with the property that row pivoting is not needed to compute the LU decomposition. It also returns a small perturbation of T as Tp, which has no particular structure. Use `tic` and `toc` to compare the time needed to solve $Tx = b$ with your `TDMS` code from Exercise 17 versus the time needed to solve $(Tp)\,x = b$ with \. Consider $n \times n$ matrices for which $n = 2^k$ and $k = 5 \ldots 10$.

Chapter 3
Approximation

The scientist explains the world by successive approximations. – Edwin Hubble

Data, either computed or gathered, is imperfect and/or incomplete, and having a handful of methods to display, analyze, and compress it is important. We consider three introductory methods that should be known by computational scientists. The first approximation is the method of least squares, which is a technique to optimize a particular functional form to a collection of data. If data is assumed to be stochastic, i.e. drawn from a random process, then the method of least squares supports the substantial statistical study known as linear regression. The second approximation is that of cubic splines, which creates a smooth curve to match data. The third approximation is principal component analysis, which compresses data to preserve statistical characteristics.

These techniques are selected since they broadly expose us to the overriding ideas of approximating, fitting, and compressing data. The method of least squares presumes flawed information and finds a best approximation to imperfect data. The process of fitting data with a cubic spline essentially assumes precise information that can be nicely "filled in" with a smooth function. Principal component analysis collapses information in a way that allows us to describe a large dataset's features without maintaining the entire dataset. These three methods apply and adapt to numerous computational pursuits.

3.1 Linear Models and the Method of Least Squares

The method of least squares optimally approximates disparate data with an assumed functional form, called a model. We assume that the parameters α_0,

$\alpha_2, \ldots, \alpha_n$ define an approximating model f. For instance, if we want to approximate data with a second order polynomial, then our model is

$$f(x) = \alpha_0 + \alpha_1 x + \alpha_2 x^2. \tag{3.1}$$

A model f is linear if it is linear in its defining parameters, which are the α coefficients of the quadratic. This definition can seem odd at first, especially since we regularly use the term linear to characterize functions like $T(x,y) = 3x + 2y = \langle 3, 2 \rangle \cdot \langle x, y \rangle$. In this case T is one possible manifestation of the linear model $f(x,y) = ax + by = \langle a, b \rangle \cdot \langle x, y \rangle$. When we say that T is linear we mean that it is linear in its arguments x and y, but when we say that the model f is linear we mean that it is linear in its parameters a and b. In this case the function T and the model f are both linear. The function $T(x,y) = 5\sqrt{x} + 2y^2 - xy$ is nonlinear in x and y, but it is one possible realization of the linear model

$$f(x, y) = a\sqrt{x} + by^2 + cxy = \langle a, b, c \rangle \cdot \langle \sqrt{x}, y^2, xy \rangle.$$

Hence, f is a linear model whereas its associated functions are nonlinear. Similarly, the quadratic in (3.1) is a linear model even though the resulting functions are quadratics (as long as $\alpha_2 \neq 0$).

Assume that we have data points (x_i, y_i), for $i = 1, 2, \ldots, m$. The approximating model f would fit the data exactly if it simultaneously satisfied the m equations,

$$f(x_1) = y_1, \ f(x_2) = y_2, \ \ldots, \ \text{and } f(x_m) = y_m. \tag{3.2}$$

In the common event of an inconsistent system, the method of least squares seeks coefficients α_j that minimize the aggregated discrepancies, called residuals. Although there are many ways to measure discrepancies, we restrict our development to squared residuals, and the problem defining the parameters is

$$\min_{\alpha_j} \sum_{i=1}^{m} (f(x_i) - y_i)^2. \tag{3.3}$$

Such optimization problems are called (ordinary) least squares problems.

Assuming the linear model of a quadratic function in (3.1), we are able to express the simultaneous equations in (3.2) as the following matrix equation,

$$A\alpha = \begin{bmatrix} 1 & x_1 & x_1^2 \\ 1 & x_2 & x_2^2 \\ & \vdots & \\ 1 & x_m & x_m^2 \end{bmatrix} \begin{bmatrix} \alpha_0 \\ \alpha_1 \\ \alpha_2 \end{bmatrix} = \begin{bmatrix} y_1 \\ y_2 \\ \vdots \\ y_m \end{bmatrix} = y.$$

3.1 Linear Models and the Method of Least Squares

The matrix A is called the coefficient matrix, and each element of the vector $A\alpha - y$ is one of the individual residuals, from which we have

$$\min_{\alpha_j} \sum_{i=1}^{m} (f(x_i) - y_i)^2 = \min_{\alpha} \|A\alpha - y\|^2.$$

Using the fact that the squared norm is the inner product of its argument with itself, we find

$$\begin{aligned}\|A\alpha - y\|^2 &= (A\alpha - y)^T(A\alpha - y) \\ &= \alpha^T A^T A\alpha - y^T A\alpha - \alpha^T A^T y + y^T y \\ &= \alpha^T A^T A\alpha - 2y^T A\alpha + y^T y.\end{aligned}$$

As discussed in Chap. 2, we know that $A^T A$ is positive definite as long as the columns of A are linearly independent, a property of A that we assume. Hence, we know from Theorem 6 that the optimization problem in (3.3) has a unique solution.

We solve the following linear system to calculate the optimal α vector,

$$D_\alpha \left(\alpha^T A^T A\alpha - 2\alpha^T Ay + y^T y \right) = 2A^T A\alpha - 2A^T y = 0.$$

Hence, the optimal function f is defined by the unique solution to the positive definite system

$$A^T A\alpha = A^T y. \tag{3.4}$$

These equations are known as the normal equations of the system $A\alpha = y$, which are created by pre-multiplying by A^T. From Chap. 2 we know that we can solve such a system efficiently, either by factoring A into its Cholesky factorization and then backsolving or through the iterative algorithm of conjugate directions.

As an example, suppose we have the data

i	1	2	3	4	5	6
x_i	-2	-1	0	1	2	3
y_i	2	2	1	0	0	3

(3.5)

We fit several polynomials of varying degree using the model

$$f_n(x) = \alpha_0 + \alpha_1 x + \alpha_2 x^2 + \ldots + \alpha_n x^n.$$

The systems defining the coefficients, and subsequently the approximating functions themselves, for $n = 1$ and 3 are

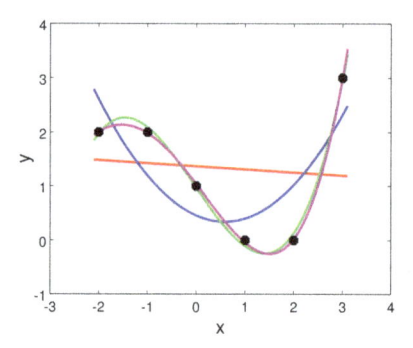

Fig. 3.1 Polynomial least squares solutions of degree 1 (red), 2 (blue), 3 (green), and 4 (magenta) for the data listed in (3.5).

$$n = 1$$

$$\begin{bmatrix} 1 & -2 \\ 1 & -1 \\ 1 & 0 \\ 1 & 1 \\ 1 & 2 \\ 1 & 3 \end{bmatrix} \begin{bmatrix} \alpha_0 \\ \alpha_1 \end{bmatrix} = \begin{bmatrix} 2 \\ 2 \\ 1 \\ 0 \\ 0 \\ 3 \end{bmatrix} \quad \text{and}$$

$$n = 3$$

$$\begin{bmatrix} 1 & -2 & 4 & -8 \\ 1 & -1 & 1 & -1 \\ 1 & 0 & 0 & 0 \\ 1 & 1 & 1 & 1 \\ 1 & 2 & 4 & 8 \\ 1 & 3 & 9 & 27 \end{bmatrix} \begin{bmatrix} \alpha_0 \\ \alpha_1 \\ \alpha_2 \\ \alpha_3 \end{bmatrix} = \begin{bmatrix} 2 \\ 2 \\ 1 \\ 0 \\ 0 \\ 3 \end{bmatrix}.$$

The defining system becomes square and invertible for $n = 5$, which means that there is a 5th degree polynomial through the 6 points. This observation extends to the fact that there is an n-th degree polynomial through $n + 1$ points, a case that we address separately in the next section. Solving the normal equations for $n = 1, \ldots, 4$ gives the following approximating polynomials, which are plotted in Fig. 3.1,

$f_1(x) = 1.361905 - 0.057143\, x,$
$f_2(x) = 0.457140 - 0.396430\, x + 0.339291\, x^2,$
$f_3(x) = 0.968254 - 1.312169\, x + 0.019841\, x^2 + 0.212963\, x^3,$ and
$f_4(x) = 1.039683 - 1.181217\, x - 0.090278\, x^2 + 0.171296\, x^3 + 0.020833\, x^4.$

The polynomial examples show that the method of least squares encompasses polynomial approximation, but the technique is broader still. For example, suppose data is known to be periodic with period $2\pi/\omega$, then an appropriate approximation could be

$$f(x) = A\cos(\omega x - \phi).$$

The parameters of the model are the amplitude, A, and the phase, ϕ. While f might be a reasonable approximating model, it is not linear in ϕ, and hence,

3.1 Linear Models and the Method of Least Squares

we cannot directly use the normal equations to solve the least squares problem. However, we can re-model f as the difference of two periodic functions to use the method of least squares. Let ϕ be such that $A_2 = A_1 \tan(\phi)$, from which we have

$$A_1 = \cos(\phi)\sqrt{A_1^2 + A_2^2} \quad \text{and} \quad A_2 = \sin(\phi)\sqrt{A_1^2 + A_2^2}.$$

Then,

$$\begin{aligned} A_1 \cos(\omega x) + A_2 \sin(\omega x) &= \sqrt{A_1^2 + A_2^2}\,(\cos(\phi)\cos(\omega x) + \sin(\phi)\sin(\omega x)) \\ &= \sqrt{A_1^2 + A_2^2}\,\cos(\omega x - \phi) \\ &= A\cos(\omega x - \phi), \end{aligned} \quad (3.6)$$

where the second equality follows from the trigonometric identity

$$\cos(\theta_1 + \theta_2) = \cos(\theta_1)\cos(\theta_2) - \sin(\theta_1)\sin(\theta_2).$$

So, while f is not linear in A and ϕ, it is linear in A_1 and A_2, and we conclude that the method of least squares can be used.

As an example, suppose the following data is collected every other hour over a 24-hour period,

t	0	2	4	6	8	10	12	14	16	18	20	22
y_t	2.01	2.41	3.05	3.15	2.80	2.62	2.21	1.43	1.05	1.23	1.18	1.30

(3.7)

If the underlying process is assumed to be periodic over the day, then a good approximation might be

$$f(t) = \alpha_1 + \alpha_2 \cos(\pi t/12) + \alpha_3 \sin(\pi t/12).$$

The over determined linear system is

$$\begin{bmatrix} 1 & \cos(0) & \sin(0) \\ 1 & \cos(\pi/6) & \sin(\pi/6) \\ 1 & \cos(\pi/3) & \sin(\pi/3) \\ \vdots & \vdots & \vdots \\ 1 & \cos(11\pi/6) & \sin(11\pi/6) \end{bmatrix} \begin{bmatrix} \alpha_1 \\ \alpha_2 \\ \alpha_3 \end{bmatrix} = \begin{bmatrix} 2.01 \\ 2.51 \\ 2.91 \\ \vdots \\ 1.50 \end{bmatrix},$$

and the solution to the corresponding normal equations gives a best approximation of

$$\begin{aligned} f(t) &= 2.02235 - 0.00169 \cos(\pi t/12) + 1.00945 \sin(\pi t/12) \\ &= 2.02235 + \sqrt{0.00169^2 + 1.00945^2}\,\cos(\pi t/12 - \tan^{-1}(-1.00945/0.00169)) \\ &= 2.02235 + 1.0095 \cos(\pi t/12 + 1.57), \end{aligned}$$

which is plotted in Fig. 3.2.

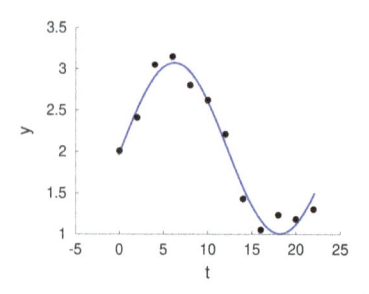

Fig. 3.2 A best periodic approximation to the tabulated data in (3.7).

We conclude by noting that $A^T A$ is invertible under the assumption that the columns of A are linearly independent, and hence, one is tempted to solve the normal equations as $\alpha = (A^T A)^{-1} A^T y$. However, calculating the inverse is ill-advised, as it is more accurate to solve $A^T A \alpha = A^T y$, say by factoring $A^T A$ and then solving for α with forward and backward substitution. Part of the concern lies with the fact that the numerical properties of $A^T A$ can only degrade from those of A, a topic we discuss in the next section.

3.1.1 Lagrange Polynomials: An Exact Fit

The case of finding an n-th degree polynomial through $n+1$ points deserves special attention. The coefficient matrix A in this case is square and (assumed) invertible. However, the $A^T A$ matrix of the normal equations is often ill-conditioned, which loosely means that diminutive errors in the system's right-hand side could lead to substantial changes in the solution. Indeed, the pre-multiplication by A^T is part of the problem, as it can be shown that A is at least as well conditioned as $A^T A$, and hence, for this case it would be better to solve $A\alpha = y$ instead of $A^T A \alpha = A^T y$. So while we know mathematically that there is a unique n-th degree polynomial through $n+1$ points, calculating it with the normal equations can be computationally difficult.

We may construct the solution with an algebraic observation instead of considering a least squares problem. Assume we want to fit a quadratic through the three points (x_1, y_1), (x_2, y_2), and (x_3, y_3). Consider

$$f(x) = \frac{(x-x_2)(x-x_3)}{(x_1-x_2)(x_1-x_3)} y_1 + \frac{(x-x_1)(x-x_3)}{(x_2-x_1)(x_2-x_3)} y_2 + \frac{(x-x_1)(x-x_2)}{(x_3-x_1)(x_3-x_2)} y_3.$$

Notice that f is a quadratic since it is the sum of three quadratics. Moreover,

3.1 Linear Models and the Method of Least Squares

$$f(x_1) = y_1, \quad f(x_2) = y_2, \quad \text{and} \quad f(x_3) = y_3.$$

There is only one quadratic through these three points, which means that f is this unique function. Hence we have calculated the best polynomial approximation and have sidestepped the possibility of magnifying errors by solving an ill-conditioned system.

The Lagrange polynomial through the points (x_i, y_i), for $i = 1, 2, \ldots, n$, is

$$f(x) = \sum_{i=1}^{n} \left(\prod_{j \neq i} \frac{x - x_j}{x_i - x_j} \right) y_i, \tag{3.8}$$

where each product is over $j = 1, 2, \ldots, n$ such that $j \neq i$. The assumption that $x_i \neq x_j$ for $i \neq j$ is important, since the polynomial is otherwise ill-defined.

We use the data from Fig. 3.3 to illustrate the numerical advantage of a Lagrange polynomial over a solution calculated from the normal equations. For $n = 6$ the condition number of $A^T A$ is 2.17×10^{16}. As a basis of comparison, a perfectly conditioned matrix has a condition number of 1. The substantial value of 2.17×10^{16} suggests that minor changes in y can be magnified on the order of 10^{16} as the normal equations are solved. Hence, even slight round-off errors can lead to noticeable effects on the α_i coefficients. MATLAB and Octave both warn that $A^T A$ is ill-conditioned upon using `A'*A\A'*y`, which should raise suspicion about the solution. The suspicion is confirmed once the polynomial from the normal equation is plotted against the Lagrange polynomial, see Fig. 3.3.

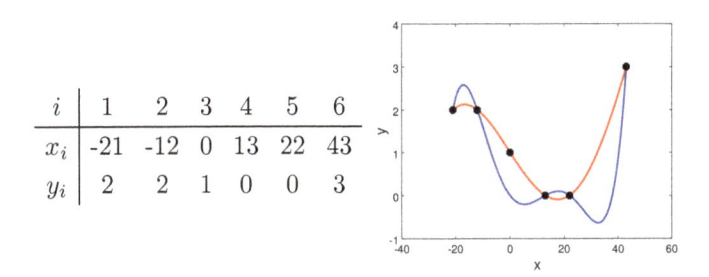

i	1	2	3	4	5	6
x_i	-21	-12	0	13	22	43
y_i	2	2	1	0	0	3

Fig. 3.3 The Lagrange polynomial is the red curve. The approximating polynomial from the normal equations is blue and is supposed to match the red curve. The numerical instability of the normal equations produces an inaccurate curve that even misses the third data point.

Lagrange polynomials demonstrate the advantage of carefully considering a linear system prior to submitting it to computational reckoning. Linear systems vary in their difficulty to solve, even if unique solutions are known to exist. The linear system associated with a Lagrange polynomial need not, and indeed should not, be solved because the solution can be calculated directly. The only possible numerical concern with calculating a Lagrange polynomial is that the distance between x_i and x_j might be exceedingly small, in which case we might have a near division by zero. However, attempting to calculate the polynomial directly is preferred to trying to solve the normal equations even in this case.

Lagrange polynomials have the shortcoming that they tend to overfit data, a concern illustrated by the next example. As such, Lagrange polynomials should be used with caution. The requirement of a single polynomial agreeing with data is uncommon, and such functions often make predictions dubious. To illustrate, consider the data and the associated Lagrange polynomial in Fig. 3.4. While the Lagrange polynomial agrees with the data, as it must, the

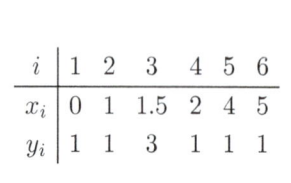

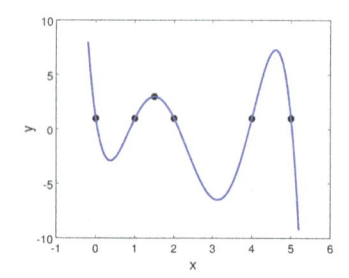

i	1	2	3	4	5	6
x_i	0	1	1.5	2	4	5
y_i	1	1	3	1	1	1

Fig. 3.4 The example data on the left is not likely well-represented by the Lagrange polynomial on the right.

requirement of a sole function doing so suggests the predictions:

if x is	3	4.5	6
then y is	-6.3143	7.000	-145.2857

All of these predictions are noticeably questionable against the data, as y only obtains the two values of 1 and 3. In particular, the high degree polynomial quickly becomes unrealistic outside of x's data range.

3.2 Linear Regression: A Statistical Perspective

Linear regression is a mainstay of statistical analysis, and having a working knowledge of the statistical interpretation of regression is important. The difference between the data mentality of the previous section and the statistical perspective of this section is that data is now assumed to represent an underlying random process. Adding a stochastic assumption provides much in the way of interpretation, and while the assumed flawed data of the previous section can be approximated optimally with the technique of least squares

3.2 Linear Regression: A Statistical Perspective

independent of any notion of randomness, the statistical perspective of this section improves our ability to analyze and exploit a model.

The stochastic interpretation of this section requires an understanding of random variables, and we begin with a succinct introduction to probability and statistics. Students with a foreknowledge of probability and statistics should feel comfortable with a cursory read.

3.2.1 Random Variables

Random variables express quantities that happen by chance, and they are regularly employed in computational science to explain random observations and/or to aid computations. We consider the situation of a random experiment. The collection of all possible outcomes of the experiment is called the sample space, denoted by S, and an event is a set of outcomes, i.e. a subset of S. Random variables, which are commonly denoted by capital letters, assign a real value to each outcome of the sample space. So if X is a random variable on the sample space S, then $X(s)$ is a real value for each $s \in S$, and the range of X is $X(S) = \{X(s) : s \in S\}$.

Suppose we flip a fair coin three times as an illustration. If we denote heads by H and tails by T, then the sample space contains eight possible outcomes,

$$S = \{(H,H,H),(H,T,H),(H,H,T),(T,H,H),(H,T,T),(T,H,T),(T,T,H),(T,T,T)\}.$$

We let X be the random variable that counts the number of tails in an outcome. So the assignments $X(H,H,H) = 0$, $X(H,T,H) = 1$, and $X(T,T,T) = 3$ illustrate how X maps outcomes to real values. In this example $X(S) = \{0, 1, 2, 3\}$.

Suppose we want to calculate the probability of having at least two tails in the coin flipping experiment. This calculation corresponds with the event

$$A = \{(H,T,T),(T,H,T),(T,T,H),(T,T,T)\} \subseteq S.$$

We express the probability as

$$P(A) = P(X \geq 2).$$

The value of $P(X \geq 2)$ is computed with a probability function f that assigns every value of $X(S)$ a real value, which means that $f : X(S) \to \mathbb{R}$. If we assume that each outcome is equally likely, i.e. that the coin is fair, then a reasonable probability function is

$f(0) = 1/8 \Leftarrow$ one of eight outcomes has no tails,
$f(1) = 3/8 \Leftarrow$ three of eight outcomes have one tail,
$f(2) = 3/8 \Leftarrow$ three of eight outcomes have two tails, and
$f(3) = 1/8 \Leftarrow$ one of eight outcomes has three tails.

Then,
$$P(A) = P(X \geq 2) = P(X \in \{2,3\}) = f(2) + f(3) = \frac{1}{2},$$
which reflects the fact that A accounts for half of the equally likely outcomes.

A random variable is discrete if its range is discrete and is (absolutely) continuous if its range is a continuous interval. If X is a discrete random variable, then f is a probability mass function, but if X is instead continuous, then f is a probability density function. Mass and density functions always satisfy $0 \leq f(x)$ for $x \in X(S)$, and the integrated or summed total over all possible x is 1, i.e.

Discrete Case $\qquad$ Continuous Case

$$\sum_{x \in X(S)} f(x) = 1 \quad \text{or} \quad \int_{X(S)} f(x)\,dx = 1.$$

The probability of event A is calculated as

Discrete Case $\qquad$ Continuous Case

$$P(A) = P(X \in X(A)) \quad \text{is} \quad \sum_{x \in X(A)} f(x) \quad \text{or} \quad \int_{X(A)} f(x)\,dx.$$

There are numerous well-studied probability distributions, and even a modest review would be a course in itself. However, a couple of the most important distributions should be among the computational scientist's immediate reserve. In particular,

$$X \sim \mathcal{U}(a,b) \quad \leftarrow \quad \text{read } X \text{ is uniformly distributed on [a,b], and}$$

$$X \sim \mathcal{N}(\mu, \sigma^2) \quad \leftarrow \quad \text{read } X \text{ is normally distributed with mean } \mu \text{ and variance } \sigma^2.$$

If $X \sim \mathcal{U}(a,b)$, then the probability density function is $f(x) = 1/(b-a)$ as long as $a \leq x \leq b$. If $X \sim \mathcal{N}(\mu, \sigma^2)$, then the probability density function for any real x is

$$f(x) = \frac{1}{\sigma\sqrt{2\pi}} e^{-(x-\mu)^2/2\sigma^2}.$$

A standard uniform variable assumes $a = 0$ and $b = 1$, and a standard normal variable assumes $\mu = 0$ and $\sigma = 1$. The iconic bell shaped curve of the standard normal is illustrated in Fig. 3.6.

Suppose Z is a standard normal variable. Then the probability of Z satisfying $-1 \leq Z \leq 2$ is

$$P(-1 \leq Z \leq 2) = \int_{-1}^{2} \frac{1}{\sqrt{2\pi}} e^{-x^2/2}\,dx \approx 0.8186.$$

3.2 Linear Regression: A Statistical Perspective

The probability of any single outcome is zero, i.e.

$$P(Z = z) = \int_z^z \frac{1}{\sqrt{2\pi}} e^{-x^2/2}\, dx = 0.$$

Similarly, if Y is a standard uniform variable, then

$$P(1/4 \leq Y \leq 5/8) = \int_{1/4}^{5/8} dx = 3/8.$$

Random variables are discussed in terms of statistical parameters such as their means, variances, and standard deviations. Suppose Y is a discrete random variable with probability mass function f. Then the mean of Y, also called the expected value of Y, is

$$\mu = \mu_Y = \mathrm{E}(Y) = \sum_{y \in Y(S)} y\, f(y).$$

If X is a continuous random variable with probability density function f, then the mean of X, or the expected value of X, is

$$\mu = \mu_X = \mathrm{E}(X) = \int_{X(S)} x\, f(x)\, dx.$$

Routine calculations show that if $X \sim \mathcal{U}(a,b)$, then $\mathrm{E}(X) = (b-a)/2$, and if $Z \sim \mathcal{N}(\mu, \sigma^2)$, then $\mathrm{E}(Z) = \mu$.

A continuous random variable's variance is

$$\sigma^2 = \sigma_X^2 = \mathrm{Var}(X) = E((X - E(X))^2) = \int_{X(S)} (x - E(X))^2 f(x)\, dx.$$

Variance measures how far outcomes range from their mean, with the variations being squared, weighted by their probability, and totaled. A useful re-expression is

$$\mathrm{Var}(X) = \int_{X(S)} x^2 p(x)\, dx - \left(\int_{X(S)} x\, p(x)\, dx\right)^2 = E(X^2) - E(X)^2. \quad (3.9)$$

If $X \sim \mathcal{U}(a,b)$, then $\mathrm{Var}(X) = (b-a)^2/12$, and if $Y \sim \mathcal{N}(\mu, \sigma^2)$, then $\mathrm{Var}(Y) = \sigma^2$. The standard deviation is the square root of variance and is denoted by σ or σ_X. If the random variable is discrete, then the integral defining the variance is replaced with an appropriate summation.

Data does not equate with perfect distributional knowledge, and instead, data is a sample drawn from an assumed distribution. We can't expect to calculate exact statistical parameters from data because we only have the sample's evidence of the true distribution. For example, the following outcomes comprise a 10-element sample drawn from the standard uniform X.

$X \sim \mathcal{U}(0,1)$

Sample	x_1	x_2	x_3	x_4	x_5
	0.9963082	0.3860629	0.3405858	0.9133750	0.8468924
Sample	x_6	x_7	x_8	x_9	x_{10}
	0.2982297	0.0073326	0.6034754	0.4430176	0.9382423

Because we know the true distribution of X, the true mean and variance are

$$\mathrm{E}(X) = 0.5 \quad \text{and} \quad \mathrm{Var}(X) = 0.0833.$$

If we had not known the distribution of X and had only seen the data, then we would have suspected

$$\mathrm{E}(X) \approx \bar{x} = \frac{1}{10} \sum_{i=1}^{10} x_i = 0.5774$$

and

$$\mathrm{Var}(X) \approx s^2 = s_X^2 = \frac{1}{9} \sum_{i=1}^{10} (x_i - \bar{x})^2 = 0.1117.$$

These calculations only approximate the true statistics. The estimate of the mean is called the sample mean, and the estimate of the variance is called the sample variance

Notice the divisor of 9 in the calculation of the sample variance, a value that is one short of the number of data points. We might have anticipated a divisor of 10 since variance is an expected value; however, dividing by 10 would have biased the calculation. Both the sample mean and variance are unbiased as calculated, meaning that if we repeatedly draw n element samples, then

$$\mathrm{E}\left(\frac{1}{n} \sum_{i=1}^{n} x_i\right) = \mathrm{E}(\bar{x}) = \mathrm{E}(X) \quad \text{and}$$

$$\mathrm{E}\left(\frac{1}{n-1} \sum_{i=1}^{n} (x_i - \bar{x})^2\right) = \mathrm{E}(s^2) = \mathrm{Var}(X).$$

The divisor of $n-1$ is called Bessel's correction, and we assume its use throughout.

Notice that $\bar{x}$ and s depend on the random process of drawing a sample, and hence, they are outcomes of the stochastic action that selects a sample. We let $\bar{X}$ be the random variable that achieves the values of $\bar{x}$ as the sample varies. The sample variance and standard deviation also associate with ran-

3.2 Linear Regression: A Statistical Perspective

dom variables, typically S^2 and S in statistics, but we forego this notation because S already denotes the sample space.

Two landmark results help us understand and interpret $\bar{X}$. The first is commonly called the Law of Large Numbers.

Theorem 8 (Law of Large Numbers). *If X_i is the i-th independent sample from a distribution with mean μ, then for any positive value ε,*

$$\lim_{n \to \infty} P\left(\left|\frac{1}{n}\sum_{i=1}^{n} X_i - \mu\right| > \varepsilon\right) = 0.$$

The assumed independence requires that the i-th outcome be decided without regard of the previous $i-1$ outcomes. For example, if we toss a coin 10 times, then the outcome of the 11th toss does not depend on the previous 10, even if all previous outcomes had been, for example, heads.

The law of large numbers instills a trust in the sample mean's ability to approximate the true mean as the sample size increases. The convergence statement reads, "the sample mean converges to the true mean in probability." The result establishes a likely agreement between the sample mean and the true mean as long as the sample size is sufficiently large. To illustrate, suppose ε is the small value of 10^{-6}. Then the likelihood that the sample mean fails to agree with the true mean in the first five decimal places is negligible as long as n is sufficiently large. There is no sense of how large n should be to reasonably guarantee the approximation, an interpretive shortcoming that is somewhat overcome by our next result.

A comment about interpreting probability statements is worthwhile before continuing. Many grade-school posters claim that zero probability is the same as having perfect certainty that an event will not occur, but this is patently false. Suppose X is a standard uniform variable, and consider the situation in which each of your classmates selects a value of x satisfying $0 \leq x \leq 1$. The probability of X achieving one of these values is zero, and indeed, this is true even if you had had an infinitely enumerated number of classmates. Each of the selected values is possible, but the collection is just overwhelmingly unlikely against the infinite number of values left unselected. A full discussion of this nuance requires an advanced look at integration, which is beyond the confines of our discussion. That said, we encourage readers to interpret probability statements with care.

The most pervasive result in statistics is arguably the central limit theorem, which is so often referenced that it is abbreviated as the CLT. Referring to "the" central limit theorem is a bit of a misnomer, as the CLT is really a family of related results, many of which are studied to this day. The general theme of the CLT is to establish a sense of normality for a random variable that is derived from other random variables, the latter of which may or may not be normal. The prevailing concept is that the normal distribution arises naturally in numerous stochastic settings. We only state the CLT with regard to the sample mean.

Theorem 9 (Central Limit Theorem (of Sample Means)). *If $\bar{X}$ is the sample mean of a random, independent, n-element sample from a distribution with mean μ and variance σ^2, then $\bar{X}$ is approximately distributed as $\mathcal{N}(\mu, \sigma^2/n)$ for sufficiently large n.*

Notice that both μ and σ^2 are assumed finite and that n needs to be sufficiently large, a case we generally assume for interpretive purposes. The approximation is perfect if X is normally distributed; however, in this case we regularly need to estimate σ^2 with s^2, which leads to an approximate distribution of $\bar{X}$ called the t-distribution. The examples in Sect. 3.2.2 demonstrate an appropriate use of the t-distribution for small n.

The CLT adds to our interpretation of the law of large numbers by estimating the proximity of the sample mean to the true mean. The approximate variance of $\bar{X}$ is inversely proportional to the size of the sample, with the proportionality constant being σ^2. We can couple this fact with the added information that any normal has 68.27% of its probability within one standard deviation, 95.45% of its probability within two standard deviations, and 99.73% of its probability within three standard deviations. For example, suppose our estimate of σ^2 is $s^2 = 12.3$ based on a random sample of size 100. Our estimated standard deviation for $\bar{X}$ is thus

$$\sqrt{\frac{s^2}{100}} = \sqrt{\frac{12.3}{100}} = 0.35071.$$

If $\bar{x} = 8.79$, then we are roughly 95% confident that the true mean is within $8.79 \pm 2 \times 0.35071$, i.e. $\mu_{\bar{X}}$ is in $[8.0886, 9.4914]$ with about 95% confidence. If we raise the sample size to 1000, then $\mu_{\bar{X}}$ is in $[8.5682, 9.0118]$ with about 95% confidence. So, while we don't exactly know how far $\bar{x}$ is from the true mean, we can suspect that it is within an interval up to a specified confidence.

The standard uniform and normal distributions are common in computational settings, and both MATLAB and Octave have standard commands to draw samples. The commands are `rand` and `randn` for, respectively, the standard uniform and standard normal distributions. Both accept dimensional arguments, so `rand(5)` is a 5×5 matrix with each element being drawn from a standard uniform and `randn(2,4)` is a 2×4 matrix with each element being drawn from a standard normal. The standard distributions can be algebraically manipulated to alter their defining parameters. If we want $X \sim \mathcal{U}(2,5)$, then we use $X = 2+3*Y$, where Y is a standard uniform. So we can draw a 10-element sample of X with `2 + 3*rand(1,10)`. Likewise, if we want $X \sim \mathcal{N}(4,2)$, then $X = 4+\sqrt{2}*Z$, where Z is a standard normal. In this case we could generate a 10-element sample with `4 + sqrt(2)*randn(1,10)`. Notice that if $X \sim \mathcal{N}(\mu, \sigma^2)$, then $Z = (X - \mu)/\sigma$ is a standard normal.

Figures 3.5 and 3.6 depict the standard uniform and standard normal distributions together with a random sample. The 100-element sample from the uniform distribution is somewhat equally distributed over the interval $[0, 1]$, as each value is equally likely. The 200-element sample of the normal

3.2 Linear Regression: A Statistical Perspective

has many more values near the mean, with values becoming less likely as the distance from the mean increases. Of the 200-element sample, about 137 (68.27% of 200) of the values are expected to be within ± 1, about 191 (95.45% of 200) of the values are expected to be within ± 2, and about 9 (4.55% of 200) of the values are expected to be outside ± 2. Outcomes beyond ± 3 are sometimes called "rare events," although this term has no precise definition and should be clarified if used.

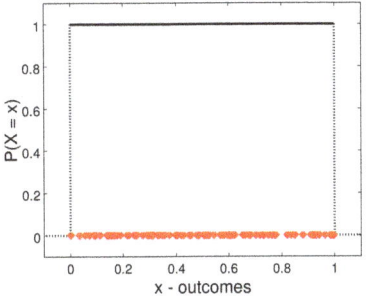

Fig. 3.5 A standard uniform distribution and a random sample of 100 values, which are the red dots.

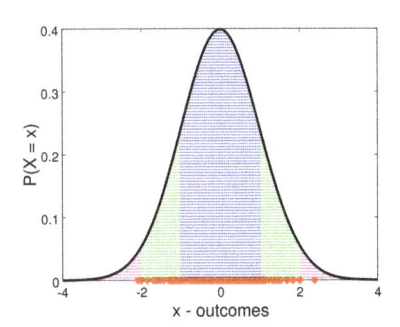

Fig. 3.6 A standard normal distribution and a random sample of 200 values, which are the red dots. The blue region over $[-1, 1]$ accounts for 68.27% of the total probability. The union of the blue and green regions over $[-2, 2]$ accounts for 95.45% of the total probability, and the union of the blue, green, and magenta regions over $[-3, 3]$ accounts for 99.73% of the probability.

MATLAB and Octave have functions to calculate sample means and variances. For example, we can generate a 100-element sample from a standard normal with x = randn(100,1). The commands mean(x) and var(x) then calculate the sample mean and variance. The general formulas are

$$\texttt{mean(x) calculates } \bar{x} = \frac{1}{n}\sum_i x_i = \frac{e^T x}{n} \quad \text{and}$$

$$\texttt{var(x) calculates } s_x^2 = \frac{1}{n-1}\sum_i (x_i - \bar{x})^2$$

$$= \frac{(x - \bar{x}\,e)^T (x - \bar{x}\,e)}{n-1}.$$

The vector e in these expressions is the vector of ones in which length is decided by the context of its use. For this example e would be a column vector of ones of length 100, i.e. e = ones(100,1) in MATLAB. Both commands also work with matrices, in which case they return a vector of the sample means and variances of the columns.

Our last statistical parameter is the covariance between two random variables, say X and Y. Covariance is a measure of how two variables simultaneously trend above and below their means. The definition is

$$\mathrm{Cov}(X,Y) = \sigma_{xy} = E((X - E(X))(Y - E(Y))).$$

Notice that $\mathrm{Cov}(X,X) = \mathrm{Var}(X)$. Estimating the covariance from a sample is similar to calculating the sample variance. If x_i and y_i comprise an n element sample from X and Y, then the sample covariance is

$$s_{xy} = \frac{1}{n-1} \sum_{i=1}^{n} (x_i - \bar{x})(y_i - \bar{y}) = \frac{(x - \bar{x}\,e)^T (y - \bar{y}\,e)}{n-1},$$

where the divisor of $n-1$ is again Bessel's correction. If x and y are vectors of the sample, then the MATLAB command cov(x,y) computes the sample covariance.

Large, positive covariances imply a tendency for X and Y to simultaneously be above or below their respective means. Large, negative covariances imply that X and Y trend negatively to each other, i.e. if one is above its mean, then the other is likely below its mean. A covariance of zero indicates that X and Y are uncorrelated, and hence, one variable's propensity to be above or below its mean has no apparent bearing on the other variable's propensity to be above or below its mean.

Scale is a bit of an interpretive issue with covariance, and it is often normalized by the standard deviations to help rectify this concern. The resulting statistical parameter is the Pearson correlation, which is

$$\rho_{XY} = \frac{\mathrm{Cov}(X,Y)}{\sigma_X \sigma_Y}.$$

The Pearson correlation always satisfies $-1 \leq \rho_{XY} \leq 1$. A value of -1 indicates perfect negative correlation, a value of 1 indicates perfect positive correlation, and a value of 0 means X and Y are linearly uncorrelated. The sample estimate from data is

$$r_{xy} = \frac{s_{xy}}{s_x s_y},$$

which can be calculated with cor(x,y) in MATLAB and Octave, where x and y are again sample vectors.

3.2.2 Stochastic Analysis and Regression

The method of linear regression relates one random variable to a collection of others. Suppose our data has $p+1$ components and is of the form

$$(x_{1i}, x_{2i}, \ldots, x_{pi}, y_i), \quad \text{for } i = 1, 2, \ldots, n.$$

There is nothing special about the last component being designated as "y," and any component could be selected.

We presume a statistical perspective that counters our earlier deterministic assumption about least squares, and we now assume that data is a random sample drawn from the random variables $X_1, X_2, \ldots, X_p$, and Y. The interpretation of the linear model is

$$\left.\begin{aligned} Y(X_1 &= x_1, X_2 = x_2, \ldots, X_p = x_p) \\ &= \alpha_0 + \alpha_1 x_1 + \ldots + \alpha_p x_p + \varepsilon \\ &= E(Y \mid X_1 = x_1, X_2 = x_2, \ldots, X_p = x_p) + \varepsilon. \end{aligned}\right\} \quad (3.10)$$

The left-hand side is the dependent, random variable Y under the assumption that all of X_1 through X_p have obtained their stated values, meaning that each random X_i has achieved the value x_i. We routinely interpret the right-hand side of the first equality as the two terms

$$\underbrace{(\alpha_0 + \alpha_1 x_1 + \ldots + \alpha_p x_p)}_{\text{prediction}} + \underbrace{\varepsilon}_{\text{error}}.$$

The model's prediction of Y for the known values of the Xs is the first term, and the error associated with the observation is the second term. The last equality of (3.10) illustrates the fundamental statistical tenet of regression, which is that model predictions are the expected values of Y for the observed values of the X variables.

The X variables are called inputs, regressors, or independent variables, and the Y variable is called the output, response, or dependent variable. A multiple regressor model is common even if data is paired as (x_i, y_i), which might appear to have a single regressor. For example, a quadratic approximation would alter (x_i, y_i) to (x_i, x_i^2, y_i) and then seek α_0, α_1, and α_2 to best satisfy

$$y_i \approx \alpha_0 + \alpha_1 x_i + \alpha_2 x_i^2.$$

In this case Y is the response variable and X and X^2 are the two regressors. Recall that the regression model is linear because it is linear in its parameters; that is, the model is linear in the three α values.

The left-hand side of Eq. (3.10) is the random variable Y, but the first terms of the following two expressions are non-random scalars satisfying

$$\alpha_0 + \alpha_1 x_1 + \ldots + \alpha_p x_p = E(Y \,|\, X_1 = x_1, X_2 = x_2, \ldots, X_p = x_p).$$

The conclusion is that the random element must be the error, and hence, the statistical model mandates random error. The error term is technically a function of $(x_1, x_2, \ldots, x_p)$, an observation we address momentarily.

Calculating the α parameters to minimize squared error isn't possible because data is only a sample of $X_1, X_2, \ldots, X_p$, and Y. For instance, we can readily conceive that a new sample might alter the parameter estimates. Indeed, the key difference between the deterministic and stochastic viewpoints is:

> If data is assumed to be a sample drawn from random variables, then the true α parameters that minimize error can only be **estimated** from data's evidence of the true distributions.

The intent of minimizing error suggests that we should calculate coefficients $a_0, a_1, \ldots, a_n$ to define the function $\hat{y}$ to best satisfy

$$\hat{y}(x_1, x_2, \ldots, x_p) = a_0 + a_1 x_1 + a_2 x_2 + \ldots + a_n x_n$$
$$\approx \alpha_0 + \alpha_1 x_1 + \alpha_2 x_2 + \ldots + \alpha_n x_n,$$

where the best approximation means that the a coefficients minimize

$$\sum_{i=1}^{n} (y_i - \hat{y}(x_{1i}, x_{2i}, \ldots, x_{pi}))^2.$$

This minimization problem uses the sample value y_i in lieu of $Y(X_1 = x_{1i}, X_2 = x_{2i}, \ldots, X_p = x_{pi})$, the former of which is data's evidence of the latter.

The difference $y_i - \hat{y}(x_{1i}, x_{2i}, \ldots, x_{pi})$ is the i-th residual, and our goal of minimizing error is supplanted by minimizing the sum of squared residuals. The substitution is reasonable since the collection of residuals is an approximate sample of the random error. The a coefficients are thus calculated by minimizing the sum of squared residuals and not the sum of squared errors. This approximation is likely the best we can do from a data perspective since we don't assume any foreknowledge about the distributions of Y or X_i.

The resulting least squares problem is like that of the previous section, and the a coefficients can be calculated by solving the normal equations in (3.4) with the $n \times p$ matrix and the right-hand side of

$$A = \begin{bmatrix} 1 & x_{11} & x_{21} & \ldots & x_{p1} \\ 1 & x_{12} & x_{22} & \ldots & x_{p2} \\ \vdots & \vdots & \vdots & & \vdots \\ 1 & x_{1n} & x_{2n} & \ldots & x_{pn} \end{bmatrix} \quad \text{and} \quad y = \begin{pmatrix} y_1 \\ y_2 \\ \vdots \\ y_n \end{pmatrix}. \tag{3.11}$$

3.2 Linear Regression: A Statistical Perspective

Hence, a satisfies $A^T A a = A^T y$, and as long as the columns of A are linearly independent, which is most common and assumed, the matrix $A^T A$ is positive definite and the vector a is unique. In this case we have

$$A^T A a = A^T y \iff a = (A^T A)^{-1} A^T y.$$

We repeat that solving the system on the left has better computational accuracy and efficiency than does the explicit calculation on the right, and the a coefficients should be calculated by solving the normal equations instead of calculating $(A^T A)^{-1}$. That said, the expression on the right is convenient mathematically.

The predicted values of the model for the data, denoted by $\hat{y}_{\text{data}}$, are

$$\hat{y}_{\text{data}} = \begin{pmatrix} \hat{y}(x_{11}, x_{21}, \ldots, x_{p1}) \\ \hat{y}(x_{12}, x_{22}, \ldots, x_{p2}) \\ \vdots \\ \hat{y}(x_{1n}, x_{2n}, \ldots, x_{pn}) \end{pmatrix} = \begin{pmatrix} a_0 + a_1 x_{11} + a_2 x_{21} + \ldots + a_p x_{p1} \\ a_0 + a_1 x_{12} + a_2 x_{22} + \ldots + a_p x_{p2} \\ \vdots \\ a_0 + a_1 x_{1n} + a_2 x_{2n} + \ldots + a_p x_{pn} \end{pmatrix}$$

$$= A a = A(A^T A)^{-1} A^T y.$$

So the $\hat{y}$ values at the regressor's sample data are efficiently and accurately calculated by first solving $A^T A a = A^T y$ for a and then multiplying the result on the left by the matrix A. The model's output at any other collection of values, say $x = (1, x_1, x_2, \ldots, x_p)^T$, is

$$\hat{y}(x_1, x_2, \ldots, x_p) = \hat{y}(x) = x^T a = x^T (A^T A)^{-1} A^T y.$$

Notice the pre-padding of x with a 1, which is needed to account for the constant term a_0. The residuals of the optimized regression are

$$y - \hat{y}_{\text{data}} = y - A a = y - A(A^T A)^{-1} A^T y = (I - A(A^T A)^{-1} A^T) y.$$

The second expression can be calculated directly from the data without an inverse by first solving for a and then making the appropriate calculation.

A simple linear regression model has a single input and output, and this case warrants some attention because it helps with interpretation. Consider the simple linear regression model for which

$$\hat{y} = a_0 + a_1 x.$$

In this case

$$A = \begin{bmatrix} 1 & x_1 \\ 1 & x_2 \\ \vdots & \vdots \\ 1 & x_n \end{bmatrix} \quad \text{and} \quad A^T A = \begin{bmatrix} n & e^T x \\ e^T x & x^T x \end{bmatrix}.$$

The inverse of the 2×2 matrix $A^T A$ is

$$(A^T A)^{-1} = \frac{1}{n\, x^T x - (e^T x)^2} \begin{bmatrix} x^T x & -e^T x \\ -e^T x & n \end{bmatrix}$$

$$= \frac{1}{x^T x - n(e^T x/n)^2} \begin{bmatrix} x^T x/n & -e^T x/n \\ -e^T x/n & 1 \end{bmatrix}$$

$$= \frac{1}{x^T x - n\bar{x}^2} \begin{bmatrix} x^T x/n & -\bar{x} \\ -\bar{x} & 1 \end{bmatrix}$$

$$= \frac{1}{(n-1) s_x^2} \begin{bmatrix} x^T x/n & -\bar{x} \\ -\bar{x} & 1 \end{bmatrix}.$$

The last equality holds because

$$\left.\begin{aligned}
(x - \bar{x}\, e)^T (x - \bar{x}\, e) &= x^T x - 2\bar{x}\, e^T x + \bar{x}^2 e^T e \\
&= x^T x - 2\bar{x}\, (e^T x/n)\, n + n\bar{x}^2 \\
&= x^T x - 2n\bar{x}^2 + n\bar{x}^2 \\
&= x^T x - n\bar{x}^2,
\end{aligned}\right\} \quad (3.12)$$

from which we have

$$s_x^2 = \frac{(x - \bar{x}\, e)^T (x - \bar{x}\, e)}{n - 1} = \frac{x^T x - n\bar{x}^2}{n - 1}.$$

So the model parameters are

$$a = (A^T A)^{-1} A^T y$$

$$= \frac{1}{(n-1) s_x^2} \begin{bmatrix} x^T x/n & -\bar{x} \\ -\bar{x} & 1 \end{bmatrix} \begin{bmatrix} 1 & 1 & \dots & 1 \\ x_1 & x_2 & \dots & x_n \end{bmatrix} \begin{pmatrix} y_1 \\ y_2 \\ \vdots \\ y_n \end{pmatrix}$$

$$= \frac{1}{(n-1) s_x^2} \begin{bmatrix} x^T x/n & -\bar{x} \\ -\bar{x} & 1 \end{bmatrix} \begin{pmatrix} e^T y \\ x^T y \end{pmatrix}.$$

If we complete the calculation for a_1, i.e. the second component of a, then

$$a_1 = \frac{-\bar{x}\, e^T y + x^T y}{(n-1) s_x^2} = \frac{x^T y - n\bar{x}\bar{y}}{(n-1) s_x^2} = \frac{(x - \bar{x} e)^T (y - \bar{y} e)/(n-1)}{s_x^2} = \frac{s_{xy}}{s_x^2},$$

3.2 Linear Regression: A Statistical Perspective

where the second to last equality follows from a calculation similar to (3.12). So the a_1 parameter is the ratio of the sample covariance of x and y to the sample variance of x. The intercept parameter a_0 can then be calculated upon recognizing that

$$\bar{y} = a_0 + a_1 \bar{x} \Rightarrow a_0 = \bar{y} - \frac{s_{xy}}{s_x^2} \bar{x},$$

a fact that follows by solving for a_0 analogous to how we solved for a_1. MATLAB and Octave can calculate a_0 and a_1 for the simple linear model with

```
a1 = cov(x,y) / var(x);
a0 = mean(y) - a1 * mean(x);
```

A comment on notation is apt. The previous section used the method of least squares to calculate α coefficients, but the statistical development of this section estimates the α coefficients with the same technique. The resulting estimates are a coefficients. Greek letters in statistics are often used to represent statistical parameters like the mean μ, the variance σ^2, and the true model parameters α. Estimates of these statistical parameters, called statistics, are often denoted by their Roman counterparts, e.g. s instead of σ and a instead of α. The regression parameters are most commonly denoted as β values in statistical texts, with the corresponding estimates then being b values. We have refrained from this notation since b is so commonly used as the right-hand side of a matrix equation in a computational setting.

The statistical interpretation of regression depends on several statistical assumptions about error.

Error Assumptions of Linear Regression

- The expected value of each $\varepsilon(x_1, x_2, \ldots, x_p)$ is zero.
- All $\varepsilon(x_1, x_2, \ldots, x_p)$ share a common variance.
- The collection of $\varepsilon(x_1, x_2, \ldots, x_p)$ is independent.
- Each $\varepsilon(x_1, x_2, \ldots, x_p)$ is normally distributed.

The validity of these assumptions can be investigated by studying the collection of residuals, which is again data's evidence of the true error. Such investigations are referred to as residual analysis, and the exercises include a few examples. Whereas some of the assumptions are paramount to a statistical interpretation, others can be violated and overcome by adapting the method of least squares. For example, statistical analysis is nearly impossible if the mean error isn't zero, but non-constant variance can be somewhat mitigated by weighting the least squares problem. The assumption of independence mandates that error be indifferent to the state of the input variables, and hence, this assumption shows that the random variable ε does not depend on $(x_1, x_2, \ldots, x_p)$. So if we adhere to the assumptions, then it is reasonable to use ε instead of $\varepsilon(x_1, x_2, \ldots, x_p)$.

The first three error assumptions give a powerful theorem named after the towering mathematicians of Carl Friedrich Gauss and Andrey Markov.

Theorem 10 (Gauss-Markov Theorem). *The first three error assumptions imply that the least squares estimates a_i are the best linear unbiased estimates of α_i.*

The interpretation of "best" is that the random variable a_i gives the smallest variance estimate of α_i over all possible unbiased linear estimators. We note that the third condition of independence can be relaxed to an assumption of uncorrelated errors, which provides a stronger version of the theorem.

The least squares estimates a_i are unbiased, meaning that $E(a_i) = \alpha_i$, under the first assumption that the errors have an expected value of zero. So, if we repeatedly sample data and calculate estimates of a_i, then the average of these estimates would approach the true α_i parameter as the number of samples increased. Moreover, the Gauss-Markov Theorem states that if we have the first three assumptions, then the sample variance of these estimates is as small as we can expect.

The error assumptions have a natural geometric interpretation in the simple linear model of $\hat{y} = a_0 + a_1 x$. Suppose the constant variance of the errors is σ^2, and hence each error ε is distributed as $\mathcal{N}(0, \sigma^2)$. The errors are also assumed to be independent of one another, so the error at one x_i has no bearing on the error at another x_j. In this case we can visualize the sole normal that is the random error as being able to "slide" along the true regression line, see Figs. 3.7 and 3.8. The residuals of our data are only a sample of the error, but our modeling framework forces us to consider the possibility that the residual's could have been elsewhere according to the distribution $\mathcal{N}(0, \sigma^2)$.

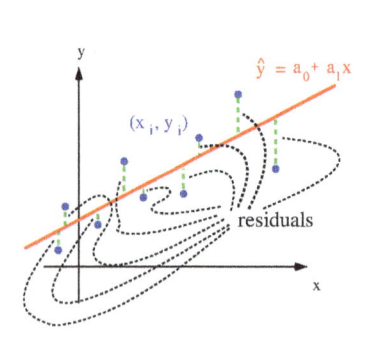

Fig. 3.7 An illustration of a regression line and its residuals.

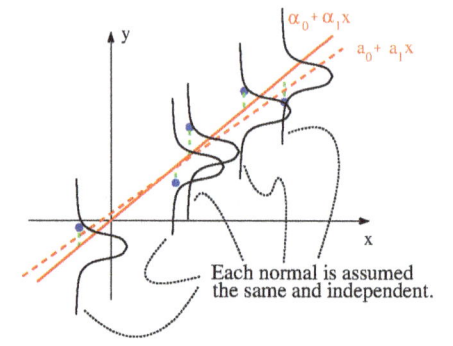

Fig. 3.8 The residuals are assumed samples from an independent normal. The solid line is the true regression line in terms of the α parameters, whereas the dashed line is an estimate based on data's sample.

3.2 Linear Regression: A Statistical Perspective

A reasonable question at this point is, what have we gained over our prior deterministic use of least squares by adopting a stochastic perspective? Here are some of the benefits:

- We can measure the percentage of the response variable's variation explained by the regressors' variations.
- We can statistically test the significance of the correlation between the response variable and the collection of regressors.
- We can build confidence intervals for predictions.
- We can quantify the error in the a_i estimates of α_i.

There are numerous other benefits related to the statistical analysis of uncertain and imperfect data, but these four provide the essential elements that help us assess and use a regression model. We consider each below.

The most popular measure of a model's efficacy is the coefficient of determination, which is the famed R^2 value. In short, this value is the percentage of the response variable's variation explained by the model. If the model perfectly accounts for the response variable's variation, then there is no error. In this case the data's evidence of Y's deviations from its mean has been perfectly explained by the regressors' deviations from their means, and the value of R^2 is 1 because all of the response's variability is accounted for by a linear combination of the regressors' variabilities. The other extreme is for none of Y's variance to be explained by the regressors' variabilities, and in this case $R^2 = 0$. The coefficient of determination is almost always the first consideration in a model's assessment, although be warned that it is an imperfect evaluation. High R^2 values don't always lead to trustworthy models, and low R^2 models can be useful in some settings.

The definition of R^2 is

$$R^2 = \frac{(\hat{y}_{\text{data}} - \bar{y}\,e)^T(\hat{y}_{\text{data}} - \bar{y}\,e)}{(y - \bar{y}\,e)^T(y - \bar{y}\,e)}$$

$$= \frac{(\hat{y}_{\text{data}} - \bar{y}\,e)^T(\hat{y}_{\text{data}} - \bar{y}\,e)/(n-1)}{(y - \bar{y}\,e)^T(y - \bar{y}\,e)/(n-1)} \quad \Leftarrow \quad \frac{\text{explained variance}}{\text{total variance,}}$$

where e is again a vector of ones and y is the vector of data. The denominator of the second expression is the sample variance s_y^2. The numerator of the second expression measures the model's variations from $\bar{y}$, and the formulaic similarity with that of a sample variance is why the numerator is referred to as the "explained variance."

There are many widely held rules-of-thumb about R^2, the most common being that a value of at least 0.8 indicates a good model. Any such claim is suspect and should be considered with caution and skepticism. First, R^2 is non-decreasing as more independent variables are added, so a model can often look good by adding lots of (potentially meaningless) independent variables. Second, we are dealing with a sample, so small improvements or degradations in R^2 are not concomitant with model quality. The value of R^2 should be calculated and reviewed, but its value should be benchmarked against the application and other model assessments.

An additional perspective on R^2 is possible after a bit of algebraic manipulation, which shows

$$R^2 = \frac{(\hat{y}_{\text{data}} - \bar{y}\,e)^T(\hat{y}_{\text{data}} - \bar{y}\,e)}{(y - \bar{y}\,e)^T(y - \bar{y}\,e)} = 1 - \frac{(y - \hat{y}_{\text{data}})^T(y - \hat{y}_{\text{data}})}{(y - \bar{y}\,e)^T(y - \bar{y}\,e)}. \quad (3.13)$$

The numerator of the last term is the total squared error that is minimized in the calculation of $\hat{y}$. Hence R^2 is as large as possible for the data. The process of calculating a regression model is thus simultaneously minimizing the squared residuals and maximizing the percentage of the response variable's variance that can be described by the collection of regressors. In other words, the regression model guarantees that the collection of regressors cannot be linearly combined to better explain the response's variation, at least as assessed through the data.

Each estimate a_i is a random variable whose expectation is the true α_i, and we can leverage this statistical relationship to conduct a hypothesis test to help decide if there is a correlation between X_i and Y. A hypothesis test is a probabilistic evaluation about the truth of a statement. In this case we want to test whether $\alpha_i = 0$, since if this is true, then X_i has no bearing on Y. The equality being tested is the null hypothesis and is denoted by $H_0 : \alpha_i = 0$. The alternative hypothesis is $H_a : \alpha_i \neq 0$. A rejection of the null hypothesis depends on how likely it is that such a rejection will coincide with the hypothesis being true. Rejecting the null hypothesis in the case that it is true is called a Type I error, and the goal is to be somewhat certain that we do not make such a mistake.

A hypothesis test does not require us to decide if the null hypothesis is, in fact, true or false. Instead, a hypothesis test is asking how sure we are about the falsehood of the null hypothesis, which in this case is the equality $\alpha_i = 0$. However, we don't know α_i and instead only know its estimate a_i. We can show that the standard error of a_i is

$$\begin{aligned} s_{a_i} &= \sqrt{\frac{(y - \hat{y}_{\text{data}})^T(y - \hat{y}_{\text{data}})}{n - p - 1} \left(e_i^T (A^T A)^{-1} e_i\right)} \\ &= s_{\text{est}} \sqrt{\left(e_i^T (A^T A)^{-1} e_i\right)}, \end{aligned} \quad (3.14)$$

where e_i is the vector of zeros except for a 1 in the i-th position. The multiplier

$$s_{\text{est}} = \sqrt{\frac{(y - \hat{y}_{\text{data}})^T(y - \hat{y}_{\text{data}})}{n - p - 1}} \quad (3.15)$$

3.2 Linear Regression: A Statistical Perspective

is the estimated standard deviation of ε, which is the square root of the sum of squared residuals divided by $n - p - 1$. Recall that p is the number of regressors.

Calculation (3.14) isn't as opaque as it might seem. Recall that the coefficients of $\hat{y}$ are the least squares solution to $Aa = y$, where A and y have the form in (3.11) and a is the vector of our estimated parameters. So the optimal parameters satisfy

$$A^T A a = A^T y \quad \text{and} \quad a = (A^T A)^{-1} A^T y,$$

provided that the columns of A are linearly independent. The second equality suggests that variations in y cascade to variations in a through the matrix $(A^T A)^{-1}$, and Eq. (3.14) states this cascade effect precisely.

The random variable a_i follows a t-distribution with $n - p - 1$ degrees of freedom. The t-distribution is a parameterized collection of distributions used to describe how an estimate of a mean is distributed. If $n - p - 1 \geq 30$, which is routine, then the t-distribution is approximately normal, and in this case it is appropriate to use the normal distribution in lieu of the t-distribution. We generally assume the appropriate use of the normal, but if $n - p - 1 < 30$, then both MATLAB and Octave have functions to accommodate the t-distribution. The approximating normal distributing of a_i is $\mathcal{N}(\alpha_i, s_{a_i}^2)$, and hence,

$$\frac{a_i - \alpha_i}{s_{a_i}} \sim \mathcal{N}(0, 1)$$

The distribution we are testing against has $\alpha_i = 0$, so $a_i/s_{a_i} \sim \mathcal{N}(0, 1)$.

The specific value of a_i from the regression is one possibility drawn from the collection of all possible values of a_i. We further have a_i/s_{a_i} as one of the possible values of the standard normal after calculating s_{a_i}. This ratio is called the test statistic. As an example, suppose the estimated parameter for X_i is $a_i = 3$ and the standard error is $s_{a_i} = 1.5$. Then our test statistic is $a_i/s_{a_i} = 3/1.5 = 2$. Let Z be a standard normal variable, and let d satisfy

$$P(-d < Z < d) = \int_{-d}^{d} \frac{e^{-x^2/2}}{\sqrt{2\pi}} \, dx = 0.9. \tag{3.16}$$

Then 90% of the standard normal lies between $\pm d$. Any value within this interval is sufficiently close to zero that we might be unwilling to claim with adequate confidence that α_i is not zero. So the result of the hypothesis test rests on calculating d and asking if our test statistic of 2 is within the interval $[-d, d]$. If the test statistic is not in $[-d, d]$, then we claim with at least 90% confidence that $\alpha_i \neq 0$. The null hypothesis is rejected in this case. If the test statistic is instead inside $[-d, d]$, then we are unable to claim with 90% confidence that $\alpha_i \neq 0$, and we fail to reject the null hypothesis. Failing to reject the null hypothesis is not the same as accepting it, and if we fail to reject the hypothesis, then we are not claiming that α_i is indeed zero.

Instead, we are indicating a lack of sufficient statistical evidence to dismiss our concern that it might be zero.

The last question is how do we calculate d? Such values had to be approximated from tables not so long ago, but thankfully MATLAB and Octave have tools to facilitate this calculation. In particular, the command `norminv(p)` calculates d satisfying

$$P(Z < d) = \int_{-\infty}^{d} \frac{e^{-x^2/2}}{\sqrt{2\pi}}\, dx = p.$$

The symmetry of the normal about its mean implies the use of $p = 0.1/2 = 0.05$ to satisfy Eq. (3.16), as this value of p will symmetrically place half of the 10% to the far right and half to the far left of the density function. So we calculate `norminv(0.05)` = -1.6449 or `norminv(0.95)` = 1.6449, as either provides the correct value of d. Since the test statistic of 2 is outside the interval $[-1.6449, 1.6449]$, we reject the null hypothesis and claim with 90% confidence that $\alpha_i \neq 0$. The statistical result indicates that the response variable Y is somewhat explained by the regressor X_i, or alternatively, that the variations of X_i help predict the variations in Y. Importantly, this inference holds against the collection of independent variables, and the conclusion might change if this collection is altered.

Hypothesis testing necessitates the selection of a confidence probability, θ, prior to conducting the test, but such sequential decision making can be awkward. For instance, suppose we a priori decide to use $\theta = 0.9$ to test the significance of a_i. Suppose further that the calculation of the test statistics results in $a_i/s_{a_i} = 1.62$. We then fail to reject the null hypothesis and conclude that there is a lack of statistical evidence to declare a relationship between X_i and Y; although we emphasize, again, that this conclusion is couched against the collection of regressors. We might decide to remove X_i as a regressor since we failed to reject the hypothesis $\alpha_1 = 0$. However, we would have had the opposite result if we had instead selected only a slightly smaller value of θ, a fact clarified by the calculation

$$1 - \int_{-1.62}^{1.62} \frac{e^{-x^2/2}}{\sqrt{2\pi}}\, dx = 0.1052. \tag{3.17}$$

A reasonable argument is that while a_i fails to be significant at a 90% level, we remain highly confident that $\alpha_i \neq 0$ and that there is a likely relationship between X_i and Y. Indeed, what gives 90% so much more interpretive ability than does 89.48%, especially in light of random data?

A more metered approach is to report the integral calculations of (3.17) for each a_i to aid our analysis, a scheme that has become the contemporary standard. The p-value associated with each a_i is

3.2 Linear Regression: A Statistical Perspective

$$p_{\text{val}} = 1 - \int_{-|a_i|/s_{a_i}}^{|a_i|/s_{a_i}} \frac{e^{-x^2/2}}{\sqrt{2\pi}}\, dx,$$

where we again assume a sufficiently large sample to use the normal instead of the t-distribution. Note that we denote a p-value by p_{val}, whereas p remains to be the number of regressors. The absolute values in the limits of integration are required because a_i can be negative. The command `2*(1-normcdf(1.62))` computes the appropriate p-value in both MATLAB and Octave for the test statistic above. Modern model evaluation would be to compute the p-values for each a_i and then use these values to see if there is evidence against α_i being zero. If so, then a modeler might decide to keep X_i in the model to help explain the response.

We can use a model to make new predictions once we have gained trust in its efficacy to represent Y. New predictions require some care with regard to what is being predicted. The issue here is that $\hat{y}$ has two interpretations:

1. $\hat{y}(x_1, x_2, \ldots, x_p)$ is the expected value of Y under the assumption that $X_1 = x_1, X_2 = x_2, \ldots, X_p = x_p$.
2. $\hat{y}(x_1, x_2, \ldots, x_p)$ is a potential value of Y under the assumption that $X_1 = x_1, X_2 = x_2, \ldots, X_p = x_p$.

In the first case we seek the average or expected response if the regressors achieve their stated values. In the second case we instead seek the response itself. Suppose as an illustrative example that Y is the purchase price of a house and that X_1 is a purchaser's annual income. Historical data from your real estate firm over the last decade results in a regression model of

$$\hat{y}(x_1) = 30{,}000 + 0.8 x_1, \tag{3.18}$$

with $R^2 = 0.83$ and the p-value of a_1 being 0.06. So this model seems to explain 83% of the variability of your historic sales in terms of your customers' annual incomes. Moreover, we are 94% confident that annual income and purchase price are correlated. Here are two related, but different, questions,

1. What is the expected purchase price among individuals whose income is $100,000?
2. What is the predicted purchase price of any particular customer whose income happens to be $100,000?

Our best estimate of both is $\hat{y}(100{,}000) = 110{,}000$. The difference lies with our trust in this answer relative to its interpretation.

The standard error of the average customer given the values in x is

$$s_{\text{est}} \sqrt{x^T (A^T A)^{-1} x}, \tag{3.19}$$

whereas the standard error of any particular customer with the same x is

$$s_{\text{est}} \sqrt{1 + x^T (A^T A)^{-1} x}. \tag{3.20}$$

The first expression becomes the standard error of a_i in (3.14) if $x = e_i$, which follows immediately since our model's prediction at e_i is a_i. In the simple case of model (3.18), $A^T A$ is a 2×2 matrix and $x = (1, 100000)^T$. Suppose $x^T(A^T A)^{-1}x = 1.2$ and $s_{\text{est}} = 16,000$. Then a 95% confidence interval for the average customer is (approximately),

$$110{,}000 \pm 2\,(16{,}000\,\sqrt{1.2}) = 110{,}000 \pm 35{,}054.24.$$

So, if our company helps numerous clients with an income of $100,000, then we are 95% confident that the average purchase price among these customers will be somewhere between $74,945.76 and $145,054.24, assuming that historical data is reflective of the near future. Alternatively, the standard error for any individual customer implies a 95% chance of purchasing a house within

$$110{,}000 \pm 2\,(16{,}000\sqrt{1+1.2}) = 110{,}000 \pm 47{,}463.67.$$

So with 95% confidence individual purchases should be within the interval [62,536.33, 157,463.67], which is broader than the interval for the average purchase of [74,945.76, 145,054.24] with the same confidence. This wider interval is called a "prediction interval" instead of a confidence interval because it is not germane to a statistic of the model but rather to a prediction of the model. The larger standard error of the prediction always implies that we are less certain about individual predictions than we are of estimated average predictions.

The standard error values in (3.19) and (3.20) can be used to generate confidence and prediction intervals for percentages other than 95%. Suppose we want to calculate θ confidence and prediction intervals. Then calculating d so that

$$\frac{1-\theta}{2} = 1 - \int_{-\infty}^{d} \frac{e^{-x^2/2}}{\sqrt{2\pi}}\,dx,$$

the intervals are

Confidence Interval		Prediction Interval
$\hat{y}(x) \pm d\,s_{\text{est}}\sqrt{x^T(A^T A)^{-1}x}$	and	$\hat{y}(x) \pm d\,s_{\text{est}}\sqrt{1 + x^T(A^T A)^{-1}x}.$

These intervals are predicated on the assumption that errors are normally distributed, and hence, this supposition should be assessed, see Exercise 18. The value of d is readily calculated in MATLAB and Octave with the command `norminv((1+theta)/2)`.

We consider two examples to illustrate the interpretive ability associated with regression. The first is the groundbreaking work in astronomy by S. Faber and R. Jackson in the 1970s. Their research established a power law between a galaxy's luminosity and its velocity, a theory that has been extended and confirmed since its introduction. The primary analytical tool was

3.2 Linear Regression: A Statistical Perspective

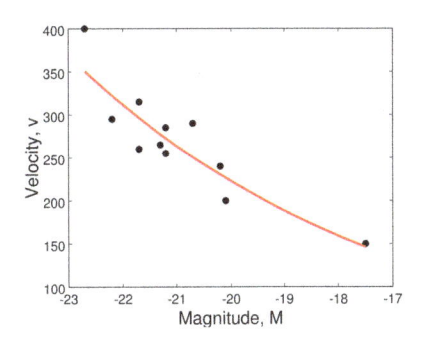

Fig. 3.9 The untransformed Faber-Jackson data and its exponential best fit.

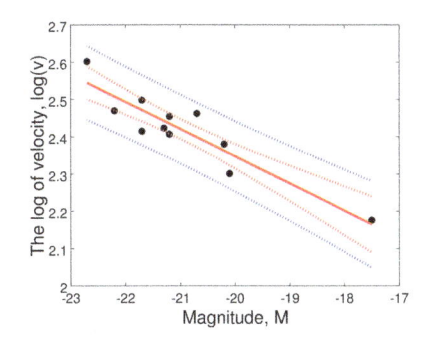

Fig. 3.10 The transformed Faber-Jackson data and its best affine approximation. The red dashed curves show the 95% confidence intervals for the mean, and the blue dashed curves define the 95% prediction intervals.

linear regression, and in particular, the study established a relationship between a galaxy's line-of-sight velocity, v, in units of km/s, and its absolute magnitude, M, in units of brightness. The data is below.

Galaxy ID	3379	4374	4406	4459	4472	4486	4486B	4552	4649	4889	5846
M	-20.2	-21.2	-21.3	-20.1	-22.2	-21.7	-17.5	-20.7	-21.7	-22.7	-21.2
v	240	285	265	200	295	315	150	290	260	400	255

The posited model is exponential,

$$v = k_1 10^{k_2 M} \Leftrightarrow \log(v) = \log(k_1) + k_2 M. \tag{3.21}$$

The re-formulation on the right is linear, and in terms of our regression notation we have $y = \log(v)$, $x = M$, $a_1 = \log(k_1)$, and $a_2 = k_2$. The least squares solution is

$$\log(v) = 0.888346 - 0.072962\, M \Leftrightarrow v = 7.330 \times 10^{-0.072962\, M}.$$

The original and transformed data are plotted with the least squares solution in Figs. 3.9 and 3.10. The value of R^2 is 0.8529, which suggests that about 85% of the variation in $\log(v)$ can be explained by the variation of M.

The test statistic for a_2 is $a_2/s_{a_2} = -7.2238$, and the resulting p-value is 0.00005. The p-value is calculated with the true t-distribution and not its normal approximation because we are shy of having enough data to use the normal with only 11 galaxies. We need to provide a parameter called the degrees-of-freedom, which is $n - p - 1 = 11 - 1 - 1 = 9$, to identify the correct t-distribution. In this case the p-value is calculated with `2*(1-tcdf(7.2238,9))`. In general, the t distribution is always preferred, although the difference is nearly indistinguishable if $n - p - 1 > 30$.

The small p-value of 0.00005 indicates a high degree of trust in rejecting the null hypothesis, $H_0 : \alpha_2 = 0$. In other words, we suspect $\alpha_2 \neq 0$. The true value of the parameter α_2 is the ratio of the covariance of $\log(v)$ and M to the variance of M, and the only way this ratio can be zero is if $\text{Cov}(\log(v), M) = 0$. Since correlation is a scaled version of covariance, the ratio is zero if and only if $\log(v)$ and M are uncorrelated. So in terms of correlation, our small p-value suggests strong evidence of a true correlation.

The combination of a relatively high R^2 and a small p-value for a_2 promotes the application of the model. Suppose we want to predict the average velocity of a galaxy with a brightness of $M = -19.71$. Our prediction of this average is

$$\log(v(M = -19.71)) = 0.888346 - 0.072962(-19.71) = 2.3264.$$

The standard error of this estimate as calculated by (3.19) is 0.01835, and a 95% confidence interval is

$$2.2905 = 2.3264 - 0.0359$$
$$\leq E(\log(v(M = -19.71))$$
$$\leq 2.3264 + 0.0359 = 2.3624$$

This range is illustrated by the red dashed curves in Fig. 3.10 for the value of $M = -19.71$. So, if we were to observe several galaxies with a brightness of $M = -19.71$, then we would predict with about 95% confidence that the average value of $\log(v)$ would be in $[2.2905, 2.3624]$.

We could instead ask about the collection of velocities for all galaxies with a brightness of $M = -19.71$. Our predicted average velocity satisfies $\log(v(M = -19.71)) = 2.3264$, but the standard error of this prediction as calculated by (3.20) is 0.04796, and in this case we expect that 95% of the galaxies with a brightness of $M = -19.71$ satisfy

$$2.2324 = 2.3264 - 0.0940$$
$$\leq \log(v(M = -19.71)$$
$$\leq 2.3264 + 0.0940 = 2.4204.$$

This predication interval is determined in Fig. 3.10 by the broader blue dashed lines. We should generally expect about 95% of our observations to fall within these blue prediction intervals, and all observations for this relatively small dataset are favorably within them.

Our second example considers a civil engineering study on bridge failures from 1987 through 2008. The following data was collected from various samples over this period to assess a possible relationship between population and bridge failures.

Population (X)	17,300	24,200	24,400	3900	13,300	5140	6560
Number of failures (Y)	92	74	97	13	79	21	19

3.2 Linear Regression: A Statistical Perspective

The simple linear model $\hat{y} = a_0 + a_1 x$ was considered, and the resulting least squares solution was

$$\hat{y} = 5.1601092 + 0.0037856\, x.$$

The data and the regression line are shown in Fig. 3.11.

The R^2 value is relatively high at 0.79347, which suggests that about 79% of the variation in bridge failures can be explained by the variation in population. Moreover, the positive a_1 value of 0.0037856 suggests an expected increase of about four bridge failures for every 1000 people added to a population, which is quite the daunting prediction. The test statistic for a_1 is $a_1/s_{a_1} = 4.3829$, and the resulting p-value is 0.0071, which is computed with the t distribution as `2*(1-tcdf(4.3829,5))` since $n - p - 1 = 7 - 1 - 1 = 5 < 30$. This small p-value suggests that we reject the null hypothesis of $H_0 : \alpha_1 = 0$, and hence, we have reasonable evidence to conclude that $\alpha_1 \neq 0$. Moreover, the data provides evidence of a positive correlation between population size and bridge failure.

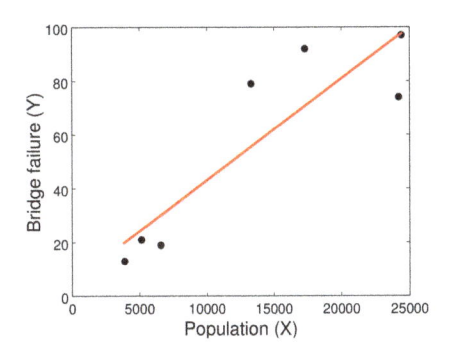

Fig. 3.11 Bridge failure against population. The red line is the least squares approximation.

We finish this section with a reminder about computational concerns. Many of the calculations associated with linear regression are easily expressed in terms of $(A^T A)^{-1}$, but in all cases we should solve $A^T A y = x$ to calculate $(A^T A)^{-1} x$. The repeated use of $A^T A$ suggests that the matrix be factored once and that different solutions then be calculated for different right-hand sides with backsolve. So a computational perspective encourages the use of something like the inner product algorithm to calculate the Cholesky factorization of $A^T A$ instead of an iterative algorithm like that of conjugate directions. This rationale is sound if the collection of regressors is fixed, although iterative solution procedures can make sense as different models are explored. We consider model building in the second portion of this text.

3.3 Cubic Splines

A spline is a piecewise polynomial curve through a set of points. Our perspective on data changes from that of regression, and we now assume that data is accurate. Estimating the underlying continuous process is then accomplished by fitting a curve through the data. Splines somewhat avoid the concern associated with Lagrange polynomials of overfitting the data by constructing an approximating curve piecewise.

Outside a spline's mathematical definition, it is a curve drawn by a draftsman through a set of points with the aid of a flexible object. The result is a smooth curve, and while connecting data linearly, say in a connect-the-dots mentality, is technically a spline, we restrict our development to the most frequently used cubic splines to gain the aforementioned smoothness. We assume the spline should pass through the points (x_i, y_i), and we design an approximating cubic for each pair of consecutive points of the form

$$f_i(x) = a_i + b_i(x - x_i) + c_i(x - x_i)^2 + d_i(x - x_i)^3,$$

where x is in the interval $[x_i, x_{i+1}]$ and i ranges from 0 to n. See Fig. 3.12 for an example through four points.

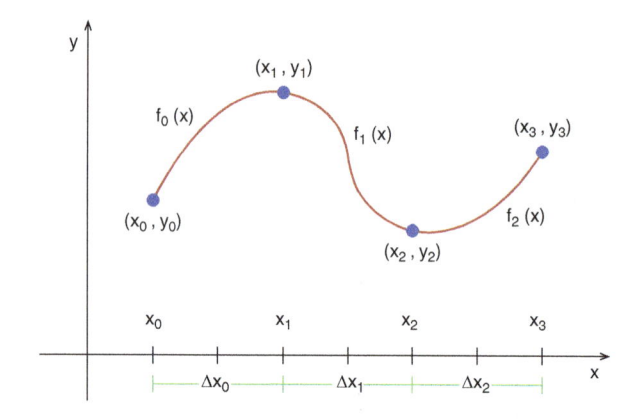

Fig. 3.12 A cubic spline through four points.

The first and second derivatives of the interval functions f_i and f_{i+1} are required to agree at x_{i+1} to achieve the smoothness of the overall curve. We let $\Delta x_i = x_{i+1} - x_i$ so that

3.3 Cubic Splines

$$\left.\begin{array}{ll} f_i(x_i) = a_i & f_i(x_{i+1}) = a_i + b_i\Delta x_i + c_i\Delta x_i^2 + d_i\Delta x_i^3 \\ f'_i(x_i) = b_i & f'_i(x_{i+1}) = b_i + 2c_i\Delta x_i + 3d_i\Delta x_i^2 \\ f''_i(x_i) = 2c_i & f''_i(x_{i+1}) = 2c_i + 6d_i\Delta x_i. \end{array}\right\} \quad (3.22)$$

The spline is defined by a collection of parameters a_i, b_i, c_i, and d_i so that

$$\left.\begin{array}{l} f_i(x_i) = y_i \\ f_i(x_{i+1}) = y_{i+1} \end{array}\right\} \text{ ensure the spline passes through the data}$$

and

$$\left.\begin{array}{l} f'_i(x_i) = f'_{i-1}(x_i) \\ f''_i(x_i) = f''_{i-1}(x_i) \end{array}\right\} \text{ ensure the spine is smooth.}$$

These equalities combine with (3.22) to give a set of linear equations, which for the example in Fig. 3.12 has the matrix equation $Aw = v$, where the matrix is

$$A = \left[\begin{array}{cccc|cccc|cccc} 1 & & & & & & & & & & & \\ 1 & \Delta x_0 & \Delta x_0^2 & \Delta x_0^3 & & & & & & & & \\ & 1 & 2\Delta x_0 & 3\Delta x_0^2 & -1 & & & & & & & \\ & & 2 & 6\Delta x_0 & & -2 & & & & & & \\ \hline & & & & 1 & & & & & & & \\ & & & & 1 & \Delta x_1 & \Delta x_1^2 & \Delta x_1^3 & & & & \\ & & & & & 1 & 2\Delta x_1 & 3\Delta x_1^2 & -1 & & & \\ & & & & & & 2 & 6\Delta x_1 & & -2 & & \\ \hline & & & & & & & & 1 & & & \\ & & & & & & & & 1 & \Delta x_2 & \Delta x_2^2 & \Delta x_2^3 \end{array}\right]$$

and the variable vector w and the right-hand side v are

$$w = \left(\begin{array}{cccc|cccc|cccc} a_0 & b_0 & c_0 & d_0 & a_1 & b_1 & c_1 & d_1 & a_2 & b_2 & c_2 & d_2 \end{array}\right)^T \text{ and}$$

$$v = \left(\begin{array}{cccc|cccc|cc} y_0 & y_1 & 0 & 0 & y_1 & y_2 & 0 & 0 & y_2 & y_3 \end{array}\right)^T.$$

The top two equations associated with each horizontal partition of A guarantee that the spline passes through the data, and the last two impose the equality of the first and second derivatives. The last horizontal partition lacks two equations that would equate the derivatives since the spline does not extend beyond the last data point. The system thus lacks two equations to precisely define the coefficients, a topic we re-visit shortly.

The linear system as presented has several shortcomings. The top equations of each horizontal partition can be removed since they simply state that $a_i = y_i$. Removing these equations necessitates that the right-hand sides of the second equations become $y_{i+1} - y_i$. This reduction leaves each row with three or four nonzero elements that cluster around the diagonal. However, it is possible to re-model the system so that the coefficient matrix is tridiagonal, i.e. the only nonzero elements are either on the main diagonal or on the immediate sub/super diagonal. Each row has two or three nonzero elements in this case, which can expedite the solution procedure. A final numerical concern is that the values Δx_i, Δx_i^2, and Δx_i^3 can have varied magnitudes, which can challenge numeric stability.

The concerns just noted can be overcome by re-expressing the problem in terms of the second derivatives $f_i''(x)$. For $i = 0, 1, \ldots, n-1$, we let s_i satisfy

$$s_i = f_i''(x_i) = 2c_i \implies c_i = \frac{s_i}{2}.$$

Moreover, assume for convenience that $s_n = f_{n-1}''(x_n)$. Then,

$$s_{i+1} = f_i''(x_{i+1}) = 2c_i + 6d_i \Delta x_i = s_i + 6d_i \Delta x_i \implies d_i = \frac{s_{i+1} - s_i}{6 \Delta x_i}.$$

Recall that we already know $y_i = a_i$, so the only coefficient of f_i not defined in terms of y_i and s_i is b_i. Using the fact that

$$f_i(x_{i+1}) = a_i + b_i \Delta x_i + c_i \Delta x_i^2 + d_i \Delta x_i^3,$$

we have upon substituting for a_i, c_i, and d_i that

$$y_{i+1} = y_i + b_i \Delta x_i + \frac{s_i}{2} \Delta x_i^2 + \frac{s_{i+1} - s_i}{6 \Delta x_i} \Delta x_i^3.$$

Solving for b_i gives

$$b_i = \frac{y_{i+1} - y_i}{\Delta x_i} - \frac{s_{i+1} + 2s_i}{6} \Delta x_i.$$

With a_i, b_i, c_i, and d_i expressed in terms of s_i, s_{i+1}, y_i, and y_{i+1}, the requirement that $f_i'(x_i) = f_{i-1}'(x_i)$ is

$$\begin{aligned}
f_i'(x_i) &= b_i \\
&= \frac{y_{i+1} - y_i}{\Delta x_i} - \frac{s_{i+1} + 2s_i}{6} \Delta x_i \\
&= \left(\frac{y_i - y_{i-1}}{\Delta x_{i-1}} - \frac{s_i + 2s_{i-1}}{6} \Delta x_{i-1} \right) + 2 \left(\frac{s_{i-1}}{2} \right) \Delta x_{i-1} + 3 \left(\frac{s_i - s_{i-1}}{6 \Delta x_{i-1}} \right) \Delta x_{i-1}^2 \\
&= b_{i-1} + 2c_{i-1} \Delta x_{i-1} + 3d_{i-1} \Delta x_{i-1}^2 \\
&= f_{i-1}'(x_i).
\end{aligned}$$

3.3 Cubic Splines

The third equality is the same as

$$\Delta x_{i-1} s_{i-1} + 2(\Delta x_i + \Delta x_{i-1})s_i + \Delta x_i s_{i+1}$$
$$= 6\left(\frac{y_{i+1} - y_i}{\Delta x_i} - \frac{y_i - y_{i-1}}{\Delta x_{i-1}}\right), \quad (3.23)$$

where i ranges from 1 to $n-1$. This linear equation relates s_i with its predecessor s_{i-1} and its successor s_{i+1}, which provides the tridiagonal structure. Moreover, squared and cubic powers of Δx_i or Δx_{i-1} are no longer required.

Equality (3.23) does not overcome the problem of there being only $n-1$ equations but $n+1$ variables. Several typical options are available to add two conditions to uniquely define the s_i variables. The most common is to create a natural spline by assuming that there is no curvature at the starting and ending point,

$$f_0''(x_0) = s_0 = 0 = s_n = f_{n-1}''(x_n).$$

The matrix form of the equations that define a natural spline for the example in Fig. 3.12 is now reduced to

$$\begin{bmatrix} 2(\Delta x_1 + \Delta x_0) & \Delta x_1 \\ \Delta x_1 & 2(\Delta x_2 + \Delta x_1) \end{bmatrix} \begin{pmatrix} s_1 \\ s_2 \end{pmatrix}$$
$$= 6 \begin{pmatrix} (y_2 - y_1)/\Delta x_1 - (y_1 - y_0)/\Delta x_0 \\ (y_3 - y_2)/\Delta x_2 - (y_2 - y_1)/\Delta x_1 \end{pmatrix}.$$

The initial system had $3n = 9$ equations with $a_i = y_i$ removed and an additional two equations to specify the second derivatives at x_0 and x_3. The re-expressed system accomplishes the same goal in $n - 1 = 2$ equations, and it does so in way that improves computational stability.

The general form of the coefficient matrix defining a natural spline is

$$\begin{bmatrix} 2(\Delta x_1 + \Delta x_0) & \Delta x_1 & & & & \\ \Delta x_1 & 2(\Delta x_2 + \Delta x_1) & \Delta x_2 & & & \\ & \Delta x_2 & 2(\Delta x_3 + \Delta x_2) & \Delta x_3 & & \\ & & & \ddots & & \\ & & & & \ddots & \\ & & & & \Delta x_{n-2} & 2(\Delta x_{n-1} + \Delta x_{n-2}) \end{bmatrix}.$$

This matrix has $3n - 5$ nonzero elements and lends itself nicely to sparse matrix arithmetic. Moreover, its tridiagonal structure means that the system can be solved efficiently. Indeed, row reducing the matrix to an upper triangular form only requires $n-2$ steps, and the process of back substitution similarly benefits from the tridiagonal structure.

The coefficients a_i, b_i, c_i, and d_i are calculated after the values of s_i are known:

$$a_i = y_i,$$
$$b_i = \frac{y_{i+1} - y_i}{\Delta x_i} - \frac{s_{i+1} + 2s_i}{6}\Delta x_i,$$
$$c_i = \frac{s_i}{2}, \text{ and}$$
$$d_i = \frac{s_{i+1} - s_i}{6\Delta x_i}.$$

The spline is then

$$f(x) = \begin{cases} a_0 + b_0(x - x_0) + c_0(x - x_0)^2 + d_0(x - x_0)^3, & x_0 \leq x < x_1 \\ a_1 + b_1(x - x_1) + c_1(x - x_1)^2 + d_1(x - x_1)^3, & x_1 \leq x < x_2 \\ \vdots \\ a_{n-1} + b_{n-1}(x - x_{n-1}) \\ \quad + c_{n-1}(x - x_{n-1})^2 + d_{n-1}(x - x_{n-1})^3, & x_{n-1} \leq x < x_n. \end{cases}$$

Consider the data tabulated below as an example,

$$\begin{array}{c|cccccc} x & -4.9 & -1.7 & 0.4 & 0.7 & 1.4 & 4.4 \\ \hline y & 9.1 & 7.7 & 2.8 & 9.1 & 6.3 & 0.8 \end{array} \quad (3.24)$$

The system defining the s_i variables is

$$\begin{bmatrix} 10.6 & 2.1 & 0 & 0 \\ 2.1 & 4.8 & 0.3 & 0 \\ 0 & 0.3 & 2.0 & 0.7 \\ 0 & 0 & 0.7 & 7.4 \end{bmatrix} \begin{pmatrix} s_1 \\ s_2 \\ s_3 \\ s_4 \end{pmatrix} = \begin{pmatrix} -11.375 \\ 140.000 \\ -150.000 \\ 13.000 \end{pmatrix}.$$

The solution with the embedded natural conditions is

$$(s_0, s_1, s_2, s_3, s_4, s_5)$$
$$= (0, -8.6422, 38.2058, -84.1312, 9.7151, 0).$$

Calculating a_i, b_i, c_i, and d_i results in the spline,

3.4 Principal Component Analysis

$$f(x) = \begin{cases} 9.1 + 4.1717(x+4.9) + 0(x+4.9)^2 - 0.45011(x+4.9)^3, \\ \qquad \text{if } -4.9 \leq x < -1.7 \\[1ex] 7.7 - 9.6558(x+1.7) - 4.3211(x+1.7)^2 + 3.7181(x+1.7)^3, \\ \qquad \text{if } -1.7 \leq x < 0.4 \\[1ex] 2.8 + 21.386(x-0.4) + 19.103(x-0.4)^2 - 67.965(x-0.4)^3, \\ \qquad \text{if } 0.4 \leq x < 0.7 \\[1ex] 9.1 + 14.497(x-0.7) - 42.066(x-0.7)^2 + 22.344(x-0.7)^3, \\ \qquad \text{if } 0.7 \leq x < 1.4 \\[1ex] 6.3 - 11.548(x-1.4) + 4.8576(x-1.4)^2 - 0.53973(x-1.4)^3, \\ \qquad \text{if } 1.4 \leq x \leq 4.4, \end{cases}$$

which is illustrated in Fig. 3.13.

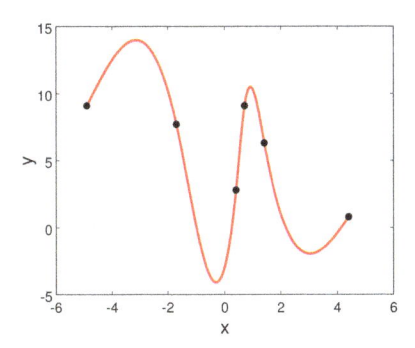

Fig. 3.13 The natural cubic spline through the data in (3.24).

3.4 Principal Component Analysis

Principal component analysis (PCA) is a computational tool used in the analysis and compression of data. We illustrate the technique with a simple example to motivate the general case. Suppose an experiment is assessed by the two characteristics x and y, and that repeating the experiment ten times results in the following data.

$$\begin{array}{c|cccccccccc}
\multicolumn{11}{c}{\text{Experiment}} \\
 & 1 & 2 & 3 & 4 & 5 & 6 & 7 & 8 & 9 & 10 \\
\hline
x & -0.2 & -0.1 & 0.8 & 0.0 & 0.8 & -0.4 & -0.1 & 0.6 & -0.5 & -0.6 \\
y & 0.1 & 0.2 & 2.1 & 0.6 & 2.9 & -2.0 & -0.7 & 1.8 & -1.3 & -2.5
\end{array} \quad (3.25)$$

PCA is premised on calculating a description of the data in terms of its variability, and the underlying assumption is that x and y are sample characteristics dependent on an unknown random process.

We tabulate the experimental data to form a matrix X, in which the first column contains the characteristic x and the second contains the characteristic y,

$$X = \begin{bmatrix} -0.2 & 0.1 \\ -0.1 & 0.2 \\ 0.8 & 2.1 \\ \vdots & \vdots \\ -0.6 & -2.5 \end{bmatrix}.$$

The data is "centered" by subtracting the sample means, and if we denote the vector of samples means by $\bar{\mu} = (\bar{x}, \bar{y}) = (0.03, 0.12)$, then the centered data is

$$W = X - e\bar{\mu} = \begin{bmatrix} -0.2 - 0.03 & 0.1 - 0.12 \\ -0.1 - 0.03 & 0.2 - 0.12 \\ 0.8 - 0.03 & 2.1 - 0.12 \\ \vdots & \vdots \\ -0.6 - 0.03 & -2.5 - 0.12 \end{bmatrix},$$

where e is a column vector of ones. The data is centered because the sample means of the columns of W are zero, i.e.

$$\frac{1}{m} e^T W = \frac{1}{m} e^T X - \frac{1}{m}\left(e^T e\right) \bar{\mu} = \bar{\mu} - \frac{m}{m}\bar{\mu} = 0, \quad (3.26)$$

where m is the number of data points, i.e. the number of rows of X.

The sample variances and covariances can now be succinctly expressed as components of the following matrix computation,

$$Q = \frac{1}{9}(X - e\bar{\mu})^T(X - e\bar{\mu}) = \frac{1}{9} W^T W$$

$$= \frac{1}{9}\begin{bmatrix} s_x^2 & s_{xy} \\ s_{yx} & s_y^2 \end{bmatrix} = \begin{bmatrix} 0.27344 & 0.89156 \\ 0.89156 & 3.19511 \end{bmatrix}.$$

The matrix Q is called the covariance matrix, and it can be computed in MATLAB or Octave with any of

- Q = cov(X),

3.4 Principal Component Analysis

- W = X - ones(10,1)*mean(X); Q = (1/9)*W'*W;, or
- W = X - repmat(mean(X),10,1); Q = (1/9)*W'*W.

While the first is most succinct, learning to use all three will broaden one's MATLAB repertoire.

The factored form $(1/9)\,W^T W$ shows that Q is positive definite as long as the columns of W are linearly independent, an assumption we make. From Theorem 6 in Chap. 2 we know that there is a diagonal matrix D and a matrix V such that
$$Q = VDV^T \text{ and } V^T = V^{-1}.$$

The property that $V^T = V^{-1}$ is the same as the columns of V being orthonormal, which means that the norm of each column is 1 and that the inner product of any two different columns is zero. For this example we have

$$Q = \begin{bmatrix} -0.96270 & -0.27057 \\ 0.27057 & -0.96270 \end{bmatrix} \begin{bmatrix} 0.022873 & 0 \\ 0 & 3.445682 \end{bmatrix} \begin{bmatrix} -0.96270 & -0.27057 \\ 0.27057 & -0.96270 \end{bmatrix}.$$

Multiplying on the right by V gives

$$QV = \begin{bmatrix} Q \begin{bmatrix} -0.96270 \\ 0.27057 \end{bmatrix}, \ Q \begin{bmatrix} -0.27057 \\ -0.96270 \end{bmatrix} \end{bmatrix}$$
$$= \begin{bmatrix} 0.022873 \begin{bmatrix} -0.96270 \\ 0.27057 \end{bmatrix}, \ 3.445682 \begin{bmatrix} -0.27057 \\ -0.96270 \end{bmatrix} \end{bmatrix}$$
$$= VD.$$

The second equality shows that the columns of V are eigenvectors of Q with corresponding eigenvalues along the diagonal of D. So the factorization $Q = VDV^T$ decomposes the covariance matrix, which explains the variability exhibited by the data, in terms of eigenvectors and eigenvalues. The MATLAB command

[V, D] = eig(cov(X))

computes the eigenvectors and eigenvalues of the covariance matrix. The factorization VDV^T is called the spectral factorization or spectral decomposition of Q.

PCA is especially beneficial if it can exploit the eigenvector and eigenvalue decomposition of the covariance matrix. Consider Fig. 3.14, which plots the centered data with the eigenvectors of Q and illustrates how such an advantage might exist. The axes through the eigenvectors are eigenspaces, and the eigenspace for the largest eigenvalue of 3.445682 trends along the central direction of the data. Indeed, the data for this example would maintain much of its information if it was projected onto this eigenspace. The fact that this behavior is common across numerous applications is what makes PCA so useful.

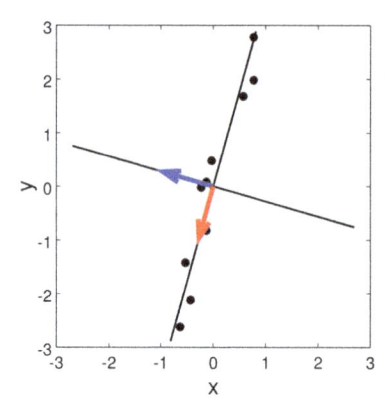

Fig. 3.14 The centered data of (3.25) plotted with the eigenvectors of the covariance matrix. The red vector is an eigenvector for the largest eigenvalue of 3.445682, and the blue vector is an eigenvector for the smallest eigenvalue of 0.022873. The data clusters around the axis through the eigenvector with the largest eigenvalue.

We seek coefficients that express the centered data in terms of the eigenvectors. The first experimental outcomes of the centered data are the elements of $(-0.23, -0.02)$, and we want coordinates $z_{1,1}$ and $z_{1,2}$ so that

$$(-0.23, -0.02) = z_{1,1}(-0.96270, 0.27057) + z_{1,2}(-0.27057, -0.96270).$$

The second centered data would similarly provide $z_{2,1}$ and $z_{2,2}$ by satisfying

$$(-0.13, 0.18) = z_{2,1}(-0.96270, 0.27057) + z_{2,2}(-0.27057, -0.96270).$$

Allowing Z to be the matrix whose (i,j) element is $z_{i,j}$, we have the following matrix equation for the entire dataset,

$$W = ZV^T \iff Z = WV = (X - e\bar{\mu})V.$$

The equalities on the right express the fundamental relationship of PCA, which maps the data in X to the transformed data of Z by multiplying the centered data by V. The i-th row of Z contains the transformed coordinates of the i-th experiment, which for our example are tabulated below (up to two decimals).

	\multicolumn{10}{c}{Experiment}
	1	2	3	4	5	6	7	8	9	10
$z_{\cdot,1}$	0.22	0.15	−0.21	0.16	0.01	−0.16	−0.10	−0.09	0.13	−0.10
$z_{\cdot,2}$	0.08	−0.04	−2.11	−0.45	−2.88	2.15	0.82	−1.77	1.51	2.69

3.4 Principal Component Analysis

Each eigenvalue measures data's variance along its associated eigenspace. For example, suppose the data points in Fig. 3.14 were projected onto the eigenspace of the red eigenvector, creating a new set of points that were all scalar multiples of the red vector. The sample variance of these new points would be the eigenvalue for the red eigenvector divided by Bessel's divisor of $m - 1 = 9$, which is $3.445682/9$. Likewise, if the data were projected onto the eigenspace of the blue eigenvector, then the sample variance of the transformed data would be the associated eigenvalue of 0.022873 divided by 9. So each eigenvalue informs us of data's geometric spread along the associated eigenspace. In particular, small eigenvalues suggest little variation, and hence, a reasonable approximation might ignore these spacial directions. For example, the variation along the blue vector in Fig. 3.14 is small enough that we might project the data onto the red vector and remove the blue vector. Doing so would allow us to compress the data by representing each projected point as a single scalar of the red vector instead of the original two characteristics x and y.

We recall the projection formula from calculus to substantiate the interpretation of the eigenvalues as projected sample variances,

$$\text{proj}_v(w) = \left(\frac{w^T v}{v^T v}\right) v = \left(\frac{w^T v}{\|v\|^2}\right) v = \left(\frac{w^T v}{\|v\|}\right) \frac{v}{\|v\|}.$$

So if w_i^T is the i-th row of W, which is our i-th centered sample, then its projection onto the normalized vector $v/\|v\|$ is of length $w_i^T v/\|v\|$. The lengths of the projected vectors remain centered because the sample mean is

$$\frac{1}{m} \sum_{i=1}^{m} \frac{w_i^T v}{\|v\|} = \frac{1}{m\|v\|} e^T (Wv) = \frac{1}{m\|v\|} \left(e^T W\right) v = 0^T v = 0,$$

where $e^T W = 0$ from (3.26). Hence the sample variance of the projected lengths is the sum of the squares of the lengths themselves divided by Bessel's divisor, which is

$$\frac{1}{m-1} \sum_{i=1}^{m} \left(\frac{w_i^T v}{\|v\|}\right)^2 = \frac{1}{m-1} \left(\frac{v^T W^T W v}{v^T v}\right). \tag{3.27}$$

We can visually think of the black dots in Fig. 3.14 as being projected onto any vector v, of which the red and blue eigenvectors are candidates. If we drew a line through v and thought of it as our sole axis, then the new data would collapse into a string of scalars, and the sample variance would thus be the sum of these squared scalars divided by Bessel's divisor of $m - 1$.

The projected variance calculation in (3.27) shows how to use the covariance matrix to calculate data's spacial variation over a vector v, and we hope to find v for which the projected variance is as large as possible. Hence, we want to solve

$$\max_v \frac{v^T W^T W v}{v^T v}.$$

The ratio being maximized is called the Rayleigh-Ritz quotient, and the solution to the optimization problem is a unit length eigenvector for the largest eigenvalue. The maximum value is the largest eigenvalue. Hence, a unit length eigenvector for the largest eigenvalue is a direction along which the most variation occurs, and the eigenvalue itself is the sample variance multiplied by $m-1$ in this direction. For our example we have

$$v_1 = \begin{pmatrix} -0.27057 \\ -0.96270 \end{pmatrix} \text{ solves } 3.445682 = \max_v \frac{v^T W^T W v}{v^T v}.$$

This eigenvector is called the first principal component because it captures the maximum variation of the data.

The process continues by computing a second principal component that explains the second most variation in the data. In this case we seek a vector orthogonal to the first v_1, and the problem being solved is

$$\max_v \frac{v^T W^T W v}{v^T v} \text{ such that } v^T v_1 = 0.$$

The problem's solution is a unit eigenvector for the second largest eigenvalue, and hence,

$$v_2 = \begin{pmatrix} -0.96270 \\ 0.27057 \end{pmatrix} \text{ solves } 0.022873 = \max_v \frac{v^T W^T W v}{v^T v} \text{ such that } v^T v_1 = 0.$$

We conclude that the principal components are the eigenvectors for the eigenvalues in decreasing magnitude and that the interpretation of maximizing variation shows that the eigenvalues measure the variances of the data projected onto the eigenspaces. The first principal component in Fig. 3.14 maximizes the spread of the data along its eigenspace, and because the variation along the second principal component is minor, we know that the data can be reasonably collapsed onto the first principal component. Doing so gives the compressed data corresponding to the first principal component, which is tabulated below (the subscript c indicates that we have "compressed" the collection of principal components).

3.4 Principal Component Analysis

$$\begin{pmatrix} x - \bar{x} \\ y - \bar{y} \end{pmatrix} \begin{pmatrix} -0.23 \\ -0.02 \end{pmatrix} \begin{pmatrix} -0.13 \\ 0.08 \end{pmatrix} \begin{pmatrix} 0.77 \\ 1.98 \end{pmatrix} \begin{pmatrix} -0.03 \\ 0.48 \end{pmatrix} \begin{pmatrix} 0.77 \\ 2.78 \end{pmatrix}$$

$$\begin{pmatrix} x_c \\ y_c \end{pmatrix} \begin{pmatrix} -0.02 \\ -0.08 \end{pmatrix} \begin{pmatrix} 0.01 \\ 0.04 \end{pmatrix} \begin{pmatrix} 0.57 \\ 2.04 \end{pmatrix} \begin{pmatrix} 0.12 \\ 0.44 \end{pmatrix} \begin{pmatrix} 0.78 \\ 2.78 \end{pmatrix}$$

$$\begin{pmatrix} x - \bar{x} \\ y - \bar{y} \end{pmatrix} \begin{pmatrix} -0.43 \\ -2.12 \end{pmatrix} \begin{pmatrix} -0.13 \\ -0.82 \end{pmatrix} \begin{pmatrix} 0.57 \\ 1.68 \end{pmatrix} \begin{pmatrix} -0.53 \\ -1.42 \end{pmatrix} \begin{pmatrix} -0.63 \\ -2.62 \end{pmatrix}$$

$$\begin{pmatrix} x_c \\ y_c \end{pmatrix} \begin{pmatrix} -0.58 \\ -2.08 \end{pmatrix} \begin{pmatrix} -0.22 \\ -0.79 \end{pmatrix} \begin{pmatrix} 0.48 \\ 1.71 \end{pmatrix} \begin{pmatrix} -0.41 \\ -1.45 \end{pmatrix} \begin{pmatrix} -0.73 \\ -2.59 \end{pmatrix}$$

Figure 3.15 plots the original centered data along with its compressed counterpart. The compressed data lies along the eigenspace of the largest eigenvector. We note that the original data can be approximated by adding the sample means back to the compressed data.

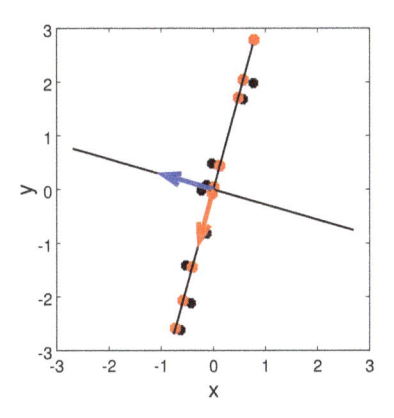

Fig. 3.15 The original centered data (black) and the compressed data (red).

The term "compressed" might seem odd at first since we have replaced 20 experimental outcomes with 20 less-exact values. However, since all compressed values are scalar multiples of a single two-dimensional eigenvector, each of the 10 experimental pairs is compressed to a single scalar in addition to the eigenvector. For example, the compressed data is

$$\begin{pmatrix} -0.02 \\ -0.08 \end{pmatrix} = 0.08 \begin{pmatrix} -0.27057 \\ -0.96270 \end{pmatrix}, \ldots, \begin{pmatrix} -0.73 \\ -2.59 \end{pmatrix} = 2.692734 \begin{pmatrix} -0.27057 \\ -0.96270 \end{pmatrix}.$$

This reduces the total number of values from 20 to 10+2 = 12, a savings of 40%.

PCA extends analogously to higher dimensions. Generally, we let X be a matrix with m rows and n columns so that the n outcomes of the i-th experiment fall across the i-th row of X. Again, factoring the covariance matrix so that
$$W = \frac{(X - e\bar{\mu})^T(X - e\bar{\mu})}{m-1} = VDV^T,$$
we select a percentage, ρ, of the total variation that we would like to maintain. A typical value of ρ is 0.95. We order the eigenvalues as
$$\lambda_1 >= \lambda_2 >= \ldots >= \lambda_n$$
and find the first k so that
$$\sum_{i=1}^{k} \lambda_i \geq \rho \sum_{i=1}^{n} \lambda_i.$$

The vector $(\lambda_1, \lambda_2, \ldots, \lambda_k)$ is called a feature vector. The compressed, transformed data is
$$Z_c = (X - e\bar{\mu})V_c,$$
where columns of V_c are unit length eigenvectors of the first k eigenvalues. The compressed version of the original data is
$$X_c = Z_c V_c^T + e\bar{\mu},$$
which can be calculated from the $m \times k$ matrix Z_c, the $n \times k$ matrix V_c, and the n-vector $\bar{\mu}$. Pseudocode for the process is in Algorithm 10.

Algorithm 10 The method of principal component analysis. Input is an $m \times n$ data matrix X and a percentage ρ

Calculate the n sample means $\bar{\mu}$.
Center the data $W = X - e\bar{\mu}$.
Factor the covariance matrix $(1/(m-1))W^TW = VDV^T$.
Sort the eigenvalues in descending order.
Find the first k eigenvalues so that $\sum_i^k \lambda_i \geq \rho \sum_i^n \lambda_i$.
Construct the submatrix V_c of V whose columns correspond with
    the feature vector $(\lambda_1, \ldots, \lambda_k)$.
Compute $Z_c = (X - e\bar{\mu})V_c$.
return Return V_c, Z_c, and $\bar{\mu}$.

PCA compression reduces the storage of the experimental data from $m \times n$ elements to $(m+n)k+n$ elements, which is substantial if k is much less than n. As an example, if $m = 1000$, $n = 100$, and $k = 10$, then the original data requires the storage of 10^5 elements. The compressed data requires the storage of 11,100 elements, a savings of 88.9%. The size of k depends on how quickly the largest eigenvalues accrue the desired percentage of the overall sum. In favorable cases only a small percentage of the eigenspaces are needed and the compression is substantial.

PCA compression returns the original data if and only if $\rho = 1$. In this case the original data is stored in a way that requires $(m+n)n + m$ values. This is not a compression because Z_c, V_c, and $\bar{\mu}$ require the storage of an additional $n^2 + m$ elements over the $m \times n$ elements of the original data. So realizing a meaningful compression requires an approximation of the original data. Any compression algorithm for which the original data cannot be exactly reproduced is called lossy, and PCA is lossy because it provides an approximate reconstruction of the original data as it achieves compression.

3.4.1 Principal Component Analysis and the Singular Value Decomposition

We have already observed that matrix factorizations play an important role in computational science. For instance, the LU and Cholesky factorizations support applications such as splines and linear regression. The decomposition of the covariance matrix W, i.e.

$$W = \frac{(X - e\bar{\mu})^T (X - e\bar{\mu})}{m-1} = VDV^T,$$

is called the spectral decomposition since it factors the covariance matrix in terms of its eigenvalues and eigenvectors. The term "spectral" is used because the collection of eigenvalues is called the spectrum of a matrix.

The use of the spectral decomposition associated with PCA depends on the fact that the covariance matrix is positive definite, but there is an alternative factorization from which PCA can be conducted directly from the centered data. This factorization is called the singular value decomposition, which exists for any matrix.

Theorem 11. *If A is an $m \times n$ matrix, then there are matrices U, V, and Σ such that*

$$A = U\Sigma V^T.$$

The matrices U, V, and Σ satisfy

- *U is $m \times m$ such that $U^T = U^{-1}$,*
- *V is $n \times n$ such that $V^T = V^{-1}$, and*
- *Σ is an $m \times n$ diagonal matrix.*

A full study of the singular value decomposition and its calculation exceeds the introductory nature of this text, but with regards to PCA, notice that if

$$X - e\bar{\mu} = U\Sigma V^T,$$

then

$$\begin{aligned}(X - e\bar{\mu})^T(X - e\bar{\mu}) &= (U\Sigma V^T)^T(U\Sigma V^T)\\ &= V\Sigma^T(U^TU)\Sigma V^T\\ &= V\Sigma^T\Sigma V^T\\ &= VDV^T,\end{aligned}$$

where D is a diagonal $n \times n$ matrix. Accounting for the Bessel correction of $m - 1$, we see that the singular value decomposition leads to the covariance matrix by dividing the diagonal elements of D by $m - 1$. This is especially convenient since the transformed data must satisfy

$$Z = (X - e\bar{\mu})V = U\Sigma.$$

For any particular feature vector, the compressed, transformed data is thus

$$Z_c = (X - e\bar{\mu})V_c = U\Sigma_c.$$

The singular value decomposition of the centered data can be computed in MATLAB and Octave with

$$[\text{U}, \text{S}, \text{V}] = \text{svd}(\text{X} - \text{ones}(\text{m}, 1) * \text{mean}(\text{X}));$$

from which the compressed, transformed data can be calculated once a feature vector is established.

3.5 Exercises

1. Assume an approximating polynomial of degree n in the method of least squares, and suppose that the points (x_i, y_i), for $i = 1, 2, \ldots, m$ with $m > n$, have the property that $x_i \neq x_j$ for $i \neq j$. Show that the columns of the coefficient matrix A are linearly independent, where the normal equations defining the polynomial are $A^T A a = A^T y$.
2. Prove the re-expression of variance in (3.9).
3. Show that if X is a standard uniform variable, then $Y = a + (b-a)X \sim \mathcal{U}(a, b)$.
4. Show that if $X \sim \mathcal{N}(\mu, \sigma^2)$, then $Z = (X - \mu)/\sigma$ is a standard normal variable.
5. Write a function with declaration

    ```
    [alpha]=LeastSquares(x,y,n)
    ```

3.5 Exercises

that returns the coefficients of the least-squares polynomial of degree n for approximating the data in x and y.

The input argument are:

x - A vector containing the x-coordinates of the data.

y - A vector containing the y-coordinates of the data. You may assume that x and y have the same number of elements. Your code should work if x and y are row vectors or if they are column vectors.

n - The degree of the polynomial to be returned.

The output argument is:

alpha - A vector with $n+1$ entries specifying the coefficients of $p(x)$, the least-squares approximating polynomial of degree n, in order from highest to lowest degree.

6. Establish Eq. (3.13).
7. Get the daily average temperatures from the National Oceanic and Atmospheric Administration (NOAA) for a location near yourself for the last 30 years. Use the method of least squares to find a best approximation of the form

$$f(d) = \alpha_0 + \alpha_1 d + \alpha_2 \cos\left(\frac{2\pi d}{365.25} - \phi\right),$$

where the days are indexed by d. Be sure to use a linear model based on (3.6). Does the α_1 coefficient of the best approximation indicate that temperatures are increasing or decreasing? Plot the data from NOAA with your least squares approximation.
8. The graph on the left of Fig. 3.16 depicts data from a simulated tornado. The horizontal axis is the radial distance from the center of the tornado's rotation. The vertical axis is the tangential velocity v of the rotation in meters per second. The data is in the text file Tornado.txt available at http://www.springer.com. The i-th velocity entry in Tornado.txt gives a velocity measurement at position $r_i = 2i/2500$ m from the center of the tornado. Velocity measurements of 0 m/s should be ignored. The graph on the right depicts a least squares approximation to the data. The functional form of the approximation is

$$f(r) = \begin{cases} \dfrac{\alpha_2}{a-r+1}, & b \leq r < a \\ \alpha_1(r-a) + \alpha_2, & a \leq r \leq d. \end{cases}$$

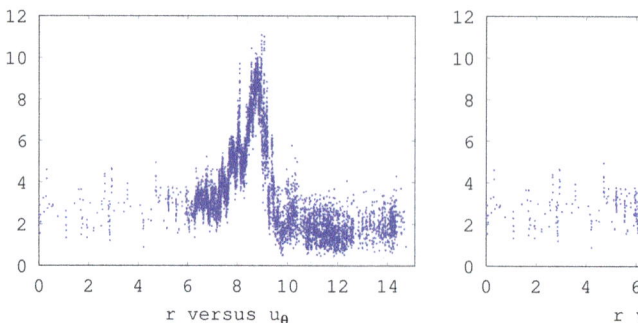

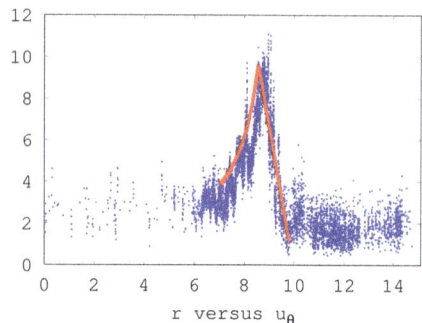

Fig. 3.16 A plot of radial distance versus radial velocity for the simulated tornado data in Exercise 8 is on the left, the same data with an approximating function is on the right.

The value of a is unknown and is to be searched for over the interval satisfying $r_{\max} - 0.75 \leq a \leq r_{\max} + 0.75$, where $v(r_{\max})$ is the largest velocity in the data. For each value of a, the parameter d is to be searched for over the interval satisfying $2/3 \leq d - a \leq 5/4$ and the parameter b is to be searched for over the interval satisfying $3/2 \leq a - b \leq 3$.

Use the method of least squares for each parameter triple (a, b, d) to calculate f as follows:

a. **Step 1:** Find a best fit approximation to f over $a \leq r \leq d$,
b. **Step 2:** Use the value of α_2 from Step 1 to define f over $b \leq r < a$.

Find the parameter triple (a, b, d) that (approximately) minimizes the average squared error over your search. That is, if there are $k_{a,d}$ data points between a and d, then the error is

$$e = \sum_{a \leq r_i \leq d} \frac{(f(r_i) - v_i)^2}{k_{a,d}}.$$

Your code should create a plot like that on the right showing how your best approximation fits the data. The values of a, b, d should all be searched for with a step size of 0.05 units. The value $1.1455\, f^2(a)$, for the best a, estimates the drop in pressure at the center of the vortex, and this value should be reported.

9. Some engineering applications seek a best fit circle of the form

$$(x - a)^2 + (y - b)^2 = r^2.$$

The parameters of this nonlinear model are a, b, and r, and the goal is to find parameters that minimize

$$\sum_i \left((x_i - a)^2 + (y_i - b)^2 - r^2\right)^2,$$

3.5 Exercises

where the collection of data is (x_i, y_i). Show that this least squares problem can be reformulated as a linear model. Use your reformulation to calculate the least squares solution through the points,

i	1	2	3	4	5	6
x_i	5.31	1.90	2.48	-4.60	5.28	-2.11
y_i	-0.34	2.33	-2.05	-1.28	-0.54	-2.20

10. Write a function with declaration

    [z] = LagrangePoly(x,y,v)

    that creates a Lagrange polynomial interpolating the data in x and y and evaluates the polynomial at the points in v.

    The input arguments are:

    x - A vector of length n that contains the x-coordinates of the data being interpolated; may be either a row or a column vector.

    y - A vector of length n that contains the y-coordinates of the data being interpolated; may be either a row or a column vector.

    v - A vector containing points at which $p(x)$ is to be evaluated, where $p(x)$ is the interpolating polynomial; may be either a row or column vector.

    The output argument is:

    z - A vector such that $z_i = p(v_i)$, where $p(x)$ is the interpolating polynomial.

    Your method should *not* solve a linear system of equations, but rather it should build the Lagrange polynomials from (3.8).

11. Write a function with declaration

    [z]=CubicSpline(x,y,v)

    that computes a natural cubic spline.

    The input arguments are:

    x - A vector of length n, either as a row or a column, that specifies the x-coordinates of the data. Do not assume that the data in this vector is sorted in increasing order.

    y - A vector of length n, either as a row or a column, that specifies the y-coordinates of the data.

    v - An optional argument, with default value x. The cubic spline should be evaluated at all the values in v.

The output argument is:

z - A vector containing $f(v_i)$, where $f(x)$ is the natural cubic spline for the data described in x and y.

The function should calculate the cubic spline by solving the tridiagonal system developed in Sect. 3.3.

12. Generate an N element sample from the standard uniform with x = rand(N,1), where N is some suitably large value. Let $\bar{x}_n$ be the sample mean of the first n elements of x as n progresses from 1 to N. Plot $\bar{x}_n$ against n and note that the sample mean progresses toward the true mean of a standard uniform variable, i.e. toward 1/2. Repeat the exercise with the standard uniform.

13. Repeatedly draw an N element sample from a distribution of your choice and calculate the sample mean for each sample. Store these values in a vector, e.g. sampleMeans, and then use the command hist(sampleMeans) to generate a histogram of sampleMeans. A histogram discretely bins the values of its argument and then counts the number of occurrences in each bin. Interpret the histogram with regard to the Central Limit Theorem.

    As an example, the command hist(mean(rand(75,2000))) produces a histogram of 2000 sample means, each calculated from a 75-element sample drawn from the standard uniform distribution.

14. Write a function with declaration

    [a,Rsqr,pVal,yHat,cInt,pInt] = LinearRegression(X,y,V,conf)

    that calculates the linear regression model,

    $$\hat{y}(x) = a_0 + a_1 x_1 + a_2 x_2 + \ldots a_n x_n,$$

    where the random variables X_i and Y_i have been sampled m times.

    The input arguments are:

    X - An $m \times n$ matrix such that each of the columns corresponds with an independent variable, with X_{ij} being the i-th sample of the j-th regressor.

    y - A column vector y such that y_i is the i-th sample of the response variable.

    V - An optional $q \times n$ matrix V, containing hypothetical observations.

    conf - an optional $q \times 1$ vector conf of values between 0 and 1.

    The output arguments are:

3.5 Exercises 117

a - A row vector whose components are the model parameters from the method of linear regression.
Rsqr - The coefficient of determination R^2.
pVal - A vector of p-values for the model parameters.
yHat - A vector of model predictions $\hat{y}$, from the hypothetical observations in V. If V is included, then yHat should be a vector of model predictions so that the i-th component is

$$\hat{y}(X_1 = V_{i1}, X_2 = V_{i2}, \ldots, X_n = V_{in}).$$

In MATLAB syntax, yHat[i] == a * [1, V(i,:)]' should be true for each row of V.
cInt - A vector to construct confidence intervals.
pInt - A vector to construct prediction intervals.

Should conf be included as an input argument, then cInt and pInt should be returned so that

yHat[i] ± cInt[i] and yHat[i] ± pInt[i]

are the (estimated) confidence and prediction intervals, respectively, for the i-th model prediction corresponding to conf[i]. If conf is omitted but V is not, then conf should default to a vector for 95% confidence and prediction intervals, i.e. the default should be conf = 0.95 * ones(q,1). Should V and conf be omitted, then only a, Rsqr, and pVal should be returned (the others can be returned as empty).

15. Use your solution to Exercise 14 to analyze the Faber-Jackson and bridge datasets at the end of Sect. 3.2.2. In particular, generate graphs similar to Fig. 3.10 for both data sets.
16. Calculate and interpret a 95% confidence interval on a_1 for the bridge dataset.
17. Repeat Exercise 7 but use your linear regression code from Exercise 14. Interpret the efficacy of the model by interpreting the coefficient of determination and the parametric p-values. Do you trust the model, and in particular, do you trust the estimated trend parameter a_1?
18. A regression model should be validated prior to its use. Since the statistical assumptions are imposed on the errors, a model is validated by finding evidence that the residuals are consistent with an independent sample drawn from a normal distribution with constant variance. Model validation is a substantial study in itself, but the common techniques below are routine. We assume the simple linear model $\hat{y} = a_0 + a_1 x$, where the parameters are decided by the n points (x_i, y_i). The notation $\hat{y}_i$ is succinctly used for $\hat{y}(x_i)$.

    (a) **Residual Plots:** A common graphical method to investigate the residuals is to plot them against the independent observations x_i, creating a scatter plot. This plot should be inspected and evaluated. For example, the "spread" of the residuals should roughly agree with the statistics from a normal distribution with a constant variance. An estimate of the standard deviation is

$$s_{\text{err}} = \sqrt{\frac{1}{n-1} \sum_i^n (y_i - \hat{y}_i)^2}.$$

The majority of residuals, roughly 68% of them, should be within $\pm s_{\text{err}}$, and nearly all of them, roughly 95% of them, should be within $\pm 2\, s_{\text{err}}$. Residuals whose magnitudes are greater then $3\, s_{\text{err}}$ should be (exceedingly) rare. Moreover, there should not be a pattern in the plot, and the "spread" of the residuals should remain constant. Patterns or trends raise concerns.

Plot the residuals from Exercise 17 and evaluate how they agree or disagree with the underlying assumptions of regression. Include on your plot horizontal lines for $\pm s_{\text{err}}$ and $\pm 2\, s_{\text{err}}$ to help your assessment.

(b) **Probability Plots:** While scatter plots of residuals help us identify patterns and trends, they do not otherwise aid our ability to identify agreement with the iconic bell curve of the normal. A visual test in this regard is to use a probability plot, which graphically benchmarks the occurrence pattern of the residuals against what would be expected from a standard normal distribution. This plot can be generated in MATLAB with `probplot`, and a command like `probplot(y - yHat)` will produce the desired graph. Should the residuals be distributed normally, then all data points should lie on the diagonal line from the bottom left to the top right—indicating that the occurrence pattern matches that of the normal distribution. However, deviations from the diagonal line indicate disagreement with the normal distribution.

Try the following computational test to help understand the nature of a probability plot. Let `w = rand(1000,1)` be a 1000-element sample drawn from a standard uniform distribution. Investigate how `probplot(w)` illustrates a non-normal sample. The deviations at the ends of the diagonal line indicate a lack of agreement with the tails of the normal. Explain why this makes sense. Repeat the process with `w = randn(1000,1)` to assess agreement with a normally distributed variable. Finally, evaluate the normality of the residuals from Exercise 17.

(c) **Autocorrelation:** If data is explained in terms of time, i.e. data forms a time series, or is otherwise naturally ordered, then autocorrelation is a concern. This means that there is a relationship between consecutive residuals, which subsequently suggests a violation of the independence assumption on the errors. A common statistic is the Durbin-Watson test statistic,

$$d = \frac{\sum_{i=2}^{n}\left((y_i - \hat{y}_i) - (y_{i-1} - \hat{y}_{i-1})\right)^2}{\sum_{i=1}^{n}(y_i - \hat{y}_i)^2}.$$

The value of d always lies between 0 and 4. Values closer to 0 indicate positive autocorrelation, meaning that consecutive residuals tend to be the same. Values closer to 4 indicate negative autocorrelation, meaning that consecutive residuals tend to be the negatives of each other. Values near 2 suggest no autocorrelation.

Calculate the Durbin-Watson test statistic for the residuals in Exercise 17. If we consider d being outside $[0.5, 3.5]$ as problematic, then should we be concerned with autocorrelation?

19. Sometimes a regression model is unduly dependent on a few data points, possibly caused by errors during data entry or by experimental inaccuracies. An outlier is a data point whose dependent observation is unusually outside the expected scope. Such data points can skew a simple linear regression model since the underlying optimization problem "pulls" the regression line toward the outlier to limit the outlier's contribution to the sum of squared residuals. Outliers are commonly discovered by considering the standardized residuals (sometimes called studentized residuals)

$$\frac{y_i - \hat{y}_i}{\sqrt{\frac{\sum_{i=1}^{n}(y_i - \hat{y}_i)^2}{n-2}}\sqrt{1 - \left(\frac{1}{n} + \frac{(x_i - \bar{x})^2}{\sum_{i=1}^{n}(x_i - \bar{x})^2}\right)}}.$$

Values exceeding ± 3 are suspect, increasingly so as the magnitude of this value grows.

The point (x_i, y_i) is influential if the x_i value surpasses a reasonable scope. Influential points are identified by calculating a data point's leverage,

$$h_i = \frac{1}{n} + \frac{(x_i - \bar{x})^2}{\sum_{i=1}^{n}(x_i - \bar{x})^2}.$$

Many software packages flag a data entry as influential if $h_i \geq 6/n$.

Are there concerns with either outliers or influential points in Exercise 17? If so, remove these points from the data and re-assess the resulting regression model.

20. Numerous online resources report gun ownership and homicide rates by country. Build a regression model to evaluate the possible relationship between gun ownership and homicide. Feel free to experiment with different models, and base your final model on a sound evaluation that includes consideration of R^2, the p-values of the coefficients, a regression analysis, and the impact of influential points and outliers (see Exercises 18 and 19).

21. Write a function with declaration

    ```
    [Zc,Vc,muBar] = PCA(X,p)
    ```

    that performs principal component analysis on the data in X.

    The input arguments are:

    X - A matrix describing the data, with one observation per row.

    p - A scalar, $0 < p \leq 1$, giving a lower bound for the variation in data to be retained.

    The output arguments are:

    Zc   - The data, in its compressed form.

    Vc   - The eigenvector basis in which the compressed data is expressed.

    muBar - the row vector of column means $\bar{\mu}$.

    As an example,

    ```
    [Zc, Vc, muBar] = PCA(X, 0.75);
    ```

    will return matrices Zc and Vc and a row vector muBar so that

    ```
    Xapprox = Zc * Vc' + ones(m,1) * muBar;
    ```

    calculates the approximated data matrix Xapprox to X that accounts for 75% of the variability in X.

22. MATLAB has several commands to read and convert images. Write a script that reads an image and converts it to grayscale. The resulting image will be encoded as a matrix of values between 0 and 255. Use this matrix as X and compress the data with PCA. The result will be a compressed version of the image, say X_c. Display and compare these images as ρ varies. How small can ρ become before you lose reasonable image quality, and what is the savings in space for this ρ?

23. PCA can be used as a filter. The value of ρ controls the percentage of variation maintained in the compressed data, and hence, the value $1 - \rho$ expresses how much variation is lost. If data is scrambled with lots of small amplitude noise, then PCA with a small value of ρ might recapture the original signal by removing the variation caused by the noise.

3.5 Exercises

The file PCAsignal, available at http://www.springer.com, contains a data matrix A (use load('PCAsignal') to read the data). Each row is a sample of a function, say $f(t)$, where the columns index the values of t. The first column is for $t = 0$, the second for $t = 0.1$, and so on in steps of $\Delta t = 0.1$. Plot the data with blue dots over the time domain. Then perform PCA on the data to remove much of the variation—you may want to experiment with different values of ρ. Create a second plot that superimposes a plot of the filtered data over the original data. Again plot the original data as blue dots but use red for the filtered data. Label each graph and include a caption to explain the figure. Does the signal identified by PCA appear periodic, and if so, then what is the period approximately?

Chapter 4
Optimization

Premature optimization is the root of all evil. – Donald Knuth

...in fact, the great watershed in optimization isn't between linearity and nonlinearity, but convexity and nonconvexity. – R. Tyrrell Rockafellar

People say that you should not micro-optimize. But if what you love is micro-optimization ... that's what you should do. – Linus Torvalds

Numerous computational problems are appropriately modeled in terms of an optimal property. Indeed, we have already seen two examples in linear regression and principal component analysis, both of which describe a computational and modeling intent in terms of an optimal quality. Any earnest taxonomy of the field of optimization would demonstrate a wide girth and a sizable depth of application, solution procedure, and mathematical analysis. In this chapter we restrict our attention to the general themes that underlie many of the algorithms used to solve optimization problems. The chapter is divided into three parts, those being unconstrained, constrained, and global optimization. The algorithms contained herein provide a suite of computational procedures to solve a variety of optimization problems.

4.1 Unconstrained Optimization

We assume functions of the form $f : \mathbb{R}^n \to \mathbb{R}$, and we further assume derivatives of any desirable order. The function being real-valued is important since $\mathbb{R}$ is linearly ordered, i.e. for any real u and v we have exactly one of $u < v$, $u > v$, or $u = v$. Hence for x and y in $\mathbb{R}^n$ we can establish if $f(x) < f(y)$, $f(x) > f(y)$, or $f(x) = f(y)$. If the function had instead mapped into $\mathbb{R}^2$, then we would not have been so fortunate, e.g. is $(1, -1) > (-1, 1)$

or $(1, -1) < (-1, 1)$? Mapping $\mathbb{R}^n$ into $\mathbb{R}$ allows us to confirm a preference between two elements of $\mathbb{R}^n$ by evaluating f. So if we seek to find x that minimizes f, then x is preferred to y if $f(x) < f(y)$.

We consider the basic question of solving

$$\min_x f(x).$$

The sense of optimal being minimization generally includes maximization since we can replace f with $-f$. Our use of min, while standard, is less pedantic than would be a more rigorous development. To illustrate, consider the two unconstrained problems,

$$\frac{\text{Problem A}}{\min x^2} \quad \text{and} \quad \frac{\text{Problem B}}{\min e^{-x}}.$$

Problem A has a minimum value of 0 at $x = 0$, but Problem B does not have a minimum since every possible x can be bested by another. For instance, the claim that $x = 10^{10}$ is a solution to Problem B is immediately false because $x = 10^{11}$ gives yet a smaller value of e^{-x}, which of course could be further improved. In this case we would formally use the infimum instead of the minimum, and the infimum of Problem B is 0 even though no possible x can achieve this value. The use of min and max presupposes the possibility of an optimal solution, but the use of inf (infimum) and sup (supremum) does not. The differences are important mathematically but are less so computationally, after all, our computers have a maximum value, and hence, Problem B has a solution computationally. The use of max and min is more intuitive and germane to most applications, and we bypass any further discussion of infimums and supremums for this reason.

Nearly all our motivation lies with results from introductory calculus. The first is a general form of the first derivative test, and the second is an interpretation of the gradient.

First Order Necessary Condition of Optimality If $f(x^*)$ is a minimum or maximum of the differentiable function f, then

$$D_x f(x^*) = \nabla f(x^*) = 0.$$

Steepest Search Direction The gradient $\nabla f(x)$ (negative gradient $-\nabla f(x)$) is the direction of steepest ascent (descent) at x.

The first statement establishes a necessary condition since it states a property that is shared by all optimal solutions. However, the property might hold at non-optimal solutions, and hence, the condition lacks the sufficiency to conclude optimality. The second statement states that from any x in the domain, the direction from x along which the function marginally increases (decreases) most rapidly is the (negative) gradient. Since we seek to minimize f, searching a direction related to $-\nabla f(x)$ makes sense.

4.1 Unconstrained Optimization

A second condition is needed to verify that an x^* satisfying $\nabla f(x^*) = 0$ does indeed solve the minimization problem. In calculus we learn a second derivative test for single variable functions, which states that if $f'(x^*) = 0$ and $f''(x^*) > 0$, then x^* is a (local) minimizer of f. There is a direct extension of this result for multi-variate functions.

Second Order Sufficient Condition of Optimality If $\nabla f(x^*) = 0$ and $D_x^2 f(x^*)$ is positive definite, then x^* is a (local) minimizer of f.

The second derivative $D_x^2 f(x^*)$ is an $n \times n$ matrix called the Hessian of f at x^*. The Hessian of f is the Jacobian of the gradient $\nabla f(x)$, and as with the Jacobian, notation is not uniform across disciplines. The Hessian is commonly denoted by H, H_f, and $\nabla^2 f$. We use $D_x^2 f$ to notationally indicate the fact that the Hessian is being used as an extension of the second derivative. If f is a function of $x = (x_1, x_2, x_3)^T$, then we have as an example that

$$D_x f(x) = \nabla f(x) = \begin{pmatrix} \partial f/\partial x_1 \\ \partial f/\partial x_2 \\ \partial f/\partial x_3 \end{pmatrix}$$

and

$$D_x^2 f(x) = D_x \nabla f(x) = \begin{bmatrix} \partial^2 f/\partial x_1 \partial x_1 & \partial^2 f/\partial x_2 \partial x_1 & \partial^2 f/\partial x_3 \partial x_1 \\ \partial^2 f/\partial x_1 \partial x_2 & \partial^2 f/\partial x_2 \partial x_2 & \partial^2 f/\partial x_3 \partial x_2 \\ \partial^2 f/\partial x_1 \partial x_3 & \partial^2 f/\partial x_2 \partial x_3 & \partial^2 f/\partial x_3 \partial x_3 \end{bmatrix},$$

where we have assumed that all derivatives exist.

Functions whose Hessians are positive definite for any x are strictly convex, a mathematical quality that we advance in the next section. For strictly convex functions it can be shown that a local minimizer is a global minimizer, see, for instance, the last conclusion of Theorem 6. Moreover, if a minimizer exists, which is not generally guaranteed, then it is unique. Recall that we can think of a positive definite matrix as a diagonal matrix with positive diagonal elements, see the fourth property of Theorem 6. To illustrate the sufficiency condition and the strict convexity, consider the quadratic function

$$f(x,y) = 2x^2 + y^2 = \frac{1}{2} (x \ y) \begin{bmatrix} 4 & 0 \\ 0 & 2 \end{bmatrix} \begin{pmatrix} x \\ y \end{pmatrix}.$$

The Hessian is the constant matrix,

$$D_{x,y}^2 f(x,y) = \begin{bmatrix} 4 & 0 \\ 0 & 2 \end{bmatrix} = Q.$$

The eigenvalues of the Hessian are $\lambda_1 = 4$ and $\lambda_2 = 2$, which have corresponding eigenvectors $v = (1,0)^T$ and $w = (0,1)^T$. If we evaluate f over

any scalar multiple of v, say kv, then f reduces to the single dimensional quadratic
$$f(kv) = f(k,0) = 2k^2.$$
Likewise, $f(kw) = k^2$ for any scalar k. Obviously these simple quadratics share the unique minimum of zero at $k = 0$. Moreover, any vector can be written as a linear combination of v and w, e.g. $(2,-3)^T = 2v - 3w$. So, for any unit vector $u = (u_1, u_2)^T$ we have

$$\begin{aligned}
f(ku) &= f(ku_1v + ku_2w) \\
&= \frac{1}{2}(ku_1v + ku_2w)^T Q(ku_1v + ku_2w) \\
&= \frac{1}{2}(ku_1v + ku_2w)^T (ku_1 Qv + ku_2 Qw) \\
&= \frac{1}{2}(ku_1v + ku_2w)^T (ku_1 \lambda_1 v + ku_2 \lambda_2 w) \\
&= \frac{1}{2}\left(k^2 u_1^2 \lambda_1 v^T v + k^2 u_1 u_2 \lambda_1 w^T v + k^2 u_1 u_2 \lambda_2 v^T w + k^2 u_2^2 \lambda_2 w^T w\right) \\
&= \frac{1}{2} k^2 (u_1^2 \lambda_1 + u_2^2 \lambda_2) \\
&= k^2 (2u_1^2 + u_2^2),
\end{aligned}$$

where the fourth equality uses the fact that w and v are eigenvectors and the sixth equality uses the fact that the eigenvectors are orthogonal. Notice that

$$1 \le (2u_1^2 + u_2^2) \le 2$$

because u is a unit vector. Hence f reduces to a single dimensional quadratic over the general line ku, and any such quadratic is bounded by the quadratics over the eigenspaces because for any k we have

$$k^2 = f(kw) \le f(ku) = k^2(2u_1^2 + u_2^2) \le f(kv) = 2k^2.$$

The second derivatives of the left and right extremes are $\lambda_2 = 2$ and $\lambda_1 = 4$, from which we see that the eigenvalues of the Hessian bound the curvature of f along any line. Since the minimum eigenvalue is positive, we have the important geometric intuition that f forms a "bowl" whose sides are parabolas. In higher dimensions this geometric perspective still holds with similar algebraic support as long as the Hessian is positive definite, which leads to a formal proof of the second order sufficient conditions. The geometry for this example is illustrated in Fig. 4.1.

The optimization algorithms we consider are iterative, and starting with an initial x we search along a direction for an improvement. Mathematically, the iteration is
$$x_{k+1} = x_k + \alpha_k d_k, \ k = 0, 1, 2, \ldots.$$

4.1 Unconstrained Optimization

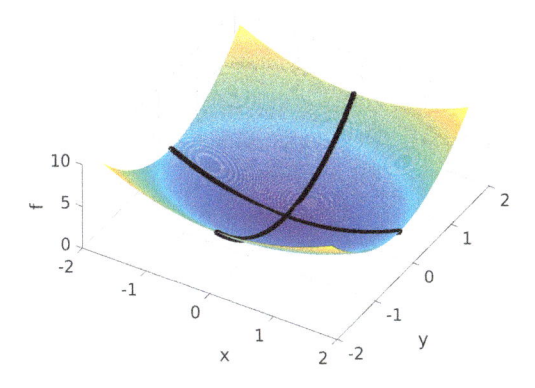

Fig. 4.1 A plot of $f(x,y) = 2x^2 + y^2$. The parabolas k^2 and $2k^2$ over the eigenspaces of the Hessian are plotted in black.

The two primary design questions of an optimization algorithm are, what is the search direction, d_k, and what is the step size, α_k?

4.1.1 The Search Direction

An obvious search direction for a minimization problem is $-\nabla f(x)$, and while this leads to an important algorithm, the negative gradient has shortcomings. As an illustration, consider $f(x,y) = x^2 + 3y^2$, which obviously has a minimum of 0 at the unique solution $(x^*, y^*) = (0,0)$. Suppose we initiate an iterative algorithm at $(1,1)$. Since

$$-\nabla f(1,1) = -\binom{2}{6},$$

we have for any α that an update along the negative gradient is

$$\binom{1}{1} + \alpha \binom{-2}{-6} = \binom{1-2\alpha}{1-6\alpha}.$$

No α leads to the solution since it is impossible to simultaneously make both components zero. A reasonable and important question is, can the negative gradient be improved so that the search direction more accurately points to the unique solution?

As with many numerical and computational problems, we learn much from a Taylor approximation. The second order approximation of $f(x)$ expanded about the n-vector v is

$$f(x) \approx f(v) + (D_x f(v))^T (x - v) + \frac{1}{2}(x - v)^T (D_x^2 f(v))(x - v),$$

where $(D_x f(v))$ is the gradient $\nabla f(v)$ and $(D_x^2 f(v))$ is the $n \times n$ Hessian of f at v. If we replace f with its second order Taylor approximation, then the first order necessary condition becomes

$$0 = D_x \left(f(v) + (D_x f(v))^T (x - v) + \frac{1}{2}(x - v)^T (D_x^2 f(v))(x - v) \right)$$
$$= D_x f(v) + (D_x^2 f(v))(x - v).$$

Setting $\Delta v = x - v$, we see that

$$(D_x^2 f(v))\Delta v = -D_x f(v),$$

which is exactly the Newton iteration from Chap. 2 related to the system $D_x f(x) \Delta v - \nabla f(x) = 0$. If the Hessian is invertible, then we have

$$\Delta v = -(D_x^2 f(v))^{-1}(D_x f(v)) = -(D_x^2 f(v))^{-1} \nabla f(v),$$

although we warn that the inverse should not be calculated directly from the Hessian. A linear solver should be used instead.

If we assume as before that $f(x,y) = x^2 + 3y^2$, then

$$D_{(x,y)} f(x,y) = \begin{pmatrix} 2x \\ 6y \end{pmatrix} \quad \text{and} \quad D_{(x,y)}^2 f(x,y) = \begin{bmatrix} 2 & 0 \\ 0 & 6 \end{bmatrix}.$$

Expanding about $(x,y) = (1,1)$, we have

$$f(x,y) \approx f(1,1) + (x-1, y-1) \begin{pmatrix} 2 \\ 6 \end{pmatrix} + \frac{1}{2}(x-1, y-1) \begin{bmatrix} 2 & 0 \\ 0 & 6 \end{bmatrix} \begin{pmatrix} x-1 \\ y-1 \end{pmatrix}$$

$$= 4 + (2(x-1) + 6(y-1)) + ((x-1)^2 + 3(y-1)^2)$$

$$= x^2 + 3y^2.$$

This calculation illustrates the, not so surprising, fact that the second order Taylor polynomial of a quadratic is itself. We have already observed that the negative gradient does not point to the unique minimizer, but the Newton direction at $(1,1)$ is

$$\Delta v = -\begin{bmatrix} 2 & 0 \\ 0 & 6 \end{bmatrix}^{-1} \begin{pmatrix} 2 \\ 6 \end{pmatrix} = -\begin{pmatrix} 1 \\ 1 \end{pmatrix}.$$

So for $\alpha = 1$ we have

4.1 Unconstrained Optimization

$$v + \alpha \Delta v = \begin{pmatrix} 1 \\ 1 \end{pmatrix} + \alpha \begin{pmatrix} -1 \\ -1 \end{pmatrix} = \begin{pmatrix} 0 \\ 0 \end{pmatrix}.$$

We conclude that if the search direction is the Newton direction, then the optimal solution is calculated in one step, which is an improvement over the negative gradient.

The directions of steepest descent and Newton's method are part of a family of unconstrained optimization algorithms that stem from variations of the search direction d satisfying

$$Hd = -\nabla f(x),$$

where H is an $n \times n$ matrix. The two cases already presented are:

Steepest Descent The search direction d is the negative gradient, and H is the $n \times n$ identity matrix.

Newton's Method The search direction d is the Newton direction, and H is the Hessian matrix.

Many other algorithms, called Quasi-Newton methods, lie somewhere in the middle by iteratively updating H so that it better approximates the Hessian. One of the most successful Quasi-Newton methods is due to Broyden, Fletcher, Goldfarb, and Shanno, called the BFGS algorithm. Starting with an initial guess at the Hessian, say H_0, the algorithm iteratively updates the approximation so that

$$H_{k+1} = H_k + \frac{\Delta Df_k (\Delta Df_k)^T}{(\Delta Df_k)^T \Delta x_k} - \frac{H_k \Delta x_k (\Delta x_k)^T H_k}{(\Delta x_k)^T H_k \Delta x_k}, \tag{4.1}$$

where

$$\Delta Df_k = \nabla f(x_{k+1}) - \nabla f(x_k) \text{ and}$$
$$\Delta x_k = x_{k+1} - x_k.$$

The concept of approximating the Hessian is an extension of the secant method from Chap. 1. Recall that the secant method builds from the mean value theorem to replace the first derivative with

$$f'(x) \approx \frac{f(b) - f(a)}{b - a},$$

where $a \leq x \leq b$ and the approximation is perfect for at least one such x. If we increase the order of the derivative by one, then this approximation is subsequently

$$f''(x) \approx \frac{f'(b) - f'(a)}{b - a}.$$

Letting $b = x_{k+1}$ and $a = x_k$ so that $b - a = \Delta x_k$, we have the multidimensional extension,

$$f''(x) \approx \frac{f'(b) - f'(a)}{b - a} \quad \Leftrightarrow \quad H_{k+1}\Delta x_k = \nabla f(x_{k+1}) - \nabla f(x_k) = \Delta Df_k.$$

The condition that $H_{k+1}\Delta x_k = \Delta Df_k$ is often referred to as the secant condition, and it is this property that is desired of an iterative scheme that updates the Hessian. This prompts the question of whether or not the BFGS update satisfies the secant condition. The following calculation shows that it does,

$$\begin{aligned} H_{k+1}\Delta x_k &= H_k \Delta x_k + \left(\frac{\Delta Df_k(\Delta Df_k)^T}{(\Delta Df_k)^T \Delta x_k}\right)\Delta x_k - \left(\frac{H_k \Delta x_k (\Delta x_k)^T H_k}{(\Delta x_k)^T H_k \Delta x_k}\right)\Delta x_k \\ &= H_k \Delta x_k + \Delta Df_k \left(\frac{(\Delta Df_k)^T \Delta x_k}{(\Delta Df_k)^T \Delta x_k}\right) - H_k \Delta x_k \left(\frac{(\Delta x_k)^T H_k \Delta x_k}{(\Delta x_k)^T H_k \Delta x_k}\right) \\ &= H_k \Delta x_k + \Delta Df_k - H_k \Delta x_k \\ &= \Delta Df_k. \end{aligned}$$

The BFGS method has a property with regard to its inverse that is not shared by its Newton counterpart. Newton's method requires the solution to $D_x^2 f(x)d = -\nabla f(x)$ at each iteration, and the BFGS method as developed thus far similarly requires the solution to $H_k d = -\nabla f(x)$ per iteration. However, we can use a special algebraic expression, called the Sherman-Morris formula, to calculate the inverse of H_k at each iteration, which means that we can compute the BFGS direction as $-H_k^{-1}\nabla f(x)$. The computation of a matrix inverse is not typically recommended in numerical computation, but this advice comes from the concern of computing an inverse from the original matrix. In the case of the BFGS algorithm we are **not** suggesting that we calculate H_k^{-1} from H_k, but rather, it is possible to calculate H_{k+1}^{-1} from H_k^{-1}.

The Sherman-Morris formula for the $n \times n$ matrix A is

$$[A + uv^T]^{-1} = A^{-1} - \frac{A^{-1}uv^T A^{-1}}{1 + v^T A^{-1} u}, \tag{4.2}$$

where u and v are n-vectors and all inverses are assumed to exist. The rank of the $n \times n$ matrix uv^T is one, and the Sherman-Morris formula provides a straightforward way to compute the inverse of a rank one update to A if we already know A^{-1}. The BFGS algorithm is a rank two update since it adds two rank one matrices to H_k to construct H_{k+1}. If we cleverly apply the Sherman-Morris formula to the two rank one updates, then we find

$$\begin{aligned} H_{k+1}^{-1} = H_k^{-1} &+ \left(1 + \frac{\Delta f_k^T H_k^{-1} \Delta Df_k}{\Delta Df_k^T \Delta x_k}\right)\frac{\Delta x_k \Delta x_k^T}{\Delta x_k^T \Delta Df_k} \\ &- \frac{\Delta x_k \Delta Df_k^T H_k^{-1} + H_k^{-1} \Delta Df_k \Delta x_k^T}{\Delta x_k^T \Delta Df_k}. \end{aligned} \tag{4.3}$$

4.1 Unconstrained Optimization   131

So if we initiate the BFGS algorithm with $H_0 = I$, then $H_0^{-1} = I^{-1} = I$. We can then calculate H_1^{-1} and compute the search direction $d = -H_1^{-1}\nabla f(x)$ instead of calculating H_1 and then solving $H_1 d = -\nabla f(x)$. The process repeats at each iteration since we know the inverse of H_k.

There are numerous alternatives to the BFGS update, with the most common alternatives using a single rank one correction at each iteration to satisfy the secant condition. However, years of numerical experience have established the BFGS update as a consistent and efficient default.

4.1.2 The Line Search

The method used to decide the step size once a search direction is selected is called the line search. Importantly, the multi-dimensional problem reduces to the following single variable problem once a search direction is decided,

$$\min_\alpha f(x + \alpha d), \qquad (4.4)$$

where α is a scalar. The first order necessary condition for this problem is

$$0 = D_\alpha f(x + \alpha d) = (\nabla f(x + \alpha d))^T d, \qquad (4.5)$$

and we seek an α for which the directional derivative is zero. If gradient calculations can be had efficiently, then we might query the directional derivative to search for an $\hat{\alpha}$ so that the signs of $\nabla f(x)^T d$ and $(\nabla f(x + \hat{\alpha}d))^T d$ are different. We could then use, for example, the method of bisection or linear interpolation to search the interval $[0, \hat{\alpha}]$ to find an α satisfying (4.5).

Another possibility is to approximate $f(x + \alpha d)$ and then calculate an approximate solution to (4.4). Two approximations are the Taylor expansion

$$f(x + \alpha d) \approx f(x) + \alpha \nabla f(x)^T d + \frac{\alpha^2}{2} d^T (D_x^2 f(x)) d$$

and the Lagrange polynomial

$$f(x + \alpha d) \approx \frac{(\alpha - \alpha_2)(\alpha - \alpha_3)}{(\alpha_1 - \alpha_2)(\alpha_1 - \alpha_3)} f(x + \alpha_1 d)$$

$$+ \frac{(\alpha - \alpha_1)(\alpha - \alpha_3)}{(\alpha_2 - \alpha_1)(\alpha_2 - \alpha_3)} f(x + \alpha_2 d) + \frac{(\alpha - \alpha_1)(\alpha - \alpha_2)}{(\alpha_3 - \alpha_1)(\alpha_3 - \alpha_2)} f(x + \alpha_3 d),$$

where α_1, α_2, and α_3 are reasonable step lengths. Common values are to let α_1 be 0, α_3 be the largest permissible step size, and α_2 be $\alpha_3/2$. If f is a quadratic, then the Lagrange and Taylor polynomials are different algebraic expressions for the same function.

If the Taylor expansion is used, then (4.5) is approximated as

$$0 = D_\alpha f(x + \alpha d)$$
$$\approx \nabla f(x)^T d + \alpha\, d^T \left(D_x^2 f(x)\right) d,$$

from which we have that

$$\alpha = -\left(\frac{\nabla f(x)^T d}{d^T \left(D_x^2 f(x)\right) d}\right). \tag{4.6}$$

The downside to this calculation is that it requires an evaluation of the Hessian, which can be inefficient for large complicated problems. Using this α equates to performing a single iteration of Newton's method to solve $f(x + \alpha d) = 0$ for α. We could of course continue to perform further Newton iterations, but this tactic is less common since each iteration requires an additional evaluation of the Hessian.

The Lagrange polynomial removes the need to calculate the Hessian. If we differentiate the Lagrange polynomial with respect to α and solve for the root, then

$$\alpha = \frac{(\alpha_2^2 - \alpha_3^2)f(x') + (\alpha_3^2 - \alpha_1^2)f(x'') + (\alpha_1^2 - \alpha_2^2)f(x''')}{2(\alpha_2 - \alpha_3)f(x') + 2(\alpha_3 - \alpha_1)f(x'') + 2(\alpha_1 - \alpha_2)f(x''')}, \tag{4.7}$$

where $x' = x + \alpha_1 d$, $x'' = x + \alpha_2 d$, and $x''' = x + \alpha_3 d$. Although this expression is a bit tedious, it provides a numerical benefit. The downside of the Lagrange polynomial is deciding which values to use for α_1, α_2, and α_3. We emphasize that if f is (approximately) quadratic, then the two calculations are (approximately) the same.

Optimization routines typically bound the step size and search within a predefined range. Algorithms that stringently restrict their search so that $\|\alpha d\|$ is small are called small-step algorithms, whereas algorithms that permit a broader search are called large-step algorithms. There is unfortunately no canonical line search, and if an algorithm is not performing as expected, then an alternative could lead to marked improvement.

4.1.3 Example Algorithms

Pseudocode for a general unconstrained optimization algorithm is listed in Algorithm 11. The simplicity of the overriding theme of selecting a search direction and a step size somewhat masks the number of decisions that need to be considered for any specific method. We illustrate the design process by completing a few algorithms and demonstrating their effectiveness.

One of the most straightforward algorithms is gradient (steepest) descent with the step size being decided by the second order Taylor expansion in (4.7). The H matrix in this case is the identity. This algorithm requires the calculation of the Hessian at each iteration, and hence, is only applicable if the

4.1 Unconstrained Optimization

Hessian is efficiently available. In this case the pseudocode of Algorithm 11 becomes that of Algorithm 12. The termination criteria are to decide if

- the first order necessary condition is sufficiently satisfied,
- the iteration count has become excessive, or
- the step size has diminished below an acceptable standard, i.e. the algorithm has stalled.

Algorithm 11 General pseudocode for solving an unconstrained optimization problem

---
$k = 1$
while unmet termination criteria **do**
    calculate a search direction d by solving $Hd = -\nabla f(x)$
    calculate a step size α with a line search
    let $x = x + \alpha d$ and $k = k + 1$
end while
return best estimate of solution and termination status

Algorithm 12 Gradient descent with the step size being decided by a second order Taylor polynomial

---
$k = 1$
while $\|\nabla f(x)\| > \delta_{\nabla f}$ and $k < k_{\max}$ and $\|\alpha d\| > \delta_x$ **do**
    let $d = -\nabla f(x)$
    let
$$\alpha = -\left(\frac{\nabla f(x)^T d}{d^T \left(D_x^2 f(x)\right) d}\right)$$
    let $x = x + \alpha d$
    let $k = k + 1$
end while
return x and $f(x)$

Another example is to combine Newton's method with the line search associated with the Lagrange polynomial in (4.7). If we allow $\alpha_1 = 0$, $\alpha_2 = 1/2$, and $\alpha_3 = 1$, then (4.7) reduces to

$$\alpha = \frac{3f(x) - 4f(x + (1/2)d) + f(x + d)}{4f(x) - 8f(x + (1/2)d) + 4f(x + d)}. \tag{4.8}$$

The result is Algorithm 13.

Consider the problem of minimizing

$$f(x, y) = x(3x + \sin(y)) + y(3y + \cos(x)) + xy, \tag{4.9}$$

Algorithm 13 Newton's method with the step size being decided by the Lagrange polynomial with $\alpha_1 = 0$, $\alpha_2 = 1/2$, and $\alpha_3 = 1$

$k = 1$
while $\|\nabla f(x)\| > \delta_{\nabla f}$ and $k < k_{\max}$ and $\|\alpha d\| > \delta_x$ **do**
  solve $D_x^2 f(x)\, d = -\nabla f(x)$ for d
  set
$$\alpha = \frac{3f(x) - 4f(x + (1/2)d) + f(x+d)}{4f(x) - 8f(x + (1/2)d) + 4f(x+d)}$$

  let $x = x + \alpha d$
  let $k = k + 1$
end while
return x and $f(x)$

for which
$$\nabla f(x,y) = \begin{pmatrix} 6x + \sin(y) - y\sin(x) + y \\ 6y + \cos(x) + x\cos(y) + x \end{pmatrix}$$
and
$$D_{x,y}^2 f(x,y) = \begin{bmatrix} 6 - y\cos(x) & 1 + \cos(y) - \sin(x) \\ 1 + \cos(y) - \sin(x) & 6 - x\sin(y) \end{bmatrix}.$$

The iterations for Algorithms 12 and 13 are listed in Table 4.1. The gradient descent algorithm requires six iterations to satisfy the first order condition with a tolerance of 10^{-3}. The Newton algorithm bests the gradient descent algorithm at every iteration and satisfies the same convergence criterion in three iterations.

	Algorithm 12		Algorithm 13	
Iteration	(x,y)	$\|\nabla f(x,y)\|$	(x,y)	$\|\nabla f(x,y)\|$
0	(1.000, 3.000)	19.70	(1.000, 3.000)	19.70
1	(−0.281,−0.591)	4.31	(−0.635, 0.383)	3.39
2	(0.080,−0.218)	0.16	(0.075,−0.183)	0.11
3	(0.066, −0.185)	0.04	(0.060,−0.186)	8.3e−5
4	(0.061, −0.188)	7.3e−3		
5	(0.060, −0.186)	2.0e−3		
6	(0.160, −0.186)	3.4e−4		

Table 4.1 Iterations for Algorithms 12 and 13 to minimize function (4.9).

4.1 Unconstrained Optimization

We conclude this section by comparing the BFGS algorithm to Newton's method. Since the BFGS algorithm emulates Newton's method without calculating the Hessian, this comparison tests if calculating the Hessian is worthwhile (at least in terms of iterations). We use a step size of $\alpha = 1$ in the comparison, which is called the full Newton step in Newton's method (sometimes called a pure Newton algorithm). The pseudocodes for the comparison are found in Algorithms 14 and 15. Iterations for function (4.9) are listed in Table 4.2. The reduced number of iterations for Newton's method illustrates the value of having accurate Hessian evaluations. That said, numerous computational studies have shown that the BFGS algorithm is often the algorithm of choice because it removes the calculation of $D^2 f(x)$, which is generally costly and often unavailable in closed form.

The four algorithms of this section are only a small selection of the design possibilities. Selecting the search direction and step size should generally be decided by the ease and accuracy of the second order calculations of the Hessian. If evaluating the Hessian is efficient and trustworthy, then Newton directions and step sizes are good choices. If Hessians are time consuming or subject to numerical errors, then algorithms like gradient descent and BFGS are appropriate with a line search that does not depend on the Hessian. No single algorithm works well in all cases, and experimentation is encouraged.

Algorithm 14 The BFGS algorithm with unit step size

$k = 1$
while $\|\nabla f(x)\| > \delta_{\nabla f}$ and $k < k_{\max}$ and $\|\alpha d\| > \delta_x$ **do**
   solve $H\,d = -\nabla f(x)$ for d
   let $\Delta Df = \nabla f(x+d) - \nabla f(x)$
   let $x = x + d$
   let $H = H + \frac{\Delta Df (\Delta Df)^T}{(\Delta Df)^T d} - \frac{H d d^T H}{d^T H d}$
   let $k = k + 1$
end while
return x and $f(x)$

Algorithm 15 Newton's method with a unit step size

$k = 1$
while $\|\nabla f(x)\| > \delta_{\nabla f}$ and $k < k_{\max}$ and $\|\alpha d\| > \delta_x$ **do**
   solve $D_x^2 f(x)\,d = -\nabla f(x)$ for d
   let $x = x + d$
   let $k = k + 1$
end while
return x and $f(x)$

	Algorithm 14		Algorithm 15	
Iteration	(x, y)	$\|\nabla f(x,y)\|$	(x, y)	$\|\nabla f(x,y)\|$
0	(1.000, 3.000)	19.70	(1.000, 3.000)	19.70
1	(−5.617, −15.550)	100.78	(−1.171, −0.474)	9.60
2	(0.881, −0.441)	4.77	(0.093, −0.270)	0.44
3	(−3.229, 1.263)	17.43	(0.061, −0.187)	3.8e−3
4	(0.006, −0.125)	0.34	(0.060, −0.186)	4.9e−7
5	(0.060, −0.191)	0.03		
6	(0.060, −0.186)	2.6e−3		
7	(0.060, −0.186)	5.2e−5		

Table 4.2 Iterations for Algorithms 14 and 15 to minimize function (4.9).

4.2 Constrained Optimization

Optimization in most applications is restricted by conditions placed on the variables, and in this section we consider problems of the form

$$\min_x \ f(x)$$
such that
$$g(x) = 0$$
$$h(x) \geq 0. \tag{4.10}$$

As in the unconstrained case, x is generally a vector with n components, and we seek a vector that renders f as small as possible. The objective function f is real-valued so that $f(x)$ is a scalar for all valid x. The functions g and h need not be scalar valued, and we generally assume that $g : \mathbb{R}^n \to \mathbb{R}^m$ and $h : \mathbb{R}^n \to \mathbb{R}^p$. Both the "=" and "≥" of the model are componentwise, and hence, there are m equality constraints and p inequality constraints. The feasible set is the collection of $x \in \mathbb{R}^n$ such that $g(x) = 0$ and $h(x) \geq 0$.

Consider the following example to demonstrate the flexibility of the model,

$$\left. \begin{aligned} \min_x \ & x_1 + x_2^2 + \sin(x_3) \\ \text{such that} \ & \\ & x_1(1 - x_1) = 0 \\ & x_3 \geq x_1^2 + x_2^2 \\ & x_3 \leq 1.2. \end{aligned} \right\} \tag{4.11}$$

4.2 Constrained Optimization

If we let

$$f(x) = x_1 + x_2^2 + \sin(x_3),$$
$$g(x) = x_1(1 - x_1), \text{ and}$$
$$h(x) = \begin{pmatrix} -x_1^2 - x_2^2 + x_3 \\ 1.2 - x_3 \end{pmatrix},$$

then our model fits the description in (4.10). The constraint $g(x) = 0$ forces x_1 to be either 0 or 1, which means combinatorial, and even integral, restrictions conform to the general paradigm.

The unconstrained case of the previous section benefited from the straightforward goals of searching for an x^* so that $\nabla f(x^*) = 0$ (the first order necessary condition of optimality) and $D_x^2 f(x^*) \succ 0$ (the second order sufficient condition of optimality). Satisfying these two conditions promises a local minimizer. Constrained problems are often more difficult because these conditions are not as meaningful in the presence of constraints. As an example, consider the optimization problem

$$\min_x \; x_1 + x_2 \text{ such that } x_1^2 + x_2^2 \leq 1. \tag{4.12}$$

In this case $f(x) = x_1 + x_2$, from which we have

$$\nabla f(x) = \begin{pmatrix} 1 \\ 1 \end{pmatrix} \text{ and } D_x^2 f(x) = \begin{bmatrix} 0 & 0 \\ 0 & 0 \end{bmatrix}.$$

So neither the first order necessary condition nor the second order sufficiency condition can be satisfied. Nonetheless, the optimization problem is well-posed and has a finite optimal solution. We solve this problem later, but you might want to try on your own first.

The study of constrained optimization circles around the work of Joseph Lagrange and John von Neumann—although many others have played substantial roles. Associated with any constrained optimization problem is a function bearing the name of the former, and the Lagrangian of (4.12) is

$$\mathcal{L}(x, \lambda, \rho) = f(x) - \lambda^T g(x) - \rho^T h(x).$$

The elements of the m-vector λ and the p-vector ρ are called Lagrange multipliers, and each constraint receives its own multiplier. The original optimization problem is stated in terms of n dimensions since each of f, g, and h has the n-vector x as its argument. The Lagrangian is instead a function of x, λ, and ρ, and hence, $\mathcal{L}$ acts on a combined vector of length $n + m + p$. The output of the Lagrangian is a scalar, making $\mathcal{L}$ a function from $\mathbb{R}^{n+m+p}$ into $\mathbb{R}$.

The brilliance of the Lagrangian is that it re-casts constraint satisfaction as an implied outcome of optimality. So instead of feasibility being an externally mandated requirement against which all candidate solutions must be vetted, the requirement of feasibility is naturally satisfied by an optimal condition. Consider the following unconstrained optimization problem to initiate our study,

$$\min_x \left(\max_{\lambda, \rho \geq 0} \mathcal{L}(x, \lambda, \rho) \right) = \min_x \left(\max_{\lambda, \rho \geq 0} f(x) - \lambda^T g(x) - \rho^T h(x) \right). \quad (4.13)$$

The function in the parentheses is a function of x. So if we set

$$T(x) = \max_{\lambda, \rho \geq 0} \mathcal{L}(x, \lambda, \rho) = \max_{\lambda, \rho \geq 0} f(x) - \lambda^T g(x) - \rho^T h(x),$$

then we could evaluate T at any x of our choice. Once x is selected, the values of f, g, and h become set, and the resulting value of T is decided by selecting λ and ρ to make $f(x) - \lambda^T g(x) - \rho^T h(x)$ as large as possible. In the case that λ and ρ can be chosen to make $T(x)$ arbitrarily large, we assign $T(x)$ the "value" of ∞.

Suppose x is selected so that

$$f(x) = 5.4, \quad g(x) = \begin{pmatrix} 0 \\ 0 \\ -0.2 \end{pmatrix} \quad \text{and} \quad h(x) = \begin{pmatrix} 0 \\ 1.4 \\ 0 \\ -0.01 \end{pmatrix}.$$

Then,

$$T(x) = \max_{\lambda, \rho \geq 0} \left(5.4 - \begin{pmatrix} \lambda_1 \\ \lambda_2 \\ \lambda_3 \end{pmatrix}^T \begin{pmatrix} 0 \\ 0 \\ -0.2 \end{pmatrix} - \begin{pmatrix} \rho_1 \\ \rho_2 \\ \rho_3 \\ \rho_4 \end{pmatrix}^T \begin{pmatrix} 0 \\ 1.4 \\ 0 \\ -0.01 \end{pmatrix} \right)$$

$$= \max_{\lambda, \rho \geq 0} (5.4 + 0.2\,\lambda_3 - 1.4\,\rho_2 + 0.01\,\rho_4).$$

Two terms are particularly troublesome with regard to the minimization of T. The first is $0.2\,\lambda_3$ and the second is $0.01\,\rho_4$. Both of these can become arbitrarily large as either, or both, $\lambda_3 \to \infty$ or $\rho_4 \to \infty$. The term $-1.4\,\rho_2$ is incapable of increasing without bound because all elements of ρ must be nonnegative. Hence the largest possible value of $-1.4\,\rho_2$ is 0 for $\rho_2 = 0$. We conclude that $T(x) = \infty$ because at least one, and in this case two, of the terms can increase without bound.

Suppose x had instead been selected so that

4.2 Constrained Optimization

$$T(x) = \max_{\lambda,\, \rho \geq 0} \left(6.1 - \begin{pmatrix} \lambda_1 \\ \lambda_2 \\ \lambda_3 \end{pmatrix}^T \begin{pmatrix} 0 \\ 0.3 \\ 0 \end{pmatrix} - \begin{pmatrix} \rho_1 \\ \rho_2 \\ \rho_3 \\ \rho_4 \end{pmatrix}^T \begin{pmatrix} 0 \\ 2.9 \\ 1.8 \\ 0 \end{pmatrix} \right)$$

$$= \max_{\lambda,\, \rho \geq 0} \left(6.1 - 0.3\,\lambda_2 - 2.9\,\rho_2 - 1.8\,\rho_3 \right).$$

In this case only $-0.3\,\lambda_2$ can increase without bound by allowing $\lambda_2 \to -\infty$ (notice that λ has no sign restriction). Neither $-2.9\,\rho_2$ nor $-1.8\,\rho_3$ can increase without bound because ρ is nonnegative, so the largest these terms can be is zero. In any event, we again set $T(x) = \infty$ due to the λ_2 term.

We might begin to ask which, if any, value of x leaves T finite. This is an important question since the optimization problem in (4.13) seeks to minimize T. If every x renders T infinite, then seeking a minimum is futile. The examples identify how T can be infinite, but they also show how a judicious selection of x could thwart T's maximization from returning an infinity. If any component of $g(x)$ is nonzero, then T becomes infinity because the corresponding λ multiplier could approach either ∞ or $-\infty$ as needed. However, only negative components of $h(x)$ result in T being infinity due to the sign restriction on the ρ multipliers. We conclude that

$$T(x) = f(x) < \infty \quad \text{if and only if} \quad g(x) = 0 \text{ and } h(x) \geq 0.$$

The reasons why $T(x)$ is $f(x)$ for a feasible x are (1) the equality $g(x) = 0$ removes the influence of λ, and (2) either the equality $h_i(x) = 0$ similarly removes the influence of ρ_i or the inequality $h_i(x) > 0$ forces $\rho_i = 0$ due to T's maximization. Hence, the feasibility of x is identical to a finite value of T, and in this case $T(x) = f(x)$ if x is feasible.

The outer minimization in (4.13) forces the selection of a feasible x if possible since otherwise the quantity being minimized achieves the worst possible outcome of infinity. Moreover, the equality of $T(x)$ and $f(x)$ for feasible x shows that

$$\min_{x} \left(\max_{\lambda,\, \rho \geq 0} \mathcal{L}(x, \lambda, \rho) \right) = \min_{x} \{ f(x) : g(x) = 0,\ h(x) \geq 0 \}.$$

This equality lays bare the relationship between an optimization problem and its Lagrangian. The problem on the right is a classically constrained problem, whereas the problem on the left is nearly an unconstrained problem—the only constraint being the sign restriction on ρ. The equality permits us to "optimize" the Lagrangian without directly enforcing the constraints $g(x) = 0$ and $h(x) \geq 0$. The constraints are instead enforced through the optimization of $\mathcal{L}$.

Reversing the sense of optimization provides a related problem called the (Lagrangian) dual, which is

$$\max_{\lambda,\,\rho\geq 0}\left(\min_x \mathcal{L}(x,\lambda,\rho)\right).$$

The original problem is called the primal, and there is a primal–dual relationship at the bedrock of constrained optimization. The primal problem is intuitively the smallest of the big-values of the Lagrangian, and the dual is the largest of the small values. Hence we should expect

$$\max_{\lambda,\,\rho\geq 0}\left(\min_x \mathcal{L}(x,\lambda,\rho)\right) \leq \min_x\left(\max_{\lambda,\,\rho\geq 0} \mathcal{L}(x,\lambda,\rho)\right). \tag{4.14}$$

This relationship follows from the general inequality,

$$\max_u \min_v f(u,v) \leq \min_v \max_u f(u,v),$$

where we recall our assumption that the maximums and minimums exist. The proof is left as an exercise.

Inequality (4.14) is called weak duality, and any value of the dual automatically bounds the primal from below and any value of the primal automatically bounds the dual from above. If the primal and dual problems exhibit the same value, then optimality is ensured. Unfortunately not all problems can be solved by establishing equality, but those that do are said to satisfy strong duality. While the lack of strong duality is a persistent nuisance in many fields of optimization, e.g. in combinatorial and integer programming, the important and widely applicable class of convex optimization problems is guaranteed to satisfy strong duality. In particular, the two important subclasses of linear and quadratic programming are discussed in a forthcoming section.

The optimization problems that set upper and lower bounds on the Lagrangian promote a general framework with which to solve constrained problems. Consider problem (4.11) to illustrate how a Lagrangian method might proceed. The Lagrangian of this problem is

$$\mathcal{L}(x,\lambda,\rho) = x_1 + x_2^2 + \sin(x_3) - \lambda\,x_1(1-x_1) - \begin{pmatrix}\rho_1\\\rho_2\end{pmatrix}^T \begin{pmatrix}-x_1^2 - x_2^2 + x_3\\1.2 - x_3\end{pmatrix}.$$

If we initiate with $\hat{\lambda} = 1$ and $\hat{\rho} = (1,1)^T$, then we can solve

$$\min_x \mathcal{L}(x,\hat{\lambda},\hat{\rho})$$

with any of the unconstrained algorithms of the previous section. We use gradient descent with a bisection line search initiated with $x = (1,1)^T$. The result is $\hat{x} = (0, 0, -1.5708)^T$, for which

$$f(\hat{x}) = -1.0,\ g(\hat{x}) = 0,\ h(\hat{x}) = (-1.5708, 2.7708)^T,\ \text{and } \mathcal{L}(\hat{x},\hat{\lambda},\hat{\rho}) = -2.2.$$

4.2 Constrained Optimization

Since we have solved the inner minimization problem for a fixed collection of Lagrange multipliers, we have calculated a lower bound on the dual problem. With regards to weak duality we currently have

$$-2.2 \leq \max_{\lambda, \rho \geq 0} \left(\min_x \mathcal{L}(x, \lambda, \rho) \right) \leq f(x^*) = \min_x \left(\max_{\lambda, \rho \geq 0} \mathcal{L}(x, \lambda, \rho) \right),$$

and hence the solution to the problem cannot drop below -2.2.

We have not yet solved our problem, a fact that follows from either of two observations. First, $\hat{x}$ is infeasible because $h(\hat{x}) \not\geq 0$, and hence we do not yet have a feasible solution. Second, the Lagrangian at $(\hat{x}, \hat{\lambda}, \hat{\rho})$ would have been equal to $f(\hat{x})$ if and only if we had had a feasible $\hat{x}$. The discrepancy between $\mathcal{L}(\hat{x}, \hat{\lambda}, \hat{\rho}) = -2.2$ and $f(\hat{x}) = -1$ means that we have not yet solved the problem.

There are numerous possible continuations as we proceed with a search for optimality. One option is to establish an upper bound by seeking a feasible solution. Calculating a feasible solution is regularly as difficult as solving the optimization problem itself, but for this problem we can readily construct a host of feasible solutions by inspection. After a few moments thought, a good candidate is $x' = (0, 0, 0)^T$, which is feasible and gives an objective value of $f(x') = 0$. The feasibility of x' ensures that

$$f(x^*) = \min_x \left(\max_{\lambda, \rho \geq 0} \mathcal{L}(x, \lambda, \rho) \right) \leq \max_{\lambda, \rho \geq 0} \mathcal{L}(x', \lambda, \rho) = f(x') = 0,$$

and hence,

$$-2.2 \leq \max_{\lambda, \rho \geq 0} \left(\min_x \mathcal{L}(x, \lambda, \rho) \right) \leq f(x^*) = \min_x \left(\max_{\lambda, \rho \geq 0} \mathcal{L}(x, \lambda, \rho) \right) \leq 0.$$

The question remains if we can select a better feasible solution for which the objective function falls below zero. The combined upper and lower bounds guarantee that the optimal value is somewhere between -2.2 and 0.

Instead of attempting to find a better feasible solution we could work to improve the lower bound by updating the Lagrange multipliers. If we fix $\hat{x}$, then we want to consider how we might update $\hat{\lambda}$ and $\hat{\rho}$ to increase our current lower bound of -2.2. The Lagrangian for $\hat{x}$ is

$$\mathcal{L}(\hat{x}, \lambda, \rho) = f(\hat{x}) - \lambda^T g(\hat{x}) - \rho^T h(\hat{x}) = -1 - \lambda \cdot 0 - \begin{pmatrix} \rho_1 \\ \rho_2 \end{pmatrix}^T \begin{pmatrix} -1.5708 \\ 2.7708 \end{pmatrix}.$$

The goal with respect to λ and ρ is to maximize the Lagrangian, and this suggests a search along $\nabla_{\lambda, \rho} \mathcal{L}(\hat{x}, \hat{\lambda}, \hat{\rho})$, i.e. use gradient ascent to maximize $\mathcal{L}$ with respect to the Lagrange multipliers. Since

$$\nabla_{\lambda,\rho}\mathcal{L}(\hat{x},\hat{\lambda},\hat{\rho}) = \begin{pmatrix} -g(\hat{x}) \\ -h(\hat{x}) \end{pmatrix} = \begin{pmatrix} 0 \\ 1.5708 \\ -2.7708 \end{pmatrix},$$

the suggested line search seeks to replace λ and ρ with

$$\begin{pmatrix} \hat{\lambda} \\ \hat{\rho} \end{pmatrix} + \alpha \begin{pmatrix} -g(\hat{x}) \\ -h(\hat{x}) \end{pmatrix} = \begin{pmatrix} 1 \\ 1 \\ 1 \end{pmatrix} + \alpha \begin{pmatrix} 0 \\ 1.5708 \\ -2.7708 \end{pmatrix}, \qquad (4.15)$$

where α is decided by the line search.

The steepest ascent iteration suggested in (4.15) is well intended, but it is notoriously difficult to use. The problem lies with the Lagrange multipliers for the constraints that are already satisfied. We argued earlier that the maximization of $\mathcal{L}$ over λ and ρ forces ρ_i to be zero if $h_i(\hat{x}) > 0$, and hence we suspect that ρ_2 should be zero since $h_2(\hat{x}) = 2.7708 > 0$. For this reason we set $\rho_2 = 0$ and select α according to the other components.

One interpretation of the Lagrange multipliers is that they penalize violations in feasibility. Indeed, recall that if any component of the constraints is violated, then the maximization over the multipliers returns an infinity. The goal of updating the multipliers is to provide just enough penalty to encourage the minimization of $\mathcal{L}$ over x to satisfy the constraints. However, what if a constraint is naturally satisfied for all optimal solutions? In this case there should be no penalty since none is needed to cajole the Lagrangian into satisfying the constraint. In our problem $h_2(\hat{x}) = 2.7708 > 0$, and hence we suspect that there is no need to penalize violations in $1.2 - x_3 \geq 0$. This rationale again suggests that ρ_2 should be zero.

We adapt gradient ascent so that ρ_i is set to zero if $h_i(\hat{x}) > 0$ and update the remaining Lagrange multipliers with $\alpha = 10^{-6}$. With each new collection of Lagrange multipliers we solve

$$\min_x \mathcal{L}(x, \lambda, \rho)$$

with gradient descent using a bisection line search initiated with our current value of x. The resulting solution x is then used to re-calculate the direction of steepest ascent for λ and ρ, i.e. $\nabla_{\lambda,\rho}\mathcal{L}$ is re-computed. The process terminates once the values of the Lagrangian and the objective function are within 10^{-6} of each other. This problem requires 27 iterations, some of which are tabulated in Table 4.3. The result is a numerical proof of the optimality of $x' = (0,0,0)^T$.

Lagrangian solution methods often require problem specific tailoring even with an adaption of gradient ascent to help decide the multipliers. One of the complications rests with the fact that solving the unconstrained problem $\min_x \mathcal{L}(x,\lambda,\rho)$ per iteration can be computationally challenging. The use of gradient descent on our small problem often terminated with a nonzero status, from which we know that the algorithm regularly failed to numeri-

4.2 Constrained Optimization

Initialized with $x = (1,1,1)^T$, $\lambda = 1$, and $\rho = (1,1)^T$			
Iteration	Best value of f(x)	g(x)	The minimum of $h_1(x)$ and $h_2(x)$
1	−0.0279*	0.0000	−0.0279
2	−0.0147*	0.0000	−0.0147
3	−0.0101*	0.0000	−0.0101
⋮			
26	−0.0001*	0.0000	−0.0001
27	0.0000	0.0000	0.0000

Table 4.3 Results of a Lagrangian method solving problem (4.11). The objective values denoted with a * indicate that x is infeasible, which is substantiated in the fourth column. The last iteration verifies the optimal value is zero

cally satisfy the first order condition $\nabla_x \mathcal{L}(x, \lambda, \rho) = 0$. Another complication rests with the gradient ascent step used to decide the new Lagrange multipliers. If the current x exhibits any violation, then gradient ascent will seek to increase the multipliers without bound. However, the goal is to set the Lagrange multipliers to assert just enough penalty to enforce the constraints. These two apposing goals can complicate convergence. The general theme is to use a small step as we did with $\alpha = 10^{-6}$. However, small steps lead to small adjustments, and the result could be a huge number of iterations if the multipliers require substantial updates. The general framework of a Lagrangian method is listed in Algorithm 16.

Algorithm 16 General pseudocode of a Lagrangian algorithm

$k = 1$
while unmet termination criteria **do**
    update λ and ρ, e.g. with a modified gradient ascent method
    solve the unconstrained problem $\min_x \mathcal{L}(x, \lambda, \rho)$ for x with an algorithm of choice
    let $k = k + 1$
end while
return x and $f(x)$

Saddle Points, KKT Conditions, and Constraint Qualifications

John von Neumann showed us how to use the relationship of weak duality to help characterize optimality in terms of saddle points.

Definition 2. The point $(\hat{u}, \hat{v})$ is a saddle point of $f(u, v)$ if $\hat{u}$ minimizes $f(u, \hat{v})$ and $\hat{v}$ maximizes $f(\hat{u}, v)$.

An illustration of a saddle point is in Fig. 4.2.

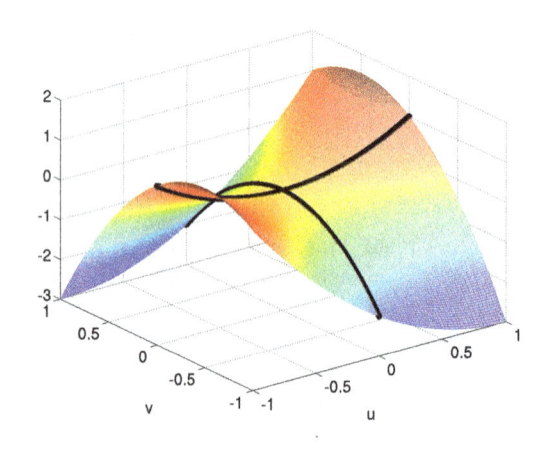

Fig. 4.2 The function $f(u, v) = u^2 + 2uv - 2v^2$. The saddle point is $(0, 0)$, and the black curves demonstrate the maximum and minimum qualities upon fixing one of u or v to be zero.

The weak duality property enforced by the Lagrangian naturally agrees with the concept of a saddle point because we want to minimize with respect to x and maximize with respect to (λ, ρ). We say that (x^*, λ^*, ρ^*) is a saddle point of $\mathcal{L}$ if

$$x^* \text{ minimizes } \mathcal{L}(x, \lambda^*, \rho^*), \text{ and}$$
$$(\lambda^*, \rho^*) \text{ maximizes } \mathcal{L}(x^*, \lambda, \rho).$$

The relationship between saddle points of the Lagrangian and optimal solutions of the primal problem was established by von Neumann.

Theorem 12. *Consider the optimization problem*

$$\min\{f(x) : g(x) = 0, \ h(x) \geq 0\}$$

and its associated Lagrangian $\mathcal{L}(x, \lambda, \rho)$. Then,

1. *If (x^*, λ^*, ρ^*) is a saddle point of $\mathcal{L}$, then x^* solves the (primal) optimization problem, and*

4.2 Constrained Optimization

2. The primal–dual point (x^*, λ^*, ρ^*) *is a saddle point of* $\mathcal{L}$, *minimizing in* x *and maximizing in* (λ, ρ), *if and only if*

   a. $g(x^*) = 0$ *and* $h(x^*) \geq 0$,
   b. $(\rho^*)^T h(x^*) = 0$ *with* $\rho^* \geq 0$, *and*
   c. x^* *solves* $\min_x \mathcal{L}(x, \lambda^*, \rho^*)$.

Theorem 12 establishes that saddle points of the Lagrangian solve the optimization problem, and moreover, saddle points are characterized by conditions 2a, 2b, and 2c. The power of these three conditions is that the first two can be verified directly, and the third is an unconstrained optimization problem, which gives us leverage in establishing necessary and sufficient conditions. Thus collectively the theorem motivates the translation of an optimization problem into an algebraic system.

Let us return to problem (4.12) to illustrate Theorem 12. The Lagrangian of this problem is

$$\mathcal{L}(x_1, x_2, \rho) = x_1 + x_2 - \rho(1 - x_1^2 - x_2^2),$$

where we have removed λ due to the absence of equality constraints. The first order necessary condition of the optimization problem in 2c requires $D_x \mathcal{L}(x, \lambda, \rho) = \langle 1 + 2\rho x_1, 1 + 2\rho x_2 \rangle = \langle 0, 0 \rangle$, which means that the necessary conditions of a saddle point are

$$1 - x_1^2 - x_2^2 \geq 0$$
$$\rho(1 - x_1^2 - x_2^2) = 0$$
$$1 + 2\rho x_1 = 0$$
$$1 + 2\rho x_2 = 0.$$

The 2nd equation implies either $\rho = 0$ or $1 - x_1^2 - x_2^2 = 0$. However, the 3rd and 4th equations are impossible if $\rho = 0$, and we conclude that $1 = x_1^2 + x_2^2$. From the 3rd and 4th equalities we have

$$\left. \begin{array}{r} x_1 = -1/2\rho \\ x_2 = -1/2\rho \end{array} \right\} \quad \Rightarrow \quad \left\{ \begin{array}{l} x_1^2 = 1/4\rho^2 \\ x_2^2 = 1/4\rho^2. \end{array} \right.$$

Hence $1/4\rho^2 + 1/4\rho^2 = 1/2\rho^2 = 1$, and $\rho = \pm 1/\sqrt{2}$. The negative value of ρ is not allowed, from which we know that $\rho = 1/\sqrt{2}$, and subsequently that $x_1 = x_2 = -1/\sqrt{2}$.

The optimality of the candidate solution $(-1/\sqrt{2}, -1/\sqrt{2})$ is not yet established because only conditions 2a and 2b have been verified. The necessary condition associated with 2c is also confirmed, but we still need to establish that our solution actually minimizes $\mathcal{L}$. Indeed, the first order necessary condition would have held if we had instead maximized $\mathcal{L}$. We can check to see if the second order sufficient condition indicates a minimum by calculating the Hessian of the Lagrangian, which is

$$D_x^2 \mathcal{L}(x_1, x_2, \rho) = \begin{bmatrix} 2\rho & 0 \\ 0 & 2\rho \end{bmatrix}.$$

The Hessian is positive (semi)definite for all $\rho > 0$ ($\rho \geq 0$), and we conclude that our candidate solution minimizes $\mathcal{L}$. We further comment that this conclusion is developed in our introduction to convexity momentarily. So
$$(x_1^*, x_2^*, \rho^*) = (-1/\sqrt{2}, -1/\sqrt{2}, 1/\sqrt{2})$$
is a saddle point of $\mathcal{L}$ and (x_1^*, x_2^*) solves problem (4.12).

We need to read Theorem 12 carefully so as not to make the mistake of thinking that all constrained problems can be solved by finding a saddle point of the Lagrangian. The theorem states that a saddle point of the Lagrangian suffices to establish optimality, but this statement leaves open the possibility that some problems might have solutions that are unassociated with saddle points. An example of such a problem is

$$\min\{-x_1 : -x_1^3 - x_2^2 \geq 0\}. \tag{4.16}$$

Since $x_2^2 \geq 0$, the smallest value of $-x_1$ is (obviously) 0, and the optimal solution is $(0, 0)$. The Lagrangian and its gradient are

$$\mathcal{L}(x_1, x_2, \rho) = -x_1 - \rho(-x_1^3 - x_2^2) \quad \text{and} \quad \nabla_x \mathcal{L}((x_1, x_2, \rho) = \begin{pmatrix} -1 + 3\rho x_1^2 \\ 2\rho x_2 \end{pmatrix}.$$

Since $\nabla_x \mathcal{L}(0, 0, \rho) = \langle -1, 0 \rangle$, we know that $(0, 0)$ is incapable of satisfying the necessary condition required by condition 2c of Theorem 12. Hence the only solution to this problem does not associate with a saddle point.

Saddle points can be theoretically guaranteed by establishing what is called a constraint qualification. There are many such qualifications, and any of them will establish the existence of a saddle point. In some cases one will work whereas others will not. How to choose an appropriate constraint qualification for any particular problem isn't obvious. Moreover, the lack of a saddle point can often be overcome by reformulating the model, which is the case for problem (4.16) if we replace x_1^3 with x_1 (but leave the objective as $-x_1$).

The classical constraint qualification of Lagrange applies to problems with equality constraints, and it requires $\nabla g(x^*)$ to have full row rank, where x^* is an optimal solution. This condition can be extended to inequalities by including the inequalities that hold with equality at x^*. While a study of constraint qualifications is of theoretical interest, they are of little practical use. The problem is that we first need to know an optimal solution to test a qualification, but if we already knew an optimal solution, then why would we need to test it? Indeed, validating a constraint qualification is generally ignored, and the aim is to develop algorithms that attempt to satisfy conditions 2a,

2b, and 2c of Theorem 12. If a technique converges, then we have a solution, but if not, then we might want to consider reformulating the model.

Condition 2c of Theorem 12 is often supplanted with its necessary condition of $\nabla_x \mathcal{L}(x, \lambda^*, \rho^*) = 0$. This replacement yields the Karush, Kuhn, Tucker (KKT) conditions.

Karush, Kuhn, Tucker Conditions

$$g(x) = 0, \ h(x) \geq 0$$
$$\rho^T h(x) = 0, \ \rho \geq 0$$
$$\nabla_x \mathcal{L}(x, \lambda, \rho) = 0.$$

Satisfying the KKT conditions is almost always the intent of a constrained optimization algorithm. Indeed, the Lagrangian algorithm in (Algorithm 16) seeks a solution to these equations. The first requirement is that x be feasible. The second is called complementarity, and it forces either ρ_i or $h_i(x)$ to be zero. Complementarity indicates that there is no need of a penalty if the corresponding constraint is naturally satisfied at optimality. The third condition is the necessary condition of the required optimality.

If we satisfy the KKT conditions, then we generally lack a proof of optimality. The Lagrangian techniques differ in this regard because they can work toward strong duality. The loss of a guarantee of optimality follows because satisfying $\nabla_x \mathcal{L}(x, \lambda, \rho) = 0$ does not necessarily mean that we have solved $\min_x \mathcal{L}(x, \lambda, \rho)$. So, while the KKT conditions favorably reduce the goal of optimization to an algebraic system of (in)equalities, they do so in a way that is neither necessary nor sufficient without additional structure. So the KKT conditions are not necessary without a constraint qualification as exemplified by problem (4.16), and they are insufficient because satisfying $\nabla_x \mathcal{L}(x, \lambda, \rho) = 0$ does not guarantee that x solves $\min_x \mathcal{L}(x, \lambda, \rho)$.

Here is an illustration of our ability to explain optimal solutions of constrained problems.

$$\boxed{\begin{array}{c}\text{A Saddle Point} \\ \text{of the} \\ \text{Lagrangian}\end{array}} \Rightarrow \boxed{\begin{array}{c}\text{An} \\ \text{Optimal} \\ \text{Solution}\end{array}} \Rightarrow \boxed{\begin{array}{c}\text{Satisfaction} \\ \text{of the} \\ \text{KKT Conditions}\end{array}} \qquad (4.17)$$

$$\uparrow$$
$$\boxed{\begin{array}{c}\text{Requires a Constraint Qualification} \\ \text{and differentiable properties}\end{array}}$$

The first implication can't generally be reversed since some optimal solutions do not associate with saddle points as demonstrated by problem (4.16). Moreover, the second implication is only conditionally valid and is not generally reversible.

The lack of reversible, or even guaranteed, implications provides a bit of a computational conundrum. Searching for a saddle point would guarantee optimality if the search were successful, but the search could unfortunately seek something that doesn't exist. It's a bit like knowing that a unicorn could end all suffering, and thus we should seek a unicorn. If we are fortunate enough to find one, well that would be great, but our search might otherwise continue forever. Similarly, if we compute a solution to the KKT conditions, then we know that our solution shares characteristics with an (assumed) optimal solution that also satisfies a constraint qualification. A brazen analogy here would be that the authors of this text might enjoy claiming that their shared love of mathematics with Carl Friedrich Gauss ensures an equally prodigious talent, but even the grandest of bravado's is shamed by such a falsehood.

Convexity and Its Consequences

There is thankfully a widely common and applicable collection of problems for which the KKT conditions are both necessary and sufficient. The property that places a problem in this class is called convexity, which is defined below.

Definition 3. A set X is convex if for any x and y in X we have

$$(1 - \theta)x + \theta y \in X,$$

where $0 \leq \theta \leq 1$. A real valued function f is convex if its epigraph,

$$\{(x, y) : y \geq f(x) \text{ for some } x\}$$

is convex.

The essence of convexity is that set membership is maintained along the line segment between any two members of the set. Examples of convex and nonconvex sets and functions are illustrated in Figs. 4.3, 4.4, 4.5 and 4.6.

The definition of convexity is set theoretical and requires no inference to topics from calculus. However, the property imposes analytic properties such as continuity and (near) differentiability. If we are willing to assume twice differentiable functions, then we can classify convexity in terms of a positive semidefinite Hessian. The definition of a positive semidefinite matrix is a relaxation of Definition 1.

Definition 4. The $n \times n$ symmetric matrix A is positive semidefinite, denoted $A \succeq 0$, if for any n-vector x we have $x^T A x \geq 0$.

The main difference between $A \succ 0$ and $A \succeq 0$ is that the latter permits zero eigenvalues, and hence a semidefinite matrix can map nonzero vectors to zero. So $A \succeq 0$ does not imply invertibility whereas $A \succ 0$ does. The definition of convexity equates with the weaker property of semidefiniteness.

4.2 Constrained Optimization

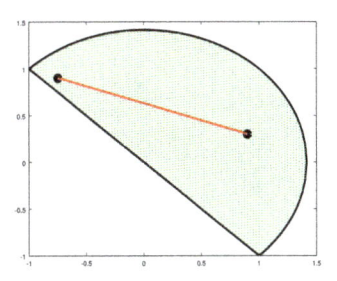

Fig. 4.3 An example of a convex set. The red line segment remains in the set for any two points of the set.

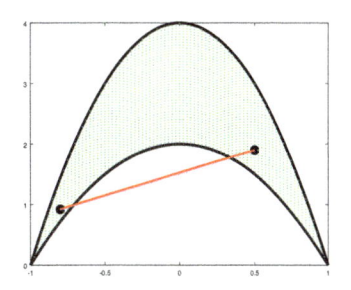

Fig. 4.4 An example of a nonconvex set. The red line segment leaves the set for some points in the set.

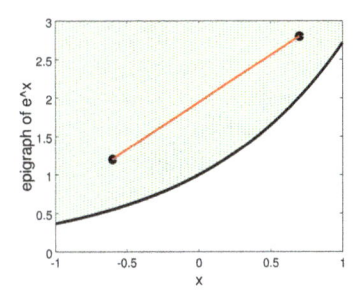

Fig. 4.5 An example of a convex function, e^x in this case. The red line segment remains in the epigraph for any two points in the epigraph.

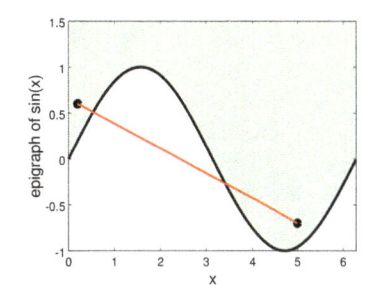

Fig. 4.6 An example of a nonconvex function, $\sin(x)$ in this case. The red line segment leaves the epigraph for some points in the epigraph.

Theorem 13. *A twice continuously differentiable function f is convex if and only if its Hessian $D_x^2 f(x)$ is positive semidefinite.*

Calculus courses regularly refer to a function being "concave up," i.e. convex, if its second derivative is positive, and Theorem 13 extends this idea to higher dimensional problems.

The optimization problem

$$\min f(x) \text{ such that } g(x) = 0, \; h(x) \geq 0$$

is convex if $f(x)$ is convex, $g(x)$ is affine, and $h(x)$ is concave. The function g being affine means that $g(x) = Ax - b$ for some matrix A and vector b. So convex optimization problems require linear equality constraints. A function is concave if its negative, in this case $-h(x)$, is convex. The definition is componentwise, and if $h(x)$ is a p-vector, then we are imposing that each $h_i(x)$ be concave by stating that $h(x)$ is concave.

For convex optimization problems we have for any vector v that

$$v^T \left(D_x^2 \mathcal{L}(x, \lambda, \rho)\right) v = v^T \left(D_x^2 (f(x) - \lambda^T g(x) - \rho^T h(x))\right) v$$
$$= v^T \left(D_x^2 f(x) - \sum_{i=1}^m \lambda_i D_x^2 g_i(x) - \sum_{i=1}^p \rho_i D_x^2 h_i(x)\right) v$$
$$= v^T D_x^2 f(x) v - \sum_{i=1}^p \rho_i \left(v^T D_x^2 h_i(x) v\right).$$

The second derivatives of g are zero because g is affine. The convexity of f ensures that $v^T D_x^2 f(x) v \geq 0$, and the concavity of h ensures that $v^T D_x^2 h_i(x) v \leq 0$ for all i. Since $\rho_i \geq 0$, we conclude that

$$v^T \left(D_x^2 \mathcal{L}(x, \lambda, \rho)\right) v \geq 0,$$

and it follows that the Lagrangian of a convex optimization problem is a convex function in x.

The reason we are interested in convexity is that it assures many favorable qualities with regards to optimization. In particular, the first order necessary condition is sufficient if the function being minimized is convex. We conclude that the convexity with respect to x of the Lagrangian of a convex problem consequently implies that condition 2c of Theorem 12, which is that x^* solves $\min_x \mathcal{L}(x, \lambda^*, \rho^*)$, can be replaced with the first order condition $\nabla_x \mathcal{L}(x^*, \lambda^*, \rho^*) = 0$. The result of this substitution is that the KKT conditions are necessary and sufficient to identify a saddle point of the Lagrangian, and hence, for convex problems there is an implication from the far right block to the far left block in the schematic depicted in (4.17). This fact nearly reverses the implications of this diagram, with the only remaining hindrance being the constraint qualification. This hindrance is also overcome by convexity, although the argument to this fact surpasses the scope of this text. Theorem 14 states some of the properties of convex problems.

Theorem 14. *Suppose the optimization problem*

$$\min f(x) \text{ such that } g(x) = 0, \ h(x) \geq 0$$

is convex, where all functions are assumed to be smooth. Then,

- *The KKT conditions are necessary and sufficient for optimality.*
- *The solution set is convex.*
- *Any locally optimal solution is globally optimal.*

Optimization problems with only equality constraints nicely lend themselves to the algebraic structure of the KKT conditions. In this case the conditions reduce to

$$\left. \begin{array}{r} g(x) = 0 \\ \nabla_x \mathcal{L}(x, \lambda) = 0 \end{array} \right\} \Leftrightarrow \left\{ \begin{array}{r} g(x) = 0 \\ \nabla_x f(x)^T = \lambda^T \nabla g(x). \end{array} \right.$$

4.2 Constrained Optimization

If the optimization problem is convex, then solving the system on the right guarantees an optimal solution. The equality $\nabla_x f(x) = \lambda^T \nabla g(x)$ shows that the gradient of the objective function is a linear combination of the gradients of the equality constraints at optimality.

Consider the following convex problem as an example,

$$\min_x x_1^2 + x_2^4 + e^{x_3} \text{ such that } x_1 + x_2 + x_3 = 1, \; x_1 - x_2 - x_3 = 10.$$

The linear constraints in matrix form are

$$g(x) = \begin{bmatrix} 1 & 1 & 1 \\ 1 & -1 & -1 \end{bmatrix} \begin{pmatrix} x_1 \\ x_2 \\ x_3 \end{pmatrix} - \begin{pmatrix} 1 \\ 10 \end{pmatrix} = \begin{pmatrix} 0 \\ 0 \end{pmatrix},$$

and the gradients of the f and g are

$$\nabla f(x) = \begin{pmatrix} 2x_1 \\ 4x_2^3 \\ e^{x_3} \end{pmatrix} \quad \text{and} \quad \nabla g(x) = \begin{bmatrix} 1 & 1 & 1 \\ 1 & -1 & -1 \end{bmatrix}.$$

The KKT conditions are the nonlinear system

$$\begin{bmatrix} x_1 + x_2 + x_3 - 1 \\ x_1 - x_2 - x_3 - 10 \\ 2x_1 - \lambda_1 - \lambda_2 \\ 4x_2^3 - \lambda_1 + \lambda_2 \\ e^{x_3} - \lambda_1 + \lambda_2 \end{bmatrix} = \begin{pmatrix} 0 \\ 0 \\ 0 \\ 0 \\ 0 \end{pmatrix},$$

where the top two equations ensure feasibility and the last three ensure that $\nabla_x \mathcal{L}(x, \lambda) = 0$. If we solve this system with Newton's method initiated with $x = (1, 1, 1)^T$ and $\lambda = (1, 1)^T$, then convergence is reached in ten iterations with a tolerance of 10^{-8}. The resulting solution is

$$x^* = (5.50000, 0.13440, -4.63440)^T \quad \text{and} \quad \lambda^* = (5.50486, 5.49514)^T.$$

The solution is guaranteed to be globally optimal due to the convexity of the optimization problem. This example highlights that convex optimization problems with only equality constraints can be directly solved by calculating a solution to a system of equations.

4.2.1 Linear and Quadratic Programming

Numerous applications reside in the areas of linear and quadratic programming, and these sub-fields warrant special attention. We have already seen that linear regression is a widely used application of quadratic optimization in Sect. 3.2, and applications in finance and radiotherapy are developed in Part II of this text.

We consider linearly constrained problems of the form

$$\min f(x) \text{ such that } Ax = b, \ x \geq 0,$$

where f is either linear or quadratic, i.e.

linear		quadratic
$f(x) = c^T x$	or	$f(x) = \frac{1}{2} x^T Q x + c^T x.$

We assume A is an $m \times n$ matrix, and we further assume $Q \succeq 0$ to ensure convexity, although please note that the class of quadratic programs more broadly includes non-convex quadratics and, potentially, quadratic constraints. The reason for factoring $1/2$ from Q is to easily express $\nabla f(x)$ as $Qx + c$ in the quadratic case.

The KKT conditions are both necessary and sufficient for linear and convex quadratic programs. In both instances the conditions of feasibility and complementarity are the same,

feasibility		complementarity
$Ax = b, \ x \geq 0$	and	$\rho^T x = 0, \ \rho \geq 0.$

The third condition of

$$\nabla_x \mathcal{L}(x, \lambda, \rho) = \nabla f(x) - A^T \lambda - \rho = 0,$$

is different between the two cases and becomes

linear program		quadratic program
$c - A^T \lambda - \rho = 0$	and	$Qx + c - A^T \lambda - \rho = 0.$

Traditional notation replaces λ with y and ρ with s if studying linear or quadratic programs, and the KKT conditions with these replacements are:

	linear program		quadratic program
primal feasibility:	$Ax = b, \ x \geq 0$		$Ax = b, \ x \geq 0$
dual feasibility:	$A^T y + s = c, \ s \geq 0$	and	$A^T y + s = Qx + c, \ s \geq 0$
complementarity:	$x^T s = 0$		$x^T s = 0.$

The primal–dual terminology of linear programming can be motivated by recalling the weak duality relationship, which for linear programming is

4.2 Constrained Optimization

$$\min\{c^T x : Ax = b,\ x \geq 0\} = \min_x \left(\max_{y,\ s \geq 0} c^T x - y^T(Ax - b) - s^T x \right)$$

$$\geq \max_{y,\ s \geq 0} \left(\min_x c^T x - y^T(Ax - b) - s^T x \right)$$

$$= \max_{y,\ s \geq 0} \left(\min_x (c - A^T y - s)^T x + b^T y \right)$$

$$= \max\{b^T y : A^T y + s = c,\ s \geq 0\}.$$

The last equality follows because the inner minimization is negative infinity for any y and s for which a single component of $c - A^T y - s$ is nonzero. Recall that the first linear program is called the primal problem and that the last linear program is called the dual, and together they form the primal–dual pair,

primal linear program		dual linear program	
$\min_x c^T x$		$\max_x b^T y$	
such that	and	such that	(4.18)
$Ax = b,\ x \geq 0$		$A^T y + s = c,\ s \geq 0.$	

So the dual of a linear program is itself a linear program.

Much of the duality theorem of linear programming is confirmed by weak duality. Each of the primal and dual has three possible outcomes. A linear program can be unbounded if, for example, the primal objective $c^T x$ can decrease without bound for a sequence of primal feasible solutions. A somewhat trivial example is

$$\min -x_1 \text{ such that } x_1 = x_2,\ x_1 \geq 0,\ x_2 \geq 0.$$

A second possibility is for a linear program to be infeasible, meaning that there is no nonnegative x such that $Ax = b$. A trivial example in this case would be to have constraints requiring $x_1 = 1$ and $x_1 = 2$. The third, and canonical, option is for a linear program to have a finite solution, and in this case the necessity and sufficiency of the KKT conditions mandate that the dual has an optimal solution with the primal and dual objectives having the same value. In this third option there is no duality gap. The duality theorem of linear programming is stated below.

Theorem 15. *Exactly one of the following possibilities holds for the primal and dual linear programs in* (4.18)*:*

- *Both the primal and dual have optimal solutions, say x^* for the primal and y^* and s^* for the dual, so that $c^T x^* = b^T y^*$.*
- *The primal is unbounded, which subsequently ensures that the dual is infeasible.*

- *The dual is unbounded, which subsequently ensures that the primal is infeasible.*
- *Both the primal and dual are infeasible.*

The second and third possibilities of Theorem 15 follow directly from weak duality. For example, if the primal is unbounded, then it is impossible for the dual to be feasible since any feasible solution would establish a lower bound on the primal. The first possibility of the theorem is nearly established by the necessity and sufficiency of the KKT conditions. What is missing is a theoretical argument that if both the primal and the dual are feasible, then both have optimal solutions. Nonetheless, the result is true and is suggested by the primal and dual relationship. The equality of the primal and dual objective values guarantees strong duality as long as one of the problems has an optimal solution. The last case is mathematically possible but is almost never, if ever, observed in practice.

Solution Algorithms for Linear and Quadratic Programming

Simplex methods are famous algorithms to solve linear programs. These methods were invented in the late 1940s by the "Father of Linear Programming," George Dantzig. Simplex algorithms have long been stalwarts of linear programming and its sister studies of combinatorial and integer programming, and they continue to play important roles in both theory and practice. The practical efficiency of a carefully implemented simplex algorithm is impressive, and several studies have indicated that its run time is approximately linear in the number of constraints over a wide range of applications. The theoretical efficiency is less impressive, and all known simplex procedures are exponential in the size of A in the worst case due to variants of a contrived example that doesn't practically arise. We do not develop a simplex algorithm here because they principally rely on combinatorial aspects, and hence, they do not immediately align with our perspective on solving general optimization problems. Interested readers should consider taking a course in operations research or linear programming.

The theoretical nuisance of the simplex method's possible exponential run time motivated the question of whether there was an algorithm with an improved worst case complexity. The question was answered in the affirmative by the Russian mathematician Leonid Khachiyan in 1979, and his seminal work spawned a multi-decade study into algorithms whose worst case complexity bested that of the simplex algorithm. Maybe the most important work was that of Narendra Karmarkar, who first demonstrated both practical and theoretical improvements over the simplex method in the 1980s. Some of the most successful algorithms are called interior-point methods, and it is their development that we outline.

4.2 Constrained Optimization

All interior-point methods relax the complementarity constraint, and most use Newton's method to search for optimality as the relaxation becomes increasingly stringent. We start by recognizing that complementarity and nonnegativity, which are

$$x^T s = x_1 s_1 + x_2 s_2 + \ldots + x_n s_n = 0, \ x \geq 0, \text{ and } s \geq 0,$$

hold if and only if

$$x \geq 0, \ s \geq 0, \text{ and } x_i s_i = 0 \text{ for all } i = 1, 2, \ldots, n.$$

So, if we let X be the diagonal matrix of x, then the complementarity condition can be re-expressed as

$$Xs = \begin{bmatrix} x_1 & 0 & 0 & \cdots & 0 \\ 0 & x_2 & 0 & \cdots & 0 \\ 0 & 0 & x_3 & \cdots & 0 \\ \vdots & \vdots & \vdots & \ddots & \vdots \\ 0 & 0 & 0 & \cdots & x_n \end{bmatrix} \begin{pmatrix} s_1 \\ s_2 \\ s_3 \\ \vdots \\ s_n \end{pmatrix} = \begin{pmatrix} x_1 s_1 \\ x_2 s_2 \\ x_3 s_3 \\ \vdots \\ x_n s_n \end{pmatrix} = \begin{pmatrix} 0 \\ 0 \\ 0 \\ \vdots \\ 0 \end{pmatrix} = 0.$$

The relaxation of an interior-point method assumes

$$Xs = \mu e \text{ instead of } Xs = 0,$$

where $\mu > 0$ is a scalar measuring how close we are to optimality. Notice that $Xs = \mu e$ with $\mu > 0$ requires that x_i and s_i agree in sign for each i, and since nonnegativity is required, we assume that $x > 0$ and $s > 0$. This strict positivity is what provides the algorithm's name because it ensures that iterates are in the strict interior of the feasible region. The general theme of an interior-point method is to partially solve the KKT conditions for some positive μ, reduce μ, and then repeat. So the algorithm approaches optimality by reducing μ toward zero.

Relaxing complementarity gives

$$\underline{\text{linear program}} \qquad\qquad \underline{\text{quadratic program}}$$

$$\begin{pmatrix} Ax - b \\ A^T y + s - c \\ Xs - \mu e \end{pmatrix} = \begin{pmatrix} 0 \\ 0 \\ 0 \end{pmatrix} \text{ and } \begin{pmatrix} Ax - b \\ A^T y + s - Qx - c \\ Xs - \mu e \end{pmatrix} = \begin{pmatrix} 0 \\ 0 \\ 0 \end{pmatrix}.$$

In both cases we can use Newton's method to define a search direction. The Jacobians with respect to x, y, and s are

linear program	quadratic program
$\begin{bmatrix} A & 0 & 0 \\ 0 & A^T & I \\ S & 0 & X \end{bmatrix}$ and	$\begin{bmatrix} A & 0 & 0 \\ -Q & A^T & I \\ S & 0 & X \end{bmatrix}$.

The system defining the search direction for a linear program is

$$\begin{bmatrix} A & 0 & 0 \\ 0 & A^T & I \\ S & 0 & X \end{bmatrix} \begin{pmatrix} \Delta x \\ \Delta y \\ \Delta s \end{pmatrix} = - \begin{pmatrix} Ax - b \\ A^T y + s - c \\ Xs - \mu e \end{pmatrix}, \qquad (4.19)$$

and the system for a quadratic program is

$$\begin{bmatrix} A & 0 & 0 \\ -Q & A^T & I \\ S & 0 & X \end{bmatrix} \begin{pmatrix} \Delta x \\ \Delta y \\ \Delta s \end{pmatrix} = - \begin{pmatrix} Ax - b \\ A^T y + s - Qx - c \\ Xs - \mu e \end{pmatrix}. \qquad (4.20)$$

While search directions can be calculated directly from these defining systems, some algebra leads to smaller systems that improve computational speed and accuracy. If we multiply the middle equation of (4.19) by X, then

$$\begin{aligned} XA^T \Delta y &= X(c - A^T y - s) - X\Delta s \\ &= X(c - A^T y - s) + S\Delta x - \mu e + Xs, \end{aligned}$$

where the second equality follows from the third equation of (4.19). If we multiply on the left by AS^{-1}, then

$$\begin{aligned} AS^{-1}XA^T \Delta y &= AS^{-1}X(c - A^T y - s) + AS^{-1}S\Delta x - \mu AS^{-1}e + AS^{-1}Xs \\ &= AS^{-1}X(c - A^T y - s) + A\Delta x - \mu AS^{-1}e + AS^{-1}Sx \\ &= AS^{-1}X(c - A^T y - s) + b - Ax - \mu AS^{-1}e + Ax \\ &= AS^{-1}X(c - A^T y - s) + b - \mu AS^{-1}e, \end{aligned}$$

where the third equality follows from the first equation of (4.19).

Let $\sqrt{S^{-1}X}$ be the positive diagonal matrix whose diagonal elements are the square roots of the diagonal of $S^{-1}X$. Then,

$$AS^{-1}XA^T = \left(A\sqrt{S^{-1}X}\right)\left(\sqrt{S^{-1}X}A^T\right) = \left(\sqrt{S^{-1}X}A^T\right)^T \left(\sqrt{S^{-1}X}A^T\right),$$

from which we have that $AS^{-1}XA^T \succ 0$ as long as the columns of $\sqrt{S^{-1}X}A^T$ are linearly independent. The columns of A^T are the rows of A, and multiplication by the invertible matrix $S^{-1}X$ does not affect rank. So the positive definite property holds as long as A has full row rank. This rank condition is mathematically tacit since if the rows of A are not linearly independent, then

4.2 Constrained Optimization

we could row reduce the augmented matrix $[A \mid b]$ and replace A and b with their reduced counterparts. There are of course possible numerical concerns with this reduction, but for now we assume that A has full row rank.

We can solve the positive definite system

$$AS^{-1}XA^T \Delta y = AS^{-1}X(c - A^T y - s) + b - \mu AS^{-1}e \qquad (4.21)$$

for Δy with, for instance, an inner product form of the Cholesky factorization. We can then calculate Δx and Δs with

$$\Delta s = c - A^T y - s - A^T \Delta y \quad \text{and} \quad \Delta x = S^{-1}\left(\mu e - Xs - X\Delta s\right).$$

Reducing (4.19) to first solve Eq. (4.21) for Δy provides substantial computational gains, as we only need to solve one $m \times m$ positive definite system per iteration. System (4.19) is instead $(2n + m) \times (2n + m)$.

A reasonable concern of the smaller systems is the regular appearance of S^{-1}. In general we do not want to calculate matrix inverses due to concerns with accuracy and speed. However, the use of S^{-1} in these calculations presents little, if any, concern because the matrix S^{-1} is multiplied either by X or by μe in all cases. As the algorithm converges the iterates satisfy

$$Xs \approx \mu e \quad \Leftrightarrow \quad Sx \approx \mu e \quad \Leftrightarrow \quad x \approx \mu S^{-1}e.$$

So the vector $\mu S^{-1}e$ approaches x in the cases in which S^{-1} is multiplied by μe. The cases in which S^{-1} appears as $S^{-1}X$ require a more advanced study to dismiss concerns, but the result is that the computational errors in Δy caused by the ill-conditioning of $S^{-1}X$ are uniformly bounded and most often negligible. In all cases S^{-1} need not, and should not, be calculated as an entire matrix, and calculations should instead be conducted with the vectors x and s. For example, Xs and $S^{-1}X$ in MATLAB and Octave are, respectively, `x./s` and `diag(x./s)`.

Reductions of system (4.20) in the quadratic case have fewer benefits than the linear case. One option is to reduce the system to

$$\begin{bmatrix} A & 0 \\ Q + SX^{-1} & -A^T \end{bmatrix} \begin{pmatrix} \Delta x \\ \Delta y \end{pmatrix} = \begin{pmatrix} b - Ax \\ A^T y + \mu X^{-1}e - Qx - c \end{pmatrix},$$

and then solve for Δx and Δy with a solver of choice. The direction Δs is then calculated as

$$\Delta s = Qx + c - A^T y - s - A^T \Delta y + Q \Delta x.$$

Once Δx, Δy, and Δs are calculated for a linear or quadratic model, we proceed with a Newton step that ensures the nonnegativity restrictions. This means that we want to decide positive scalars α_x and α_s so that

$$x + \alpha_x \Delta x \geq 0 \text{ and } s + \alpha_s \Delta s \geq 0.$$

A typical rule is to step 95% of the maximum allowed distance or to use the full Newton step if there is no restriction. This rule sets

$$\alpha_x = \begin{cases} 0.95 \min\{x_i/|\Delta x_i| : \Delta x_i < 0\} & \text{if } \Delta x \not\geq 0 \\ 1 & \text{if } \Delta x \geq 0 \end{cases}$$

and

$$\alpha_s = \begin{cases} 0.95 \min\{s_i/|\Delta s_i| : \Delta x_i < 0\} & \text{if } \Delta s \not\geq 0 \\ 1 & \text{if } \Delta s \geq 0. \end{cases}$$

The new values of x, y, and s are

$$x + \alpha_x \Delta x, \ y + \alpha_s \Delta y, \ \text{and} \ s + \alpha_s \Delta s.$$

The last algorithmic question is how to update μ so that progress toward optimality is continued. The new iterate could be close to optimal, so the update of μ should depend on the new values of x, y, and s. Otherwise we might decrease μ insufficiently, in which case the next iteration would degrade our progress. A common update is to set

$$\mu = \frac{x^T s}{n^2}.$$

If the new iterate exactly satisfies $Xs = \mu e$ for the previous μ, then $x^T s = n\mu$. So while this update appears to be a division by n^2, it is more like a division of the previous μ by n. This update can be overly aggressive for large n, in which case n^2 is replaced with something like $n\sqrt{n}$.

Example algorithms for linear and quadratic programming are listed in Algorithms 17 and 18. There are numerous variations, and an algorithm's performance can vary depending on the linear solver used to calculate the Newton direction, the step size, and the decrease in μ. Most solvers allow the user to adjust various parameters and settings, and if convergence is difficult with default settings, then adjustments often lead to improvements.

We illustrate how Algorithms 17 and 18 solve a problem in the petroleum industry. Suppose a refinery can purchase four types of crude, two of which are sweet with less than 0.5% sulfur, and the other two are sour with greater than 0.5% sulfur. The sulfur contents, octane ratings, and costs per barrel are found in Table 4.2.1. The crude oils must be blended to make two products with the characteristics in Table 4.5.

We let x_{ij} be the number of barrels of crude type j to use in product blend i. The sulfur restriction on the first product blend requires

4.2 Constrained Optimization

Algorithm 17 An example interior point method for linear programming

$k = 1$
while $x^T s > \delta$ **do**
  Set $\mu = x^T s / n^2$.
  Solve the positive definite system
  $$AS^{-1}XA^T \Delta y = AS^{-1}X(c - A^T y - s) + b - \mu AS^{-1}e.$$
  Set
  $$\Delta s = c - A^T y - s - A^T \Delta y, \text{ and}$$
  $$\Delta x = S^{-1}(\mu e - Xs - X\Delta s).$$
  Set
  $$\alpha_x = \begin{cases} 0.95 \min\{x_i/|\Delta x_i| : \Delta x_i < 0\}, & \text{if } \Delta x \not\geq 0 \\ 1, & \text{if } \Delta x \geq 0, \end{cases}$$
  and
  $$\alpha_s = \begin{cases} 0.95 \min\{s_i/|\Delta s_i| : \Delta s_i < 0\}, & \text{if } \Delta s \not\geq 0 \\ 1, & \text{if } \Delta s \geq 0. \end{cases}$$
  Update x, y, and s with
  $$x = x + \alpha_x \Delta x$$
  $$y = y + \alpha_s \Delta y$$
  $$s = s + \alpha_s \Delta s.$$
  Let $k = k + 1$
end while
return x, y, and s.

	Sweet crude		Sour crude	
Crude index	1	2	3	4
Sulfur	0.2%	0.15%	1.5%	4.5%
Octane	81	93	78	72
$/barrel	68.53	82.31	56.81	54.25

Table 4.4 Characteristics of four different crude oils.

Product index	Minimum octane	Maximum sulfur	Demand
1	85	1.0%	5000
2	80	1.5%	15,000

Table 4.5 Characteristics of the products that are to be produced from the crude oils described in Table 4.2.1.

Algorithm 18 An example interior point method for quadratic programming

$k = 1$
while $x^T s > \delta$ **do**
    Set $\mu = x^T s/n^2$.
    Solve the system
$$\begin{bmatrix} A & 0 \\ Q + SX^{-1} & -A^T \end{bmatrix} \begin{pmatrix} \Delta x \\ \Delta y \end{pmatrix} = \begin{pmatrix} b - Ax \\ A^T y + \mu X^{-1} e - Qx - c \end{pmatrix}.$$
    Set
$$\Delta s = Qx + c - A^T y - s + Q\Delta x - A^T \Delta y.$$
    Set
$$\alpha_x = \begin{cases} 0.95 \min\{x_i/|\Delta x_i| : \Delta x_i < 0\}, & \text{if } \Delta x \not\leq 0 \\ 1, & \text{if } \Delta x \leq 0, \end{cases}$$
    and
$$\alpha_s = \begin{cases} 0.95 \min\{s_i/|\Delta s_i| : \Delta s_i < 0\}, & \text{if } \Delta s \not\leq 0 \\ 1, & \text{if } \Delta s \leq 0. \end{cases}$$
    Update x, y, and s with
        $x = x + \alpha_x \Delta x$
        $y = y + \alpha_s \Delta y$
        $s = s + \alpha_s \Delta s$.
    let $k = k + 1$
end while
return x, y, and s.

$$\frac{0.2x_{11} + 0.15x_{12} + 1.5x_{13} + 4.5x_{14}}{x_{11} + x_{12} + x_{13} + x_{14}} \leq 1,$$

which we rewrite as

$$-0.80x_{11} - 0.85x_{12} + 0.50x_{13} + 3.50x_{14} \leq 0.$$

The octane requirement of the first product is

$$\frac{81x_{11} + 93x_{12} + 78x_{13} + 72x_{14}}{x_{11} + x_{12} + x_{13} + x_{14}} \geq 85,$$

which is the same as

$$-4.0x_{11} + 8.0x_{12} - 7.0x_{13} - 13.0x_{14} \geq 0.$$

The analogous constraints for the second product are

$$-1.30x_{21} - 1.35x_{22} + 0.00x_{23} + 1.00x_{24} \leq 0.$$

and

$$1.0x_{21} + 13.0x_{22} - 2.0x_{23} - 8.0x_{24} \geq 0.$$

4.2 Constrained Optimization

The constraints need to be modeled as $Ax = b$ with $x \geq 0$. We let the variable vector x be

$$x = (x_{11}, x_{12}, x_{13}, x_{14}, x_{21}, x_{22}, x_{23}, x_{24}, s_1, s_2, s_3, s_4)^T,$$

where s_1 and s_2 are slack variables and s_3 and s_4 are surplus variables. These additional variables account for any discrepancies in the inequalities. If we let

$$A = \begin{bmatrix} -0.8 & -0.85 & 0.5 & 3.5 & 0.0 & 0.00 & 0.0 & 0.0 & 1 & 0 & 0 & 0 \\ 0.0 & 0.00 & 0.0 & 0.0 & -1.3 & -1.35 & 0.0 & 1.0 & 0 & 1 & 0 & 0 \\ -4.0 & 8.00 & -7.0 & -13.0 & 0.0 & 0.00 & 0.0 & 0.0 & 0 & 0 & -1 & 0 \\ 0.0 & 0.00 & 0.0 & 0.0 & 1.0 & 13.00 & -2.0 & -8.0 & 0 & 0 & 0 & -1 \\ 1.0 & 1.00 & 1.0 & 1.0 & 0.0 & 0.00 & 0.0 & 0.0 & 0 & 0 & 0 & 0 \\ 0.0 & 0.00 & 0.0 & 0.0 & 1.0 & 1.00 & 1.0 & 1.0 & 0 & 0 & 0 & 0 \end{bmatrix}$$

and

$$b = (0, 0, 0, 0, 5000, 15000)^T,$$

then our constraints are $Ax = b$ with $x \geq 0$. The top two rows of A express our sulfur limitations, the second two mandate our minimum octane requirements, and the last two establish supply.

The objective coefficients are

$$c = (68.53, 82.31, 56.81, 54.25, 68.53, 82.31, 56.81, 54.25, 0, 0, 0, 0)^T,$$

which completes the model as

$$\min_x c^T x \text{ such that } Ax = b, \ x \geq 0.$$

Algorithm 17 converges in 15 iterations if x is initiated with $1000e$, and y and s are initiated with e, assuming a convergence tolerance of 10^{-8}. The reason x is initialized with $1000e$ instead of just e is because we know that thousands of barrels will be needed to meet demand, and hence, $1000e$ is likely a better, but still infeasible, guess at a solution. The algorithm requires an additional six iterations if x is instead initialized with e. The first ten iterations in Table 4.6 show that convergence to about four decimal places is reached at about the same time as the primal and dual iterates become feasible.

The optimal solution from Algorithm 17 creates product 1 with 2333.33 barrels of crude 2 and 2666.67 barrels of crude 3. This produces 5000 barrels with a sulfur content of 0.87% and an octane rating of 85. Product 2 is created with 2000 barrels of crude 2 and 13,000 barrels of crude 4. The result is 15,000 barrels with a sulfur content of 1.32% and an octane rating of 80. The total cost is $1,246,700.00.

Iteration	Gap ($x^T s$)	μ	$\|Ax - b\|_\infty$	$\|c - A^T y - s\|_\infty$
1	10608.1100	1000.0000	10811.6746	64.4993
2	7791.6155	73.6674	9627.4187	55.3529
3	5034.0850	54.1084	7460.3933	40.9628
4	2930.9356	34.9589	5667.0933	22.4316
5	800.8621	20.3537	1893.9450	4.8717
6	102.6245	5.5615	86.1408	0.0720
7	9.2184	0.7127	2.9208	0.0016
8	0.7267	0.0640	0.1607	0.0000
9	0.0462	0.0050	0.0121	0.0000
10	0.0026	0.0003	0.0008	0.0000

Table 4.6 The first ten iterations of Algorithm 17 as it solves the petroleum blending problem. The columns headed with $\|Ax - b\|_\infty$ and $\|c - A^T y - s\|_\infty$ contain the maximum violations in primal and dual feasibility.

Suppose management desires to blend the different types of crude as uniformly as possible to help mitigate concerns of purchasing from a small number of suppliers. If possible, management would like the solution to be $\hat{x}_{1j} = 5000/4 = 1250$ and $\hat{x}_{2j} = 15{,}000/4 = 3750$. However, this solution is infeasible. A reasonable adaptation of our previous model would be to solve

$$\min_x \sum_{ij} (x_{ij} - \hat{x}_{ij})^2 \text{ such that } Ax = b,\ x \geq 0, \qquad (4.22)$$

which would find a nearest feasible solution to the desired infeasible goal of management.

If we let

$$Q = \begin{bmatrix} 2 & 0 & 0 & \cdots & 0 & 0 & 0 & 0 & 0 \\ 0 & 2 & 0 & \cdots & 0 & 0 & 0 & 0 & 0 \\ 0 & 0 & 2 & \cdots & 0 & 0 & 0 & 0 & 0 \\ \vdots & \vdots & & \ddots & & \vdots & \vdots & \vdots & \vdots \\ 0 & 0 & 0 & \cdots & 2 & 0 & 0 & 0 & 0 \\ \hline 0 & 0 & 0 & \cdots & 0 & 0 & 0 & 0 & 0 \\ 0 & 0 & 0 & \cdots & 0 & 0 & 0 & 0 & 0 \\ 0 & 0 & 0 & \cdots & 0 & 0 & 0 & 0 & 0 \end{bmatrix} \text{ and } c = \begin{pmatrix} -2\hat{x}_{11} \\ \vdots \\ -2\hat{x}_{14} \\ -2\hat{x}_{21} \\ \vdots \\ -2\hat{x}_{24} \\ 0 \\ \vdots \\ 0 \end{pmatrix},$$

then

4.2 Constrained Optimization

$$\sum_{ij}(x_{ij} - \hat{x}_{ij})^2 = \frac{1}{2}x^T Q x + c^T x + \sum_{ij}\hat{x}_{ij}^2.$$

The last term is constant and can be removed from the optimization problem. The resulting quadratic problem is

$$\min_{x} \frac{1}{2}x^T Q x + c^T x \text{ such that } Ax = b, \ x \geq 0.$$

The Hessian of the objective is Q, which is positive semidefinite but not positive definite. The problem is nonetheless convex.

Algorithm 18 converges in 17 iterations to a gap tolerance of 10^{-8} if initiated like the linear case. Table 4.7 lists the first 12 iterations, the last of which achieves convergence to two decimal places. The solution and its characteristics are below.

	Crude 1	Crude 2	Crude 3	Crude 4	Total	Sulfur	Octane
Product 1	1250.00	2275.64	993.59	480.77	5000	0.85%	85.0
Product 2	3895.90	3901.16	3759.20	3443.74	15,000	1.5%	81.3

The cost of this production blend is \$1,343,971.53, which is \$97,271.53 more than the cheapest possibility calculated with the previous linear model (about an 8% increase in cost).

Iteration	Gap ($x^T s$)	μ	$\|Ax - b\|_\infty$	$\|Qx + c - A^T y - s\|_\infty$
1	9872.8419	1000.0000	16132.1838	5228.6566
2	8052.9612	68.5614	15877.9001	5145.9907
3	4399.2063	55.9233	13966.8287	4527.2038
4	2205.1676	30.5500	9808.1342	3179.4833
5	926.7357	15.3137	9446.6903	3146.3998
6	502.4013	6.4357	6581.8457	2227.7160
7	105.1425	3.4889	1415.2394	495.6178
8	11.2446	0.7302	104.6938	37.1110
9	1.2518	0.0781	6.7604	2.4037
10	0.1114	0.0087	0.6828	0.2424
11	0.0072	0.0008	0.0638	0.0226
12	0.0004	0.0001	0.0029	0.0010

Table 4.7 The first 12 iterations of Algorithm 18 as it solves the quadratic blending problem. The columns headed with $\|Ax - b\|_\infty$ and $\|Qx + c - A^T y - s\|_\infty$ contain the maximum violations of these quantities being zero.

There are numerous solvers for linear and quadratic programs, and both MATLAB and Octave support several. The optimization toolbox in MATLAB has a native solver called `linprog`, and several commercial solvers, most notably Gurobi© and Cplex©, work well in MATLAB. Octave works with the GNU-solver GLPK©. Model development can also be facilitated with several modeling languages such as AMPL©, GAMS©, and Pyomo©.

4.3 Global Optimization and Heuristics

The term "global optimization" is most often used to classify a problem as being difficult to solve due to a lack of known structure. If an optimization problem could instead be classified as a problem with a more well-defined structure, then that structure could likely be exploited to help solve the problem. As an example, the least squares problem in (4.22) can, and should, be classified as a convex quadratic program instead of a global optimization problem even though a global solution is desired. The more narrow classification in this case indicates that we can use the KKT conditions to globally solve the problem.

The general lack of structure for the class of global optimization problems challenges algorithm design. Global problems tend to have multiple local optima, and they can further be challenged with binary or integer constraints. Indeed, we generally have little information about the objective function, and it need not exhibit any analytic properties such as continuity or differentiability. So algorithms like gradient descent or Newton's method don't always make sense, although if they do, then one tactic is to initiate one of the earlier solvers at a host of different starting locations. The algorithm will hopefully converge for many of the initial points, and the reported minimum would be the best of these solutions. This method is often referred to as multi-start.

Another theme is that of heuristic search. The idea is to initialize a search with a feasible solution that is thus a candidate for optimality. The best solution calculated during the search is called the incumbent, and the heuristic's intent is to seek improved incumbents, i.e. if the new iterate improves the objective, then it becomes the new incumbent. Otherwise the search continues or the algorithms ceases. This general theme gives rise to a multitude of algorithms, and the design, implementation, and study of heuristics is a lively and important area of computational science. Some heuristics provide theoretical guarantees that help instill confidence. For example, some heuristics can be proven to converge to within a certain percentage of optimality for some problems, while others can be shown to converge to global optimality if algorithmic parameters are selected just so. However, we don't typically know these parameters for the problem being studied. This fosters a bit of frustration, as we know that some parameters will work, but we don't always know which ones.

4.3 Global Optimization and Heuristics 165

Two of the most common heuristics are simulated annealing and genetic algorithms, and these are the global optimization procedures we motivate. Simulated annealing is arguably the computational workhorse of global optimization, and it is regularly the algorithm against which others are compared. Genetic algorithms are also long-standing, proven solution methods for numerous problems.

4.3.1 Simulated Annealing

Simulated annealing is a random metaheuristic that seeks a global optima. The algorithm has been successfully applied to an extremely large collection of problems, and it is essentially the default solution methodology when faced with a daunting optimization problem. The term metaheuristic rather than heuristic means that simulated annealing is a general framework with which to solve problems. This framework defines an overarching search process that requires detail per implementation, and once these details are established for any particular search on any particular problem, the result is a heuristic for that problem. Thus the "meta" of metaheuristic indicates that simulated annealing is a way to generate heuristics. Which heuristic might best solve a problem is unknown, and users of simulated annealing should expect to experiment with several implementations per problem.

The annealing moniker comes from the algorithm's intent to mimic the annealing process in which a substance is heated and then cooled, often to alter physical properties. The state transitions of the material depend on the material's temperature and the rate at which it is cooled. Simulated annealing as an optimization technique similarly depends on a cooling schedule, and different solutions can arise from different schedules just as different material properties can arise by varying the rate at which the temperature decreases.

A generic simulated annealing search is initiated with a feasible solution, say x_0, which immediately becomes our first incumbent solution. A heuristic is then used to select a new candidate, say $\hat{x}$. If the objective value $f(\hat{x})$ is less than $f(x_0)$, then we accept $\hat{x}$ as our new incumbent and set $x_1 = \hat{x}$. If the objective value $f(\hat{x})$ is greater than or equal to $f(x_0)$, then $\hat{x}$ might still become our next iterate x_1 depending on the outcome of a random experiment. In this case the algorithm accepts a non-improving iteration, typically called "hill climbing," and it is these non-improving steps that allow simulated annealing to move away from nearby local optima. The possibility of escaping the draw of a local optima is what gives simulated annealing its condignity with regard to global optimization. Because the new iterate $\hat{x}$ might be less preferable than the incumbent, the incumbent solution would remain to be the best solution yet identified.

The likelihood of accepting uphill moves depends on the iterate, x_k, the candidate, $\hat{x}$, and a parameter called the temperature, T. The acceptance probability of $\hat{x}$ is most commonly

$$P(x_k, \hat{x}, T) = \begin{cases} 1, & f(\hat{x}) < f(x_k) \\ e^{(f(x_k)-f(\hat{x}))/T}, & f(\hat{x}) \geq f(x_k). \end{cases} \quad (4.23)$$

The random experiment that decides if $\hat{x}$ is accepted as the the new iterate x_{k+1} depends on a sample from a standard uniform variable $Y \sim \mathcal{U}(0, 1)$. If the sample is y and $P(x_k, \hat{x}, T) \geq y$, then the candidate $\hat{x}$ is accepted as the new iterate x_{k+1}. Otherwise the current iterate becomes the new iterate, i.e. $x_{k+1} = x_k$. Notice that all improving moves are automatically accepted for this choice of P. Also, in the non-improving case of $f(\hat{x}) > f(x_k)$, we have

$$P(x_k, \hat{x}, T') < P(x_k, \hat{x}, T'') \Leftrightarrow T' < T''$$

and

$$\lim_{T \to 0^+} P(x_k, \hat{x}, T) = 0 \text{ and } \lim_{T \to \infty} P(x_k, \hat{x}, T) = 1.$$

The conclusion is that we are more likely to accept uphill moves for large temperatures and less likely to do so for small temperatures. The first limit suggests that we define $P(x_k, \hat{x}, 0)$ to be 0, which we do.

True to simulated annealing's intent to mimic the annealing process, the tactic is to start with a large T to agree with the heating processes. The temperature is then decreased at each iteration, and the method by which this decrease occurs is called the cooling schedule. The initial iterations are more likely to be uphill, but the heuristic increasingly dismisses uphill moves as T decreases toward zero. In fact as T reaches 0 the heuristic only accepts improving iterates.

The algorithmic ingredients of simulated annealing are now in place. Any particular implementation requires a method to decide candidates, a cooling structure, a manner by which to calculate acceptance probabilities, and a termination criterion. A common termination criterion is to continue until an iteration bound is reached, but the search can also terminate once the objective function falls below a preordained value (should such a value be known) or if improvements become insufficient after a number of iterations. The pseudocode of simulated annealing is listed in Algorithm 19.

Simulated annealing with $T = 0$ often behaves like a greedy neighborhood search, which is often "trapped" by the catchment region of a local optima. A greedy neighborhood search is a heuristic that iteratively seeks the best solution from among the feasible solutions within some distance of the current iterate. The set of possible updates is called a neighborhood of the iterate, and its definition is often dependent on the problem being solved. A greedy neighborhood heuristic only considers improving iterates since it terminates once there is no improvement within the neighborhood. If $T = 0$ in simulated annealing and the candidate $\hat{x}$ is selected within a neighborhood of x_k, then simulated annealing resembles a greedy neighborhood search because only improving candidates are accepted. Simulated annealing with $T = 0$ often mirrors the behavior of a greedy neighborhood search even if $\hat{x}$ is not selected

4.3 Global Optimization and Heuristics

Algorithm 19 Pseudocode for simulated annealing
$k = 1$
while unmet termination criteria **do**
    Calculate a candidate $\hat{x}$ by an appropriate rule
    Calculate the acceptance probability
$$P(x_k, \hat{x}, T) = \begin{cases} 1, & f(\hat{x}) < f(x_k) \\ e^{(f(x_k) - f(\hat{x}))/T}, & f(\hat{x}) \geq f(x_k) \end{cases}$$
    Draw a random outcome y_s from $Y \sim \mathcal{U}(0, 1)$
    If $P(x_k, \hat{x}, T) \geq y_s$, then set $x_k = \hat{x}$
    Update the temperature T
    let $k = k + 1$
end while
return x

to be the best candidate within the neighborhood. In this case several updates of simulated annealing often result in an approximate single iteration of a greedy neighborhood search.

We use two examples to illustrate how simulated annealing can be used to solve an optimization problem. Suppose a complicated laboratory experiment is sensitive to atmospheric pressure. The procedure is commissioned prior to experimentation by comparing outcomes to a known standard for a specific instance of the experiment. The sensitivity to the atmospheric pressure suggests that the pressure should be adjusted to minimize error in the standard case. A laboratory runs the benchmarking experiments with varying pressures, measured in bars, and finds the following errors,

$$\begin{array}{c|c} \text{Pressure} & 0.75 \;\; 0.80 \;\; 0.90 \;\; 0.95 \;\; 1.00 \;\; 1.05 \;\; 1.10 \;\; 1.15 \;\; 1.20 \;\; 1.25 \\ \hline \text{Error} & 0.32 \;\; 0.28 \;\; 0.23 \;\; 0.62 \;\; 0.71 \;\; 0.22 \;\; 0.47 \;\; 0.38 \;\; 0.20 \;\; 0.54 \end{array}. \tag{4.24}$$

Error fails to be convex, and minimizing it seems to require a global solver. The laboratory decides to assume that error is smooth and to minimize the (natural) cubic spline that passes through the data.

Two design steps are needed to use simulated annealing, those being how to select candidate incumbents and how to decrease the temperature. Both should promote the orthodox progression of having a varied and mostly random initial search that transitions into a narrow and more deterministic search as the algorithm terminates. Candidates are selected to minimize experimental error so that

$$\hat{x} = x_k \pm \left(\frac{k_{\max} - k}{10 \, k_{\max}} \right) (1.25 - 0.75),$$

where $k_{\max}$ is the maximum number of iterations. The sign of the update is decided randomly with the probability of moving left or right both being a half. The value of $\hat{x}$ is set to either 0.75 or 1.25 if the update exceeds either of these bounds, i.e. the search is not allowed to leave the interval $[0.75, 1.25]$. Candidate iterates early in the search have larger updates and thus tend to sample widely across the feasible region. The step size decreases as the iteration count increases so that later candidates better sample local regions.

The cooling schedule should consider the entity being minimized, as the acceptance probability depends on the ratio $(f(x_k) - f(\hat{x}))/T$. If T remains high relative to the change in the objective, then the acceptance probabilities remain high and most uphill moves are accepted. If T is small relative to the change in the objective, then the acceptance probabilities are low and uphill moves are more likely dismissed. The standard idea is to encourage a more random search initially and a more deterministic search toward the end, and hence, T should start large but trend small compared to $f(x_k) - f(\hat{x})$. We consider the two cooling schedules of

$$\underline{\text{linear decrease}} \qquad \underline{\text{cubic decrease}}$$
$$T_k = 2 - \frac{2k}{k_{\max}} \quad \text{and} \quad T_k = \left(2 - \frac{2k}{k_{\max}}\right)^3.$$

The second cooling schedule exaggerates the linear decrease of the first, and it has larger initial temperatures and smaller ending temperatures.

Figures 4.7 and 4.8 depict the favorable search with the second (cubic) cooling schedule. The termination criterion was to stop after $k_{max} = 600$ iterations. The red dots in both figures are accepted uphill moves and the green dots are dismissed uphill moves. The black dots in Fig. 4.7 are the automatically accepted downhill moves. The search starts by randomly wondering throughout the range of the input variable with numerous uphill moves, and it then progresses to a more localized, deterministic, and downhill search as the algorithm terminates. The acceptance probabilities in Fig. 4.8 show the favorable progression of having high initial values that decline as the algorithm converges.

Figures 4.9 and 4.10 illustrate the same algorithm but with the linearly declining temperature schedule. The acceptance probabilities remain high throughout the algorithm and most uphill moves are accepted. The outcome is less favorable because the iterations never transition toward a downhill search. So while the distance between iterates decreases, the algorithm maintains a random navigation of the search space without regard of the objective function. The behavior of the algorithm is difficult to predict in this case, and multiple runs show disparate terminal iterates.

Our second example is that of a famous problem called the traveling salesperson (TSP) problem. The idea is to traverse through a set of positions

4.3 Global Optimization and Heuristics

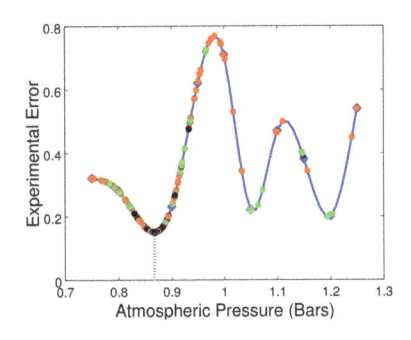

Fig. 4.7 Iterations of simulated annealing to minimize the approximated experimental error of (4.24). Red, black, and green dots are, respectively, uphill, downhill, and dismissed uphill moves. The cubically decreasing cooling schedule is used, and the final iterate is indicated by the vertically dashed line.

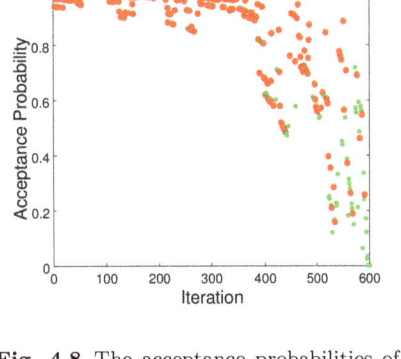

Fig. 4.8 The acceptance probabilities of the 326 iterates, out of the 600 total, that were potentially uphill. Nearly all the early possible uphill moves were accepted as indicated by the red dots. The majority of possible uphill moves were dismissed, as shown by the green dots, as the algorithm converged due to the rapid decrease in the temperature.

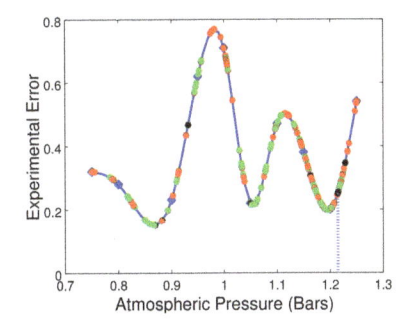

Fig. 4.9 The same simulated annealing algorithm as in Fig. 4.7 except that a linearly decreasing cooling schedule is used. In this case the algorithm tends to act like random search, and the termination is less favorable.

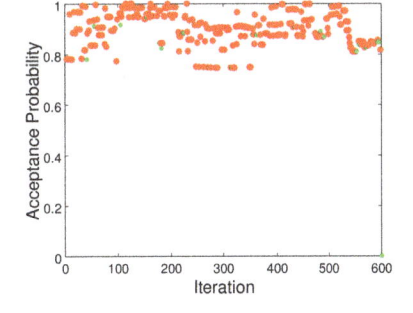

Fig. 4.10 The linearly decreasing temperature schedule maintains a high acceptance probability throughout the search. The vast majority of the 310 possible uphill moves is accepted.

(cities) and then return to the initial position after traveling the shortest distance. While the problem is traditionally stated in terms of transportation, applications of the TSP are vast and include, e.g., problems in manufacturing and computer science.

We solve the ten city problem depicted in Fig. 4.11. The city coordinates are:

Fig. 4.11 An example TSP problem.

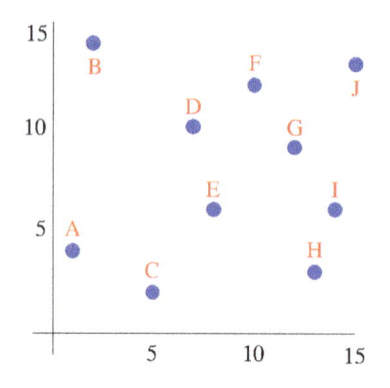

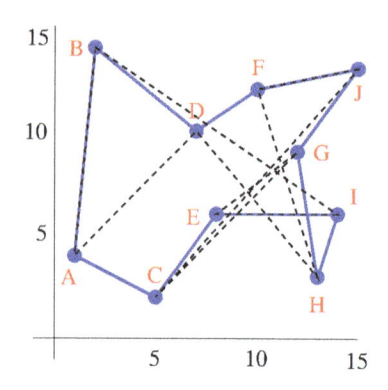

Fig. 4.12 The dashed black line is a random initial tour, and the blue tour is calculated by simulated annealing.

City	A	B	C	D	E	F	G	H	I	J
x coordinate	1	2	5	7	8	10	12	13	14	15
y coordinate	4	14	2	10	6	12	9	3	6	18

(4.25)

A tour is a permutation of the city labels. For example, the tour

$$(B, A, D, C, E, G, J, F, I, H)$$

starts at city B, and then travels to city A, then city D, etc., ending with a return from city H to city B. We assume for illustrative purposes that there is a road between every pair of cities whose length is the Euclidean distance between the cities.

Candidate iterates are generated by randomly swapping the position of two cities. So, if positions 3 and 8 are randomly selected during an iteration, then

$$\underbrace{(B,A,D,C,E,G,J,F,I,H)}_{\text{the current iterate}} \text{ would result in } \underbrace{(B,A,F,C,E,G,J,D,I,H)}_{\text{the candidate}}.$$

If the tour length of the candidate is less than the tour length of the current iterate, then the candidate is automatically accepted as the new iterate. If not, then the candidate is accepted probabilistically relative to the temperature. The initial tour is randomly generated, and the cooling schedule starts with 1/10 of the initial tour's length, after which it decreases linearly toward zero. We assume $k_{\max} = 200$ iterations.

An exemplar solution is depicted in blue in Fig. 4.12. The length of the initial (random) solution (B,I,E,G,C,J,F,H,D,A) is 99.241, and the length of the final solution (G,H,I,E,C,A,B,D,F,J) is 62.073. The last iterate is not glob-

4.3 Global Optimization and Heuristics   171

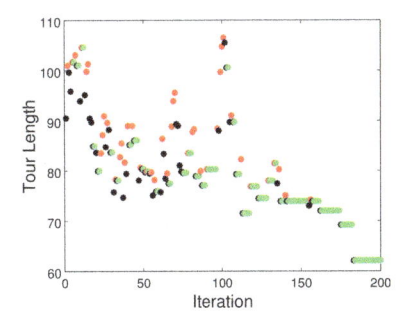

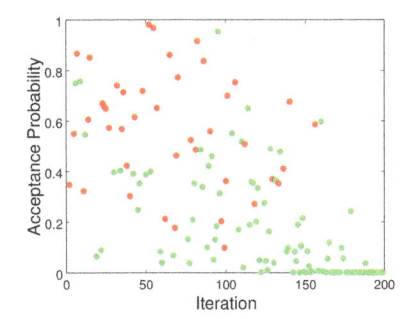

Fig. 4.13 The tour length progression of 200 iterations of simulated annealing used to solve the TSP in (4.25). Red, black, and green dots show uphill, automatically accepted downhill, and dismissed uphill moves.

Fig. 4.14 The acceptance probabilities for 200 iterations of simulated annealing used to solve the TSP in (4.25). Red and green dots are for accepted and dismissed uphill moves.

ally optimal, a fact which is easily seen by replacing the segment (G,H,I,E) with (G,I,H,E). Otherwise the final tour is optimal. The search space has 10! = 3,628,800 different tours, and to be so close to global optimality after a mere 200 iterations, which is less than 0.006% of all possible tours, is a good outcome. The progression of tour length and the acceptance probabilities are shown in Figs. 4.13 and 4.14. Red dots are for accepted uphill moves, green dots are for dismissed uphill candidates, and black dots are for downhill moves.

4.3.2 Genetic Algorithms

Genetic algorithms differ from our previous optimization algorithms because they maintain a population of solutions at each iteration. The iterative theme is to adapt biological tenets to describe how one population leads to the next. Each member of a population is awarded a "fitness," and a member's likelihood of influencing the next population increases with its fitness. Hence favorable solutions tend to propagate through iterations until better ones are found.

Experimental migration away from the best performing solutions is accomplished by crossover and mutation. Crossover is the way that current members of the population combine to generate offspring for the next generation. The biological motivation of crossover is the way that chromosomes combine to produce offspring in, for example, humans. There are numerous crossover strategies that can be singularly or collectively employed depending on what seems to best address the problem at hand. We explore a couple of the more common crossover techniques, but tailoring a genetic algorithm to a particular problem often benefits from a study of numerous crossover possibilities.

Mutations are random changes to solutions. The intent of mutation is to stochastically alter a solution to see if there is a fitness benefit. If so, then the mutation is favorable and the solution will heighten its influence on future generations. If not, then the solution will lessen or remove any forthcoming influence. The most common strategy corresponds with single nucleotide polymorphisms, which are single locations on our genome at which individuals vary among the population. If a mutation propagates to a meaningful proportion of future generations, then the mutation has created, in essence, a single nucleotide polymorphism. The Pseudocode for a genetic algorithm is listed in Algorithm 20.

Algorithm 20 Pseudocode for a genetic algorithm

$k = 1$
while unmet termination criteria **do**
   Select a mating cohort
   Use crossover to create a new population from the mating cohort
   Mutate randomly selected individuals in the population
   Calculate the fitness of the new population
   let $k = k + 1$
end while
return the incumbent

Consider the problem from the previous section of minimizing experimental error as a function of atmospheric pressure, where error is again measured by the cubic spline of the data in (4.24). The initial step of using a genetic algorithm is to encode the search space into finite sequences that match the concept of a chromosome. One way to do this for a continuous problem is to recognize that digits themselves are a sequential encoding.

Binary encodings are regularly promoted, but we choose standard base 10 for this example. Our search space is all 6 digit values in the unit interval of $[0, 1]$. So one possible solution is 0.450689, whereas 0.5589326 is not possible because it is a 7 digit sequence. Our search space is mapped to the natural feasible region by

$$x \mapsto 0.75 + x(1.25 - 0.75) = 0.75 + 0.5x,$$

where x is an element of our search space.

At each iteration a population of search space elements is maintained. The inherit competitive intent of natural selection is based on the fitness levels of the individuals within the population. Fitness is considered a good quality that should be maximized, and since our problem is naturally a minimization problem, we define the fitness of the individual x to be

4.3 Global Optimization and Heuristics

$$f(x) = 1 - g(0.75 + 0.5x),$$

where g is the cubic spline through the data in (4.24). The spline g doesn't exceed 1 on the interval of $[0.75, 1.25]$, so fitness is always positive, and maximizing fitness minimizes error. Fitness could have been $-g(x)$, but negative values of fitness are less intuitive within the genetic algorithm paradigm.

A mating cohort is drawn from a population based on the likelihood of an individual's ability to spawn an improved fitness to its offspring. Deciding the mating cohort is called selection, and numerous strategies exist. The most common is called proportionate selection, which assumes that an individual's likelihood of mating is proportionate to the individual's percentage of the population's total fitness. Assume the population is the set $\{x_j : j = 1, \ldots, J\}$, where fitness is a non-increasing function of j, i.e. $f(x_j) \geq f(x_{j+1})$. The fitness value of the i-th individual is normalized as

individual	normalized fitness
x_i	$f_N(x_i) = f(x_i) / \sum_{j=1}^{J} f(x_j).$

$\Rightarrow$

A random draw is used to decide the mating cohort. Assuming that $Y \sim \mathcal{U}(0, 1)$, we select for each sample y the first $\hat{j}$ satisfying

$$y \leq \sum_{j=1}^{\hat{j}} f_N(x_j).$$

As an example, suppose the normalized fitness scores are

$$(0.45, 0.25, 0.10, 0.10, 0.05, 0.025, 0.025).$$

Then the cumulative totals, which can be calculated with `cumsum`, are

$$(0.45, 0.70, 0.80, 0.90, 0.95, 0.975, 1.00).$$

If the random draw of y satisfies $y \leq 0.45$, then the most fit individual is selected to mate. If $0.95 < y \leq 0.975$, then the 6th individual is selected to mate. Notice that more fit individuals have a higher probability of mating.

Offspring of the new population are generated by the process of crossover, by which parental sequences are combined similar to how chromosomes combine to form offspring, e.g. in humans. A simple one point crossover is used here. A random position for each sequential mating pair is selected, and two offspring are produced by swapping their encoded representations past this location. Suppose the randomly selected digit is 2. Then an example parent and offspring relationship is

parents		offspring
$0.387429 = 0.38\|7429$		$0.38\|5310 = 0.385310$
	$\Rightarrow$	
$0.045310 = 0.04\|5310$		$0.04\|7429 = 0.047429.$

The random mating process continues until a new population is created.

The final biological adaption is that of mutation. The concept is to allow individuals to mutate to see if there is a fitness benefit. If so, then the characteristics of the improved solution gain a mating advantage and can alter future generations. If not, then it is likely that the mutation will fade from the population due to mating ineptness. Mutation is accomplished by selecting a proportion of the newly created population and then altering a randomly selected digit to a random value. So if the offspring 0.385310 is selected for mutation, then we might, for example, change the 3rd digit to a 2 creating 0.382310.

We achieve results representative of the 30 iterations depicted in Fig. 4.15, assuming the aforementioned 6 digit representations of our search space, a population size of 6, a random mutation rate of 40%, and an initial random population. The forerunning populations at the bottom of the graph do not contain a solution near the global minimum, but individuals near the minimum are introduced through mutation. Near optimal solutions died out in their first several introductions, probably due to the established competitive advantage of the solutions with atmospheric pressures near 1.1 that already had multiple copies in the population with relatively good fitness. However, a mutated solution near the atmospheric pressure of 0.88 was introduced on the twelfth iteration, and the heightened fitness of this individual leads to a maintained advantage in subsequent populations. The vertical column of red and blue dots below the global minimum illustrates this behavior and shows that nearby solutions were tested through crossover and mutation as the algorithm progressed. The population collapses around the global minimum after 30 iterations, and the best solution of the final iteration had an atmospheric pressure of 0.872551 and an experimental error of 0.152607.

The progression in Fig. 4.15 was selected to illustrate the motivating theme of a genetic algorithm. The population was relatively small, and the mutation rate was a bit high with regard to standard practice. While the outcomes varied among algorithmic runs, the outcomes were nearly always favorable in that some of the three local minimums were identified. An illustrative counterpart is seen in Fig. 4.16, which had a population size of 50 and a mutation rate of 2%. In this case the original population had many good solutions that out competed their less fit brethren as the algorithm progressed. The population collapses onto three main solutions after 30 iterations, two of which are near local minima. The random experimentation of mutation had less of an

4.3 Global Optimization and Heuristics

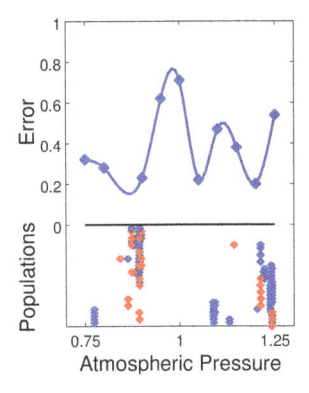

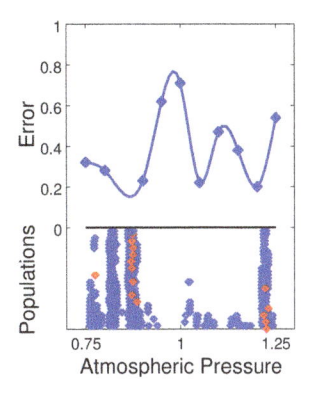

Fig. 4.15 Thirty iterations of a genetic algorithm with a population of size 6 and a mutation rate of 40%. The first iterative populations are at the bottom of the graph, and later iterations move upward. Red dots indicate mutated individuals.

Fig. 4.16 Thirty iterations of a genetic algorithm with a population size of 50 and a mutation rate of 2%. The first iterative populations are at the bottom of the graph, and later iterations move upward. Red dots indicate mutated individuals.

affect than did our other example. The final best solution had an atmospheric pressure of 0.869324 and an experimental error of 0.151098.

Our second example is the ten city traveling salesperson problem in (4.25). The genetic encoding is straightforwardly the permutation of a tour. The crossover method must maintain the offspring as permutations, and hence, a simple one-cross method like that of the previous example won't suffice. A long-standing adaption called partially mapped crossover is used instead, which like our earlier crossover method swaps a contiguous segment of the parental sequences. The adaption is illustrated with a 6 element example in which the crossover point follows the 3rd element. The first of three steps is

$$\begin{array}{ll} \text{parent A} & (\,4,2,1,6,5,3\,) \\ & \downarrow \\ \text{parent B} & (\,5,3,6,2,1,4\,) \end{array} \implies \begin{array}{l} (\,4,2,1,6,5,3\,) \\ \\ (\,4,3,6,2,1,5\,). \end{array}$$

The correction that maintains the permutation is to reinsert parent B's first element of a 5 in the second location of the two 4s that would be forced by the first element's crossover. The procedure continues as follows:

$$\begin{array}{l}(\,4,2,1,6,5,3\,)\\ \downarrow \\ (\,4,3,6,2,1,5\,)\end{array} \implies \begin{array}{l}(\,4,2,1,6,5,3\,)\\ \downarrow \\ (\,4,2,6,3,1,5\,)\end{array} \implies \begin{array}{l}(\,4,2,1,6,5,3\,)\\ \\ (\,4,2,1,3,6,5\,).\end{array}$$

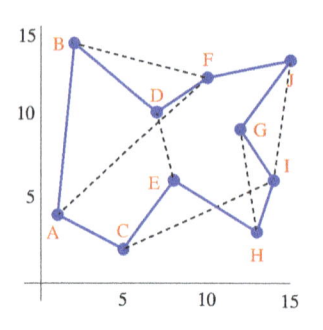

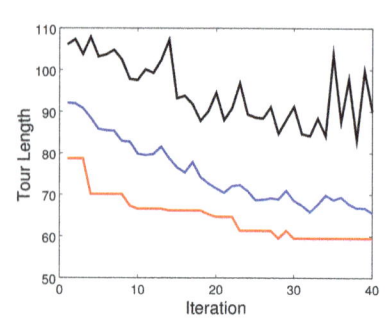

Fig. 4.17 The dashed black line is a random initial tour, and the blue tour is calculated by a genetic algorithm.

Fig. 4.18 The worst (black), average (blue), and best (red) tour lengths of 40 iterations of a genetic algorithm used to solve the TSP example in (4.25).

The first offspring is the final sequence $(4, 2, 1, 3, 6, 5)$ at the bottom right. A second offspring is created by reversing the roles of the parents, which in this case results in $(5, 3, 6, 1, 4, 2)$. The crossover point is randomly selected per mating pair.

The fitness of a tour is its length subtracted from the longest tour in the population, which equates large positive fitnesses to low tour lengths, i.e. high, positive fitness is good. The mating population is again selected proportionately with one exception. The first two elements of the mating cohort are identically the tour with the highest fitness, which up to mutation guarantees that the subsequent population's best tour is no worse than the previous population's best tour prior to mutation. Tours mutate by swapping two random elements.

The best solution of the final population after 40 iterations on the ten city problem is illustrated in Fig. 4.17. The population at each iteration was 30 tours, and the mutation rate was to have no more than 40% of the tours undergo a random, two city swap. The initial population was randomly selected with `randperm`, and the best tour of the initial population was (J, I, C, A, F, B, D, E, H, G) with a tour length of 78.577. Figure 4.18 depicts the progressions of the best, average, and worst tours of each population over the search. The best tour of the final iteration is (F, J, G, I, H, E, C, A, B, D), which has a globally optimal length of 59.426551.

The result of the genetic algorithm is impressive from the perspective that less than 1200 unique tours were generated from the $10! = 3,628,800$ possibilities. The genetic algorithm found the global optimal solution by observing less than 0.034% of all possible tours. The performance wasn't uniformly this good after multiple runs, but it was consistently this good.

Genetic algorithms have many adaptations, and our examples only illustrate a couple of the numerous options. Many other crossover and mutation methods are common, and there are multiple ways to select the mating cohort and the number of offspring. For example, our one-point crossover examples easily extend to multiple-point crossovers in which the parental sequences swap numerous segments. The population size need not be static and could instead fluctuate per iteration. Only the most fit individuals of a population, say the top 10–20%, could be selected to mate. All of these options are under the computational umbrella of a genetic algorithm, and like other stochastic search processes such as simulated annealing, designing a genetic algorithm for a particular problem is an experimental endeavor that is often based on experience and computational ability.

4.4 Exercises

1. Find a function $g : \mathbb{R} \to \mathbb{R}$ such that $g''(x) > 0$ for all x but that does not have a minimizer.
2. Show that if $\hat{x}$ is an eigenvector of the positive definite matrix Q, then the method of steepest descent initiated with $\hat{x}$ and applied to $f(x) = (1/2)x^T Q x + c^T x$ converges in a single iteration with an appropriate line search.
3. Assume $f : \mathbb{R}^n \to \mathbb{R}$ is convex. Show that x^* solves $\min_x f(x)$ if and only if $\nabla f(x^*) = 0$. You may assume whatever differential information you need to obtain the result. Note that the proposition is generally false as demonstrated by $f : \mathbb{R} \to \mathbb{R} : x \mapsto |x|$, so some differentiable assumption is required.
4. Show that any matrix of the form uv^T, where u and v are n-element column vectors, is rank one. Use this to show that both

$$\frac{(\Delta Df_k)^T (\Delta Df_k)}{(\Delta Df_k)^T \Delta x_k} \quad \text{and} \quad \frac{H_k \Delta x_k (\Delta x_k)^T H_k}{(\Delta x_k)^T H_k \Delta x_k}$$

from (4.1) are rank one.
5. Verify the Sherman-Morris formula (4.2).
6. Verify the inverse formula (4.3).
7. Let f be a function from $\mathbb{R}^{m+n}$ into $\mathbb{R}$. Show that

$$\max_u \min_v f(u,v) \le \min_v \max_u f(u,v),$$

where the maximums and minimums are assumed to exist. The vectors u and v are of respective lengths m and n.
8. Show that the only solution of

$$\min\{-x_1 : -x_1 - x_2^2 \ge 0\}$$

associates with a saddle point of the Lagrangian even though the only solution of problem (4.16) does not.

9. Assume the constrained optimization problem

$$\min f(x) \text{ such that } g(x) = 0, \ h(x) \geq 0$$

is convex.

   a. Prove that the set of optimal solutions is convex.
   b. Prove that any locally optimal solution is globally optimal.

10. Solve the convex problem

$$\min_x x_1 + x_2^2 + e^{x_1+x_2+x_3} \text{ such that } 6x_1 + 2x_2 - 5x_3 = 4$$

by solving the KKT conditions with Newton's method.

11. Assume the systems $Ax = b$, $x \geq 0$ can be satisfied for some $x > 0$, a condition known as Slater's interiority condition, and consider the following optimization problem,

$$\min c^T x - \mu \sum_{i=1}^{m} \ln(x_i) \text{ such that } Ax = b, \ x \geq 0.$$

   a. Show that the Hessian of the objective function is positive definite for all $x > 0$, and conclude that the objective is strictly convex as long as $x > 0$.
   b. Establish that there is a unique solution for each positive μ under the assumption that there is a finite minimum with $\mu = 0$.
   c. Further show that the KKT conditions of this problem agree with those of the relaxed complementarity condition used to solve a linear program with an interior point algorithm.

12. Write a function with declaration

    ```
    [xbest,fbest,itrcnt,stat]=GradDesc(f,Df,x0,
        gradTol,xTol,itrBound,s)
    ```

    that minimizes a function with the gradient descent method.

    The input arguments are:

f	-	A filename or function handle that evaluates the function to be minimized.
Df	-	A filename or function handle that evaluates the gradient of f.
x0	-	An initial iterate.
gradTol	-	A termination criterion on $\|\nabla f(x_{\text{best}})\|$.
xTol	-	A termination criterion on $\|x_k - x_{k-1}\|$.

4.4 Exercises

itrBound - A maximum number of iterations to perform.

s   - A line search parameter encoded as follows:

    s = 1: Use $\alpha = 1$ for all iterations.

    s = 2: Decide α with a Lagrange polynomial with $\alpha_1 = 0$, $\alpha_2 = 1$, and $\alpha_3 = 2$.

    s = 3: Decide α with bisection over the interval $0 \leq \alpha \leq 1$; if $\nabla f(x)$ and $\nabla f(x+d)$ have the same sign, then set $\alpha = 1$.

The algorithm should terminate if any of the termination tolerances are satisfied or if the iteration bound is exceeded.

The output arguments are:

xbest - The best calculated solution.

fxbest - The function value at **xbest**.

itrcnt - The number of iterations performed, with x_0 counting as iteration 0.

stat - A status variable, encoded as follows:

    s = 0: The algorithm succeeded and $\|\nabla f(x_{\text{best}})\|$ is less than gradtol.

    s = 1: The algorithm failed, in which case x_{best} is the best solution that was encountered.

As an example, the command

```
[xSol, fSol, itrCnt, stat] = GradDesc(@f, @Df, x0, 10^-6, 10^-8, 100, 2);
```

will attempt to minimize the function defined by the handle @f whose gradient is @Df. It will initialize the search at x0 and will terminate if the norm of the gradient is smaller than 10^{-6}, if the norm of the difference between consecutive iterates is less than 10^{-8}, or if more than 100 iterations are performed. The stat variable should return as zero if and only if the norm of the gradient is less than gradTol, otherwise it should be returned as one.

13. Write a function called **BFGS** that minimizes a function with the BFGS algorithm. Update the Hessian at each step and calculate the search direction by solving $H_k d = -\nabla f(x)$, with $H_0 = I$. The input and output arguments are the same as in Exercise 12.

14. Write a function called **BFGSinv** that minimizes a function with the BFGS algorithm. Update the Hessian inverse at each iteration and compute the

search direction by calculating $d = -H_k^{-1}\nabla f(x)$, with $H_0 = I$. The input and output arguments are the same as in Exercise 12.

15. Write a function called NewtonOpt with declaration

    [xbest,fbest,itrcnt,stat]=NewtonOpt(f,Df,Df2,
        x0,gradTol,xTol,itrBound,s)

    that minimizes a function with Newton's method.

    The input arguments are:

f	-	A filename or function handle that evaluates the function to be minimized.
Df	-	A filename or function handle that evaluates the gradient of f.
Df2	-	A filename or function handle that evaluates the Hessian of f.
x0	-	An initial iterate.
gradTol	-	A termination criterion on $\|Df_x(x_{best})\|$.
xTol	-	A termination criterion on $\|x_k - x_{k-1}\|$. The algorithm should stop when either $\|Df_x(x_{best})\|$ is sufficiently small or successive iterates are sufficiently close.
itrBound	-	A maximum number of iterations to perform.
s	-	A line search parameter encoded as follows:

    s = 1: Use $\alpha = 1$ for all iterations.

    s = 2: Decide α with a Lagrange polynomial with $\alpha_1 = 0$, $\alpha_2 = 1$, and $\alpha_3 = 2$.

    s = 3: Decide α with bisection over the interval $0 \leq \alpha \leq 1$; if $\nabla f(x)$ and $\nabla f(x+d)$ have the same sign, then set $\alpha = 1$.

    s = 4: Decide α by solving $g'(s) = 0$, where $g(s)$ is the second order Taylor polynomial for $f(x + sd)$.

    The output arguments are the same as in Exercise 12.

16. Some well-known test functions for unconstrained solvers are listed in Table 4.8. Design and conduct a computational experiment to compare the ability of gradient descent, BFGS (with and without Hessian inverses), and Newton's method to solve these problems. The computational design should explore different starting points and convergence tolerances, and both iteration count and computational speed should be considered.

4.4 Exercises

Name	Minimum	Search region
Ackley's function	0 at $(0,0)$	$-5 \leq x \leq 5$ $-5 \leq y \leq 5$
$f(x,y) = -20e^{-0.2\sqrt{(x^2+y^2)/2}} - e^{(\cos(2\pi x)+\cos(2\pi y))/2} + e + 20$			
Rosenbrock function	0 at $x = e$	no restriction
$f(x) = \sum_{i=1}^{n-1} \left(100\left(x_{i+1} - x_i^2\right)^2 + (x_i - 1)^2 \right)$			
Beale's function	0 at $(3, 0.5)$	$-4.5 \leq x \leq 4.5$ $-4.5 \leq y \leq 4.5$
$f(x,y) = (1.5 - x + xy)^2 + (2.25 - x + xy^2)^2 + (2.625 - x + xy^3)^2$			
Goldstein-Price function	3 at $(0, -1)$	$-2 \leq x \leq 2$ $-2 \leq y \leq 2$
$f(x,y) = (1 + (x+y+1)^2(19 - 14x + 3x^2 - 14y + 6xy + 3y^2))$ $\quad \times (30 + (2x - 3y)^2(18 - 32x + 12x^2 + 48y - 36xy + 27y^2))$			
Matyas function	0 at $(0,0)$	$-10 \leq x \leq 10$ $-10 \leq y \leq 10$
$f(x,y) = 0.26(x^2 + y^2) - 0.48xy$			
McCormick function	-1.9133 at $(-0.54719, -1.54719)$	$-1.5 \leq x \leq 4$ $-3 \leq y \leq 4$
$f(x,y) = \sin(x+y) + (x-y)^2 - 1.5x + 2.5y + 1$			
Styblinski-Tang function	-39.1661 at $-2.903534\,e$	$-5 \leq x_i \leq 5$
$f(x) = (1/2) \sum_{i=1}^{n} (x_i^4 - 16x_i^2 + 5x_i)$			

Table 4.8 Some standard test functions for optimization solvers.

17. Write a function with declaration

    [xbest,fxbest,ybest,sbest,itrCnt,stat]=LPSolve(
        A,b,c,x0,gapTol,itrBound)

    that solves the linear program

    $$\min c^T x \text{ such that } Ax = b, x \geq 0$$

    with an interior point method.

    The input arguments are:

    A   - The coefficient matrix, A.
    b   - The right-hand side vector, b.
    c   - The objective vector, c.
    x0   - An optional initial primal solution, with default value e.
    gapTol - An optional terminal gap tolerance, with default value 10^{-8}.
    itrBound - An optional iteration limit, with default value 1000.

    Initiate the dual s0 with e and calculate an initial y0 by solving $A^T y = c - s_0$.

    The output arguments are:

    xbest   - The best calculated solution.
    fxbest - The best calculated objective value.
    ybest   - The best calculated dual solution y.
    sbest   - The corresponding value of s.
    itrCnt - The number of iterations performed.
    stat   - A status variable encoded as follows.

    stat = 0: Success.
    stat = 1: Failure, for any reason.

    As an example, the command

    [xbest, fxbest, ybest, sbest, itrCnt, stat] = LPsolve(A,b,c,[],10^{-4});

    will try to solve the linear program starting with the default x0 but with a termination tolerance of 10^{-4}. The status variable should return as 0 only if the solution achieves the gap tolerance; otherwise, it should be returned as 1.

4.4 Exercises

18. Write a function with declaration

    ```
    [xbest,fxbest,ybest,sbest,itrCnt,stat]=QPSolve(
        A,b,Q,c,x0,gapTol,itrBound)
    ```

    that attempts to solve the convex quadratic program

    $$\min \frac{1}{2}x^T Q x + c^T x \quad \text{such that} \quad Ax = b,\ x \geq 0$$

    with an interior point algorithm.

    The input arguments are:

A	-	The coefficient matrix, A.
b	-	The right-hand side vector, b.
Q	-	The objective matrix, Q.
c	-	The objective vector, c.
x0	-	An optional initial primal solution, with default value e.
gapTol	-	An optional terminal gap tolerance, with default value 10^{-8}.
itrBound	-	An optional iteration limit, with default value 1000.

    Initiate the dual s0 with e and calculate y0 by solving $A^T y = Qx_0 + c - s_0$.

    The output arguments are:

xbest	-	The best calculated solution.
fxbest	-	The best calculated objective value.
ybest	-	The best calculated dual solution y.
sbest	-	The corresponding value of s.
itrCnt	-	The number of iterations performed.
stat	-	A status variable encoded as follows.

    stat = 0: Success.
    stat = 1: Failure, for any reason.

    As an example, the command

    ```
    [xSol,fSol,ySol,sSol,itrCnt,stat] = QPSolve(A,b,Q,c,ones(n,1),10^-10,20);
    ```

    will try to solve the quadratic problem initiated with the primal vector of ones to a gap tolerance of 10^{-10} in no more than 20 iterations. The status variable should only return as zero if the gap tolerance is achieved.

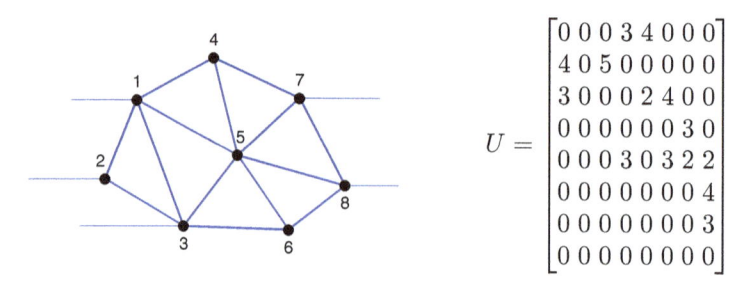

$$U = \begin{bmatrix} 0 & 0 & 0 & 3 & 4 & 0 & 0 & 0 \\ 4 & 0 & 5 & 0 & 0 & 0 & 0 & 0 \\ 3 & 0 & 0 & 0 & 2 & 4 & 0 & 0 \\ 0 & 0 & 0 & 0 & 0 & 0 & 3 & 0 \\ 0 & 0 & 0 & 3 & 0 & 3 & 2 & 2 \\ 0 & 0 & 0 & 0 & 0 & 0 & 0 & 4 \\ 0 & 0 & 0 & 0 & 0 & 0 & 0 & 3 \\ 0 & 0 & 0 & 0 & 0 & 0 & 0 & 0 \end{bmatrix}$$

Fig. 4.19 A capacitated network and an associated capacity matrix for Exercise 19.

19. A standard problem in hydraulic engineering is the design of piping networks to maximize throughput. Calculating the maximum flow through a system is accomplished by solving a linear program. As an example, suppose we wish to compute the maximum flow through the capacitated network depicted in Fig. 4.19. The matrix U on the right identifies the flow capacities, say per hour, on each of the edges. For example, there is no flow from node 1 to node 3 because $U_{13} = 0$. Alternatively, $U_{31} = 3$ means that there is a maximum flow of 3 units per hour from node 3 to node 1. Hence, the edge between nodes 1 and 3 is uni-directional. Assume the maximum, simultaneous flows into nodes 1, 2, and 3 from outside the network, i.e. from the left, are, respectively, 4, 5, and 5. The design question is, what flow capacities are needed of the pipes out of nodes 7 and 8 so that throughput is not restricted by these pipes? Notice that the answer may not be the total input flow, which is 14, due to restrictions on the intervening network.

Let x_{ij} be the flow from node i to node j. Then conservation at each node k requires

$$\sum_i x_{ik} + s_k = \sum_j x_{kj} + d_k,$$

where s_k is the supply at node k and d_k is the demand at node k. In our example only nodes 1, 2, and 3 have external supply, and nodes 7 and 8 have external demands. Three illustrative constraints are

Node 2: $s_2 = x_{21} + x_{23}$,
Node 4: $x_{14} + x_{54} = x_{47}$, and
Node 8: $x_{58} + x_{68} + x_{78} = d_8$.

The upper bound on each variable is the capacity restrictions. We have as examples that $x_{21} \leq U_{21} = 4$ and $s_1 \leq 4$. The objective is to maximize $d_7 + d_8$. Model this problem as the linear program

$$\min c^T x \text{ such that } Ax = b, \ x \geq 0,$$

4.4 Exercises

and then solve the problem with LPSolve from Exercise 17. You might also want to solve the problem with related solvers in MATLAB or Octave.

20. The least squares model developed in Sect. 3.2.2 has many favorable properties, but other regression calculations can be preferred. One adaptation is to alter the norm that is used to measure the residuals. Three standard measures are

$$\text{(1-norm)} \quad \min_x \|Ax - b\|_1 = \sum_{i=1}^{m} |(Ax - b)_i|,$$
$$\text{(2-norm)} \quad \min_x \|Ax - b\|_2^2 = \sum_{i=1}^{m} (Ax - b)_i^2, \text{ and}$$
$$\text{(inf-norm)} \quad \min_x \|Ax - b\|_\infty = \max_i |(Ax - b)_i|.$$

The 2-norm model is the method of least squares, and a minimum can be computed by solving the normal equations. The 1- and inf-norms, respectively, diminish and exaggerate the importance of deviant data, and hence, these norms can be beneficial in some settings. Calculating minimums for the 1- and inf-norms can be accomplished by solving linear programs. As an example exercise, suppose your grades for the last six assignments are those below.

Assignment	1	2	3	4	5	6
Score	89	65	70	60	100	64

Construct and solve linear programs to find the best fit line in the 1- and inf-norm cases, with grade being the response variable and assignment index being the regressor. Also calculate a best fit line with the 2-norm. Which model would you prefer to forecast your next grade?

21. Consider the multiple regression problem of calculating parameters a_1, a_2, and a_3 so that the least squares solution through a set of data is

$$\hat{y} = a_1 x_1 + a_2 x_2 + a_3 x_3.$$

The intercept term a_0 is assumed to be zero. Suppose we seek parameters such that for each i we have $0 \leq a_i \leq 0.5(a_1 + a_2 + a_3)$. Formulate the resulting least squares problem as a constrained quadratic optimization problem of the form

$$\min \frac{1}{2} x^T Q x + c^T x \text{ such that } Ax = b, \, x \geq 0.$$

Use QPSolve from Exercise 18 to solve the problem for the data below.

x_1	x_2	x_3	y
1.73	0.66	4.64	12.98
0.12	0.10	3.66	3.23
1.28	0.11	4.44	11.49
0.12	0.62	3.42	6.74
0.91	0.45	1.50	5.39

How different is the restricted solution from the unrestricted solution, and in particular, how different are the predicted rates dy/dx_i between these two models?

22. Write a function with declaration

    [tbest,lenbest,itrCnt,stat]=TSPSA(A,itrBnd,x0,T)

    that uses simulated annealing to solve a TSP problem.

    The input arguments are:

    A   - An $n \times n$ matrix A such that A_{ij} is the distance between cities i and j.
    itrBnd   - An optional iteration bound, with default value $50n$.
    x0   - An optional initial tour x0. If x0 is not provided, then the initial tour should be a random permutation of $1 \ldots n$.
    T   - An optional vector T of temperatures so that T_k is the temperature for iteration k. The default value of T should be a linearly declining temperature schedule starting with the length of the initial tour divided by n.

    The algorithm should terminate if either the iteration limit is reached or if $5n$ consecutive trial iterates are discarded.

    The return arguments are:

    tbest   - The best tour of the entire search, note that this may not be the final tour.
    lenbest - The best tour length of the entire search.
    itrCnt   - The number of iterations performed.
    stat   - A status variable encoded as follows:

       stat = 0: The algorithm terminated, and the last iterate is also the best solution found.

       stat = 1: The algorithm terminated, and the last iterate is not the best solution found.

4.4 Exercises

For example, the command

```
[tbest, lenbest, itrCnt, stat] = TSPSA(A, 100, [], T);
```

will use no more than 100 iterations of simulated annealing with the cooling schedule T after being initiated with a random tour. An example cooling structure in this case could be something like

```
T = [linspace(n*max(max(A)), 10^(-8), 100)].^(1.5);
```

23. Write a function with declaration

```
[tbst,lenbst,itrCnt,stat]=TSPGA(A,itrBnd,pSz,p)
```

that solves the TSP problem with a genetic algorithm.

The inputs arguments are:

A	-	An $n \times n$ matrix such that A_{ij} is the distance between cities i and j.
itrBnd	-	An optional iteration bound with default value $10n$.
pSz	-	An optional parameter that gives the number of tours in a population. The default value should be $4n$.
p	-	An optional parameter that gives the maximum mutation percentage. The default value should be 0.4.

The mating cohort should be selected proportionately except that it should always include a mating pair of the best tour from the previous population. Use the one-point partially mapped crossover method developed in the chapter, and assume mutations are random two-city swaps. Tours should be selected for mutation with a command similar to

```
unique(round(0.5+pSize*rand(1,round(p*pSize))));
```

If the tours in the population are indexed by 1, 2, ..., pSize, then the result will be a vector of unique indices of length no greater than $\lceil p*pSize \rceil$. The search should terminate once the iteration count is reached or there is no improvement in $2n$ iterations.

The output arguments are:

tbst	-	The best tour of the entire search, which may not be in the final population due to mutation.
lenbst	-	The best tour length of the entire search.
itrCnt	-	The number of iterations performed.

188   4 Optimization

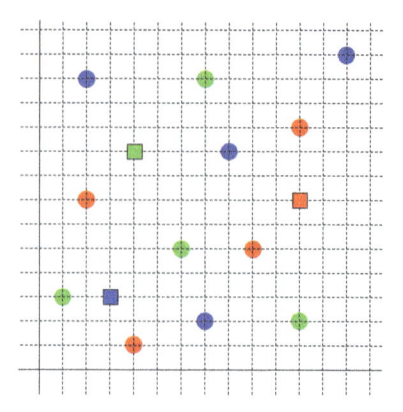

Fig. 4.20 An example configuration of the table for Exercise 24.

stat   - A status variable encoded as follows:

stat = 0: The last population contains the best tour.

stat = 1: The last population does not contain the best tour.

As an example, the command

[tBst,lenbst,itrCnt,stat] = TSPGA(A,200,8,0.2);

will use no more than 200 iterations of the genetic algorithm with a population size of 8 tours and a maximum mutation percentage of 0.2.

24. A robotic arm is only capable of moving horizontally and vertically over a table. The robot's task is to stack colored discs on top of a square marker of the same color, a job that should be accomplished with the least amount of movement. An example configuration is in Fig. 4.20. Assume that the robotic arm can hold at most a single disc and that the arm starts at the $(0,0)$ location (bottom left). Show that if the robotic arm must return to its original location, then this problem is a traveling salesperson problem. Solve this problem in three ways:

   a. Use TSPSA from Exercise 22; experiment with different settings.
   b. Use TSPGA from Exercise 23; experiment with different settings.
   c. Use the deterministic heuristic called 2-opt initiated with all possible greedy (nearest neighbor) solutions.

The 2-opt heuristic is based on a search that attempts to "untangle" tours. Candidate updates are generated for each index pair (i,j), with $i < j$, by reversing the order of the sequence from i to j. The following example illustrates the search.

4.4 Exercises

Current Iterate: (A, C, D, B, E)

$i=1, j=2$	(C,A,D,B,E)	$i=2,j=3$	(A,D,C,B,E)
$i=1, j=3$	(D,C,A,B,E)	$i=2,j=4$	(A,B,D,C,E)
$i=1, j=4$	(B,D,C,A,E)	$i=2,j=5$	(A,E,B,D,C)
$i=1, j=5$	(E,B,D,C,A)		
$i=3, j=4$	(A,C,B,D,E)	$i=4,j=5$	(A,C,D,E,B)
$i=3, j=5$	(A,C,E,B,D)		

The best of these tours becomes the new iterate, and the process continues until no improvement is found. If distances are symmetric, meaning that the distance from location i to j is the same as the distance from location j to i, then the lengths of these tours can be calculated efficiently (see if you can figure out the formula). Initiate 2-opt with nearest multiple neighbor solutions. For example, the first initial tour would be constructed by starting at city 1, from which you would move to the closest city. The process continues to generate a tour by iteratively selecting the nearest city not yet visited. The second initiation is constructed similarly but starts at city 2, the third would start with city 3, and so on. Of the three solution techniques, which seems best?

# Chapter 5
Ordinary Differential Equations

> In order to solve this differential equation you look at it until a solution occurs to you. – George Polya

> If you assume continuity, you can open the well-stocked mathematical toolkit of continuous functions and differential equations, the saws and hammers of engineering and physics for the past two centuries (and the foreseeable future). – Benoit Mandelbrot

> Man's destiny is to know, if only because societies with knowledge culturally dominate societies that lack it. Luddites and anti-intellectuals do not master the differential equations of thermodynamics or the biochemical cures of illness. They stay in thatched huts and die young. – E. O. Wilson

Integral calculus provides the mathematical tools to solve differential equations; however, integration isn't always straightforward, and even symbolic software can be challenged in routine situations. Moreover, many applications are complicated, and closed-form solutions are either impractical or impossible to compute. The methods of this chapter instead develop a class of widely used algorithms to iteratively approximate a solution.

Throughout this chapter we consider differential equations of the form

$$\frac{d}{dt} y = f(t, y).$$

The dependent variable y is a function of t, and it need not be single valued. For example, the second order equation

$$u'' + 2u' + u = \sin(t)$$

can be re-modeled by introducing a new function, v, and requiring that $u' = v$. We then define the vector valued function y as $y = (u, v)^T$ and notice that the equation above may be re-written as

$$\frac{d}{dt} y = \frac{d}{dt} \begin{pmatrix} u \\ v \end{pmatrix} = \begin{pmatrix} v \\ \sin(t) - 2v - u \end{pmatrix} = f(t, y). \qquad (5.1)$$

The solution to this differential equation with an initial condition of $y(0) = (0, 1)^T$ is

$$y(t) = \begin{pmatrix} u(t) \\ v(t) \end{pmatrix} = \frac{1}{2} \begin{pmatrix} -\cos(t) + e^{-t} + 3te^{-t} \\ \sin(t) + 2e^{-t} - 3te^{-t} \end{pmatrix},$$

which can be verified by substituting the derivatives of u and v into (5.1).

Our computational development is divided into three parts, the first of which develops three Euler methods that foreshadow the development of more advanced techniques. The second part develops the most celebrated algorithms used to solve ordinary differential equations, those being the Runge-Kutta methods. We motivate the overriding scheme of a Runge-Kutta method by developing a second order technique, after which higher order methods are introduced. The third part considers adaptive solution methods that regularly increase computational efficiency.

5.1 Euler Methods

One of the most pervasive themes in calculus is to estimate a function by its best linear approximation,

$$y(t + \Delta t) \approx y(t) + \Delta t\, D_t y(t),$$

where we expect the approximation to be reasonable if Δt is sufficiently small. A solution method associated with this linear approximation is referred to as an Euler method, and the central theme is that the differential equation

$$\frac{d}{dt} y = f(t, y)$$

provides the linear approximation,

$$y(t + \Delta t) \approx y(t) + \Delta t\, f(t, y). \tag{5.2}$$

Using (5.2) to approximation $y(t+\Delta t)$ is the forward Euler method, which is so common that it is frequently called *the* Euler method. Consider the differential equation in (5.1) to illustrate the technique. Assuming the initial condition $y(0) = y_0 = (0, 1)^T$ with $\Delta t = 1/10$, we have

$$y(0 + \Delta t) = y(0 + 0.1) \approx y(0) + \frac{1}{10} D_t\, y(0)$$

$$= \begin{pmatrix} 0 \\ 1 \end{pmatrix} + \frac{1}{10} \begin{pmatrix} 1 \\ \sin(0) - 2 \end{pmatrix}$$

$$= \begin{pmatrix} 0.1 \\ 0.8 \end{pmatrix} = y_1,$$

5.1 Euler Methods

This approximation is then used to calculate an estimate of $y(0+2\Delta t)$,

$$y(0+2\Delta t) = y(0+2(0.1)) \approx y(0.1) + \frac{1}{10} D_t f(0.1, y(0.1))$$

$$\approx \begin{pmatrix} 0.1 \\ 0.8 \end{pmatrix} + \frac{1}{10} \begin{pmatrix} 0.8 \\ \sin(0.1) - 2(0.8) - 0.1 \end{pmatrix}$$

$$= \begin{pmatrix} 0.18 \\ 0.66 \end{pmatrix} = y_2.$$

The resulting approximations generated by continuing in this fashion through $t = 1$ are listed in Table 5.1. Pseudocode is in Algorithm 21. We notationally set t_k to be the k-th value of t at which we approximate y, and we denote the approximation as y_k. So $y(t_k)$ is the actual solution at t_k, and y_k is its approximation. Notice that this notation was used in the estimates above. If the time step is the uniform value of Δt, then $t_k = t_0 + k\Delta t$, but if not, then we generally have $t_k = t_0 + \Delta t_1 + \Delta t_2 + \ldots + \Delta t_k$.

k	0	1	2	3	4	5
t_k	0	0.1	0.2	0.3	0.4	0.5
$y_k = \begin{pmatrix} u_k \\ v_k \end{pmatrix}$	$\begin{pmatrix} 0.00 \\ 1.00 \end{pmatrix}$	$\begin{pmatrix} 0.10 \\ 0.80 \end{pmatrix}$	$\begin{pmatrix} 0.18 \\ 0.64 \end{pmatrix}$	$\begin{pmatrix} 0.24 \\ 0.51 \end{pmatrix}$	$\begin{pmatrix} 0.30 \\ 0.42 \end{pmatrix}$	$\begin{pmatrix} 0.34 \\ 0.34 \end{pmatrix}$

k	6	7	8	9	10
t_k	0.6	0.7	0.8	0.9	1.0
$y_k = \begin{pmatrix} u_k \\ v_k \end{pmatrix}$	$\begin{pmatrix} 0.37 \\ 0.29 \end{pmatrix}$	$\begin{pmatrix} 0.40 \\ 0.25 \end{pmatrix}$	$\begin{pmatrix} 0.43 \\ 0.22 \end{pmatrix}$	$\begin{pmatrix} 0.45 \\ 0.21 \end{pmatrix}$	$\begin{pmatrix} 0.47 \\ 0.20 \end{pmatrix}$

Table 5.1 Iterates of forward Euler's method from $t = 0$ to $t = 1$ with a step size of $\Delta t = 0.1$ for the ordinary differential equation in (5.1).

An iterative solution procedure such as forward Euler's method produces discrete estimates of the actual solution, and these estimates can be joined with a spline to visualize the approximate solution. Natural splines of the approximations in Table 5.1 for problem (5.1) are graphed against the real solutions in Figs. 5.1 and 5.2.

Algorithm 21 Pseudocode for forward Euler's method

$k = 0$
while t_k is in range **do**
 set $y_{k+1} = y_k + \Delta t\, f(t_k, y_k)$
 set $t_{k+1} = t_k + \Delta t$
 set $k = k + 1$
end while

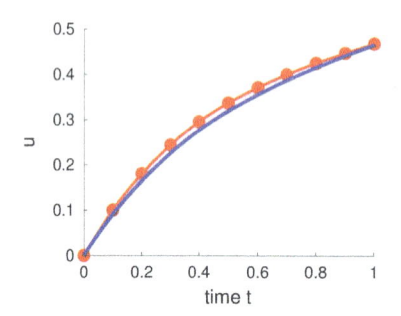

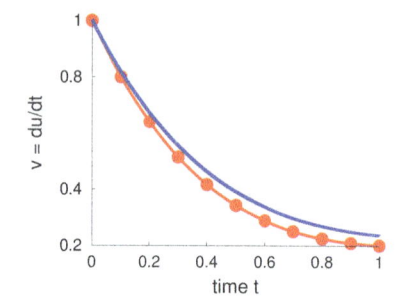

Fig. 5.1 The solution to problem (5.1) is graphed in blue, and the approximations from forward Euler's method are red dots. The red curve is a natural cubic spline of the discrete estimates from forward Euler's method.

Fig. 5.2 The derivative of the solution to problem 5.1 is graphed in blue, and the approximations from forward Euler's method are red dots. The red curve is a natural cubic spline of the discrete estimates from forward Euler's method.

The derivative, dy/dt, varies over the interval from t to $t + \Delta t$ unless the solution y is affine. Forward Euler's method estimates dy/dt at the leftmost endpoint t, but an alternative, called backward Euler's method, estimates dy/dt at the rightmost endpoint $t + \Delta t$. The motivating update in (5.2) then becomes
$$y(t + \Delta t) \approx y(t) + \Delta t\, f(t + \Delta t, y(t + \Delta t)),$$
which suggests that y_k and y_{k+1} satisfy
$$y_{k+1} = y_k + \Delta t\, f(t_{k+1}, y_{k+1}). \tag{5.3}$$

Unlike the iterates of forward Euler's method, backward approximations do not provide an explicit, forward calculation to estimate $y(t + \Delta t)$ from $y(t)$. The backward update instead needs to be deduced at each iteration, and for this reason, backward Euler's method is implicit. Forward Euler's method is instead explicit since each iteration only depends on information that has already been calculated.

5.1 Euler Methods

We illustrate the implicit calculation by noting that the estimate y_1 of $y(0.1)$ for problem (5.1) satisfies

$$y_1 = \begin{pmatrix} u_1 \\ v_1 \end{pmatrix} = \begin{pmatrix} 0 \\ 1 \end{pmatrix} + \frac{1}{10} \begin{pmatrix} v_1 \\ \sin(0.1) - 2v_1 - u_1 \end{pmatrix} = y_0 + \Delta t\, f(t_1, y_1).$$

Recognizing that the second equality is the same as

$$\begin{pmatrix} u_1 - 0.1 v_1 \\ 0.1 u_1 + 1.2 v_1 - 1 - 0.1\sin(0.1) \end{pmatrix} = \begin{pmatrix} 0 \\ 0 \end{pmatrix},$$

we see that a general system solver, such as Newton's method, can be used to calculate the unknowns u_1 and v_1—although a linear solver would suffice for this example. A natural initial guess at the solution is $y_0 = (u_0, v_0)^T = (0, 1)^T$, from which Newton's method calculates the solution $y_1 \approx (0.08347, 0.83470)^T$. We could then repeat to estimate $y(0.2)$ by solving

$$y_2 = \begin{pmatrix} u_2 \\ v_2 \end{pmatrix} = \begin{pmatrix} 0.08347 \\ 0.83470 \end{pmatrix} + \frac{1}{10} \begin{pmatrix} v_2 \\ \sin(0.2) - 2v_2 - u_2 \end{pmatrix}.$$

Continuing the process through $t = 1$ gives the results in Table 5.2. The solution is depicted in Figs. 5.3 and 5.4.

Equation (5.3) exemplifies the general requirement of solving the following equation at each iteration of backward Euler's method,

$$G(z) = z - y_k - \Delta t\, f(t_{k+1}, z) = 0.$$

The solution, say $\hat{z}$, is our computed estimate of $y(t_{k+1})$, i.e. $\hat{z} = y_{k+1}$. Notice that

$$D_z G(z) = I - \Delta t\, D_z f(t_{k+1}, z).$$

So if we know $D_z f(t, z)$, which is the same as $D_y f(t, y)$ with y renamed as z, then we can solve $G(z) = 0$ with Newton's method. Algorithm 22 assumes $D_z f(t, z)$ is known.

Algorithm 22 Pseudocode for backward Euler's method

$k = 0$
  while t_k is in range **do**
    $t_{k+1} = t_k + \Delta t$
    set $G(z) = z - y(t_k) - \Delta t\, f(t_{k+1}, z)$
    set $DG(z) = I - \Delta t D_z f(t_{k+1}, z)$
    solve $G(z) = 0$ with Newton's method initiated with y_k
    set y_{k+1} equal to the solution of $G(z) = 0$
    set $k = k + 1$
  end while
---

k	0	1	2	3	4	5
t_k	0	0.1	0.2	0.3	0.4	0.5
$y_k = \begin{pmatrix} u_k \\ v_k \end{pmatrix}$	$\begin{pmatrix} 0.00 \\ 1.00 \end{pmatrix}$	$\begin{pmatrix} 0.08 \\ 0.83 \end{pmatrix}$	$\begin{pmatrix} 0.15 \\ 0.70 \end{pmatrix}$	$\begin{pmatrix} 0.21 \\ 0.59 \end{pmatrix}$	$\begin{pmatrix} 0.26 \\ 0.50 \end{pmatrix}$	$\begin{pmatrix} 0.31 \\ 0.43 \end{pmatrix}$

k	6	7	8	9	10
t_k	0.6	0.7	0.8	0.9	1.0
$y_k = \begin{pmatrix} u_k \\ v_k \end{pmatrix}$	$\begin{pmatrix} 0.34 \\ 0.38 \end{pmatrix}$	$\begin{pmatrix} 0.38 \\ 0.34 \end{pmatrix}$	$\begin{pmatrix} 0.41 \\ 0.31 \end{pmatrix}$	$\begin{pmatrix} 0.44 \\ 0.29 \end{pmatrix}$	$\begin{pmatrix} 0.46 \\ 0.27 \end{pmatrix}$

Table 5.2 Iterates of backward Euler's method from $t = 0$ to $t = 1$ with a step size of $\Delta t = 0.1$ for the ordinary differential equation in (5.1).

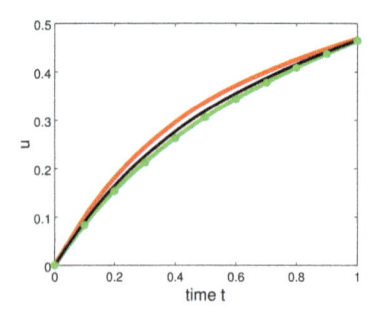

Fig. 5.3 The solution to the problem in (5.1) is graphed in black. The discrete approximation from the backward Euler method is shown as green dots. The green curve is a cubic spline of the discrete estimates from the backward Euler method. A cubic spline of the approximations from the forward Euler method is graphed in red.

Fig. 5.4 The derivative of the solution to the problem in (5.1) is graphed in black. The discrete approximation from the backward Euler method is shown as green dots. The green curve is a cubic spline of the discrete estimates from the backward Euler method. A cubic spline of the approximations from the forward Euler method is graphed in red.

The solutions in Fig. 5.3 suggest that the backward and forward Euler methods might combine, say by averaging, to improve accuracy. This observation leads to other methods and foreshadows the Runge-Kutta algorithms of the next section. Moreover, a close discernment of Fig. 5.3 hints that the backward, implicit algorithm provides a better approximation than does the

5.1 Euler Methods

forward, explicit algorithm. Indeed, the maximum errors of the backward method are 0.014 and 0.041, for u and v, respectively, whereas the maximum errors of the forward method are slightly larger at 0.019 and 0.05. As a general rule implicit methods require additional computational effort per iteration due to the need to solve an equation, but this effort is often rewarded with improved accuracy over an explicit method with the same step size.

A third Euler method, called improved Euler's method, estimates the derivative of y over the interval from t to $t + \Delta t$ by averaging estimates of the derivative at the end points. The idea is to estimate

$$D_t y(s) \approx \frac{1}{2}\left(D_t y(t) + D_t y(t + \Delta t)\right) = \frac{1}{2}\left(f(t, y(t)) + f(t + \Delta t, y(t + \Delta t))\right),$$

where s is between t and $t + \Delta t$. Notice that such an update would be implicit as stated, and therefore each forward step could take a fair amount of work. Improved Euler's method instead estimates the value of $y(t + \Delta t)$ using one step of forward Euler's method prior to performing the averaging above.

Specifically, at any point s between t and $t + \Delta t$ we estimate $D_t y(s)$ as

$$D_t y(s) \approx \frac{1}{2}\left(f(t, y(t)) + f(t + \Delta t, y(t) + \Delta t\, f(t, y(t)))\right),$$

which leads to y_{k+1} being calculated explicitly as

$$y_{k+1} = y_k + \frac{\Delta t}{2}\left(f(t_k, y_k) + f(t_{k+1}, y_k + \Delta t\, f(t_k, y_k))\right).$$

Averaging derivative estimates often improves the accuracy of the solution, and indeed, the iterates of improved Euler's method for problem (5.1) essentially lie on the actual solution depicted in Fig. 5.3. Improved Euler's method is explicit, and the iterates are calculated directly from the current iterate without solving an equation. Pseudocode for improved Euler's method is in Algorithm 23. We will see in the next section that improved Euler's method is an instance of a Runge-Kutta algorithm.

Algorithm 23 Pseudocode for improved Euler's method

$k = 0$
   while t_k is in range **do**
      set $t_{k+1} = t_k + \Delta t$
      set $y_{k+1} = y_k + (\Delta t/2)\, (f(t_k, y_k) + f(t_{k+1}, y_k + \Delta t\, f(t_k, y_k)))$
      set $k = k + 1$
   end while

5.2 Runge-Kutta Methods

Carl Runge and Martin Kutta developed numerical algorithms near the start of the twentieth century to solve differential equations, and an entire family of related algorithms now bears their names. The core idea is to extend Euler's method to add analytic information about the solution. We develop the family of second order Runge-Kutta methods to illustrate the overriding theme and then review the traditional fourth order method that is commonly employed. The concept and definition of order is further developed in Sect. 5.3.

A second order Runge-Kutta algorithm, or equivalently an RK2 method, has the form

$$y_{k+1} = y_k + b_1 k_1 + b_2 k_2, \tag{5.4}$$

where

$$k_1 = \Delta t\, f(t_k, y_k) \text{ and} \tag{5.5}$$
$$k_2 = \Delta t\, f(t_k + c_2 \Delta t,\, y_k + a_2 k_1). \tag{5.6}$$

Any particular algorithm is defined by the selection of b_1, b_2, c_2, and a_2. These parameters are chosen so that the update aligns with the second order Taylor polynomial,

$$y(t + \Delta t) \approx y(t) + D_t y(t)\, \Delta t + \frac{1}{2} D_t^2 y(t)\, \Delta t^2. \tag{5.7}$$

To equate the RK2 iteration with the second order Taylor polynomial of y, first notice that the linear approximation of f in expression (5.6) is

$$f(t_k + c_2 \Delta t,\, y_k + a_2 k_1) \approx f(t_k, y_k) + \nabla f(t_k, y_k)^T \begin{pmatrix} c_2 \Delta t \\ a_2 k_1 \end{pmatrix}$$

$$= f(t_k, y_k) + \nabla f(t_k, y_k)^T \begin{pmatrix} c_2 \Delta t \\ a_2 \Delta t\, f(t_k, y_k) \end{pmatrix}$$

$$= f(t_k, y_k) + \Delta t\, \nabla f(t_k, y_k)^T \begin{pmatrix} c_2 \\ a_2\, f(t_k, y_k) \end{pmatrix},$$

where the first equality follows from (5.5). We have from the chain rule that the second derivative in the Taylor polynomial of y is

$$D_t^2 y(t) = \nabla f(t, y(t))^T \begin{pmatrix} dt/dt \\ dy/dt \end{pmatrix} = \nabla f(t, y(t))^T \begin{pmatrix} 1 \\ f(t, y(t)) \end{pmatrix}.$$

These last two observations along with the motivating approximation in (5.7) indicate that the parameters of an RK2 method should be selected so that

5.2 Runge-Kutta Methods

$$y(t_k + \Delta t) \approx y_{k+1} = y_k + b_1 k_1 + b_2 k_2$$

$$\approx y_k + b_1 \Delta t\, f(t_k, y_k) + b_2 \Delta t \left(f(t_k, y_k) + \Delta t\, \nabla f(t_k, y_k)^T \begin{pmatrix} c_2 \\ a_2\, f(t_k, y_k) \end{pmatrix} \right)$$

$$= y_k + (b_1 + b_2)\Delta t\, f(t_k, y_k) + \Delta t^2\, \nabla f(t_k, y_k)^T \begin{pmatrix} b_2\, c_2 \\ b_2\, a_2\, f(t_k, y_k) \end{pmatrix}$$

$$\approx y(t_k) + \Delta t\, f(t_k, y(t_k)) + \Delta t^2\, \nabla f(t_k, y(t_k))^T \begin{pmatrix} (1/2) \\ (1/2)\, f(t_k, y(t_k)) \end{pmatrix}$$

$$= y(t_k) + D_t y(t_k)\, \Delta t + \frac{1}{2}\, D_t^2 y(t_k)\, \Delta t^2.$$

The last approximation in red is exact if $y(t_k) = y_k$ and if

$$b_1 + b_2 = 1, \quad b_2\, c_2 = 1/2, \quad \text{and} \quad b_2\, a_2 = 1/2.$$

The parameters b_1, b_2, a_2, and c_2 are not uniquely defined by these conditions, and any parameters satisfying these equations defines an RK2 algorithm. Hence the RK2 method is a family of algorithms instead of a unique technique.

The parameters c_2 and a_2 must be the common value of $1/(2\,b_2)$ in an RK2 method, and one important member of this class has $a_2 = c_2 = 1$ and $b_1 = b_2 = 1/2$. In this case the RK2 method becomes the improved Euler update,

$$y_{k+1} = y_k + \frac{\Delta t}{2}\Big(f(t_k, y_k) + f(t_{k+1}, y_k + \Delta t\, f(t_k, y_k))\Big).$$

So the RK2 family includes improved Euler's method.

One of the most successful methods is a fourth order RK method whose update is

$$y_{k+1} = y_k + \frac{1}{6}\left(k_1 + 2k_2 + 2k_3 + k_4\right),$$

where

$$k_1 = \Delta t\, f(t_k, y_k),$$
$$k_2 = \Delta t\, f(t_k + (1/2)\Delta t,\ y_k + (1/2)\, k_1),$$
$$k_3 = \Delta t\, f(t_k + (1/2)\Delta t,\ y_k + (1/2)\, k_2), \text{ and}$$
$$k_4 = \Delta t\, f(t_k + \Delta t,\ y_k + k_3).$$

This update is *the* Runge-Kutta method, and it has remained one of the most celebrated solution algorithms since its original design. The method is explicit since the update is directly calculated from the current iterate. Pseudocode for RK4 is in Algorithm 24. The development of the RK4 family is tedious, but the mathematical goal of matching the RK update to the fourth order Taylor polynomial is a direct counterpart of our RK2 development.

Algorithm 24 Pseudocode for RK4

$k = 0$
while t_k is in range **do**
    Calculate in order,
    $k_1 = \Delta t \, f(t_k, y_k)$,
    $k_2 = \Delta t \, f(t_k + (1/2)\Delta t, \, y_k + (1/2) k_1)$,
    $k_3 = \Delta t \, f(t_k + (1/2)\Delta t, \, y_k + (1/2) k_2)$, and
    $k_4 = \Delta t \, f(t_k + \Delta t, \, y_k + k_3)$.

    set $y_{k+1} = y_k + (1/6) (k_1 + 2k_2 + 2k_3 + k_4)$
    set $t_{k+1} = t_k + \Delta t$
    set $k = k + 1$
end while

Returning to problem (5.1) with an initial condition of $y_0 = (0,1)^T$, we have that the first iterate of the RK4 method with $\Delta t = 1$ is

$$y_1 = y_0 + (1/6)(k_1 + 2k_2 + 2k_3 + k_4)$$
$$= \begin{pmatrix} 0 \\ 1 \end{pmatrix} + \frac{1}{6}\left(\begin{pmatrix} 1.00 \\ -2.00 \end{pmatrix} + \begin{pmatrix} 0.00 \\ -0.02 \end{pmatrix} + \begin{pmatrix} 0.99 \\ -1.50 \end{pmatrix} + \begin{pmatrix} -0.50 \\ 0.85 \end{pmatrix} \right) = \begin{pmatrix} 0.41 \\ 0.30 \end{pmatrix},$$

where

$$k_1 = (1)f\left(0, \, y_0\right) = f\left(0, \, (0,1)^T\right)$$
$$= (1)\begin{pmatrix} 1 \\ \sin(0) - 2(1) - 0 \end{pmatrix} = \begin{pmatrix} 1.00 \\ -2.00 \end{pmatrix},$$
$$k_2 = (1)f\left(0 + (1/2)(1), \, y_0 + (1/2)k_1\right) = f\left(1/2, (1/2, 0)^T\right)$$
$$= \begin{pmatrix} 0 \\ \sin(1/2) - 2(0) - 1/2 \end{pmatrix} = \begin{pmatrix} 0.00 \\ -0.02 \end{pmatrix},$$
$$k_3 = (1)f\left(0 + (1/2)(1), \, y_0 + (1/2)k_2\right) = f\left(1/2, \, (0, 0.99)^T\right)$$
$$= \begin{pmatrix} 0.99 \\ \sin(1/2) - 2(0.99) - 0 \end{pmatrix} = \begin{pmatrix} 0.99 \\ -1.50 \end{pmatrix}, \text{ and}$$
$$k_4 = (1)f\left(0 + 1, \, y_0 + k_3\right) = f\left(1, \, (0.99, -0.50)^T\right)$$
$$= \begin{pmatrix} -0.50 \\ \sin(1) + 2(0.50) - 0.99 \end{pmatrix} = \begin{pmatrix} -0.50 \\ 0.85 \end{pmatrix}.$$

5.2 Runge-Kutta Methods

The iterates in Table 5.3 are the result of continuing the RK4 algorithm over the interval from $t = 0$ to $t = 10$. Figure 5.5 depicts the RK4 solution along with those of the improved and forward Euler methods. The RK4 algorithm is reasonably accurate even with the exaggerated step size of $\Delta t = 1$. The accuracy of the improved Euler method increases as the solution settles into its steady state, whereas the forward Euler method fails to mimic the solution with the large step size.

$k = t_k$	0	1	2	3	4	5
$y_k = \begin{pmatrix} u_k \\ v_k \end{pmatrix}$	$\begin{pmatrix} 0.00 \\ 1.00 \end{pmatrix}$	$\begin{pmatrix} 0.41 \\ 0.30 \end{pmatrix}$	$\begin{pmatrix} 0.66 \\ 0.21 \end{pmatrix}$	$\begin{pmatrix} 0.75 \\ -0.12 \end{pmatrix}$	$\begin{pmatrix} 0.45 \\ -0.48 \end{pmatrix}$	$\begin{pmatrix} -0.10 \\ -0.51 \end{pmatrix}$

$k = t_k$	6	7	8	9	10
$y_k = \begin{pmatrix} u_k \\ v_k \end{pmatrix}$	$\begin{pmatrix} -0.48 \\ -0.13 \end{pmatrix}$	$\begin{pmatrix} -0.38 \\ 0.34 \end{pmatrix}$	$\begin{pmatrix} 0.08 \\ 0.49 \end{pmatrix}$	$\begin{pmatrix} 0.48 \\ 0.44 \end{pmatrix}$	$\begin{pmatrix} 0.18 \\ -0.30 \end{pmatrix}$

Table 5.3 Iterates of the RK4 method from $t = 0$ to $t = 10$ with a step size of $\Delta t = 1$, which sets $k = t_k$, for the ordinary differential equation in (5.1).

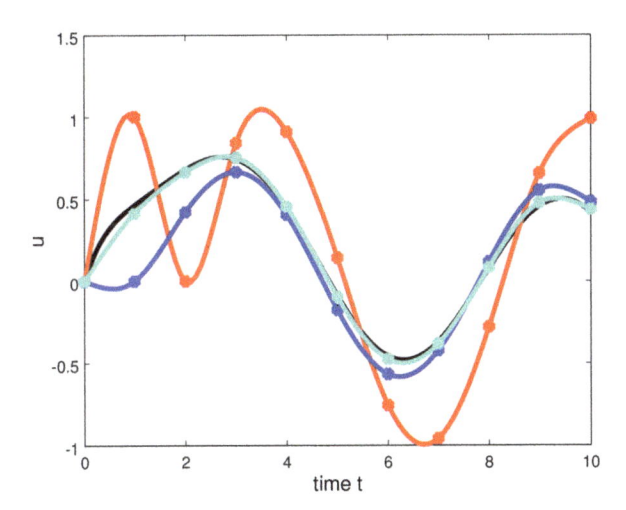

Fig. 5.5 The solution to the problem in (5.1) is graphed in black. Approximations from the forward Euler method (red), improved Euler method (blue), and RK4 (cyan) are graphed as cubic splines of their discrete approximations with $\Delta t = 1$.

There are numerous methods beyond RK4, but again, the motivation of matching a Taylor polynomial remains the central theme. Higher order methods are not always preferred, or indeed appropriate, over their lower order counterparts. Higher order methods require additional function evaluations, which can be expensive, and gains in accuracy for large step sizes may not warrant the extra time needed to calculate the higher order update.

5.3 Quantifying Error

The variety of solution methods investigated thus far invites the question of which would produce the smallest error, a question we approach by considering scalar ordinary differential equations. The local truncation error of a method is the error introduced by taking one time step under the assumption that there is no error in the current iterate. So the local truncation error is

$$e_{k+1} = y(t_{k+1}) - y_{k+1} \tag{5.8}$$

under the assumption that $y(t_k) = y_k$.

We first study the local truncation error of Euler's method, which sets

$$y_{k+1} = y_k + \Delta t\, f(t_k, y_k).$$

We have from Taylor's theorem that

$$y(t_{k+1}) = y(t_k) + \Delta t\, y'(t_k) + \frac{\Delta t^2}{2} y''(\xi_k),$$

where ξ_k is some point between t_k and t_{k+1}. So in the case of Euler's method the local truncation error is

$$e_{k+1} = y(t_k) + \Delta t\, y'(t_k) + \frac{\Delta t^2}{2} y''(\xi_k) - (y_k + \Delta t\, f(t_k, y_k)).$$

Since $y_k = y(t_k)$ by assumption and because $y'(t_k) = f(t_k, y(t_k))$, the local truncation error for Euler's method reduces to

$$e_{k+1} = \frac{\Delta t^2}{2} y''(\xi_k). \tag{5.9}$$

If we further assume that y'' is continuous, then ξ_k being between t_k and $t_k + \Delta t$ gives

$$\lim_{\Delta t \downarrow 0} |e_{k+1}| = \lim_{\Delta t \downarrow 0} \frac{\Delta t^2}{2} |y''(\xi_k)| = 0.$$

So the local truncation error associated with calculating y_{k+1} from y_k with Euler's method converges to zero as $\Delta t \downarrow 0$.

5.3 Quantifying Error

We characterize convergence by comparing error relative to its step size. Local truncation error is a function of Δt, and we say that e_{k+1} is order p if there exists a nonzero constant C_k such that

$$\frac{|e_{k+1}|}{\Delta t^p} \leq C_k,$$

where C_k is independent of Δt. Equation (5.9) shows that

$$\frac{|e_{k+1}|}{\Delta t^2} = \frac{|y''(\xi_k)|}{2},$$

where ξ_k is between t_k and $t_k + \Delta t$. Thus ξ_k depends on Δt, and $|y''(\xi_k)|/2$ cannot be C_k since $y''(\xi_k)$ depends on Δt. However, if we assume y'' is continuous, then Theorem 3 in Chap. 1 guarantees that y'' attains its maximum and minimum over any finite time interval. So if y'' is continuous and the time interval is finite, then there is a C_k independent of Δt such that

$$\frac{|e_{k+1}|}{\Delta t^2} = \frac{|y''(\xi_k)|}{2} \leq C_k.$$

So the local truncation error of Euler's method is order 2 under the assumptions that y'' is continuous and the time interval is finite. A similar analysis shows that the local truncation error for backward Euler's method is also order two, while a more involved derivation shows that the local truncation error for improved Euler's method is order 3. The local truncation error of the Runge-Kutta method in Sect. 5.2 is order 5.

Equation (5.9) shows that the ratio $e_{k+1}/\Delta t^2$ approaches $y''(t_k)$ as Δt decreases to zero, again assuming that y'' is continuous. So for small, positive Δt we have

$$e_{k+1} \approx \frac{y''(t_k)}{2} \Delta t^2.$$

The assumed continuity of y'' guarantees that $|y''(t_k) - y''(t_k + \Delta t)|$ can be made arbitrarily small as long as Δt is suitably small, and we can subsequently approximate $y''(t)/2$ with a constant if $t_k \leq t \leq t_k + \Delta t$ and Δt is sufficiently small. Calling this constant Γ_k, we have

$$e_{k+1} \approx \Gamma_k \Delta t^2 \text{ if } \Delta t \text{ is sufficiently small.} \tag{5.10}$$

This approximation aids interpretation. For instance, suppose a numerical method has local truncation error of order p, then

$$e_{k+1} \approx \Gamma_k \Delta t^p. \tag{5.11}$$

Because we consider Δt to be very small, Δt^p is much smaller for large values of p. Thus large values of p result in small errors, and in general, numerical methods with high-order converge to the exact solution more quickly than do methods with low-order as long as the exact solution is sufficiently differentiable. We thus expect the Runge-Kutta method to converge to the actual solution faster than improved Euler's method, and for improved Euler's method to converge faster than both forward and backward Euler's methods.

Once the local truncation error of a method is established, the next natural question is, how much error does a method accrue as it approximates the solution throughout the entire time domain? The local truncation error analysis is perfectly applicable to estimating the error in y_1 because we assume an exact value for y_0, that is we assume $y(0) = y_0$. However, a solver calculates y_2 from the estimate y_1, which is decidedly imperfect. The global absolute error, which we usually refer to as just the global error, is defined as the magnitude of the largest error produced by the ordinary differential equation solver. So the global absolute error is

$$\eta = \max_k |y(t_k) - y_k|. \tag{5.12}$$

We set $\eta_k = y(t_k) - y_k$ to investigate the global error in Euler's method. Notice that η_k is different from e_k since the former assumes y_{k-1} is an approximation of $y(t_{k-1})$ whereas the latter assumes $y_{k-1} = y(t_{k-1})$. So η_k is the total error that accrues in the estimate of $y(t_k)$. We have from $y(t_0) = y_0$ that

$$\eta_1 = y(t_1) - y_1 = y(t_0) + \Delta t\, f(t_0, y(t_0)) + \frac{\Delta t^2}{2} y''(\xi_0) - (y_0 + \Delta t f(t_0, y_0))$$

$$= \frac{\Delta t^2}{2} y''(\xi_0),$$

where $t_0 \leq \xi_0 \leq t_1$. We now have

$$\eta_2 = y(t_2) - y_2 = y(t_1) + \Delta t f(t_1, y(t_1)) + \frac{\Delta t^2}{2} y''(\xi_1) - (y_1 + \Delta t f(t_1, y_1))$$

$$= y(t_1) - y_1 + \Delta t\, [f(t_1, y(t_1)) - f(t_1, y_1)] + \frac{\Delta t^2}{2} y''(\xi_1)$$

$$= \eta_1 + \Delta t\, [f(t_1, y(t_1)) - f(t_1, y_1)] + \frac{\Delta t^2}{2} y''(\xi_1),$$

with $t_1 \leq \xi_1 \leq t_2$. If we assume that $\partial f/\partial y$ exists, then we can use a Taylor series expansion of f to get

$$\eta_2 = \eta_1 + \Delta t \left[f(t_1, y_1) + (y(t_1) - y_1) \frac{\partial f}{\partial y}(t_1, \omega_1) - f(t_1, y_1) \right] + \frac{\Delta t^2}{2} y''(\xi_1)$$

$$= \eta_1 + \Delta t \frac{\partial f}{\partial y}(t_1, \omega_1) \eta_1 + \frac{\Delta t^2}{2} y''(\xi_1)$$

$$= \left(1 + \Delta t \frac{\partial f}{\partial y}(t_1, \omega_1) \right) \eta_1 + \frac{\Delta t^2}{2} y''(\xi_1),$$

5.3 Quantifying Error

where ω_1 is between $y(t_1)$ and y_1. This calculation illustrates an intuitive idea; the error in y_2 is the local truncation error plus some additional error stemming from the fact that y_1 is not exactly known. Indeed, we can easily calculate for any k that

$$\eta_k = \left(1 + \Delta t \frac{\partial f}{\partial y}(t_{k-1}, \omega_{k-1})\right)\eta_{k-1} + \frac{\Delta t^2}{2} y''(\xi_{k-1}). \tag{5.13}$$

Let us assume that y'' not only exists, but that it is also bounded so that $|y''(t)| < M$ for all t. Notice that Theorem 3 in Chap. 1 guarantees this bound over any interval if y'' is continuous. We further assume that $\partial f/\partial y$ is bounded by M, i.e. $|\partial f/\partial y| < M$ for any (t, y), which can again be assured by assuming that $\partial f/\partial y$ is continuous. We can then apply Eq. (5.13) recursively to find

$$|\eta_k| < (1 + \Delta t\, M)\, |\eta_{k-1}| + \frac{\Delta t^2}{2} M$$

$$< (1 + \Delta t\, M) \left[(1 + \Delta t M)\, |\eta_{k-2}| + \frac{\Delta t^2}{2} M\right] + \frac{\Delta t^2}{2} M$$

$$= (1 + \Delta t\, M)^2 \, |\eta_{k-2}| + \frac{\Delta t^2}{2} M\left((1 + \Delta t\, M) + 1\right)$$

$$< (1 + \Delta t\, M)^3 \, |\eta_{k-3}| + \frac{\Delta t^2}{2} M\left[(1 + \Delta t\, M)^2 + (1 + \Delta t\, M) + 1\right]$$

$$\vdots$$

$$< (1 + \Delta t\, M)^k\, |\eta_0| + \frac{\Delta t^2}{2} M \sum_{k=0}^{k-1} (1 + \Delta t\, M)^k$$

$$= \frac{\Delta t^2}{2} M \sum_{k=0}^{k-1} (1 + \Delta t\, M)^k,$$

where the last equality follows from the assumption that $\eta_0 = y(t_0) - y_0 = 0$. We recognize the last expression as a geometric series and write

$$|\eta_k| < \frac{\Delta t^2}{2} M \frac{(1 + \Delta t\, M)^k - 1}{\Delta t\, M} = \frac{\Delta t}{2} \left[(1 + \Delta t\, M)^k - 1\right]. \tag{5.14}$$

Finally, let N be the last time step of our time domain $[a, b]$; that is, $t_N \leq b$ and $t_N + \Delta t > b$. We then have $N \Delta t \leq b - a$. Recalling that the exponential e^x is

$$e^x = 1 + x + \sum_{i=2}^{\infty} \frac{x^i}{i!},$$

we find that $1 + x \leq e^x$. So (5.14) gives

$$|\eta_k| < \frac{\Delta t}{2}\left[(1+\Delta t\, M)^N - 1\right]$$

$$\leq \frac{\Delta t}{2}\left(e^{M(N\Delta t)} - 1\right) \leq \frac{\Delta t}{2}\left(e^{M(b-a)} - 1\right).$$

This bound holds for all k, and hence,

$$\frac{\eta}{\Delta t} = \frac{1}{\Delta t}\max_k |y(t_k) - y_k| = \frac{1}{\Delta t}\max_k |\eta_k| \leq \frac{1}{2}\left(e^{M(b-a)} - 1\right).$$

The right-hand side is independent of Δt, and the global error of Euler's method is subsequently order 1. If the global error in a numerical method is order p, then the scheme is said to be an order p method. Euler's method is thus an order 1 method. Backward Euler's method is also order 1, while improved Euler's method is order 2. The Runge-Kutta method developed in Sect. 5.2 is an order 4 method, which is where the "4" in "RK4" comes from.

Though rigorous analysis of global error is tedious, the intuition is straightforward. At each time step the numerical method introduces an error of size Δt^p due to local truncation. By the last time step the method has made $(b-a)/\Delta t$ time steps. If the cumulative effect of the errors is additive, then we expect to have at least $(b-a)\Delta t^p/\Delta t = (b-a)\Delta t^{p-1}$ error in y_N. Indeed, the rule of thumb is that a method with order $p+1$ local truncation error is an order p method.

The global error of an order p method satisfies $\eta(\Delta t) \leq C\Delta t^p$, where we have explicitly denoted η as a function of Δt. So the ratio $\eta(\Delta t)/\Delta t^p$ is bounded above by C and below by zero, and in this case it is reasonable[1] to consider what happens if this ratio converges. Assume for the moment that

$$\lim_{\Delta t \downarrow 0} \frac{\eta(\Delta t)}{\Delta t^p} = \Gamma. \tag{5.15}$$

Then for small Δt we have $\eta(\Delta t) \approx \Gamma \Delta t^p$, and if we reduce Δt by a factor of R, then

$$\eta\left(\frac{\Delta t}{R}\right) \approx \Gamma\left(\frac{\Delta t}{R}\right)^p = \frac{\Gamma \Delta t^p}{R^p} \approx \frac{\eta(\Delta t)}{R^p}.$$

We conclude that global error reduces by an approximate factor of R^p if the step size reduces by a factor of R. For instance, if we change Δt from 10^{-1} to 10^{-2} in RK4, then we anticipate an improvement in global error on the order of 10^{-4}. So if the global error with $\Delta t = 10^{-1}$ is $\eta(10^{-1}) \approx 2^{-3}$, then $\eta(10^{-2}) \approx 2^{-7}$, which is quite the improvement.

We consider in Table 5.4 the ordinary differential equation,

---

[1] We technically have that if $\Delta t_j \downarrow 0$ as $j \to \infty$, then there is a subsequence Δt_{j_i} such that $\eta(\Delta_{j_i})/\Delta_{j_i}$ converges as $i \to \infty$. We are assuming such a subsequence for motivational purposes.

$$y' = y + 3\cos(3t)\,e^{-1/10t} - (11/10)\sin(3t)\,e^{-1/10t},\ y(0) = 0, \qquad (5.16)$$

where $0 \leq t \leq 3$. We compare the global error achieved by the order 1 Euler method, the order 2 improved Euler method, and the order 4 Runge-Kutta method. We see that reducing step size by a factor of 10 does indeed reduce global error in Euler's method by a factor of 10, error in improved Euler's method by a factor of 100, and error in RK4 by a factor of 10,000. Note that we can only compute global error because we know an exact solution to (5.16).

Δt	Euler	Improved Euler	RK4
1	8.6491e+00	7.2304e+00	7.7037e−01
10^{-1}	2.3182e+00	9.4585e−02	3.9749e−05
10^{-2}	2.7420e−01	1.0063e−03	3.6363e−09
10^{-3}	2.7914e−02	1.0126e−05	3.5305e−13
10^{-4}	2.7964e−03	1.0132e−07	1.9706e−14
10^{-5}	2.7969e−04	1.0131e−09	1.9995e−13

Table 5.4 Global error in Euler's method, improved Euler's method, and the RK4 method on example problem (5.16).

While high-order methods achieve low error with a few time steps, it is worth pointing out that each step takes more time to compute than a low order method. Further, in order to achieve the high convergence rate the true solution must be highly differentiable, which may be difficult to verify. Solution methods of order higher than 6 or 7 are rarely used for general purposes for these reasons.

5.4 Stiff Ordinary Differential Equations

Iterates from a numerical algorithm can be challenged by innate computational shortcomings, with one of the most common concerns being that of stiffness. Stiff models are common, and readers should learn to identify such systems and to know how to use alternative solvers to improve accuracy and speed. There is no formal definition of what constitutes a stiff (system of) differential equation(s), but the general idea is that if a solution makes a rapid change, then a numerical procedure might have difficulty approximating the solution. We introduce the concept with the first-order, nonlinear equation

$$y' = y^2 - y^3,\ y(0) = y_0. \qquad (5.17)$$

Since stiffness is about abrupt changes, we first calculate the maximum value of y'. Solving

$$\frac{d}{dy}\left(y^2 - y^3\right) = 2y - 3y^2 = y(2 - 3y) = 0,$$

we find that y' has potential optima at $y = 0$ and $y = 2/3$. Since

$$y(2 - 3y) \geq 0 \text{ if } 0 \leq y \leq 2/3 \quad \text{and} \quad y(2 - 3y) \leq 0 \text{ if } y \geq 2/3,$$

we conclude that y' has a maximum value of $(2/3)^2 - (2/3)^3 \approx 0.15$ at $y = 2/3$, provided that $y > 0$. We now address for what value of t this maximum is achieved.

We learn in Calculus that problem (5.17) can be solved by the technique of separation of variables, which gives

$$\int \frac{1}{y^2(1-y)} dy = \int dt.$$

The integral on the left is

$$\int \frac{1}{y^2(1-y)} dy = \int \frac{1}{y} + \frac{1}{y^2} + \frac{1}{1-y} dy = \ln|y| - \frac{1}{y} - \ln|1-y| + c.$$

Hence, the implicit, general solution is

$$\ln|y| - \frac{1}{y} - \ln|1-y| = t + c.$$

The constant in terms of y_0 is

$$c = \ln(y_0) - \frac{1}{y_0} - \ln(1 - y_0) = \frac{y_0 \ln(y_0/(1-y_0)) - 1}{y_0},$$

where we assume $0 < y_0 < 1$. The value of t at which the solution y achieves its maximum rate is

$$t_{\max} = \ln|2/3| - 3/2 + \ln|1/3| - \frac{y_0 \ln(y_0/(1-y_0)) - 1}{y_0}$$

$$= \ln(2/9) - 3/2 + \frac{1 - y_0 \ln(y_0/(1-y_0))}{y_0}. \tag{5.18}$$

We use L'Hopital's rule to find

5.4 Stiff Ordinary Differential Equations

$$\lim_{y_0 \to 0^+} y_0 \ln(y_0/(1-y_0)) = \lim_{y_0 \to 0^+} \frac{\ln(y_0/(1-y_0))}{1/y_0}$$

$$= \lim_{y_0 \to 0^+} \frac{\left(\frac{1-y_0}{y_0}\right)\left(\frac{(1-y_0)+y_0}{(1-y_0)^2}\right)}{\frac{-1}{y_0^2}}$$

$$= \lim_{y_0 \to 0^+} \frac{-y_0}{1-y_0}$$

$$= 0,$$

which in turn gives

$$\lim_{y_0 \to 0^+} \frac{1 - y_0 \ln(y_0/(1-y_0))}{y_0} = \infty.$$

We conclude that $t_{\max} \to \infty$ as $y_0 \downarrow 0$. A graph of $t_{\max}$ is in Fig. 5.6.

The problem in (5.17) has many favorable theoretical qualities. First, the solutions $y(t) = 0$ and $y(t) = 1$ are equilibrium solutions, meaning that they are constant functions that solve the differential equation. Second, the analytic properties of this equation guarantee a unique solution through any pair $(t, y(t))$. Hence the solution y satisfies $0 < y < 1$ as long as $0 < y_0 < 1$. Also, notice that $y' = y^2 - y^3 = y^2(1-y) > 0$ if $0 < y < 1$, which shows that the solution increases. If y_0 is positive but near 0, then we anticipate that y increases to 1 as $t \to \infty$, which is indeed true (this is a good exercise). Our calculation of $t_{\max}$ indicates that its value increases as the initial condition decreases to zero, and it is this horizontal shift to the right of $t_{\max}$, together with the rather abrupt change in y, that makes this differential equation stiff. For example, Figs. 5.7, 5.8 and 5.9 depict the solution over the range $0 \leq t \leq 2 t_{\max}$ for $y_0 = 0.1$, 0.01, and 0.001. The corresponding $t_{\max}$ values are approximately 9.2, 101.6, and 1003.9.

The reason this ordinary differential equation is stiff for y_0 near zero is that the solution behaves differently over various (uniform) time steps within the range of interest. For example, the solution is nearly flat if $y_0 = 0.001$ as long as t is only a relatively small distance away from $t_{\max}$. However, the solution climbs rapidly if t is near $t_{\max}$. Large step sizes make sense over much of the desired domain, but they might skip an important change and skew an answer over the entire calculation interval. Small steps are safer but require numerous function evaluations and added computational effort. Figures 5.10, 5.11, 5.12 and 5.13 illustrate some of the potential problems with the RK4 iterates. The step sizes in Figs. 5.12 and 5.13 only differ by 0.21 over a domain from 0 to 2007.8, i.e. by about 0.0001% of the total width. The behavior in Fig. 5.12 clearly illustrates a departure from what would be expected of the solution, but the RK solution in Fig. 5.13 almost neglects the erratic behavior and instead shows an abrupt climb and then a return to $y = 0$. We might suspect

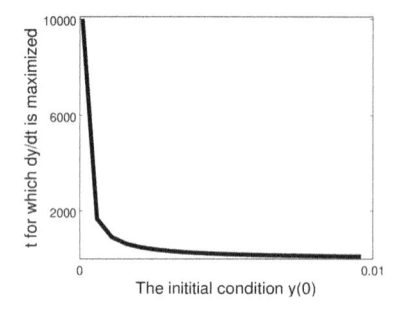

Fig. 5.6 The value t_{max} at which the solution to $dy/dt = y^2 - y^3$, $y(0) = y_0$ achieves its maximum derivative as y_0 decreases to zero.

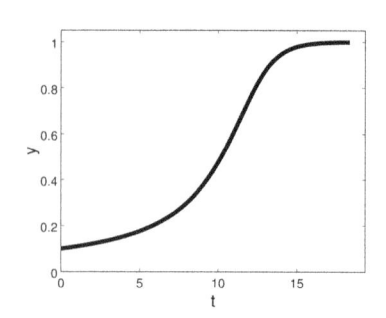

Fig. 5.7 The solution to the differential equation in (5.17) with $y_0 = 0.1$ over the range $0 \leq t \leq 2\,t_{max} = 18.4$.

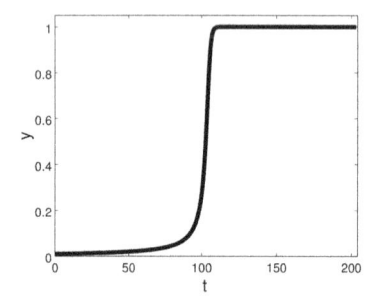

Fig. 5.8 The solution to the differential equation in (5.17) with $y_0 = 0.01$ over the range $0 \leq t \leq 2\,t_{max} = 203.2$.

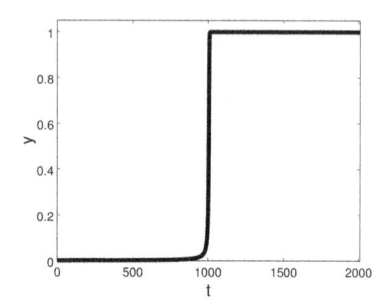

Fig. 5.9 The solution to the differential equation in (5.17) with $y_0 = 0.001$ over the range $0 \leq t \leq 2\,t_{max} = 2007.8$.

that one of these solutions is possible without the mathematical analysis of this simple differential equation.

MATLAB and Octave both have a suite of solvers for differential equations, with the standards being ode45 and ode23. These two methods are both adaptive, meaning that they automatically adjust the step size as they proceed. Adaptive methods are discussed in some detail in Sect. 5.5. Both of these solvers will solve problem (5.17) without considerable errors even in the (really) stiff case of $y_0 = 0.0001$. However, ode45 requires 216.40 s and ode23 requires 297.64 s. The solver ode23s in Octave, which is designed specifically to solve stiff differential equations, requires only 0.09 s to accurately predict the solution with $y_0 = 0.0001$. Most solvers designed to solve stiff systems, including ode23s, are implicit, see Exercise 6. One of the most telling signs of a stiff differential equation, or system of differential equations, is the time

5.5 Adaptive Methods

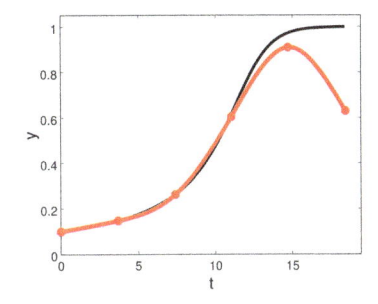

Fig. 5.10 The red dots are the RK4 solution of the problem in (5.17) with $y_0 = 0.1$ and the range $[0, 2\,t_{\max}]$ divided into 5 intervals, i.e. $\Delta t = 3.68$.

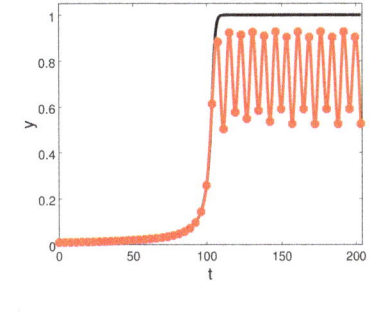

Fig. 5.11 The red dots are the RK4 solution of the problem in (5.17) with $y_0 = 0.01$ and the range $[0, 2\,t_{\max}]$ divided into 53 intervals, i.e. $\Delta t = 3.83$.

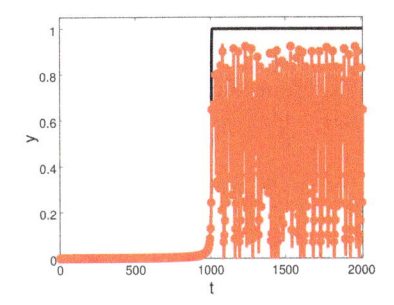

Fig. 5.12 The red dots are the RK4 solution of the problem in (5.17) with $y_0 = 0.001$ and the range $[0, 2\,t_{\max}]$ divided into 455 intervals, i.e. $\Delta t = 4.41$.

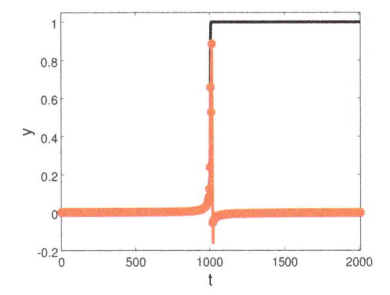

Fig. 5.13 The red dots are the RK4 solution of the problem in (5.17) with $y_0 = 0.001$ and the range $[0, 2\,t_{\max}]$ divided into 435 intervals, i.e. $\Delta t = 4.62$.

needed to calculate a solution. If a solver such as ode45 takes longer than what you might consider reasonable, then a stiff solver might be a better choice. That said, the default solver should be ode45 in most instances, as it is a well-tested default that performs well over a large range of problems.

5.5 Adaptive Methods

Selecting a proper step size is a critical decision when numerically solving an ordinary differential equation. On the one hand, using a small step should give accurate results, but on the other hand, a small step could result in unacceptably long computations. Further complicating the issue is the fact that a uniform step size may not be appropriate over the entire time domain.

If the exact solution to the ordinary differential equation is locally predictable by the solution method, then it may be possible to use large time steps. If, however, the exact solution is instead less predictable over a local region, then small steps should help maintain accuracy. Adaptive methods attempt to vary the time step as the algorithm proceeds in order to use large steps if possible, while automatically detecting if small steps are needed to improve accuracy.

The general idea of an adaptive ordinary differential equation solver is to estimate $y(t_k + \Delta t)$ from y_k and then to estimate the error in y_{k+1} with a method that is believed to be *more reliable* than the method being used to estimate $y(t_k + \Delta t)$. Increased reliability is generally assumed of higher order methods, and the techniques below use higher order methods to approximate error. If the estimated error is suitably small, then the step length Δt is accepted and the algorithm proceeds. If the estimated error is too large, then the proposed step length is reduced and the process repeats. Similarly, an adaptive method provides an avenue for a step size to increase so that large steps are used if it is safe to do so. Typically step size increases if the estimated error is sufficiently small or if a certain number of time steps have been accepted without reduction. Pseudocode for a general adaptive method is in Algorithm 25.

One way to develop an error estimate is to use Richardson extrapolation. This scheme combines two estimates of $y(t_k + \Delta t)$ made on different time scales to produce a third, more accurate estimate of $y(t_k + \Delta t)$. As an example, if our base method is Euler's method, then we begin an adaptive step by estimating $y(t_k + \Delta t)$ as

$$\hat{y} = y_k + \Delta t f(t_k, y_k).$$

We then calculate a second estimate to $y(t_k + \Delta t)$ using Euler's method with two steps of length $\Delta t/2$,

$$y^*_{1/2} = y_k + \frac{\Delta t}{2} f(t_k, y_k) \text{ and } y^* = y^*_{1/2} + \frac{\Delta t}{2} f(t_k + \Delta t/2, y^*_{1/2}).$$

We first note that it is inappropriate to approximate the error in $\hat{y}$ by comparing it to y^* since both are from the same order 1 technique of forward Euler's method, and our objective is to approximate error by comparing $\hat{y}$ to an estimate from a higher order method. Recall from (5.10) that the local truncation error of Euler's method satisfies

$$\hat{e}_{k+1} = y(t_k + \Delta t) - \hat{y} \approx \Gamma_k \Delta t^2 \text{ and } e^*_{k+1} = y(t_k + \Delta t) - y^* \approx \Gamma_k \frac{\Delta t^2}{2},$$

where Δt is sufficiently small and Γ_k is a constant approximating $y''(t)$ over the interval $t_k \leq t \leq t_k + \Delta t$. The second approximation is the result of taking

5.5 Adaptive Methods

Algorithm 25 Pseudocode for an adaptive ordinary differential equation solver

$k = 0$
while t_k is in range **do**
    initiate proposed error by setting $e_{k+1}^{\text{prop}} = \infty$
    while e^{prop} is too large **do**
        set $t_{k+1}^{\text{prop}} = t_k + \Delta t$.
        set y_{k+1}^{prop} according to some rule, using step size Δt.
        set e_{k+1}^{prop} according to error estimator
        if e_{k+1}^{prop} is acceptable **then**
            set $t_{k+1} = t_{k+1}^{\text{prop}}$
            set $y_{k+1} = y_{k+1}^{\text{prop}}$
        else
            reduce Δt
        end if
    end while
    if step length should be increased **then**
        increase Δt
    end if
    set $k = k + 1$
end while

---

two time steps, making a reasonable estimate of the error $2\Gamma_k(\Delta t/2)^2 = \Gamma_k \Delta t^2/2$.

We combine the local error estimates to calculate

$$\hat{e}_{k+1} - 2e_{k+1}^* = (y(t_k + \Delta t) - \hat{y}) - 2(y(t_k + \Delta t) - y^*)$$
$$\approx \Gamma_k \Delta t^2 - 2\left(\Gamma_k \frac{\Delta t^2}{2}\right)$$
$$= 0.$$

Rearranging the right-hand side of the first equality, we see that

$$(2y^* - \hat{y}) - y(t_k + \Delta t) \approx 0,$$

and we thus define $\bar{y}$ as the approximation,

$$\bar{y} = 2y^* - \hat{y} \approx y(t_k + \Delta t).$$

The estimate $\bar{y}$ is a weighted average of estimates from Euler's method, and it is thus similar to improved Euler's method, which is order 2. Indeed, analysis shows that $\bar{y}$ has the same local and global order of improved Euler's method, which means that $\bar{y}$ is an order 2 approximation of $y(t_k + \Delta t)$. We now have two estimates of $y(t_k + \Delta t)$, one being our original $\hat{y}$ from Euler's method, which is order 1, and the other being $\bar{y}$, which is order 2.

The perspective of an adaptive method is now in play by estimating

$$\hat{e}_{k+1} = y(t_k + \Delta t) - \hat{y} \approx \bar{y} - \hat{y}. \tag{5.19}$$

The left-hand value is the true error that we want to assess, but we cannot calculate this value because $y(t_k + \Delta t)$ is unknown. The motivation behind an adaptive scheme is to replace $y(t_k + \Delta t)$ with an estimate that is more reliable than $\hat{y}$, and $\bar{y}$ is such an estimate.

We can gauge the approximation in (5.19) by recalling that the local truncation error of $\bar{y}$ is order 3. So from (5.11) we have

$$y(t_k + \Delta t) - \bar{y} \approx \Omega_k \Delta t^3 \Leftrightarrow \bar{y} \approx y(t_k + \Delta t) - \Omega_k \Delta t^3.$$

where Ω_k is a constant and Δt is suitably small. The difference between our approximate error $\bar{y} - \hat{y}$ and the local truncation error $\hat{e}_{k+1}$ subsequently satisfies

$$\begin{aligned}(\bar{y} - \hat{y}) - \hat{e}_{k+1} &\approx \left(y(t_k + \Delta t) - \Omega_k \Delta t^3 - \hat{y}\right) - \hat{e}_{k+1} \\ &= (y(t_k + \Delta t) - \hat{y}) - \Omega_k \Delta t^3 - \hat{e}_{k+1} \\ &= \hat{e}_{k+1} - \Omega_k \Delta t^3 - \hat{e}_{k+1} \\ &= -\Omega_k \Delta t^3.\end{aligned}$$

Remember that Δt is assumed to be small, so $\Delta t > \Delta t^2 > \Delta t^3$. For instance, if $\Delta t = 10^{-2}$, then $\Delta t^2 = 10^{-4}$ and $\Delta t^3 = 10^{-6}$. The fact that Δt^3 is near zero suggests that $\bar{y} - \hat{y} \approx \hat{e}_{k+1}$, which supports our intent to estimate $\hat{e}_{k+1}$ with $\bar{y} - \hat{y}$. Moreover, we gain trust in the approximation because $\hat{e}_{k+1} \approx \Gamma_k \Delta t^2$, see (5.10). So for small Δt we have that $\hat{e}_{k+1} >> (\bar{y} - \hat{y}) - \hat{e}_{k+1}$.

Assume $\Gamma_k \approx \Omega_k \approx 1$ as a motivating case, and further assume $\Delta t = 10^{-2}$. Then the local truncation error $\hat{e}_{k+1}$ is approximately $\Delta t^2 = 10^{-4}$, but $\bar{y} - \hat{y}$ estimates $\hat{e}_{k+1}$ with an approximate error of $\Delta t^3 = 10^{-6}$. We intuitively predict agreement between $\bar{y} - \hat{y}$ and $\hat{e}$ in the first five decimal places, and since $\hat{e}$ is intuitively accurate to only three decimal places, our estimate $\bar{y} - \hat{y}$ exceeds our predicted accuracy as measured by $\hat{e}$, making $\bar{y} - \hat{y}$ a reasonable proxy to $\hat{e}$.

We use $\bar{y} - \hat{y}$ to build an adaptive method. The development above extends beyond the scalar case, and we now assume y is a vector dependent on t. Once time step k is complete, the algorithm proposes a new time step $t_k + \Delta t$.

5.5 Adaptive Methods

We compute $\hat{y}$, y^*, and $\bar{y}$ and then estimate the error in y^* by calculating $e^{\text{prop}} = \|y^* - \bar{y}\|$. If e^{prop} is sufficiently small, then the step $t_k + \Delta t$ is accepted, and we use $\bar{y}$ as our estimate of $y(t_k + \Delta t)$. Notice that we use $\bar{y}$ as the new approximation instead of $\hat{y}$ despite the fact that the error estimates are for $\hat{y}$. This tactic works well in practice since $\bar{y}$ is generally more accurate than $\hat{y}$. If e^{prop} is too large, then the proposed step is rejected, Δt is reduced, and a new step is proposed. If e^{prop} is very small, then we assume that we can take a larger time step in the future, and the initial proposed step length for the next iteration increases.

Richardson extrapolation are possible with any base scheme, for example we may use improved Euler or RK4 as the base algorithm. The only required modification is the calculation of $\bar{y}$. It can be shown that if $\hat{y}$ and y^* are calculated using the same order p method with step sizes Δt and $\Delta t/2$, respectively, then

$$\bar{y} = \frac{2^p y^* - \hat{y}}{2^p - 1}$$

is an order $p+1$ approximation of $y(t + \Delta t)$, and therefore, $\|\bar{y} - y^*\|$ is a suitable estimate of $\|y(t + \Delta t) - y^*\|$. The update is $\bar{y}$ if the approximate error is sufficiently small.

Many algorithms use two separate methods of different orders rather than use Richardson extrapolation. The most popular implementation is to use two Runge-Kutta schemes of differing order, chosen carefully so that the higher order step is cheap to calculate from the lower order case. For example, define

$$k_1 = \Delta t\, f(t, y_k),$$
$$k_2 = \Delta t\, f(t + 1/2\Delta t, y_k + 1/2\, k_1), \text{ and}$$
$$k_3 = \Delta t\, f(t + \Delta t, y_k + (1/256)\, k_1 + (255/256)\, k_2),$$

to get

$$y^* = y_k + (1/256)k_1 + (255/256)k_2, \text{ and}$$
$$\bar{y} = y_k + (1/512)k_1 + (255/256)k_2 + (1/512)k_3.$$

It can be shown that y^* is a first order approximation of $y(t_k + \Delta t)$ and that $\bar{y}$ is a second order approximation. Of special note is the fact that $\bar{y}$ only requires a single additional evaluation of f to compute k_3 once y^* is determined. This sole additional evaluation of f is a major advantage over using, say, two unrelated first and second order methods for y^* and $\bar{y}$.

The idea of using two distinct Runge-Kutta methods, one of which requires a subset of the calculations of the other, is due to Erwin Fehlberg. Such schemes are called embedded RK methods, and we often refer to them by including the order of the initial estimate y^* along with the better estimate of $\bar{y}$ in the name. Many further include an "F" for Fehlberg. For example, the scheme above would be called an RKF1(2) or RK1(2) scheme. The most famous embedded method proposed by Fehlberg in 1969 was a fourth order method with a fifth order error control, and this method is used in Octave's implementation of `ode45`. MATLAB uses a slightly different RK4(5) method in its version of `ode45`.

An embedded RK scheme proceeds exactly like a Richardson extrapolation once y^* and $\bar{y}$ are calculated. Error in y^* is approximated as $\|y^* - \bar{y}\|$, and the step is accepted if the predicted error is suitably small. Two common acceptance criteria are the absolute error criterion and the error per unit step size criterion. A step is accepted in the former as soon as the predicted error in the step falls below a predetermined threshold, i.e. as soon as

$$e^{\mathrm{prop}} < \varepsilon. \tag{5.20}$$

A step is accepted with the latter criterion relative to the step length, and Δt^{prop} is accepted as soon as

$$e^{\mathrm{prop}}/\Delta t^{\mathrm{prop}} < \varepsilon. \tag{5.21}$$

The latter criterion is obviously more stringent, but it usually produces better results. We note that neither criteria guarantees an approximate solution with global error less than ε.

Finally, an adaptive scheme must include a mechanism for increasing and decreasing step size as the algorithm proceeds. One simple scheme is to reduce/increase Δt by a constant, predetermined factor if adjustment is needed. For example, we might take

$$\Delta t = \Delta t/2 \tag{5.22}$$

if a reduction is warranted and

$$\Delta t = 1.2\Delta t \tag{5.23}$$

if a magnification is appropriate. The increase and decrease scalars should promote many possible values for Δt, and in particular, only doubling and halving should be avoided.

Adjusting step length by a constant factor is often successful. However, we might improve performance by more carefully adjusting step size. Suppose that the step for y^* has local truncation error of order p. Because both y^* and $\bar{y}$ use the same initial value of y_k, the estimated error $\|y^* - \bar{y}\|$ should be modeled as a local truncation error. So the predicted error for Δt is modeled as

$$e(\Delta t) \approx \Gamma \Delta t^p.$$

5.5 Adaptive Methods

Suppose we make a proposed step of length Δt, but the resulting estimated error is too large. We should then determine a new step, $\alpha \Delta t$, such that the estimated error arising from the new step will be acceptable. If we use absolute error as the acceptance criterion, then we seek α satisfying

$$e(\alpha \Delta t) = \Gamma \alpha^p \Delta t^p = \alpha^p e(\Delta t) \leq \varepsilon. \tag{5.24}$$

We typically choose α with a small safety factor, s, which leads to

$$\alpha = \sqrt[p]{\frac{s\varepsilon}{e(\Delta t)}}. \tag{5.25}$$

The new proposed step size is set to $\Delta t = \alpha \Delta t$. Note that this adjustment scheme is appropriate regardless of whether we are increasing or decreasing Δt. The safety factor s is usually something like $s = 0.8$.

The variety of choices involved in the authorship of an adaptive ordinary differential equation solver suggests that such methods should be used with care. Indeed, choices of method order, error tolerance criteria, and step length adjustment can significantly impact the performance of the solver. As the example in the previous section demonstrated, differences among adaptive methods can provide substantial differences in calculation times.

Consider the following problem as an example,

$$y' = -(2t - 6)\tan((t-3)^2)y, \quad y(0) = 1, \quad 0 \leq t \leq 7. \tag{5.26}$$

The exact solution is

$$y(t) = \frac{\cos\left((t-3)^2\right)}{\cos(9)},$$

see Fig. 5.14. Notice that the exact solution varies slowly near $t = 3$ and oscillates with increasing frequency as t nears 0 and 7. We might expect an adaptive method to spend more effort near the endpoints of the domain and less effort near $t = 3$. Figure 5.14 shows the result of an adaptive method on problem (5.26). The particular method is improved Euler's method with Richardson extrapolation for error estimation. The algorithm uses an absolute estimated error criterion with $\varepsilon = 10^{-3}$. The algorithm does a good job of focusing its effort on the interesting parts of the solution.

Figure 5.15 illustrates the potential benefit of an adaptive method. In this figure we compare a few different adaptive methods with varying orders, error criteria, and step size adjustment schemes to non-adaptive methods. Even a scheme as humble as Euler's method, if made adaptive and used on suitable problems, can be quite performant. Indeed, we see in Fig. 5.15 that the adaptive Euler's method achieved an error on the same scale as that of non-adaptive RK4, with nearly an order of magnitude fewer time steps!

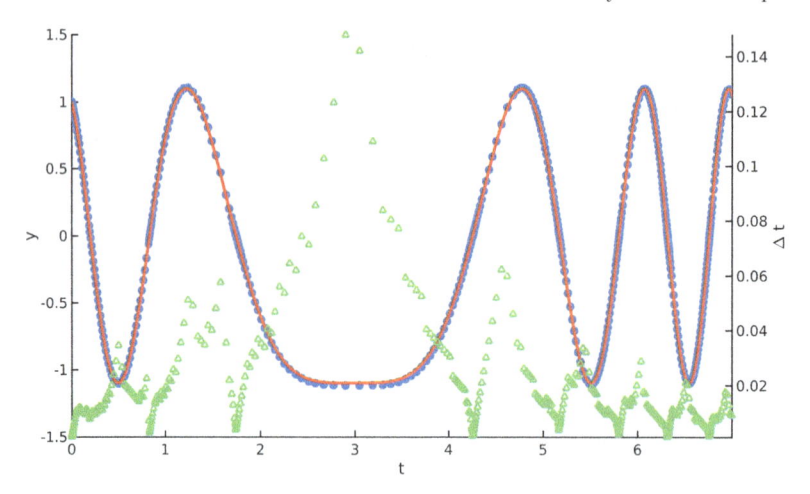

Fig. 5.14 The result of using an adaptive improved Euler method with Richardson extrapolation to solve problem (5.26). The exact solution is red, the approximated solution is blue dots, and the y-axis is on the left. The step size at each time is shown by green triangles, and the appropriate y-axis is on the right.

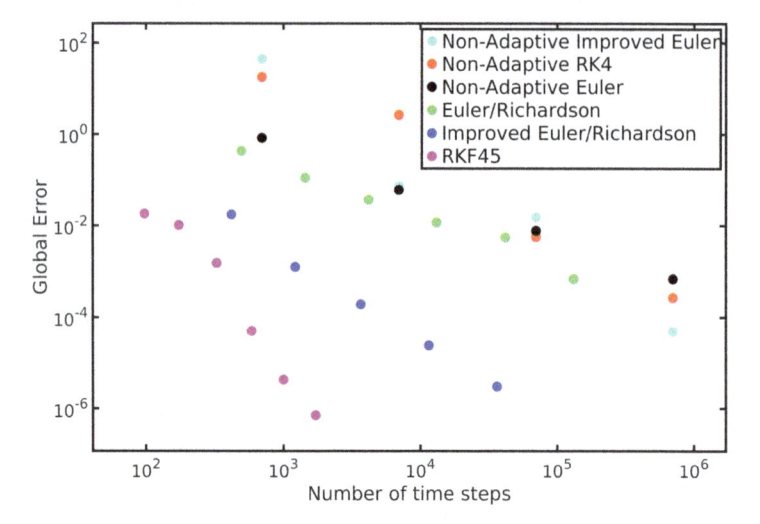

Fig. 5.15 Comparison of several adaptive methods to several non-adaptive ODE solvers applied to problem (5.26). The Euler/Richardson and RKF45 methods both used an absolute error criterion, while the Improved Euler/Richardson method used an error per unit step criterion. Euler/Richardson and Improved Euler/Richardson both used a predetermined step size adjustment, while the RKF45 method used an optimal step size adjustment following equation (5.25).

5.6 Exercises

1. One of the most common differential equations is the exponential model

$$y' = ky + a, \; y(t_0) = y_0,$$

where k and a are constants. Assume that y is a scalar valued function and show, by direct substitution, that the solution to this initial value problem is

$$y(t) = e^{k(t-t_0)}\left(y_0 + \frac{a}{k}\right) - \frac{a}{k}.$$

2. One method for solving scalar differential equations is called separation of variables. Suppose we have a differential equation of the form

$$\frac{dy}{dt} = g(y)h(t).$$

Such equations are called separable because the y and t variables on the right-hand side are separated as a product. Not all, or even most, differential equations are separable, but those that are may be rewritten as

$$\frac{1}{g(y)}\frac{dy}{dt} = h(t),$$

assuming that $g(y) \neq 0$. Suppose $G(y)$ satisfies

$$\frac{dG}{dy} = \frac{1}{g(y)}.$$

Then the chain rule shows that

$$\frac{dG}{dt} = \frac{dG}{dy}\frac{dy}{dt} = \frac{1}{g(y)}\frac{dy}{dt},$$

and our differential equation becomes

$$\frac{dG}{dt} = h(t).$$

Integrating both sides of this equation with respect to t gives

$$G(y) = \int h(t)\,dt + C.$$

We need at this point to solve for y and to use the initial conditions to calculate C.

Consider the following differential equation as an example,

$$\frac{dy}{dt} = \frac{t^2}{y}, \quad y(0) = 2.$$

This equation is separable with $g(y) = 1/y$ and $h(t) = t^2$. Separating variables gives
$$y \frac{dy}{dt} = t^2,$$
and an anti-derivative for $g(y)$ is $G(y) = y^2/2$. So,
$$\frac{dG}{dt} = t^2,$$
and integrating gives
$$\frac{y^2}{2} = G(y) = \frac{t^3}{3} + C.$$
We conclude that
$$y(t) = \pm\sqrt{2}\sqrt{\frac{t^3}{3} + C}.$$
This expression is the general solution of the ordinary differential equation, as it describes all possible solutions, regardless of initial condition. This particular initial value problem is solved by choosing the positive square root and by taking $C = 2$.

Use the method of separation of variables to solve the exponential model in Exercise 1.

3. Write a function with declaration

    [t,y,stat]=Euler(f,y0,trange,deltat,method)

    that solves
    $$\frac{dy}{dt} = f(t, y), \quad y(t_0) = y_0.$$
    using one of three variants of Euler's method.

    The input arguments are:

    f   - The function, either as a function handle or a string, that determines the right-hand side of the ordinary differential equation. If y returns vectors of length n, then $f(t, y)$ needs to take scalar values for t and column vectors of length n for y, and it needs to return column vectors of length n.
    y0   - The initial value of $y(t_0)$ as a column vector.
    trange   - A 1×2 vector defining the range of t, the first element of trange is t_0, the last element of trange is the end of the time interval.

5.6 Exercises

deltat - The size of the time step to take. Make sure that the last time step returned is at least as large as trange(2).

method - An optional parameter, with default value 0, that specifies which Euler method to use. The variable is encoded as follows:

method 0: Use forward Euler's method.

method 1: Use backward Euler's method.

method 2: Use improved Euler's method.

For backward Euler's method use `fsolve`, with appropriate options to keep its output minimal, to solve the system of equations necessary at each time step.

The output arguments are:

t - A vector containing all times at which approximations to y were calculated. Include the initial time trange(1).

y - An $n \times m$ matrix containing the approximation to $y(t_j)$ in column $j+1$. In this context m is the length of t. Include the initial condition in column 1.

stat - A status variable that should be zero if termination was successful and one otherwise. This function can fail, for instance, if `fsolve` fails on a step during backward Euler's method.

4. Write a function with declaration

[t,y] = RK4(f,y0,trange,deltat)

that uses the RK4 method described in Algorithm 24 to solve a differential equation of the form

$$\frac{dy}{dt} = f(t,y),\ y(t_0) = y_0.$$

The input arguments are:

f - The function, either as a function handle or a string, that determines the right-hand side of the ordinary differential equation. If y returns vectors of length n, then $f(t,y)$ needs to take scalar values for t and column vectors of length n for y, and it needs to return column vectors of length n.

y0 - The initial value $y(t_0)$ as a column vector.

trange   - A 1×2 vector defining the range of t, the first element of trange is t_0, the last element of trange is the end of the time interval.

deltat   - The size of the time step to take. Make sure that the last time step returned is at least as large as trange(2).

The output arguments are:

t   - A vector containing all times at which approximations to y were calculated. Include the initial time trange(1).

y   - An $n \times m$ matrix containing the approximation to $y(t_j)$ in column $j+1$. In this context m is the length of t. Include the initial condition in column 1.

5. Write a function with declaration

   [t,y,status] = AdaptiveOde(f,y0,trange,epsilon,
                              method,minstep,maxstep)

   that uses an adaptive method as described in Algorithm 25 to solve an ordinary differential equation in the form

   $$\frac{dy}{dt} = f(t,y), \; y(t_0) = y_0.$$

   The input arguments are:

   f   - The function, either as a function handle or a string, that determines the right-hand side of the ordinary differential equation. Assume that f returns scalars.

   y0   - The initial value $y(t_0)$.

   trange   - A 1×2 vector defining the range of t, the first element of trange is t_0, the last element of trange is the end of the time interval.

   epsilon - The error tolerance; an optional argument with default value 10^{-3}. Use the absolute estimated error criterion in (5.20).

   method   - An optional parameter, with default value 0, that specifies what method to use. The variable is encoded as follows:

   method = 0: Use Euler's method as the base method, with Richardson extrapolation to produce the error estimate.

   method = 1: Use the Runge-Kutta-Fehlberg 4(5) method, described below.

5.6 Exercises

minstep - The smallest time step the algorithm is allowed to make. If it is not possible to make a step that satisfies the estimated error criterion while respecting this minimum step size, then `status` should be set to 1, and the algorithm should cease. This is an optional argument with default value 10^{-14}.

maxstep - The largest time step the algorithm is allowed to make. If the algorithm attempts to use a step larger than `maxstep`, then Δt should be set to `maxstep`, `status` should be set to 2, and the algorithm should proceed. This is an optional argument with default value 1.

The output arguments are:

t - A vector containing all times at which y is approximated. Include the initial time `trange(1)`. The solver should proceed until it *exceeds* `trange(2)`, i.e.

$$\texttt{t(end-1)} < \texttt{trange(2)} \quad \text{and} \quad \texttt{t(end)} \geq \texttt{trange(2)}.$$

y - A vector containing the approximation to $y(t_j)$ in column $j+1$. Include the initial condition in the first entry of y.

status - A variable indicating the success or failure of the method. This method can fail if it is not possible to satisfy the estimated error condition while respecting the minimum step size option. The variable `status` should be encoded as follows:

status = 0: Success, and the algorithm never tried to take a step larger than `maxstep`.

status = 1: Failure, the algorithm could not simultaneously respect the estimated error criterion and the minimum step size.

status = 2: Success, but the algorithm tried to take a step larger than `maxstep`.

The algorithm should adjust the step size with the fixed ratios in (5.22) and (5.23).

The RKF4(5) estimate is calculated as follows:

$$k_1 = \Delta t \, f(t_k, y_k),$$
$$k_2 = \Delta t \, f(t_k + (1/4)\Delta t, y_k + (1/4)k_1),$$
$$k_3 = \Delta t \, f(t_k + (3/8)\Delta t, y_k + (3/32)k_1 + (9/32)k_2),$$
$$k_4 = \Delta t \, f(t_k + (12/13)\Delta t, y_k + (1932/2197)k_1 - (7200/2197)k_2 + (7296/2197)k_3),$$
$$k_5 = \Delta t \, f(t_k + \Delta t, y_k + (439/216)k_1 - 8k_2 + (3680/513)k_3 - (845/4104)k_4), \text{ and}$$
$$k_6 = \Delta t \, f(t_k + (1/2)\Delta t, y_k - (8/27)k_1 + 2k_2 - (3544/2565)k_3 + (1859/4104)k_4 - (11/40)k_5),$$

from which

$$y^* = y_k + (25/216)k_1 + (1408/2565)k_3 + (2197/4104)k_4 - (1/5)k_5 \text{ and}$$
$$\bar{y} = y_k + (16/135)k_1 + (6656/12825)k_3 + (28561/56430)k_4 - (9/50)k_5 + (2/55)k_6.$$

It turns out that y^* is a fourth order estimate of $y(t_k + \Delta t)$ and that $\bar{y}$ is a fifth order estimate of $y(t_k + \Delta t)$. The algorithm should initialize $\Delta t = 0.1$ for the first proposed step length, and it should increase step length if it accepts a step with proposed error $e^{\text{prop}} < \varepsilon/10$.

6. Compare backward, forward, and improved Euler's methods against RK4 on problem (5.17). Use $N = 5 \times 10^k$, with $k = 1, 2, 3, 4, 5$, steps on the interval $[0, 1.5\,t_{\max}]$, where $t_{\max}$ is defined in (5.18). Assume the initial value $y_0 = 10^{-5}$. Use `fzero` in backward Euler's method for solving the nonlinear equation; `fzero` is a bracketing method and is often more trustworthy than secant-based methods like `fsolve` in this case.

7. We study a system of equations in Chap. 10 known as the susceptible, infected, and recovered (SIR) model. This model expresses how individuals might progress through these health states as a disease spreads through a populace. The model is

$$S'(t) = -k_1 I(t) S(t),$$
$$I'(t) = k_1 I(t) S(t) - k_2 I(t), \text{ and}$$
$$R'(t) = k_2 I(t),$$

where $S(t)$, $I(t)$, and $R(t)$ are, respectively, the number of susceptible, infected, and recovered individuals at time t. The parameter k_1 is an infectious rate in units of per person per time, and k_2 is a recovery rate in units of per time.

5.6 Exercises

Assume the initial susceptible population is 10,000 people, the initial infected population is 100 people, and the initial recovered population is 0 people. Use ode45 to investigate the impact of different values of k_1 and k_2. Report your findings in the form of several graphs with appropriate labels.

8. Consider a combined population of rabbits and foxes for which $x_1(t)$ is the population of rabbits at time t and $x_2(t)$ is the population of foxes at time t. Foxes prey on rabbits, and an abundance of rabbits means more prey for foxes, and thus a larger fox population. However, a large fox population reduces the rabbit population, leading to fewer foxes! We posit a predator–prey model to investigate this behavior,

$$x_1'(t) = (k_1 - k_2)x_1(t) - k_3 x_1(t) x_2(t) \text{ and}$$
$$x_2'(t) = k_4 x_1(t) x_2(t) - k_5 x_2(t).$$

The parameters are:

k_1 – the birth rate of rabbits (per time)
k_2 – the natural death rate of rabbits (per time)
k_3 – the death rate of rabbits due to fox–rabbit interactions (per fox per time)
k_4 – the growth rate in the fox population due to fox–rabbit interactions (per rabbit per time)
k_5 – the natural death rate of foxes (per time).

Use your RK4 solver from Exercise 4 to investigate the dependence of the system on the parameters k_1 through k_5.

9. Newton's Law of Cooling approximates the temperature of an object within an ambient environment if the temperature of the environment is independent of that of the object. For instance, consider a coffee cup placed in a room. If we assume the coffee has uniform temperature and that the heat escaping the cup does not impact the temperature of the room, then the temperature of the coffee cup may be modeled by Newton's Law of Cooling.

Newton's Law of Cooling posits that the rate of change in the temperature of the object is proportional to the difference between the object's temperature and the ambient temperature. The model with $y(t)$ being an object's temperature and T_a being the ambient temperature is

$$\frac{dy}{dt} = k(y - T_a).$$

The constant k is a heat transfer coefficient in per time units. The model is exponential if T_a is a constant, see Exercise 1, but exact solutions can otherwise be infeasible as T_a becomes complicated.

Consider a thermos of coffee left outside for one week. The manufacturer advertises (oddly) that liquid inside the thermos would have a heat transfer coefficient of $k = -0.004$ min^{-1}. Approximate the solution with your RK4 solver from Exercise 4 assuming an ambient temperature of

$$T_a(t) = 65 + 20 \sin\left(\frac{\pi}{720}t\right).$$

Time is in minutes, and $t = 0$ corresponds with the start of the coffee being left outside.

10. The Belousov-Zhabotinsky (BZ) reaction is a famous chemical reaction, and one simple model of the dynamics is the Oregonator model, which is

$$A + Y \to X + P$$
$$X + Y \to 2P$$
$$A + X \to 2X + 2Z$$
$$X + X \to A + P$$
$$B + Z \to (1/2)\,Y.$$

If we let $[\cdot]$ be the concentration of its argument, e.g. $[X]$ is the concentration of species X, then this system can be studied by solving the following system of ordinary differential equations,

$$\left.\begin{aligned}\frac{d[X]}{dt} &= k_1[A][Y] - k_2[X][Y] + k_3[A][X] - 2k_4[X]^2 \\ \frac{d[Y]}{dt} &= -k_1[A][Y] - k_2[X][Y] + (1/2)k_5[B][Z] \\ \frac{d[Z]}{dt} &= 2k_3[A][X] - k_5[B][Z],\end{aligned}\right\} \quad (5.27)$$

where $[A]$ and $[B]$ are assumed constant and the reaction constants k_1 through k_5 are measured empirically. Chemical systems and their associated differential equations are studied in Chap. 9. Some reasonable settings and initial conditions are:

$[A] = 0.06\,M$, $[B] = 0.02\,M$, $k_1 = 2.3\,M^{-1}s^{-1}$, $k_2 = 3 \times 10^6\,M^{-1}s^{-1}$,
$k_3 = 42\,M^{-1}s^{-1}$, $k_4 = 3 \times 10^3\,M^{-1}s^{-1}$, $k_5 = 1\,M^{-1}s^{-1}$,
$[X](0) = 4.2 \times 10^{-4}\,M$, $[Y](0) = 8.4 \times 10^{-7}\,M$, and $[Z](0) = 0.11\,M$,

and a reasonable time interval is $[0, 1000]$ s. The large difference in the size of the rate constants and initial conditions indicates a likelihood of this system being stiff.

a. Use ode45 and ode23s to solve system (5.27) with the stated settings. Use odeset to specify that the adaptive methods should strive for a relative tolerance, in turn, of 10^{-k}, $k = 1\ldots 10$. Track how many steps it takes both `ode45` and `ode23s` to complete for each relative tolerance, and use `tic` and `toc` to compare the amount of time taken by each method. Report your results in the form of two graphs.

b. Use your Euler code from Exercise 3 to compare the performances of forward, backward, and improved Euler's methods. Use a step size $\Delta t = 10^{-k}$, $k = 0\ldots 4$. Make sure to pay attention to the `status` flag of your Euler code, as implicit Euler's method may fail. Why do we expect the implicit Euler method to be useful in this situation?

11. Consider a pendulum consisting of an object with mass m connected to a massless rod of length L, see Fig. 5.16. Let $\theta(t)$ be the angle between the pendulum's arm and the vertical line passing through the pivot point. The position, velocity, and acceleration of the object at time t are then, respectively, $(L\sin(\theta), -L\cos(\theta))$, $L\theta'(t)$, and $L\theta''(t)$. The force of gravity has magnitude mg, where g is a gravitational constant. However, the object is constrained to move only along a circle of radius L, and the resulting net force of gravity acting on the object is $mg\cos(\gamma(t))$. Some quick trigonometry shows that $\gamma(t) = \pi/2 - \theta(t)$, making the net magnitude of gravity's force $mg\cos(\pi/2 - \theta(t)) = mg\sin(\theta(t))$. Gravitational force acts in the direction opposite to the x-coordinate, that is, in the opposite direction of $\sin(\theta(t))$. Thus, we model the net gravitational force as $-mg\sin(\theta(t))$. We can model friction as having a magnitude proportional to the speed of the object at time t, acting in a direction opposite of $\theta'(t)$. That is, a model of friction is $-\mu L\theta'(t)$.

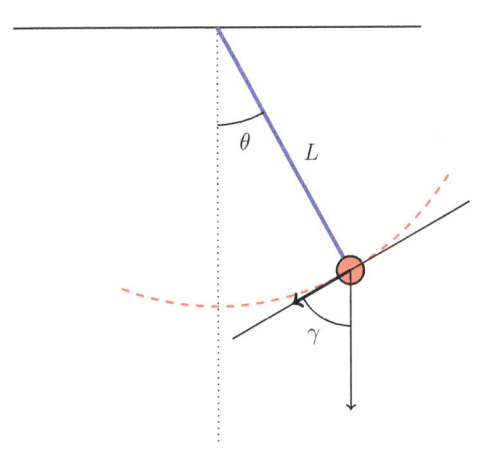

Fig. 5.16 A pendulum for Exercise 11.

Employing Newton's second law of $F = ma$ gives us

$$mL\theta''(t) = -\mu L\theta'(t) - mg\sin(\theta(t)),$$

which we rearrange to

$$\theta''(t) + \frac{\mu}{m}\theta'(t) + \frac{g}{L}\sin(\theta(t)) = 0. \tag{5.28}$$

This second-order equation is re-expressed as the first order system,

$$\theta'(t) = u(t)$$
$$u'(t) = -\frac{\mu}{m}u(t) - \frac{g}{L}\sin(\theta(t)).$$

The second-order equation is nonlinear, and a common approximation is to replace $\sin(\theta(t))$ with $\theta(t)$, which is reasonable if $\theta(t)$ is small. This results in the linear, second-order differential equation

$$\theta''(t) + \frac{\mu}{m}\theta'(t) + \frac{g}{L}\theta(t) = 0, \tag{5.29}$$

which translates to a similar system of first order differential equations.

a. Use ode45 to solve the linear and nonlinear pendulum models with $\mu = 0.1$ N·s/m, $g = 9.8$ m/s^2, $m = 10$ kg, and $L = 1.1$ m. Use the initial conditions $\theta(0) = \pi/2$ and $\theta'(0) = 0$. Produce plots of $\theta(t)$ versus time demonstrating the differences between the solutions of the models. Repeat the experiment with $\theta(0) = \pi/12$ and $\theta'(0) = 0$.
b. A rudimentary technique for creating animated figures in MATLAB and Octave is to repeatedly call plot. For example, the following code will animate a circle moving along a spiral path, including a trail behind it.

```
%list of times at which to create a frame
times=linspace(0,10);
%plot one frame per time
for i=1:length(times)
    %draw the path, including all times up to times(i)
    xcoords=times(1:i).*cos(times(1:i));
    ycoords=times(1:i).*sin(times(1:i));
    %plot; mark the last coordinate with a circle
    plot(xcoords,ycoords,xcoords(end),ycoords(end),'o','
        MarkerSize',15);
    %stop the axis from re-adjusting
    axis([-10,10,-10,10]);
    %slow down the animation
    pause(0.1);
end
```

Use this idea to create an animation of a pendulum with arm length L moving according to the solutions from Problem 11a. In order to ensure that you are plotting the positions at comparable times, first create a spline of the solutions using spline, then create each frame of the animation by taking data from the splines.

5.6 Exercises

12. A pendulum with variable arm length does not satisfy the equations of the previous problem but instead satisfies

$$\theta''(t) + \left(\frac{\mu}{m} + 2\frac{L'(t)}{L(t)}\right)\theta'(t) + \frac{g}{L(t)}\theta(t) = f(t), \qquad (5.30)$$

where $f(t)$ is an external driving force being applied to the system. We assume in the following problems that the length of the pendulum has been measured at discrete times; the measurements are found in Table 5.5.

t	0	2	4	6	8	10	12	14	16	18
L	1.101	1.100	1.100	1.099	1.097	1.096	1.094	1.091	1.088	1.085

t	20	22	24	26	28	30	32	34	36	38
L	1.081	1.078	1.073	1.068	1.064	1.059	1.054	1.049	1.044	1.038

t	40	42	44	46	48	50	52	54	56	58
L	1.031	1.025	1.019	1.013	1.007	1.000	0.994	0.988	0.982	0.975

t	60	62	64	66	68	70	72	74	76	78
L	0.970	0.963	0.958	0.953	0.946	0.942	0.937	0.932	0.927	0.924

t	80	82	84	86	88	90	92	94	96	98
L	0.920	0.916	0.913	0.910	0.908	0.905	0.903	0.902	0.901	0.901

Table 5.5 The length of the pendulum for Problem 12.

a. Approximate $L(t)$ with a natural cubic spline and plot L.
b. If we make the assumption that $L(t)$ changes slowly compared to $\theta(t)$, then we might simplify (5.30) by setting $L' = 0$. Simplify the equation above under this assumption and compare it to the models of the previous exercise.
c. Let $\mu = 0.1$ N· sec/m, $m = 10$ kg, and $f(t) = \sin(3\pi t)$ N. Solve the simplified equation from Exercise 12b with Euler's method using your approximation for $L(t)$ from Exercise 12a on the time range $0 \leq t \leq 100$ (assume $\Delta t = 0.1$). Plot your solution over the time ranges of $0 \leq t \leq 5$ and $0 \leq t \leq 50$. Be sure to label your axes and title your graph. You will need to re-express the second order equation as a system of first order equations similar to (5.1). Do you trust your solution?
d. Repeat Exercise 12c with improved Euler's method. Do you trust this solution?
e. Repeat Exercises 12c and 12d using model (5.30) directly rather than the simplification from Exercise 12b.
f. Animate your solutions to Problems 12c, 12d, and 12e as you did in Problem 11b.

13. Consider the collection of water tanks in Fig. 5.17. Suppose that salt is initially dissolved in the water in one or more of the tanks and that we wish to know how that salt will move through the system. Such models are studied in Chap. 9.

Let $a_i(t)$ be the amount of salt in T_i at time t. Then a model for the depicted system under the assumption of uniformly mixed salt is

$$\frac{da_1}{dt} = 10c_1 - 12\left(\frac{a_1}{100}\right) + 2\left(\frac{a_2}{100}\right)$$

$$\frac{da_2}{dt} = 6\left(\frac{a_1}{100}\right) - 13\left(\frac{a_2}{100}\right) + 7\left(\frac{a_3}{80}\right)$$

$$\frac{da_3}{dt} = 6\left(\frac{a_1}{100}\right) - 11\left(\frac{a_3}{80}\right) + 5\left(\frac{a_4}{120}\right)$$

$$\frac{da_4}{dt} = 11\left(\frac{a_2}{100}\right) + 4\left(\frac{a_3}{80}\right) - 15\left(\frac{a_4}{120}\right),$$

where c_1 is the concentration of salt from the exterior source to tank one and each tank is at its indicated capacity. The brine solution entering the first tank has a varying concentration, and measurements taken every five minutes are listed in Table 5.6.

t	5	10	15	20	25	30	35	40	45	50
g/ℓ	38.79	35.80	40.84	35.25	27.81	28.63	21.63	18.43	19.14	18.98
t	55	60	65	70	75	80	85	90	95	100
g/ℓ	12.14	6.73	10.27	3.71	5.49	6.68	1.32	1.76	1.64	0.51

Table 5.6 Concentration of salt in brine solution entering Tank 1 in Problem 13.

Answer the following questions assuming that there is initially no salt in any of the tanks.

a. Approximate the rate at which salt enters the first tank from the external source using three different approximations: the linear, quadratic, and cubic best fit polynomials arising from the data in Table 5.6. Rather than returning a negative inflow rate, each of your approximations should be set to zero once it becomes negative. If an approximation does not become zero, then it should be discarded.

b. Use your RK4 solver from Exercise 4 to solve the system of differential equations for each of the approximated inflow rates in Problem 13a. Estimate the time at which Tank 4 has a concentration of 1/10 of a gram per liter as the system is flushed with pure water per your approximation from Problem 13a. Plot the concentration in each tank

5.6 Exercises

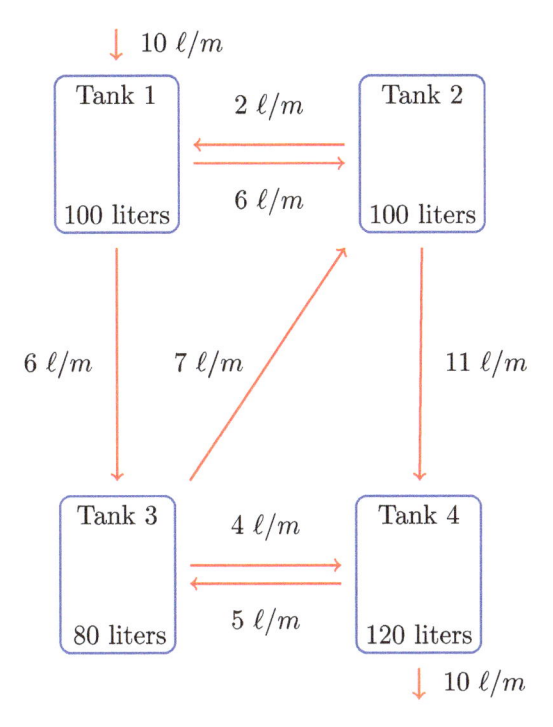

Fig. 5.17 A system of interconnected water tanks, the rate and direction of brine flow is indicated by the arrows.

on the same axes for each approximation of the entering concentration. Create a graph, or graphs, to clearly illustrate your findings, and be sure to label axes and include titles.

14. Consider a double pendulum as shown in Fig. 5.18 in which two pendulums are connected. The first pendulum has an arm length L_1 and mass m_1 connected to a pivot at the origin. The second pendulum has a pivot at the end of the first pendulum, has an arm length L_2 and mass m_2. Assume that both arms are massless and that there is no friction at the pivot points.

    The angles $\theta_1(t)$ and $\theta_2(t)$ solve the system of differential equations

$$\theta_1'' = \frac{1}{\gamma_1}\left(m_2\left(\sin\left(\Theta\right)L_1(\theta_1')^2 - \sin\left(\theta_2\right)g\right)\cos\left(\Theta\right) + m_2(\theta_2')^2 L_2 \sin\left(\Theta\right)\right.$$
$$\left. + gM\sin\left(\theta_1\right)\right) \text{ and}$$

$$\theta_2'' = \frac{1}{\gamma_2}\left(-\cos\left(\Theta\right)\sin\left(\Theta\right)L_2 m_2(\theta_2')^2 - M\sin\left(\Theta\right)L_1(\theta_1')^2 \right.$$
$$\left. - M\sin\left(\theta_1\right)\cos\left(\Theta\right)g + M\sin\left(\theta_2\right)g\right),$$

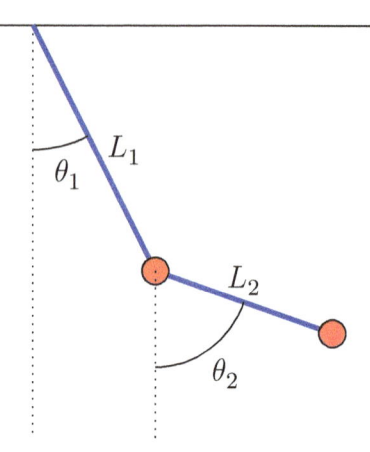

Fig. 5.18 A double pendulum.

with

$$\Theta = \theta_1 - \theta_2,$$
$$M = m_1 + m_2,$$
$$\gamma_1 = L_1 \left(\cos^2(\Theta) m_2 - M \right), \text{ and}$$
$$\gamma_2 = L_2 \left(\cos^2(\Theta) m_2 - M \right).$$

a. Use ode45 to solve the equations of motion above with initial conditions $m_1 = 2\,\text{kg}$, $m_2 = 1.5\,\text{kg}$, $L_1 = 0.6\,\text{m}$, and $L_2 = 0.8\,\text{m}$. Use the initial conditions $\theta_1(0) = 0$, $\theta_2(0) = \pi/4$, $\theta_1'(0) = 0$, and $\theta_2'(0) = 0$.

b. Animate the solution to the double pendulum problem as in Problem 11b.

c. It is easy to verify that $\theta_1(t) = 0$, $\theta_2(t) = 0$ is a solution to the double pendulum problem. Because this solution does not depend on time it is called an equilibrium solution. A second equilibrium solution is $\theta_1(t) = \pi$, $\theta_2(t) = \pi$. Use your Euler method code from Problem 3 to solve the double pendulum problem first with initial conditions $\theta_1(0) = \theta_1'(0) = \theta_2(0) = \theta_2'(0) = 0$, and then with initial conditions $\theta_1(0) = \theta_2(0) = \pi$, $\theta_1'(0) = \theta_2'(0) = 0$. Animate the results as in Problem 14b. Though they are both equilibrium solutions, the second equilibrium is called unstable; tiny perturbations to the initial condition result in dramatically different solutions. In this case, the tiny errors made by the ordinary differential equation solver is causing the numerical solution to diverge from the exact solution.

d. Alter your code from Problem 14b to trace the path of the second pendulum, see Fig. 5.19. Compare the path that results from initial conditions $\theta_1(0) = 0$, $\theta_2 = \pi/4$, $\theta_1'(0) = 0$, and $\theta_2'(0) = 0$ to the path that results from initial conditions $\theta_1(0) = 0$, $\theta_2 = \pi/4 + 10^{-1}$, $\theta_1'(0) = 0$,

5.6 Exercises

and $\theta_2'(0) = 0$. You should see that even tiny differences in initial conditions lead to very different solutions; this is called chaotic behavior.

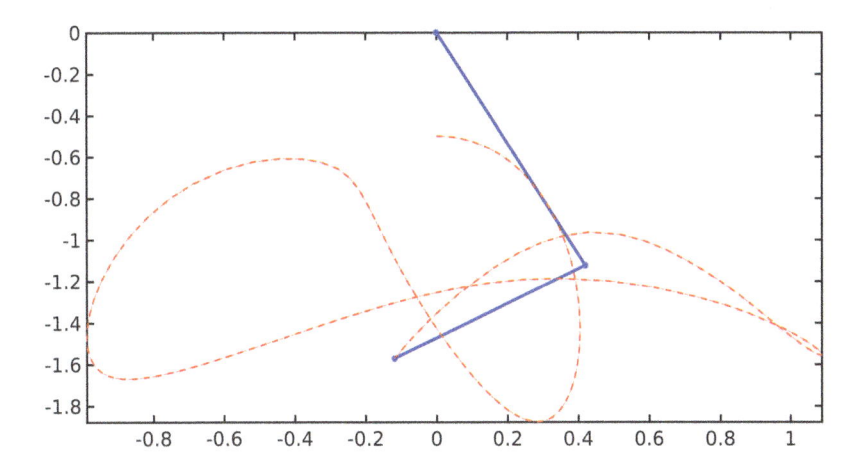

Fig. 5.19 The trace of the path of a double pendulum.

15. Consider N objects with masses $m_1, m_2, \ldots, m_N$, positions $x_1, x_2, \ldots, x_N$, and velocities $v_1, v_2, \ldots, v_N$ moving through space under the influence of gravity. Predicting the positions of these objects at some time in the future is frequently called the N-body problem (Fig. 5.20).

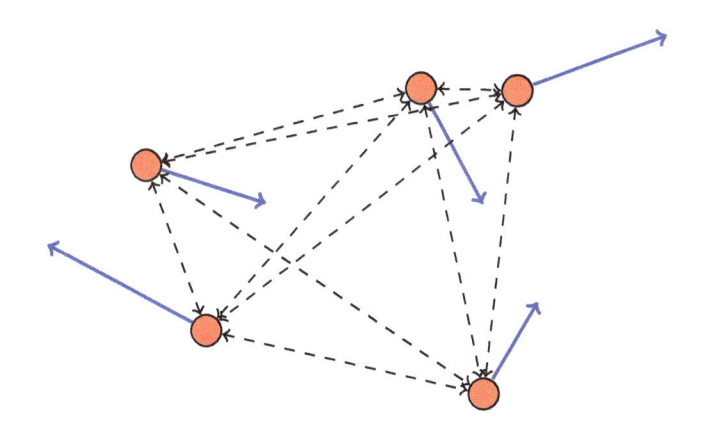

Fig. 5.20 An illustration of object in space. The velocity of each objects is represented with the solid arrow; the gravitational interaction forces are illustrated with the dashed lines.

The equations of motion for the N-body problem may be derived using $F = ma$, just as was done in Exercise 11. The positions and velocities solve

$$x_i'(t) = v_i(t) \tag{5.31}$$

$$v_i'(t) = G \sum_{j \neq i} \frac{m_j}{\|x_j(t) - x_i(t)\|^3} (x_j(t) - x_i(t)). \tag{5.32}$$

Equations (5.31) and (5.32) give us a system of $2N$ ordinary differential equations.

a. Use ode45 to solve the N-body equations modeling the solar system. Find the relevant data for as many objects in the solar system as you can find. Animate the results of your simulation.
b. The kinetic energy in the system at time t is

$$K(t) = \frac{1}{2} \sum_i m_i \|v_i(t)\|^2$$

and the potential energy at time t is

$$P(t) = \sum_{i<j} \frac{G m_i m_j}{\|x_j - x_i\|}.$$

Define the total energy in the system as

$$\mathcal{H}(t) = K(t) + P(t),$$

and show that if the system solves the system of equations (5.31) and (5.32) then the total energy of the system stays constant by showing that $\mathcal{H}'(t) = 0$.
c. Solve the N-body equations for all the bodies you can find in the solar system using ode45, Euler's method, improved Euler's method, and a non-adaptive RK4 method. Produce a plot of the total energy in the numerical solution against time for each method.
d. Forward Euler's method applied to the N-body equations is

$$x_i^{k+1} = x_i^k + \Delta t \, v_i^k$$
$$v_i^{k+1} = v_i^k + \Delta t \, G \sum_{j \neq i} \frac{m_j}{\|x_j^k - x_i^k\|^3} (x_j^k - x_i^k),$$

where x_i^k is the approximation to x_i at time step k. We saw in Exercise 15c that Euler's method does not preserve the total energy of the system well. However, a simple modification of Euler's method performs much better. The scheme

$$v_i^{k+1} = v_i^k + \Delta t\, G \sum_{j \neq i} \frac{m_j}{\|x_j^k - x_i^k\|^3} (x_j^k - x_i^k),$$

$$x_i^{k+1} = x_i^k + \Delta t\, v_i^{k+1}$$

is called symplectic Euler's method. In general, ordinary differential equation solvers that attempt to preserve total energy are called symplectic methods, and they are widely used in classical mechanics. Implement symplectic Euler's method, and compare the results to those from Exercise 15c.

Chapter 6
Stochastic Methods and Simulation

> During the 1950s, I decided, as did many others, that many practical problems were beyond analytic solution and that simulation techniques were required. – Harry Markowitz

> To dissimulate is to pretend not to have what one has. To simulate is to feign to have what one doesn't have. – Jean Baudrillard

> If God has made the world a perfect mechanism, He has at least conceded so much to our imperfect intellect that in order to predict little parts of it, we need not solve innumerable differential equations, but can use dice with fair success. – Max Born

This chapter develops stochastic computational methods that assume random elements within their computational task. The methods prior to this chapter have largely been deterministic, as they have not depended on random elements in their computation—previous exceptions being our studies of linear regression, simulated annealing, and genetic algorithms. These exceptions point to a dichotomy into how randomness leaks itself into the computational arena. One way is to use a random element to help decide a deterministic quantity. For example, solving a deterministic TSP problem with either simulated annealing or a genetic algorithm incorporates a stochastic search in the attempt to calculate the deterministic value that is the shortest tour. In this case the problem is deterministic, but the calculation method is stochastic.

Another way randomness enters a computational study is through the model itself. One case is our earlier work on linear regression, which assumes random data, and hence, the regression model itself is premised on the stochasticity of the entities being studied. Other examples of stochastic models would be differential equations with uncertain parameters, or optimization problems with random data. Solutions to such problems are random because the models themselves are random. In these cases a deterministic algorithm can be used to solve randomly selected instances, from which statistics of the solutions can then be inferred.

This chapter includes some fundamental tactics on how to accommodate and leverage uncertainty. Embedding a random element into a computational method leads to a randomized algorithm, with simulated annealing and genetic algorithms being examples. Randomized algorithms play a substantial role in many areas of modern computing, and in particular, availing random selection to address the immensity of searching and routing information on the internet has found wide acceptance. Our goal herein is more traditional, albeit still quite important, as we develop a stochastic integration technique called Monte Carlo integration to illustrate many of the stochastic themes associated with computational science. We then continue to explore how these computational themes can apply to some of our previously developed deterministic methods to infer statistical outcomes of stochastic models.

6.1 Simulation

The term simulation has a broad interpretation, and a succinct discussion about its meaning as it relates to computational science is justified. The term broadly refers to a mock event or model that mimics an underlying entity. For example, computer simulations are used daily to design radiotherapy treatments for cancer patients. The auto industry simulates proposed designs to vet them against manufacturing goals, and the efficacy of a computer network is simulated prior to its construction. A simulation can also be nonvirtual. For example, leadership teams conduct catastrophe simulations to train themselves for adverse events, and roofing shingles are exposed to harsh conditions to simulate years of weathering.

The interpretation of what is meant by simulation narrows as we enter the computational arena, where the term most commonly refers to any virtual model of an entity of interest. Although the term isn't inappropriate for nonrandom models, in most cases the term relates to a virtual model intended to study a complicated system fraught with uncertainty in its real-world setting. The source of uncertainty can be due to imperfect or random data, or it can be due to the physical, monetary, or temporal cost that precludes testing an inordinate number of cases. We assume that "simulation" imbues an intent to study a computational model with an uncertain element in either the computational method and/or the model itself. The uncertainty could arise from quantities modeled as random variables or it could arise from sampling a myriad of possible parametric states.

Some of the original virtual simulations computed quantities in atomic physics. However, the concept stemmed from predicting outcomes of the card game solitaire, and representative of this origin, the code name assigned to the technique by the military was "Monte Carlo." Thus was born the method commonly known as Monte Carlo simulation. The essential idea is to assess a quantity by randomly sampling from a population of possibilities. If the samples accurately represent the larger population, then we gain confidence in our computed quantity. Monte Carlo simulation developed side-by-side with computing because computers were able to draw random samples quickly. For instance, a computer could simulate playing thousands of games of solitaire far quicker than a person could play them. The same computational advantage is being used today to screen candidate drugs to identify those with likely potential, or to test design projects to see which are most likely to succeed. Indeed, applications seem to be everywhere once you learn the basic mentality of using efficient computations and random sampling to canvas an otherwise overwhelmingly large realm of possibilities to identify those with favorable qualities or to estimate computed results.

6.2 Numerical Integration

Integration is one of the principal studies of calculus due to its numerous and far-reaching applications. The uses of integration to solve differential equations and to calculate probabilities are themselves substantial historical pillars that have helped shape modern society. However, as any student of calculus learns, the cleverness of hand calculating integrals can be intimidating. Indeed, the industry associated with evaluating many integrals has motivated substantial catalogs of closed-form solutions so that others could bypass tedious and astute calculations.

The advent of modern computing promoted the idea of quickly and accurately estimating a definite integral's value without a closed form solution of its indefinite counterpart. The idea is to directly compute the value of a definite integral without an antiderivative. Such computational methods are referred to as numerical integration or quadrature.

Numerical integration is a long-standing study in numerical analysis, and the number of calculation algorithms is vast. We review a basic method called Simpson's rule to illustrate the prevailing theme, after which, we show how to estimate an integral's value with a Monte Carlo simulation. Comparisons between the deterministic method of Simpson's rule and the stochastic method of Monte Carlo integration motivate our presentation.

The problem we consider is that of calculating

$$\int_D f(x)\,dV,$$

where f is a real valued function of the n-vector x, D is a bounded region, and V is the (hyper)volume. The notation indicates that we are integrating f over the region D with respect to our standard concept of length, area, volume,

or hypervolume. Although not generally required, we assume for simplicity that D is defined by box constraints. For example, in 2-dimensions D could be
$$D = \{(x_1, x_2) : 5 \leq x_1 \leq 7, -3 \leq x_2 \leq 2\}. \tag{6.1}$$
In this case the "volume" of D is its area, and
$$V(D) = (7-5)(2-(-3)) = 10.$$

The cross product of two sets is all 2-tuples in which the first element belongs to the first set and the second element belongs to the second. To illustrate, assume
$$A = \{a, b, c\} \quad \text{and} \quad B = \{1, 2\}.$$
Then,
$$A \times B = \{(x, y) : x \in A, y \in B\}$$
$$= \{(a, 1), (a, 2), (b, 1), (b, 2), (c, 1), (c, 2)\}$$
and
$$B \times B = B^2 = \{(x, y) : x \in B, y \in B\}$$
$$= \{(1, 1), (1, 2), (2, 1), (2, 2)\}.$$

Conveniently, notice that D in (6.1) is the cross product of two closed intervals,
$$D = [5, 7] \times [-3, 2],$$
and that $V(D)$ is the product of the lengths of these intervals. This idea naturally extends so that
$$V([a_1, b_1] \times [a_2, b_2] \times \ldots \times [a_k, b_k]) = \prod_{j=1}^{k} (b_j - a_j).$$

The cross product notation nicely augments our experience from calculus. For example,
$$\int_2^4 x^2 \, dx = \int_{[2,4]} x^2 \, dl,$$
where dl is the 1-dimensional analog of length to dV. Similarly,
$$\int_{-1}^{2} \int_{2}^{5} f(x, y) \, dx \, dy \quad \text{and} \quad \int_{0}^{1} \int_{-2}^{3} \int_{-1}^{2} f(x, y, z) \, dx \, dy \, dz$$
are, respectively,
$$\int_R f(x, y) \, dA \quad \text{and} \quad \int_D f(x, y, z) \, dV,$$

where $R = [-1, 2] \times [2, 5]$ and $D = [0, 1] \times [-2, 3] \times [-1, 2]$. The integration measure in the 2-dimensional case is dA, indicating that we are integrating with respect to area. In general, dV is used in place of dl or dA to maintain a single, unified notation.

6.2.1 Simpson's Rule

The aim of this chapter is to introduce stochastic algorithms and models, and it is therefore ironic that our first computational method is a deterministic process to approximate an integral. However, numerical integration plays two roles in this chapter. First, any student of computational science should be comfortable with the numerical methods associated with integration, and since we have not yet reviewed such methods, we include here an introduction through the study of Simpson's rule. Second, numerical integration transitions from a deterministic to a stochastic preference as the dimension of the problem increases and as function evaluations become costly. Numerical integration is thus a paragon to illustrate the theme of using simulation to advance a computational algorithm. For these reasons we initiate our foray into stochastic algorithms and models with a review of a classical deterministic method to approximate a definite integral.

Recall that a definite integral is the limit of a Riemann sum (named after the mathematician Bernhard Riemann),

$$\int_{[a,b]} f(x)\, dx = \lim_{\|P\|\downarrow 0} \sum_{p \in P} f(p)\, \Delta p.$$

The set P is a partition of the interval $[a, b]$ into n parts, and $\|P\|$ is the largest subinterval. For example, $P = \{0, 0.75, 1, 2\}$ divides $[0, 2]$ into the three subintervals $[0, 0.75]$, $[0.75, 1]$, and $[1, 2]$. The norm of this partition is

$$\|P\| = \max\{0.75 - 0,\ 1 - 0.75,\ 2 - 1\} = \max\{0.75,\ 0.25,\ 1\} = 1.$$

A typical example in calculus is to let P contain the values

$$p_k = a + \left(\frac{b-a}{n}\right) k, \text{ for } k = 0, 1, \ldots, n,$$

which partitions the interval $[a, b]$ into n equal parts. Allowing $n \to \infty$ forces

$$\Delta p_k = p_{k+1} - p_k = \frac{b-a}{n} \to 0, \quad \text{for } k = 0, 1, \ldots, n-1.$$

Hence, $\|P\| \downarrow 0$ as $n \to \infty$.

A Riemann sum is an accumulation of weighted function evaluations, with the weights being Δp. The study of quadrature extends this sentiment to find weights w_p so that

$$\int_{[a,b]} f(x)\,dx \approx \sum_{p \in P} f(p)\,w_p.$$

The design aspects of a quadrature rule are the elements of the partition P and the subsequent weights w_p. The intent behind a rule's design is to make the approximation as good as possible.

Simpson's rule is a quadrature technique that guarantees perfect accuracy for polynomials of degree 2 or less. To illustrate, suppose we seek w_1, w_2, and w_3 so that

$$\int_{[0,1]} f(x)\,dx \approx w_1 f(0) + w_2 f(1/2) + w_3 f(1),$$

and the approximation is perfect if $f(x) = ax^2 + bx + c$. Since,

$$\int_{[0,1]} ax^2 + bx + c\,dx = a\frac{1}{3} + b\frac{1}{2} + c$$

and

$$f(0) = c,\ f(1/2) = a\frac{1}{4} + b\frac{1}{2} + c,\ \text{and}\ f(1) = a + b + c,$$

the weights must satisfy

$$\int_{[0,1]} ax^2 + bx + c\,dx = a\frac{1}{3} + b\frac{1}{2} + c$$
$$= w_1 f(0) + w_2 f(1/2) + w_3 f(1)$$
$$= w_1 c + w_2 \left(a\frac{1}{4} + b\frac{1}{2} + c\right) + w_3 (a + b + c)$$
$$= a\left(w_2 \frac{1}{4} + w_3\right) + b\left(w_2 \frac{1}{2} + w_3\right) + c(w_1 + w_2 + w_3).$$

The equality of the second and fifth expressions mandates that

$$\begin{bmatrix} 0 & 1/4 & 1 \\ 0 & 1/2 & 1 \\ 1 & 1 & 1 \end{bmatrix} \begin{pmatrix} w_1 \\ w_2 \\ w_3 \end{pmatrix} = \begin{pmatrix} 1/3 \\ 1/2 \\ 1 \end{pmatrix}.$$

The solution is $w_1 = 1/6$, $w_2 = 2/3$, and $w_3 = 1/6$, from which Simpson's rule is

$$\int_{[0,1]} f(x)\,dx \approx \frac{1}{6} f(0) + \frac{2}{3} f(1/2) + \frac{1}{6} f(1). \tag{6.2}$$

The approximation in (6.2) is perfect for polynomials of degree 2 or less, but it applies to other functions as well. This observation invites the question

6.2 Numerical Integration

of how much error we might expect for functions other than quadratics. If we assume that f has a continuous fourth derivative, then the error can be shown to satisfy

$$\left| \int_{[0,1]} f(x)\,dx - \left(\frac{1}{6} f(0) + \frac{2}{3} f(1/2) + \frac{1}{6} f(1) \right) \right| \leq \frac{1}{180} f^{(4)}_{\max}, \qquad (6.3)$$

where

$$f^{(4)}_{\max} = \max_{0 \leq x \leq 1} \left\{ \left| f^{(4)}(x) \right| \right\}.$$

The study of quantifying potential errors is a mainstay of the field of numerical analysis, and interested readers are encouraged to pursue this study to advance their computational sophistication.

The quadratic on which Simpson's rule was developed was defined by the three parameters a, b, and c, and these parameters precisely defined the three weights w_1, w_2, and w_3. An analogous development can be used to develop quadrature rules for higher degree polynomials, as a degree n polynomial precisely defines a quadrature rule with $n+1$ weights once the partition is selected. The resulting higher order methods are called Newton-Cotes rules.

Our development over the interval $[0, 1]$ extends to other intervals with a change of variable. This observation promotes partitioning an interval $[a, b]$ into subintervals and then approximating the integral over $[a, b]$ as the sum of the integrals over the subintervals. We partition $[a, b]$ into n equal length subintervals with $P = \{x_i + ih : i = 0, 1, \ldots, n\}$, where $h = (b-a)/n$. We assume n is even, $x_0 = a$, and $x_n = b$. We then have

$$\int_{[a,b]} f(x)\,dx = \int_{[x_0, x_2]} f(x)\,dx + \int_{[x_2, x_4]} f(x)\,dx + \ldots + \int_{[x_{n-2}, x_n]} f(x)\,dx.$$

We expect Simpson's rule to give better approximations over the smaller intervals, and hence, estimating the summation on the right with many applications of Simpson's rule should better approximate the integral than would a single application of the rule over $[a, b]$.

We use the change of variable

$$x = (1-t)\,x_i + t\,x_{i+2} = x_i + t\,(x_{i+2} - x_i) = x_i + 2t\,h$$

to apply Simpson's rule over the interval $[x_i, x_{i+2}]$. Then $dx = 2h\,dt$, and

$$\int_{[x_i, x_{i+2}]} f(x)\,dx = 2h \int_{[0,1]} f(x_i + 2ht)\,dt$$

$$\approx 2h \left(\frac{1}{6} f(x_i) + \frac{2}{3} f(x_i + h) + \frac{1}{6} f(x_i + 2h) \right)$$

$$= \frac{h}{3} \left(f(x_i) + 4f(x_{i+1}) + f(x_{i+2}) \right).$$

The resulting approximation is

$$\int_{[a,b]} f(x)\,dx \approx \frac{h}{3}\Big[(f(x_0) + 4f(x_1) + f(x_2)) + (f(x_2) + 4f(x_3) + f(x_4)) +$$
$$\ldots + (f(x_{n-2}) + 4f(x_{n-1}) + f(x_n))\Big]$$
$$= \frac{h}{3}\Big[f(x_0) + 4f(x_1) + 2f(x_2) + 4f(x_3) + 2f(x_4) +$$
$$\ldots + 2f(x_{n-2}) + 4f(x_{n-1}) + f(x_n)\Big]. \qquad (6.4)$$

This approximation is the composite Simpson's rule, although it is often referenced as simply Simpson's rule, and it gives a powerful, deterministic method of numerically approximating a definite integral.

The error bound for the composite rule multiplies the bound in (6.3) by $(b-a)\,h^4$, which means that the approximation in (6.4) differs from the true value of the integral by no more than

$$\left(\frac{(b-a)\,h^4}{180}\right)\max_{0\le x\le 1}\left\{\left|f^{(4)}(x)\right|\right\}.$$

To illustrate, note that $|d^4 \sin(x)/dx^4| = |\sin(x)| \le 1$ over any interval. Suppose we want to use Simpson's rule to estimate

$$\int_{[0,\pi/4]} \sin(x)\,dx$$

to within an accuracy of 10^{-8}. Then we need n to be an even number satisfying

$$\frac{\pi/4}{180}\left(\frac{\pi/4}{n}\right)^4 < 10^{-8}.$$

The smallest such n is 16, which means that we are guaranteed to achieve our desired accuracy with no more than $n + 1 = 17$ function evaluations.

The computational advantage of Simpson's rule diminishes as the dimension of our integration increases. For example, consider the two-dimensional problem

$$\int_a^b \int_c^d f(x,y)\,dx\,dy = \int_D f(x,y)\,dV,$$

where $D = [a,b] \times [c,d]$. Versions of Simpson's method for two dimensions similarly calculate a weighted sum of function evaluations over the region D. Suppose $[a,b]$ is divided into n subintervals and $[c,d]$ into m subintervals. Then, as the integral on the left suggests, we need $(m+1)(n+1)$ function evaluations, and moreover, the errors over $[a,b]$ and $[c,d]$ are cumulative. So we need many more function evaluations to have a highly accurate ap-

6.2.2 Monte Carlo Integration

Similar to a deterministic quadrature approximation like that of Simpson's rule, the stochastic estimation of Monte Carlo integration approximates an integral with a weighted sum. The difference is the interpretation of the weights. Suppose we partition $[a, b]$ with $\{x_i : i = 1, 2, \ldots, n\}$. Then

$$\int_{[a,b]} f(x)\,dx \approx \sum_{i=1}^{n} f(x_i)\,\Delta x_i = (b-a)\sum_{i=1}^{n} f(x_i)\left(\frac{\Delta x_i}{b-a}\right), \qquad (6.5)$$

where $\Delta x_i = x_i - x_{i-1}$. The ratios $\Delta x_i/(b-a)$ weight the function values, and they also sum to 1. Hence, these values can be interpreted as probabilities. Indeed, if we assume $x \sim \mathcal{U}(a, b)$, then $P(x_{i-1} \le x \le x_i) = \Delta x_i/(b-a)$. Moreover, if we approximate f over $(x_{i-1}, x_i]$ as the constant $f(x_i)$, creating the function $\hat{f}$ so that $\hat{f}(x) = f(x_i)$ if $x_{i-1} < x \le x_i$ (we assume the interval $[x_0, x_1]$ for $i = 1$), then $\hat{f}(x)$ is a random variable satisfying

$$P\left(\hat{f}(x) = f(x_i)\right) = \frac{\Delta x_i}{b-a},$$

where we assume the values of $f(x_i)$ are unique. Combining this probability with (6.5) gives

$$\int_{[a,b]} f(x)\,dx \approx (b-a)\sum_{i=1}^{n} f(x_i)\left(\frac{\Delta x_i}{b-a}\right)$$

$$= (b-a)\sum_{i=1}^{n} \hat{f}(x_i)\,P\left(\hat{f}(x) = f(x_i)\right)$$

$$= (b-a)\,E(\hat{f}). \qquad (6.6)$$

The approximating function $\hat{f}$ approaches f as the norm of the partition decreases to zero, and we subsequently define the expected value of $f(x)$ with $x \sim \mathcal{U}(a, b)$ as

$$E(f) = \frac{1}{b-a}\int_{[a,b]} f(x)\,dx.$$

This definition naturally extends to higher dimensions, and the resulting relationship between a definite integral and the expected value of the integrand is made precise by the Mean Value Theorem of Integrals.

Theorem 16 (Mean Value Theorem of Integrals). *If f is continuous on D, then for some c in D we have*

$$f(c) = E(f) = \frac{1}{V(D)} \int_D f(x)\,dx.$$

The theorem states that f achieves its average value somewhere over the integration region D.

A partition $\{x_i : i = 0, 1, \ldots, n\}$ of $[a, b]$ is a random sample of $x \sim \mathcal{U}(a, b)$, and hence, a partition provides the sample mean

$$\bar{f} = \frac{1}{n} \sum_{i=1}^{n} f(x_i).$$

Combining this observation with Theorem 16 in Chap. 3 gives

$$\lim_{n \to \infty} P\left(\left| \bar{f} - \frac{1}{b-a} \int_{[a,b]} f(x)\,dx \right| > \varepsilon \right) = 0,$$

where ε is assumed to be small and positive. This expression gives a probabilistic guarantee of approximating an integral, and for large n we expect

$$\bar{f} \cdot (b-a) \approx \int_{[a,b]} f(x)\,dx.$$

Let us return to the example of the previous section, which is to approximate

$$\int_{[0,\pi/4]} \sin(x)\,dx = 1 - \frac{1}{\sqrt{2}}. \tag{6.7}$$

The integral's approximation $\bar{f} \cdot (\pi/4 - 0)$ varies as the sample from $[0, \pi/4]$ grows. Representative progressions are illustrated in Fig. 6.1.

We argued earlier that Simpson's rule could approximate the integral in (6.7) to within a guaranteed accuracy of 10^{-8} with only 17 function evaluations. However, Fig. 6.1 suggests that we need many more evaluations to

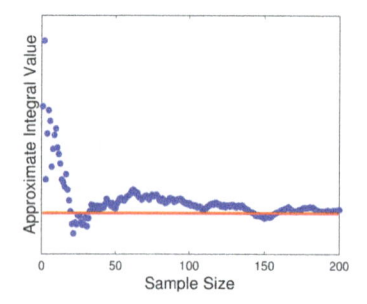

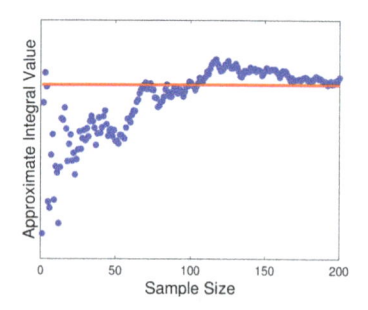

Fig. 6.1 Two representative examples of how stochastic estimates approximate the integral in (6.7) as the size of the sample in $[0, \pi/4]$ increases. The red line is the integral's true value of $1 - 1/\sqrt{2}$.

6.2 Numerical Integration

comfortably predict the integral's value with Monte Carlo integration. We might wonder if it is possible to estimate an appropriate sample size to reasonably approximate the value of an integral. However, the mathematics of Monte Carlo integration does not inform us about the required size of the sample to ensure a predefined accuracy. We might bemoan the use of Monte Carlo integration without such a guarantee, and moreover, if we decide to use Monte Carlo integration, then how do we assess the quality of our estimate? Part of the answer is that any computational advantage of Monte Carlo integration is diminished in low dimensions, especially if function evaluations are easily computed and, even more especially, if the magnitude of the fourth derivative is small. Our example of integrating $\sin(x)$ in one dimension favors Simpson's method on all fronts, and our computational outcomes substantiate the benefits of Simpson's method for this example.

We assume for the moment that function evaluations are not a concern as we address the question of interpreting an approximation from Monte Carlo integration. The analysis primarily rests on the Central Limit Theorem, see Theorem 9 in Chap. 3. The distribution of $\mathcal{I} = \bar{f} \cdot (b-a)$ is approximately normal with mean

$$\mathrm{E}(f) \cdot (b-a) = \int_{[a,b]} f(x)\,dx.$$

If we draw m, n-element, independent samples of $[a, b]$, each providing the approximation $\mathcal{I}_i = \bar{f}_i \cdot (b-a)$, with $i = 1, 2, \ldots, m$, then our sample average and standard deviation are

$$\bar{\mathcal{I}} = \frac{1}{m}\sum_{i=1}^{m} \mathcal{I}_i \quad \text{and} \quad s_{\mathcal{I}}^2 = \left(\frac{1}{m-1}\right)\sum_{i=1}^{m}\left(\mathcal{I}_i - \bar{\mathcal{I}}\right)^2.$$

We can garner confidence intervals from these sample estimates, and in particular, we have

$$P\left(\left|\bar{\mathcal{I}} - \int_{[a,b]} f(x)\,dx\right| < k\,\sigma\right) \quad \text{is approximately} \quad \begin{matrix} k{=}1 & k{=}2 & k{=}3 \\ 0.67, & 0.95, & \text{or } 0.997. \end{matrix}$$

We draw 1000 samples of size 25 from $[0, \pi/4]$ to illustrate an interpretation of the Monte Carlo method approximating the integral in (6.7). Each sample is generated with the MATLAB command (pi/4-0)*rand(1,25). An illustration of the first sample is shown in Fig. 6.2. The proximity of the green line at $\mathrm{E}(f(X))$ and the red line at the sample mean suggests that the first sample reasonably approximates the true integral. We gain additional estimates as we continue to draw the remaining 999 samples, and Fig. 6.3 is a histogram of the estimates. The figure is generated with the hist command, which by default divides the estimates into 10 equal bins and then counts how many estimates fall into each bin. The horizontal axis shows the spread of our estimates while the vertical axis denotes the number of estimates. For this example the sample mean and standard deviation of the integral's approximations are

$$\bar{\mathcal{I}} = 0.29292 \quad \text{and} \quad s_{\mathcal{I}} = 0.03376,$$

and the subsequent confidence intervals are:

67% confident	95% confident	99% confident
$\bar{\mathcal{I}} \pm s_{\mathcal{I}}$	$\bar{\mathcal{I}} \pm 2s_{\mathcal{I}}$	$\bar{\mathcal{I}} \pm 3s_{\mathcal{I}}$
$[0.25916, 0.32668]$	$[0.22540, 0.36044]$	$[0.19164, 0.39420]$.

These values change (slightly) with every simulation. However, the integral's true value of 0.29289 agrees with our estimate in the first three decimal places and is well within one standard deviation.

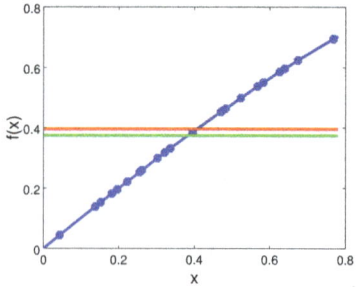

Fig. 6.2 The function $f(x) = \sin(x)$ at 25 randomly selected points in $[0, \pi/4]$. The true function average over the interval $[1, 2]$ is 0.37292 (green line). The estimated average of f over the sample is 0.39507 (red line).

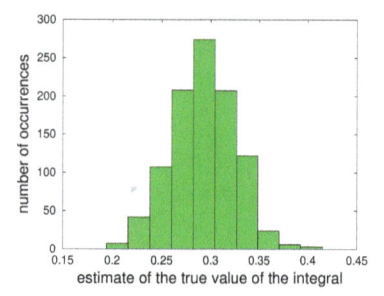

Fig. 6.3 A histogram of 1000 Monte Carlo estimates of the integral in (6.7). The mean of the distribution is 0.29292 and the standard deviation is 0.03376.

The probabilistic outcomes in Fig. 6.3 are based on 25,000 function evaluations, which is exorbitant compared with the 17 needed by Simpson's rule to achieve a near exact result. So while we understand how to conduct a stochastic interpretation, any real-world application of Monte Carlo integration is clearly going to necessitate a more judicious use of the information gained during the simulation. The next section addresses this issue and explains how the technique of bootstrapping can limit our computational burden.

6.2.3 Bootstrapping

Bootstrapping is a statistical method that imparts stochastic inference from an approximation of a random process. The approximation from which bootstrapping works is empirical, meaning that it is a collection of observed data. The method keenly applies to studies in which observational data is difficult to obtain, say due to limitations in funding, time, or ethical concerns. Restricting the acquisition of information prohibits continued sampling from the true random process of interest, sampling that would otherwise strengthen a common statistical analysis. Bootstrapping instead assumes a sufficiently meritorious approximation in support of resampling from the data in lieu of sampling from the true stochastic process.

As an example, consider the need to calculate the center-of-mass of an irregular, three-dimensional object that is being considered for a manufacturing process. The calculation would ideally make use of an accurate and continuous three-dimensional density profile, but unfortunately, density can only be approximated discretely on a coarse three-dimensional grid of a single prototype due to experimental limitations and cost concerns. The result is that the center-of-mass must be calculated from a table of density approximations of the form (x, y, z, δ). Manufacturing imperfections lead to random density variations, so our empirical observations are a sample of an assumed random process.

The estimated density profile could be used in its entirety to gain individual statistical approximations, such as a single sample mean or a single sample standard deviation. However, the goal is not to statistically analyze the empirical data of a single prototype but rather to evaluate the true stochastic phenomenon that is the manufactured product. For example, a reasonable goal is to assess the average product from the manufacturing process. Individual statistics like the sample mean of the empirical data have no guarantee of accuracy in this regard, not even if the provided density information is sizable. After all, the Law of Large Numbers only claims a likely accuracy as the number of samples increases to infinity. Figure 6.1 illustrates this concern, as the sample means waffle as they approach their true value. Moreover, even if these individual statistics accurately assess the random product, we gain little perspective on how these estimates might vary, say with additional prototypes and different density grids.

Bootstrapping necessitates that resamples of the empirical approximation be drawn with replacement. So if our empirical data has n elements and we want to draw samples of size m, then we repeatedly select and record m random elements from the entire set of empirical data to form a sample. Each addition to our sample is selected from the same set of empirical observations, which means that each time an element is selected it is then replaced prior to another selection. Samples are unlikely to match the empirical data even if $m = n$ due to the replacement strategy.

The bootstrap method applies to Monte Carlo integration since any sample of X creates an empirical sample of $f(X)$. For example, suppose we draw a 50-element, uniform sample of $[0, \pi/4]$ and evaluate $f(x) = \sin(x)$ to create an observed, empirical representation of the function over this interval. One such sample is illustrated in Fig. 6.4. The bootstrap method most commonly resamples the empirical data to create samples of the same size, although this isn't strictly required. If we generate 50 samples with replacement of size 50 from our original sample, then each sample provides a sample mean $\bar{f}_i$ for which $\bar{f}_i \cdot (b - a)$ is an estimate of

$$\int_{[0,\pi/4]} \sin(x)\,dx.$$

We thus expect the Monte Carlo approximations to satisfy

$$\int_{[0,\pi/4]} \sin(x)\,dx \approx \mathrm{E}(\bar{f}_i) \cdot (\pi/4 - 0) = \frac{\pi}{200} \sum_{i=1}^{50} \bar{f}_i.$$

For the sample illustrated in Fig. 6.4, the first Monte Carlo integration we calculated had an expected value of 0.285531, compared with the true value of $1 - 1/\sqrt{2} = 0.292893$.

Trust in the bootstrap estimate is expressed as a confidence interval. There are many ways to construct confidence intervals from bootstrap estimates, but the most basic and common technique is to list the estimates $\bar{f}_i \cdot (b-a)$ in ascending order and then remove the highest and lowest extremes to match the desired confidence. For example, if we want an 80% confidence interval for 50 estimates, then we remove the first and last 5 estimates from the ordered list, i.e. we remove half of $(1-0.8) \times 50 = 10$ elements symmetrically from the top and bottom of the sorted list. The confidence interval is then defined by the 6-th and 45-th estimates. A 90% confidence interval for our first simulation with 50 (re)samples is $[0.246501, 0.321853]$, which contains the true value of 0.292893.

The design questions associated with Monte Carlo integration and bootstrapping are the size of the sample and how many resamples to create. As already mentioned, an orthodox decision about the size of the resamples is to assume that they are the same size as the original empirical sample. Figure 6.5 depicts how bootstrap estimates and confidence intervals vary as the number of resamples changes from 30 to 100. The graph shows only slight changes in our statistical inference as the number of resamples increases, and a crude rule-of-thumb is that the number of resamples should be at least 30, although more is generally better.

Figure 6.5 illustrates that bootstrapping's dependence on the empirical data skews its predictive ability toward the empirical evidence of the integral's value. This is not surprising as the empirical evidence is all we have. The

6.2 Numerical Integration

Fig. 6.4 The blue dots are a random 50-element empirical sample of $f(x) = \sin(x)$ over $[0, \pi/4]$. The red curve is $\sin(x)$.

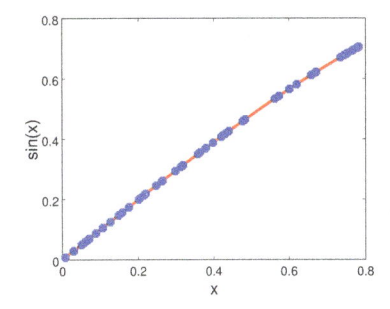

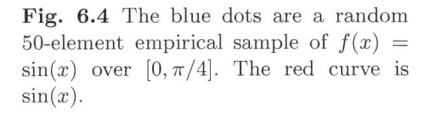

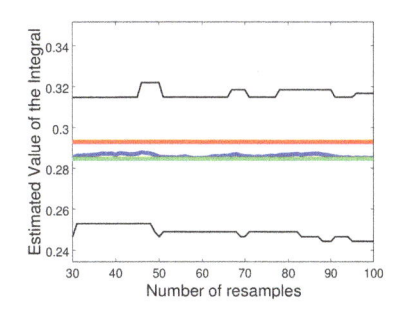

Fig. 6.5 Bootstrap estimates of the integral of $\sin(x)$ over $[0, \pi/4]$ relative to the empirical approximation in Fig. 6.4. The red and green lines are the true value and the empirical approximation of the integral. The blue curve is the bootstrap estimate of the integral as the number of samples increases. The black lines are a 90% confidence interval.

favorable outcome of the bootstrapping method is that it provides confidence in our estimate with limited function evaluations. Indeed, only 50 function evaluations were needed to calculate all estimates even though many more were assumed in the calculation of each estimate. For example, the case with 100 resamples assumed 5000 function evaluations. The trick is, of course, to randomly re-use the original 50 values. The calculation almost seems like we have created something for free, and indeed, bootstrapping was originally approached with hesitation among many statisticians for this reason. The method is widely accepted today.

6.2.4 Deterministic or Stochastic Approximation

Our illustrative examples to this point have only considered single-dimensional integrals, and a deterministic method like that of Simpson's rule is most often preferred in this case. This preference is heightened if the derivatives of the function are nicely bounded, since maximum errors can then be determined to ensure a comforting guarantee of accuracy. However, what about higher dimensional problems in which function evaluations are computationally or monetarily expensive, or what if differentiable information is impractical? In these situations neither a deterministic nor a stochastic method can give a guarantee of accuracy, although a stochastic approach will provide a confidence interval to help instill trust. Also, higher dimensional versions of deterministic quadrature rules require more function evaluations as the grid sizes multiply through the dimensions. The hope with Monte Carlo integration is to limit function evaluations by sampling, after which bootstrap estimates can be calculated to help infer statistical assurances.

We consider the following n-dimensional integral to compare Simpson's method with Monte Carlo integration beyond one dimension,

$$\int_D \sin\left(\|x\|^2\right) dV, \tag{6.8}$$

where

$$D = \{x \in \mathbb{R}^n : \|x\|_\infty \leq 1\} = [-1,1] \times [-1,1] \times \ldots \times [-1,1].$$

The value of the integral as n ranges from 1 to 6 is

n	1	2	3	4	5	6
Integral value	0.62	2.25	5.85	12.97	25.50	44.84.

Simpson's method is only used to evaluate the inner most integral, and all other integrations are approximated with a mid-point rule. The resulting approximation upon dividing each $[-1,1]$ into $2d$ equal parts is

$$\int_D \sin\left(\|x\|^2\right) dV = \int_{-1}^1 \int_{-1}^1 \cdots \int_{-1}^1 \sin\left(\|x\|^2\right) dx_1 \ldots dx_{n-1}\, dx_n$$

$$\approx \frac{1}{3d^n} \sum_{i_n=0}^{2d-1} \sum_{i_{d-1}=0}^{2d-1} \cdots \sum_{i_1=0}^{2d} w_{i_1} \sin\left(\left(-1+\frac{i_1}{d}\right)^2 + \sum_{k=2}^n \left(\frac{1-2d+2i_k}{2d}\right)^2\right),$$

where the w_{i_1} weights follow the $(1, 4, 2, 4, \ldots, 2, 4, 1)$ pattern of Simpson's rule. This approximation requires $(2d+1)(2d)^{n-1}$ function evaluations, which grows exponentially in n. There are versions of Simpson's rule for higher dimensions, and while they increase accuracy over the approximation above, they likewise require an exponential number of function evaluations in n. Moreover, the theoretical bounds on error weaken as dimension grows, limiting the value of the theoretical guarantees.

Simpson approximations are calculated for $n = 2$ and $n = 6$ as d increases from 1 to 4. The number of function evaluations for these cases is tabulated below,

	$d = 1$	$d = 2$	$d = 3$	$d = 4$
$n = 2$	6	20	42	72
$n = 6$	96	5120	54,432	294,912

We compare the approximations from Simpson's method to two variants of Monte Carlo integration. The first draws 30 uniform samples from the integration region D, each of size $10dn$. The Monte Carlo approximation is the

6.2 Numerical Integration

average of the 30 integral estimates from the individual samples, a calculation that is linear in n requiring $300dn$ function evaluations. The second Monte Carlo approximation reduces the number of evaluations by a third by drawing a single uniform sample from D of size $100dn$. The integral's approximate value is the sole sample average generated by this single sample.

Results are presented in terms of relative errors, which are plotted in Figs. 6.6 and 6.7 against the parameter d for the two comparisons. The results show that Simpson's method outperforms Monte Carlo integration in the low dimensional case of $n = 2$. For instance, Simpson's method with 20 function evaluations matches the accuracy of Monte Carlo integration with 1200 evaluations, a 60-fold increase for Monte Carlo integration. However, the story reverses for $n = 6$, and in this case Monte Carlo integration has a marked increase in performance. Simpson's method with 294,912 function evaluations is less accurate than Monte Carlo integration with as few as 1800 evaluations in the first comparison and 600 in the second. These are over 163- and 491-fold decreases in the number of function evaluations for Monte Carlo integration. The analysis depends on the randomness of the Monte Carlo process, but these outcomes are representative and clearly demonstrate a preference for Monte Carlo integration as dimension increases.

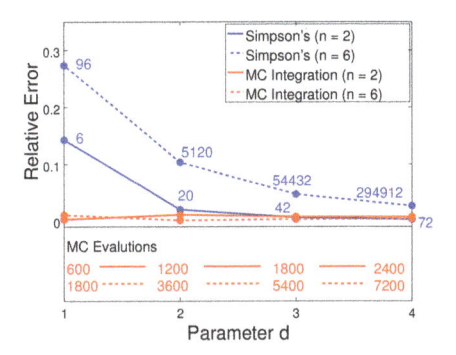

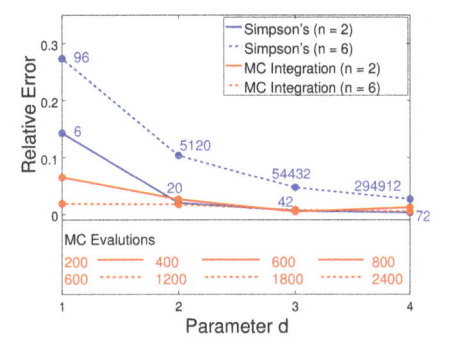

Fig. 6.6 A comparison between Simpson's method and Monte Carlo integration. The integral being evaluated is in (6.8). The Monte Carlo approximations are generated by 30 uniform samples of D of size $10dn$.

Fig. 6.7 A comparison between Simpson's method and Monte Carlo integration. The integral being evaluated is in (6.8). The Monte Carlo approximations are generated by single uniform samples of D of size $100dn$.

Both uses of Monte Carlo integration provide confidence intervals to help assess the approximation. The first estimates the standard deviation from the 30 samples, and the second calculates a bootstrap confidence interval. We calculated 100 approximations with both applications of Monte Carlo integration to measure the frequency with which an approximation failed to be within a 95% confidence interval. Notice that we should expect about 5 cases to exceed the confidence interval. All of the approximations from the first use of Monte Carlo integration were within the estimated 95% confidence interval, and hence, Monte Carlo integration exceeded the statistical expectation and gave strong assurances of quality approximations. The bootstrap approximation was less effective, with the approximations being outside the confidence interval 11 times for $d = 1$, 10 times for $d = 2$, 12 times for $d = 3$, and 9 times for $d = 4$. The reduced lack of confidence isn't surprising, after all, the function is sampled a third fewer times.

An integral's true value is unknown in most computational settings. Indeed, all that we might know is that the function can be evaluated through a lengthy computation. In such a case we have no way of guaranteeing accuracy with Simpson's rule, although we should commonly expect an improved approximation by refining the partition upon which an approximation is made. Monte Carlo integration is a worthy approach in such difficult cases because the lack of certainty is not an interpretable hindrance, and as the examples of this section illustrate, its value per evaluation can be far greater than that of Simpson's rule. So Monte Carlo integration is the method of choice for high dimensional integrals for which reasonable approximations are desired with limited function evaluations.

6.3 Random Models

The stochastic constructs introduced in our study of Monte Carlo integration extend to a much larger investigation of how simulation plays a role in computational science, and we now transition from randomized algorithms to stochastic models. Two examples are presented to promote the general theme, which is it to

1. draw a random sample of the uncertain parameters,
2. solve each random problem to create a sample of the model's outcomes, and
3. analyze the random outcomes by inferring probability statements with the bootstrap method.

The two examples we consider are that of solving a stochastic differential equation and that of solving a stochastic optimization problem.

Prefixing the word "stochastic" to any of our previously studied, deterministic methods denotes a common split in most disciplines. For example, the fields of differential equations and optimization are often divided into stochastic and deterministic sub-studies, e.g. deterministic versus stochastic differential equations and deterministic versus stochastic optimization. Our computational perspective at the moment is deterministic, and simulation essentially assumes an ability to infer stochastic properties by repeatedly solving deterministic samples. Alternatively, stochastic models can be studied mathematically as random distributions. A computational scientist is best

6.3.1 Simulation and Stochastic Differential Equations

An introductory model of projectile motion is used to illustrate a simulation of a stochastic differential equation. Suppose $x = (x_1, x_2)^T$ is the position of a projectile at time t in a two-dimensional plane, with x_1 and x_2 being the horizontal and vertical components. If the mass of the object is m kg, then a common model with air resistance is

$$m\frac{d^2 x}{dt^2} = mg - \frac{\rho\, CA}{2}\left\|\frac{dx}{dt}\right\|\frac{dx}{dt}, \tag{6.9}$$

where g is the gravitational vector $(0, -9.8)^T$ m/s^2, ρ is the density of air in kg/m^3, C is a unitless drag coefficient, and A is the cross sectional area of the projectile in units of m^2. If we let $v = dx/dt$ be the velocity vector, then a first order system of the model is

$$\frac{d}{dt}\begin{pmatrix} x \\ v \end{pmatrix} = \begin{pmatrix} v \\ g - (\rho\, CA/2m)\|v\|v \end{pmatrix}, \quad x(0) = x_0, \text{ and } v(0) = v_0.$$

The model of projectile motion is readily solved with the Runge-Kutta algorithm of Chap. 5 as long as all parameters are deterministic. However, what if some elements of the model are imprecise estimates of a random process? If model parameters are really random, then the solution itself is random, and we should use the language of statistics to describe the distribution of trajectories instead of claiming a precise trajectory. In this case we might wonder if a single model premised on average data can accurately reflect an average outcome. Answering this question motivates our study.

We consider two random cases of a baseball being launched from ground level. The first case assumes that the launch angle θ is distributed uniformly over $[30°, 50°]$, i.e. $\theta \sim \mathcal{U}(30°, 50°)$. The second case maintains the random launch angle and adds a variable launch speed of $\|v_0\| \sim \mathcal{N}(44.704, 1.7882)$ m/s, which is an average launch speed of 100 mph with a standard deviation of 2 mph. We assume

$$x_0 = (0, 0)^T \text{ and } v_0 = \|v_0\|\, (\cos(\theta), \sin(\theta))^T.$$

The mass and radius of a baseball are 0.145 kg and 0.0366 m, and hence, the cross sectional area is $A = 0.0366^2\, \pi$. The air density is assumed to be 1.22 kg/m^3, a common assumption at sea level. The drag coefficient is

$C = 0.5$, which is an appropriate value for a baseball moving at velocities commensurate with a pitch or a batted ball.

We draw a single 50 element sample of θ in radians with the command `pi*(30+20*rand(1,50))/180`. The initial speed for the first model is assumed to be certain at 44.704 m/s. Each of the sampled values of θ gives a deterministic model associated with the sampled value. Each model is solved with RK4, and the horizontal distance traveled by the baseball is calculated with bisection by searching for the time at which the vertical distance is zero. The results are illustrated in Fig. 6.8.

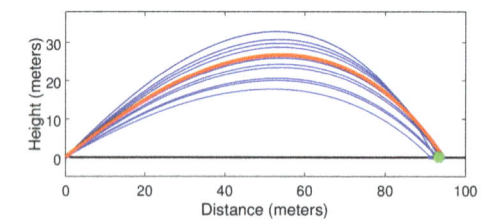

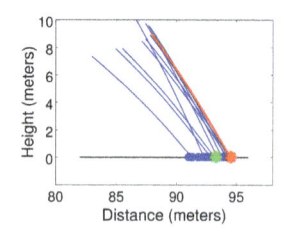

Fig. 6.8 Sample trajectories of a baseball with randomly selected launch angles. The red trajectory is launched at $40°$, and the green dot is the sample mean of the distances traveled over 50 randomly selected launch angles drawn uniformly in $[30°, 50°]$. The landing trajectories on the right illustrate a bias toward shorter horizontal distances over the sample compared with the horizontal distance with $40°$.

If we had ignored the random launch angle by assuming an average of $40°$, then the associated deterministic model would have predicted a horizontal travel of 94.55 m. A common misconception would be to assume that repeated launches with angle variations about $40°$ would result in travel distances varying about 94.55 m. In other words, it is natural but incorrect to assume that average inputs, i.e. model parameters, result in average outcomes. Although we assume distributional knowledge of the input, which in this case is the uniform distribution of θ, the distribution of the outcome, which is the horizontal travel distance, is unknown. Hence a priori assumptions about the outcome's statistics are dubious.

The landing trajectories in Fig. 6.8 suggests a shift toward shorter distances. Indeed, the sample mean of the travel distances, shown as the green dot in Fig. 6.8, is 93.31 m, and the sample standard deviation of the horizontal distances is 1.07 m. The fact that 94.55 m is outside 93.31 ± 1.07 m gives us some hesitation about the trustworthiness of 94.55 m being a reasonable estimate of the average distance. First, residing outside a standard deviation of the sample mean suggests that 94.55 m is somewhat uncommon. Second, because we want to estimate the average distance, the fact that 94.55 m is

6.3 Random Models

uncommon as an individual outcome magnifies our concern that it is an inadequate representation of the average outcome.

While the distribution of the horizontal distance is unknown, we do know the distribution of the sample mean. The distribution of the sample mean is the t-distribution, but as noted in Sect. 3.2.2, the t-distribution is nearly normal if we have enough samples, say at least 30. We bootstrap our current sample instead of repeatedly drawing new samples to gain a sufficient number of sample means. As with numerical integration, bootstrapping reduces the computational burden, which is favorable since coupling RK4 with bisection to generate Fig. 6.8 already requires several minutes. We create 40 resamples of size 50 by randomly selecting with replacement from our original population of horizontal distances. The expected value of the sample mean is 93.32 m, which nearly matches that of our original sample mean.

A 95% confidence interval on the sample mean can be approximated in two ways. One way is to rank the sample means and remove the highest and lowest value, which symmetrically removes a total of 5% of the 40 ranked samples. The resulting 95% confidence interval is [93.02, 93.60]. A second approximation relies on the near normality of the sample mean, from which a 95% confidence interval is approximately the spread within two standard deviations of the mean. The sample standard deviation of the bootstrap estimates of the mean is 0.16, which gives an approximate 95% confidence interval of $[93.32 - 2(0.16), 93.32 + 2(0.16)] = [93.00, 93.64]$. The similarity of the two intervals gives credence to our bootstrap sample. The horizontal distance of 94.55 m from the model with average data is not in either of these intervals, and we conclude that 94.55 m does not approximate the average distance.

The sample mean being nearly normal allows us to estimate the probability of the average distance being at least 94.55 m. The bootstrap statistics indicate that 94.55 m is $(94.55 - 93.32)/0.16 = 7.68$ standard deviations away from the expected mean. The probability that the average horizontal distance is at least 94.55 m is thus approximately

$$1 - \int_{-\infty}^{7.68} \frac{e^{-x^2/2}}{\sqrt{2\pi}}\, dx = 7.55 \times 10^{-15},$$

which can be calculated with `1-normcdf(7.68)`. This minute probability renders hopeless any perception of the average data's outcome being representative of the average outcome.

The difference between 94.55 m and 93.32 m could be argued as insignificant depending on the situation. If we were estimating the flight distance of a batted ball, then the difference is likely meaningless since an outfielder would almost certainly be able to account for the 1.23 m difference (note that 94.55 m is just shy of the shortest home run distances). However, if we are instead studying an outfielder's throw to home plate, then a 1.23 m difference is meaningful and could easily decide a catcher's ability to get an out. Independent of the application, the lesson is to be leery of a deterministic model built from average data, as the resulting outcome is not necessarily indicative of the true stochastic process.

The practice of simulation is essentially unchanged if other parameters are random, and our second case adds a stochastic initial speed. We assume $\|v_0\| \sim \mathcal{N}(44.704, 1.7882)$ and repeat the entire experiment with $\|v_0\|$ being random instead of static. Sample trajectories are depicted in Fig. 6.9. Several trajectories land beyond that of the average data, and so unlike the trajectories in Fig. 6.8, we might suspect better agreement between average inputs and outputs.

The horizontal distance of the model with the average data is again 94.55 m, and the sample mean over the 50 randomly selected models is 93.20 m, with a sample standard deviation of 2.64 m. In this case 94.55 m is within 93.20 ± 2.64 m, which suggests that 94.55 m is a reasonable outcome in itself. However, the expected sample mean from bootstrapping is 93.20, which matches the sample mean. The 95% confidence interval of the expected distance from the method that ranks the bootstrap estimates is [92.48, 93.80]. The sample standard deviation of the sample means is 0.41, and the associated estimate of the 95% confidence interval is [92.39, 94.01], which is fairly close to the other interval. The value of 94.55 m is outside both confidence intervals, and we again conclude that the outcome with average data is not representative of the average case. The probability of the expected distance being at least 94.55 m is estimated to be `1-normcdf((94.55-93.20)/0.41)`$= 4.96 \times 10^{-4}$. So, while the added variability in the launch speed provides a wider spread of horizontal distances, the predicted horizontal distance from the model with average data does not accurately predict the average horizontal distance.

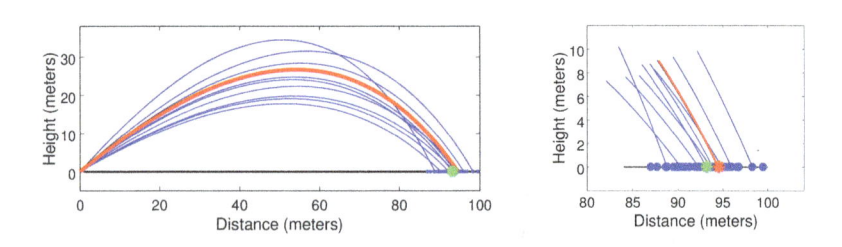

Fig. 6.9 Sample trajectories of a baseball with randomly selected launch angles and initial speeds. The red trajectory is launched at 40° with $\|v_0\| = 44.704$ m/s. The green dot is the sample mean of the distances traveled over 50 randomly selected models. The landing trajectories on the right illustrate the spread of observed horizontal distances.

6.3.2 Simulation and Stochastic Optimization Models

Optimization problems, like differential equations, depend on data, and in the vast majority of real-world applications at least some of the data is estimated due to uncertainty. A practical model with numerous applications is that of a network flow, and we illustrate how simulation and bootstrapping can be used to analyze the results of a stochastic network flow.

The problem we solve is depicted in Fig. 6.10. The goal of the problem is to find the least expensive way to route supply at node 1 to demand at node 13. We assume 5 units of supply and demand at nodes 1 and 13, respectively. The units need not be transported as a single collection, and sending 1.8 units from node 1 to node 2, 3.2 units from node 1 to node 3, and 0 units from node 1 to node 4 is feasible. An arc is a pair (i, j) that denotes the possibility of directly shipping from node i to node j. All possible arcs are shown in Fig. 6.10, so it is possible to send units directly from node 3 to node 5 but not from node 3 to node 6.

Each arc has an average cost per unit transport, which we denote by $\hat{c}_{ij}$. The average costs are illustrated by color in Fig. 6.10. For example, the average cost of shipping one unit from node 7 to node 10 is $\hat{c}_{7,10} = 1$, in whatever monetary units are being used, and the average cost of shipping one unit from node 11 to node 13 is $\hat{c}_{11,13} = 3$. All arcs have an average capacity of $\hat{w}_{ij} = 3$ units, and the flow along any arc must be below its capacity.

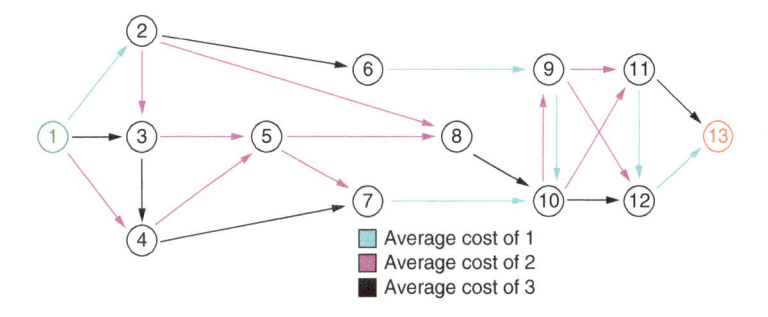

Fig. 6.10 An illustration of a 13 node network flow model. The supply node is number 1 and the demand node is number 13. The average cost per unit transport of each arc is denoted by color.

The variables decided by the optimization problem are the flows along the individual arcs, and

$$x_{ij} \text{ is the flow from node } i \text{ to node } j.$$

The variables are constrained by ensuring conservation, which means that whatever flows into a node must flow out. The exceptions to conservation are the supply and demand nodes, since flow can only enter or leave these nodes. For instance, at node 3 we ensure

$$x_{13} + x_{23} - x_{34} - x_{35} = 0,$$

whereas for nodes 1 and 13 we have

node 1 (supply)		node 13 (demand)
$x_{12} + x_{13} + x_{14} \leq 5$	and	$x_{11,13} + x_{12,13} \geq 5.$

In many situations the costs and capacities are assumed to be predetermined quantities, but we assume that they are instead random. The assumed distributions are

cost distributions		capacity distributions
$c_{ij} \sim \mathcal{N}(\hat{c}_{ij}, \hat{c}_{ij}^2/16)$	and	$w_{ij} \sim \mathcal{U}(2,4).$

We comment that the use of the normal distribution technically permits negative costs, albeit with extremely low probability with $\hat{c}_{ij}$ being positive. A truncated normal disallowing negative outcomes could have been used, but such an adaptation was unnecessary since no negative costs were observed in our simulation. The stochastic cost associated with any flow is

$$\sum_{ij} c_{ij} x_{ij},$$

and the arc capacities require $0 \leq x_{ij} \leq w_{ij}$.

The stochastic optimization model for the network is

$$\min \sum_{ij} c_{ij} x_{ij}$$

such that

$$\sum_i x_{ik} - \sum_j x_{kj} = 0, \text{ for } k = 2, 3, \ldots 12,$$

$$\sum_j x_{1j} \leq 5,$$

$$\sum_i x_{i,13} \geq 5,$$

$$0 \leq x_{ij} \leq w_{ij}, \text{ for all } (i,j).$$

6.3 Random Models

For any sample of costs and capacities the model is linear and can be expressed as
$$\min c^T x \quad \text{such that} \quad Ax = b,\ x \geq 0,$$
where A, b, and c appropriately index the arcs, see Exercise 16. We can solve a sampled instance with the algorithm in Sect. 4.2.1 once in this format.

If the network model is solved with the average data, i.e. $c_{ij} = \hat{c}_{ij}$ and $w_{ij} = 3$, then the minimum cost is 46. The question we answer with simulation is, does this cost reasonably approximate the expected cost of the stochastic model? We create a sample of 1000 models by sampling from the cost and capacity distributions, and each instance is solved to create a sample of the random minimum cost. From the sample of minimum costs we draw 200 resamples with replacement to calculate bootstrap statistics on the average minimum cost.

The sample average of the minimum cost from the 1000 samples is 45.15, with a sample standard deviation of 3.90. From this calculation the cost of 46 appears to be a reasonable outcome, as it is only $(46 - 45.15)/3.9 = 0.28$ standard deviations away from our estimate of the expected average. The expected sample mean from the bootstrap resamples is 45.14, which essentially matches the sample mean of the entire sample. Due to a relatively large sample and a large number of resamples, the approximate 95% confidence intervals from the bootstrap estimates of the expected minimum cost should be nearly identical independent of how they are created. Ranking the minimum cost estimates of the 200 resamples and removing the top and bottom 5 values gives an approximate 95% confidence interval of $[44.87, 45.38]$. The sample standard deviation of the bootstrap sample means is 0.13, and since the sample mean is approximately normal, a 95% confidence interval on the expected minimum cost is $45.14 \pm 2 \times 0.13$, which equates to $[44.89, 45.40]$. The similarity of the confidence intervals imparts trust, and since the minimum cost of 46 is outside these intervals, we have little confidence that 46 is a reasonable estimate of the average minimum cost. Indeed, 46 is 6.76 sample standard deviations away from 45.14, from which we estimate the probability that the average minimum cost is at least 46 to be `1- normcdf((46-45.14)/0.13)`$= 1.85 \times 10^{-11}$. The conclusion is that the minimum cost of 46 is an unreasonable estimate of the average cost. Instead, the average cost is about 45.14. If the monetary unit is millions of dollars, then the difference could be quite important.

The stochastic models of the last two sections demonstrate a common theme, one that should be at the forefront of your consideration if you are asked to "solve" a stochastic problem. The input to the problem is a collection of random data, and part of the modeling process is to decide which input distributions are appropriate. Once these distributions are decided, the model then translates random inputs into random outputs. In our examples,

- random launch angles were translated into random travel distances through a stochastic model of projectile motion,

- random launch angles and initial velocities were translated into random travel distances through a stochastic model of projectile motion, and
- random costs and capacities were translated into random minimum costs through a network flow model.

All three cases show that solving a single model with average data presages an ill-fated representation of the average outcome. This fact is evidenced by standard statistical arguments conducted by first simulating the outcome and then bootstrapping statistical properties about the average outcome. In some cases the added computation required by the simulation will show that the average model suffices, but without the analysis you would not know and could easily make incorrect recommendations. For example, if the network of this section was regularly used as a transportation network, and if budgets had to be decided in advance, then you might recommend a substantially larger budget than what would be needed. Such a recommendation could limit resources for other projects and subsequently restrict profits. The lesson is to study stochastic models as well as you can, and in particular, combining simulation with bootstrapping is a powerful strategy to initiate such investigations.

6.4 Exercises

1. The card game of War is a two-person game in which the goal is to collect all of the cards. Each player starts with a randomly shuffled deck of 52 cards. A play consists of:
   - Each player displays the top card from their deck.
   - If one card has a higher value, then the player with the higher value collects both cards.
   - If the displayed cards have a common value, then each player displays the next three cards from the top of their decks, after which the fourth cards are displayed. If one of the fourth cards has a higher value, then the player with the highest value collects all ten cards, otherwise this process repeats until a winner is decided or the game remains tied with all cards being displayed.

   Write a simulation of the game to estimate the expected number of plays needed to end a game. Call the simulation warSim.m. Assume that cards are placed on the bottom of the deck in the order in which they are played, with the winner's card(s) preceding the loser's card(s). The code should report the expected number of plays and display a histogram of the number plays calculated. The number of simulated card games should be an input to warSim.

2. In the card game of War, see Exercise 1, cards can be put on the bottom of the deck by whatever rule to which the players agree. Use your code

from Exercise 1 to see if there is a difference in the expected number of plays depending on how cards are recycled to the players' decks. One option is to add cards to the bottom of the deck in the order in which they are acquired. Another is to add each play's acquisition to the bottom of the winner's deck in a random order, and yet another is to place all acquired cards in a separate pile which is then shuffled and used once a player's deck is exhausted.

Compute 95% confidence intervals for the expected number of plays in two manners. First, repeatedly simulate n games to compute a sample of approximate sample means for the number of plays. Calculate the sample standard deviation and estimate a 95% confidence interval on the expected number of plays as being within two standard deviations from the mean. Second, create one n element sample and then bootstrap other samples by resampling instead of generating new samples. Estimate the 95% confidence interval by ranking the bootstrap estimates and by calculating the sample standard deviation of the bootstrap estimates. Use the `tic` and `toc` commands to time the calculation speed in each case. Which method would you promote with regard to balancing speed and accuracy?

3. Continue Exercise 2 by mixing replacement strategies for the two players. Which combination best promotes a winning strategy?
4. The "Monte Hall" question has become a chic paragon to illustrate how humans struggle to quantify probability statements. The situation is that of a game show in which a contestant is to select one of three doors. Gag gifts are behind two of the doors, but opening the third would reveal a wondrous prize, say a proof that P=NP. The contestant selects one of the three doors but is not shown what is behind it. The host then opens one of the remaining two doors, behind which is one of the gag gifts. The contestant is then offered the opportunity to change her or his choice to the closed door not originally selected. Should the contestant do so? Write a simulation called `MonteHall.m` to estimate the probability of winning the P=NP prize depending on whether or not the contestant changes.
5. The game of baseball is a statisticians dream, with statistics covering nearly all aspects of the game being readily available. The game can be thought of as a progression of state transitions from pre- to post-pitch. A game's state consists of the position of base runners, the number of runs, balls, strikes, and outs, and which team is at bat. One of the primary responsibilities of a team manager is to decide the batting order, which is a consecutive list of how players must bat. Design a baseball simulation that allows you to decide the batting order with the highest amount of expected runs. Tune your simulation to two of your favorite teams.
6. Write a function with declaration

    ```
    [I]=SimpsonInt(f,a,b,n)
    ```

that approximates an integral with Simpson's rule.

The input arguments are:

f - The function being integrated, either as a function handle or a string.
a - A vector of lower bounds describing the region D.
b - A vector of upper bounds describing the region D.
n - A vector of the number of subdivisions for each variable.

Assume the region being integrated over is

$$R = [a_1, b_1] \times [a_2, b_2] \times \ldots \times [a_k, b_k],$$

where $a = (a_1, a_2, \ldots, a_k)$ and $b = (b_1, b_2, \ldots, b_k)$. For example, if the second, third, and fourth arguments are a = [0, -1], b = [2, 0] and n = [20, 50], then the interval $[0, 2]$ is divided into 20 equal subdivisions for x_1 and the interval $[-1, 0]$ is divided into 50 subdivisions for x_2. The integral should be calculated as

$$\int_{a_k}^{b_k} \int_{a_{k-1}}^{b_{k-1}} \ldots \int_{a_2}^{b_2} \int_{a_1}^{b_1} f(x)\, dx_1\, dx_2 \ldots dx_{k-1}\, dx_k,$$

where Simpson's rule is applied to the inner most integration. All other integrals are approximated with the midpoint rule.

The return argument is:

I - The approximate value of the integral.

7. Write a function with declaration

    [Imu, Isigma] = MCInt(f, a, b, n, r)

that estimates an integral with Monte Carlo integration.

The input arguments are:

f - The function being integrated, either as a function handle or a string.
a - A vector of lower bounds describing the region R.
b - A vector of upper bounds describing the region R.
n - The sample size used to approximate the average value of the integrand, this is optional with a default value of 1000.
r - The number of samples of x to be used; optional, with a default value 100.

6.4 Exercises

Assume the region being integrated over is

$$R = [a_1, b_1] \times [a_2, b_2] \times \ldots \times [a_k, b_k],$$

where $a = (a_1, a_2, \ldots, a_k)$ and $b = (b_1, b_2, \ldots, b_k)$.

The return arguments are:

Imu   -   The sample mean of the integral's approximated values.

Isigma - The standard deviation of the integral's approximated values.

8. Add to your MCInt code from Exercise 7 by including a return argument that is an approximate 95% confidence interval. Calculate the approximation by bootstrapping new samples and ranking their estimates of the integral.

9. The Fourier series of a function $f(x)$ defined over $-p \leq x \leq p$ is

$$\frac{a_0}{2} + \sum_{n=1}^{\infty} \left[a_n \cos\left(\frac{n\pi x}{p}\right) + b_n \sin\left(\frac{n\pi x}{p}\right) \right],$$

where

$$a_n = \frac{1}{p} \int_{-p}^{p} f(x) \cos\left(\frac{n\pi x}{p}\right) dx \text{ and } b_n = \frac{1}{p} \int_{-p}^{p} f(x) \sin\left(\frac{n\pi x}{p}\right) dx.$$

The index for the a coefficients includes $n = 0$, in which case the cosine reduces to the constant 1.

Write a function with declaration

[z] = FourierSum(f,N,x)

that calculates the first N terms of the Fourier series for f and then evaluates the summation at the values of the vector x.

The input arguments are:

f - The function whose Fourier series is evaluated, either as a function handle or a string.

N - A nonnegative integer indicating the number of terms in the Fourier sum.

x - A vector of points at which to evaluate the partial Fourier series.

The return argument is:

z - the vector resulting from the Fourier series being evaluated at the elements of x.

Evaluate the integrals with Simpson's method with the number of divisions being a multiple of N. Experiment with this multiple to find a reasonable value, i.e. one that is not overly large but that is appropriate for accuracy. You may want to consider the bound in (6.3) to guide your search.

10. Add an optional argument to FourierSum in Exercise 9 to indicate that the integrals should be calculated with MCInt from Exercise 7. Simpson's rule should remain as the default. Experiment with different simulation parameters to assess how they affect accuracy. With regard to speed and accuracy, would you rather use Simpson's method or Monte Carlo integration?

11. Consider the problem of calculating the center-of-mass of a doughnut shaped ring of metal that is to spin around its center. The manufacturing process attempts to create the ring with a homogeneous density, but imperfections in the production process instead lead to a density profile that is distributed normally with a mean of 8000 kg/m^3 and a standard deviation of 125 kg/m^3.

Create a grid through the three-dimensional ring and generate a computed sample of densities over the grid. Several different sized rings are made, so your calculation scheme should work for variable radii and thicknesses. Assume the sample is empirical data with which you have been asked to approximate a 99% confidence interval for the center-of-mass. Use the technique of bootstrapping by resampling the density errors. Each grid point in the empirical sample has an error from the targeted mean of 8000 kg/m^3. The set of errors is what you draw resamples from with replacement, and so, each resample assigns randomly selected errors to the grid points, creating what is essentially a new ring from the manufacturing process. Generate enough resamples to approximate the desired 99% confidence interval, using a technique of your choice.

12. The method of bootstrapping can be used to approximate confidence intervals for the parameters of the weather model in Exercise 7 of Chap. 3. Assume each day's average temperature is the predicted value from the optimal model. The residual difference between the data and the average value is then an assumed error. Bootstrap estimates are based on new datasets by resampling from the collection of errors with replacement. So each day is assigned its average temperature and a randomly selected error. Each new dataset provides estimates of α_0, α_1, and α_2, which can then be ranked and trimmed to provide approximate confidence intervals. Compute 95% confidence intervals for α_1 and interpret your result.

6.4 Exercises

13. Conduct the following experiment. Randomly generate m polynomials of degree $n-1$ or less with coefficients having a maximum magnitude of $10n$. For example, if $m = 3$ and $n = 4$, then

$$40*\text{rand}(3,4).*\text{sign}(\text{rand}(3,4) - 0.5)$$

generates a 3×4 matrix of random numbers between -40 and 40. Each row of the matrix associates with a randomly selected polynomial of degree 3 or less. If the random matrix is

$$\begin{bmatrix} a_{11} & a_{12} & a_{13} & a_{14} \\ a_{21} & a_{22} & a_{23} & a_{24} \\ a_{31} & a_{32} & a_{33} & a_{34} \end{bmatrix},$$

then the three polynomials are

$$p_1(x_1) = a_{11}x_1^3 + a_{12}x_1^2 + a_{13}x_1 + a_{14},$$
$$p_2(x_2) = a_{21}x_2^3 + a_{22}x_2^2 + a_{23}x_2 + a_{24}, \text{ and}$$
$$p_3(x_3) = a_{31}x_3^3 + a_{32}x_3^2 + a_{33}x_3 + a_{34}.$$

Assume f is the random function of the product of these polynomials, i.e., $f(x) = p_1(x_1)p_2(x_2)\ldots p_m(x_m)$. Repeatedly sample new coefficients and integrate each $f(x)$ over

$$D = [-\hat{a}_1, \hat{a}_1] \times [-\hat{a}_2, \hat{a}_2] \times \ldots \times [-\hat{a}_m, \hat{a}_m],$$

where $\hat{a}_i = \max\{|a_{ij}| : j = 1, 2, \ldots, n\}$, with your SimpsonInt and MCInt codes from Exercises 6 and 7. Use Hörner's method to evaluate the polynomials. Investigate how the two methods perform as n increases. If possible, use a symbolic solver to calculate exact answers on which to base your study.

14. Use your simulation from Exercise 11 to consider alterations to the geometry. For instance, what if the ring's geometry is only approximately toric, or if the ring is ellipsoidal with a variable thickness? Could you use your model to approximate the center-of-mass of an automotive tire?

15. Solve the projectile motion model in (6.9) for the flight of a baseball, assuming a static initial speed of 44.704 m/s. Vary the launch angle from $\theta = 20°$ to $\theta = 60°$, and plot the resulting horizontal distances against the angles. Explain why the average distance from a randomized model is likely to be shorter than the distance from the model with average data.

16. Find matrix A and vectors b and c so that the network flow model in Fig. 6.10 is

$$\min c^T x \quad \text{such that} \quad Ax = b, \ x \geq 0.$$

Solve the resulting model with a standard solver in MATLAB or Octave or with your implementation from Exercise 17 in Chap. 4.

17. Assume the arc capacities of the network flow model in Fig. 6.10 are static at a value of 3. Simulate the solution using your model in Exercise 16 under the assumption that the arc costs are uniformly distributed about the stated averages. If $c_{ij} \sim \mathcal{U}(\hat{c}_{ij}-\theta, \hat{c}_{ij}+\theta)$, where the same value of θ is used for all distributions, then how large does θ need to be before the cost associated with the average data falls outside a 95% confidence interval for the average cost? State clearly what method you used to approximate the confidence interval.

18. Repeat Exercise 24 in Chap. 4 but with random variations in the distance traveled by the robotic arm. Instead of distance it is more intuitive to consider travel time, and so, while the distance from neighboring nodes remains one unit, the time to traverse this distance is random due to variations in the robotic controls. The goal of the TSP problem is then to minimize the total time instead of the total distance. Assume the travel time along each arc is uniformly distributed over $[0.4, 1.6]$. What are the outcomes of the expected minimum travel times from TSPSA, TSPGA, and 2-opt? Calculate 90% confidence intervals of the expected minimum travel time with a method of your choice. Which algorithm would you recommend?

Chapter 7
Computing Considerations

The purpose of computing is insight, not numbers. – Richard Hamming

Everybody who learns concurrency thinks they understand it, ends up finding mysterious races they thought weren't possible, and discovers that they didn't actually understand it yet after all. – Herb Sutter

Debugging is twice as hard as writing the code in the first place. Therefore, if you write the code as cleverly as possible, you are, by definition, not smart enough to debug it. – Brian Kernighan

To this point we have only considered mathematical and algorithmic descriptions of computational methods, but several other aspects impact the computational efficacy of a project. In particular, a vast number of computational decisions, such as which language and computing platform to use, affect the practicality of a large computational effort. There are numerous architectures on which computational tasks can be performed, and different computing platforms lend themselves to different tasks. A host of languages other than MATLAB and Octave are options, each with advantages and disadvantages. While our focus has been on MATLAB and Octave, a computational scientist requires dexterity on these fronts. After all, you may find yourself working with a team that has already spent years developing specialized codes in languages foreign to yourself. Moreover, the computing environment might use an operating system with which you are unaccustomed, and the computing architecture could differ from your experience.

This chapter introduces a few of these computing decisions. We show how to couple languages to expedite calculations, and we acquaint ourselves with parallel computing. Both of these topics amplify our computational competency and project an ability to solve large problems with alacrity.

7.1 Language Choice

MATLAB and Octave are scripting languages designed for numerical methods, especially those concerned with numerical linear algebra. Since much of science, engineering, and mathematics is expressed in terms of vectors and matrices, these languages find wide applicability in computational science. Moreover, the graphical ability of MATLAB is outstanding, and since a picture is worth a thousand words, MATLAB advances its position within computational science because its image capabilities help explain complicated observations.

A compiled language requires us to compile our code before running it. The compiler takes as input the source code written by the programmer, and it emits as output an executable program comprised of low-level machine instructions. Executable files are generally not very portable. While they can run on like processors with like resources, if you change machines, then you will likely need to recompile your code. Since compilations are not always straightforward from one machine to the next, the task of recompiling can be tedious and difficult. The advantage of a compiled language is speed, since the entirety of the source code is translated into machine-native instructions.

A scripting, or interpreted, language interprets code as it progresses through a script. As opposed to a compiled language, the source code may never be translated to machine instructions. Instead it may be transformed into some intermediate code, or it may not be transformed at all. The code is executed line by line by an interpreter, which reads each line and updates the state of the program accordingly. The act of determining how to update the state of the program adds overhead to execution time. The benefit is the pleasantry of authoring code in a simple, human-readable language that works on any platform on which an interpreter has been installed. MATLAB and Octave are both scripting languages.

Scripting languages have historically been shunned because their diminished performance has rendered them impractical for many real-world problems. The expectation has been that all serious computing should be done with compiled languages like FORTRAN and C/C++, and training in both was a prerequisite to computational studies. However, scripting language interpreters have advanced, and computing resources have gained substantially. If we couple these facts with the advantages of one-time, easy-to-write scripts that work across platforms and that can easily read different data sources and produce outstanding graphics, then we understand why the concept and applicability of a scripting language has gained favor.

An exhaustive list of scripting languages would be difficult to construct, as many have come and gone. However, several languages have lasted, if for no other reason than many useful applications have been coded in them. We mention, Perl, php, Tcl/Tk, and Python as those with widespread appeal. Of these, Tcl is less popular, although the graphical widgets in Tk, which are written in Tcl, are widely used. Perl persists in many computational settings, and it has outstanding ability with regard to string manipulation. The language php is popular for server side scripting and web design.

Python seems to be the current champion, and we give some examples below. Python's advantage with regard to computational science is worth noting. Python has a long list of modules, several of which mimic MATLAB and Octave. The NumPy and SciPy modules provide much of the same functionality as MATLAB , and modules like BioPython and Pandas are common in computational biology and data science. MATLAB has competitive toolboxes for many of the same tasks.

We do not make any claim about the superiority of one language over another, as the truth is that a coding preference is often individualistic. Just as with many elements of life, we like what we know. The point to our discussion is that there are options, and a computational scientist should be able to work with a variety of languages and select one that reasonably approaches the task at hand. Scripting languages are excellent at high-level programming that combines numerous subtasks. Many scripting languages work well with the operating system and can access, start, and stop other processes. For example, it is generally easy to get a directory listing and then process it so that other programs can access a list of files. It is also possible to "pipe" data to and from other programs already running on the system. These tasks are, at best, onerous with a compiled language. That said, scripting languages remain slow and cumbersome options for more targeted tasks that are better suited to heightened efficiency. This is where compiled languages shine, as they simply outclass scripting languages with regard to speed. Combining the two is powerful, and the next section introduces how to advance speed by linking a scripting language to a compiled language.

7.2 C/C++ Extensions

Working through the exercises of the previous sections has probably exposed a painful computational truth about MATLAB and Octave, which is that they are noticeably slow at looping through an iterative process.[1] This sluggish attribute is easily tested by computing the product Ax as A*x versus calculating the product by iterating over the elements of the matrix and vector. The function in Example Code 7.1 requires nearly 11 s on a modern laptop to calculate Ax if A is 1000×1000 and x is 1000×1, whereas A*x only requires 6.12×10^{-4} s. So A*x is about 100,000 times faster. MATLAB and Octave are, nonetheless, using functions similar to Example Code 7.1 to calculate Ax, but they are not doing so through scripts written in their own

---

[1] All timing results in this chapter were accurate as of 2015. Scripting language developers are working constantly to decrease their performance disadvantage. As such, the reader should take from this chapter general rules of thumb regarding language performance rather than any particular result.

languages. The function has instead been written in C/C++ and precompiled so that you can access it through MATLAB and Octave. Combining scripts with C/C++ allows us to write high-level scripts while accessing the speed of precompiled functions to complete specified tasks.

```
function [b] = matProd(A, x)
[m,n] = size(A);
for i = 1:m
    b(i) = 0;
    for j = 1:n
        b(i) = b(i) + A(i,j)*x(j);
    end
end
b = b(:);
```

Example Code 7.1 A MATLAB function to calculate Ax by iterating over the elements of A and x

Both MATLAB and Octave allow us to extend their functionality by authoring code in C/C++ or FORTRAN. These extensions are defined in mex-files instead of m-files, and the mex command compiles these extensions so that they become native to MATLAB or Octave. We consider how to write a new function to calculate the product of polynomials with Hörner's method.

Please note that we make no attempt to draft an introduction to C/C++ or FORTRAN, as serious introductions would themselves be substantial topics. Students who already know these languages should be able to use the forthcoming examples to embark on their own. Otherwise, the goal is to promote an ability to increase computational efficacy by linking MATLAB and Octave with a compiled language, and those interested should consider learning C/C++ or FORTRAN.

We consider the calculation of the product of polynomials f in Exercise 13 of Chap. 6. The coefficients of the polynomials are in the rows of a matrix, and the polynomial of the i-th row is evaluated at x_i. The function f is the resulting product of the polynomials. Each polynomial is evaluated with Hörner's method as developed in Sect. 1.4.1. Example Code 7.2 shows how f can be evaluated through a MATLAB or Octave function.

A mex-file written in C defining the same function as that of HornerM is in Example Code 7.3. The more involved code is due to the need to compile this function prior to its use, and hence, all variables need to be declared. Several of the variables also need to be associated with corresponding variables in MATLAB. These necessities account for over half of the code, and the real action of the function is at the bottom of the file and mirrors that of MATLAB.

Entering mex HornerC.c, assuming HornerC.c is the name of the mex-file in Example Code 7.3, compiles the code so that we can access the function

7.2 C/C++ Extensions

```
function [fval] = HornerM(Poly, x)
[numPolys, polySize] = size(Poly);
fval = 1;
for p = 1:numPolys
    polyval = Poly(p,1);
    for k = 2:polySize
        polyval = polyval*x(p) + Poly(p,k);
    end
    fval = fval*polyval;
end
```

Example Code 7.2 An m-file defining the function HornerM to evaluate a product of polynomials

HornerC in MATLAB or Octave as HornerC. If mex has not been configured, then you will need to declare a compiler. The documentation and system resources are platform specific.

We compare HornerM and HornerC by evaluating f with 10 and 100 randomly generated polynomials, tracking the computing time as the common degree of the polynomials ranges over 10^i, for $i = 1, 2, 3, 4, 5$. The coefficients and their arguments are selected from a standard uniform distribution. The results are depicted in Fig. 7.1, and the increases in computational time with the m-file imply a strong preference for the C extension of HornerC. The case with 100 polynomials of degree 100,000 requires 69.307 s with HornerM but only 0.0995 s with HornerC, which is about a 700-fold improvement in speed. The advantage is clear, and while the mex-file has a more burdensome coding protocol, the computational benefit is worth the hassle. Indeed, the mex-file approach might be required for sufficiently large calculations.

Other scripting languages can be similarly extended with C/C++ and FORTRAN, and we repeat our testing to compare Python with Python extended with C. Example Code 7.4 lists a Python function called HornerPy that performs the same task as HornerM. The similarity between the codes should give some comfort for those who know MATLAB. While the codes are nearly identical, their performances are not. The Python interpreter outpaces that of MATLAB's to evaluate f as illustrated in Fig. 7.2. This outcome further confirms the fact that MATLAB and Octave are slow with regard to looping through an iterative process.

The C code for the Python extension is considerably more direct than is the code needed for MATLAB, see Example Code 7.5. The main reason for the welcomed brevity is that we use the SWIG compiler to generate a wrapper file, which is C code used to link Python with HornerPyC. Using SWIG is platform dependent, and unlike mex-extensions in MATLAB, extending Python with SWIG requires us to compile the C code outside Python. SWIG is useful should you decide to extend a language with your own C-style functions, and it can create wrapper files for over 15 scripting languages.

```
#include <math.h>
#include <mex.h>

void mexFunction(int numInputs, mxArray *output[],int
   numOutputs, const mxArray *input[]){
  // Declare variables
  mxArray *fOut;
  double *coef, *x, *fval, polyval;
  int polySize, numPolys, sizeX;
  int p, k;
  const size_t* dimsPoly;
  const size_t* dimsX;
  // Store the inputs into a local structures
  coef = mxGetPr(mxDuplicateArray(input[0]));
  x = mxGetPr(mxDuplicateArray(input[1]));
  // Get dimensions
  dimsPoly = mxGetDimensions(input[0]);
  numPolys = (int)dimsPoly[0];
  polySize = (int)dimsPoly[1];
  dimsX = mxGetDimensions(input[1]);
  sizeX = (int)dimsX[1];
  // Associate outputs
  fOut = output[0] = mxCreateDoubleScalar(mxREAL);
  fval = mxGetPr(fOut);
  // Check dimensions
  if(sizeX != numPolys){
    mexPrintf("Dimension mismatch\n");
    return;
  }
  // Evaluate the polynomials with Horner's method and
     calculate their product
  fval[0] = 1;
  for(p=0; p<numPolys; p++){
    polyval = coef[p];
    for(k=1; k<polySize; k++){
      polyval = polyval*x[p] + coef[k*numPolys+p];
    }
    fval[0] = fval[0]*polyval;
  }
}
```

Example Code 7.3 A mex-file defining the function HornerC to evaluate a product of polynomials

7.2 C/C++ Extensions

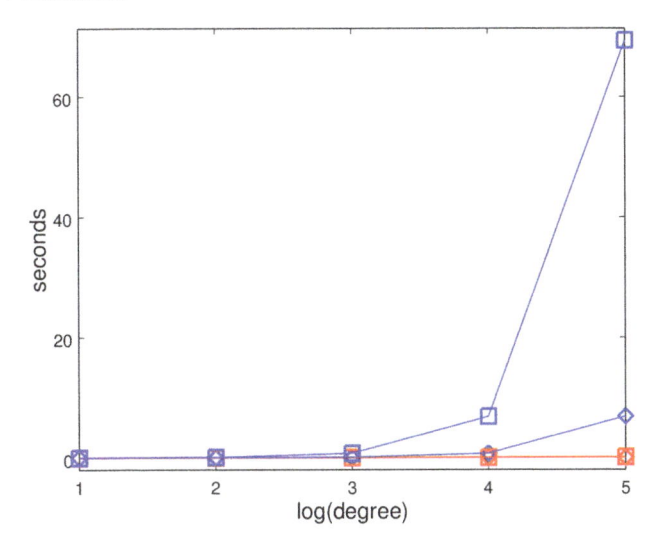

Fig. 7.1 A comparison of evaluating a product of polynomials with HornerM, in blue, and HornerC, in red. Diamonds indicate 10 polynomials, and squares indicate 100 polynomials.

```
def HornerPy(Poly, x):
    numPolys,polySize = Poly.shape;
    fval = 1;
    for p in range(numPolys):
        polyval = Poly[p,0];
        for k in range(1,polySize):
            polyval = polyval*x[p] + Poly[p,k];
        fval = fval*polyval;
    return fval
```

Example Code 7.4 A Python file defining the function HornerPy to evaluate a product of polynomials

As with MATLAB, performance substantially improves with the C extension. Figure 7.3 compares the C extensions in MATLAB and Python. The longest computation took 69.29 s to evaluate the product of 100 polynomials of degree 100,000 with HornerM. The same calculation only took 0.097 and 0.117 s in the C implementations of HornerC and HornerPyC, respectively. Unlike the native codes of HornerM and HornerPy, in which Python had a clear speed advantage over MATLAB, the Python C extension proved slightly less fast than did the MATLAB C extension. The difference is slight and changes from run to run due to perturbations in the operating system, but the slight advantage to MATLAB with C persisted over numerous runs.

```
#include <math.h>
#include "HornerPyC.h"
double HornerPyC(int numPolys, int polySize,
                 double *Poly, double *x)
{
double fval, polyval;
int p, k;
fval = 1;
for(p=0; p<numPolys; p++) {
  polyval = Poly[p];
  for(k=1;k<polySize; k++) {
    polyval = polyval*x[p] + Poly[k*numPolys+p];
  }
  fval = fval*polyval;
}
return(fval);
}
```

Example Code 7.5 C code defining the function HornerPyC to evaluate a product of polynomials as an extension to Python

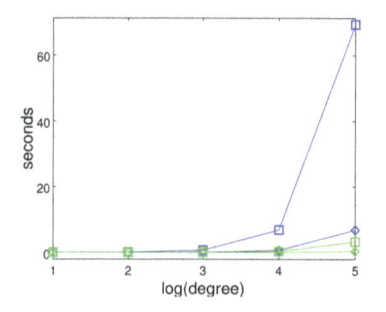

Fig. 7.2 A comparison of HornerM, in blue, and HornerPy, in green. Diamonds indicate 10 polynomials, and squares indicate 100 polynomials.

Fig. 7.3 A comparison of HornerC, in blue, and HornerPyC, in green. Diamonds indicate 10 polynomials, and squares indicate 100 polynomials.

7.3 Parallel Computing

Computational performance has been continuously improving since the outset of computational science in the 1940s. Until the early 2000s that improvement came from, among other things, increased processor speed. That is, the number of instructions (machine instructions emitted by a compiler or interpreter, not lines of MATLAB code) that could be executed by a single processor continually increased. That trend remains true today, and improvements in processor architecture and memory mean that individual processors are still getting faster. However, the largest contributor to performance since the early 2000s has been the introduction of multicore processors to the mass

7.3 Parallel Computing

market. Today nearly all machines are multicore, and some understanding of how to leverage this resource can be most useful to the computational scientist.

All modern computers possess multiple cores. A core is a piece of hardware capable of executing a machine instruction, and thus, all modern workstations are capable of executing multiple simultaneous instructions. However, we typically think of code as executing in a sequential manner, from top to bottom and left to right. This means that we must indicate which instructions may be executed simultaneously to take advantage of multiple cores. Writing code intended to run in parallel is called parallel programming, and most programming languages accommodate parallel constructs in some fashion, although ability and programming ease vary. The field of parallel programming, even just as it pertains to computational science, is large and growing. A full treatment would comprise a course in its own right, and for that reason we keep to MATLAB, with a few nods to Python. Octave did not support parallel programming particularly well at the time of this writing. In order to write parallel code in MATLAB the user must have access to the Parallel Computing Toolbox.

7.3.1 Taking Advantage of Built-In Commands

Most current computers have four or eight cores, although higher-end machines might have many more. The reader can check to see how many cores are available by looking at the task manager of the operating system and interpreting the results. As an alternative, the MATLAB command `maxNumCompThreads` returns, and possibly sets, the maximum number of cores to use. Running `maxNumCompThreads` directly after MATLAB starts is a reasonable way to determine the number of cores available on your system.

Many languages leverage parallel calculations in their built-in commands. For example, both MATLAB and Python's `numpy` use parallel implementations of matrix multiplication. To see this, first generate a large square matrix, open a system performance monitor, and then multiply the matrix by itself. During the time taken to perform the multiplication you will note that more than one core is used, indicating that the matrix multiplication is happening in parallel.

We can perform further experiments by setting the maximum number of cores that MATLAB can use. For instance, Example Code 7.6 performs the same matrix multiplication using one, two, and four cores, and speedups are achieved as the number of cores increases. On a modern laptop the multiplication with 1 core took roughly 70.6 s, with two cores about 36.4 s, and with four cores approximately 19.2 s. We note that it is possible to set `maxNumCompThreads` to a value larger than the physical number of cores on the system; however, doing so is unlikely to provide any advantage.

```
N=1e4;
A=rand(N,N);
for n=[1,2,4]
    maxNumCompThreads(n);
    tic; A*A;
    elapsed=toc;
    fprintf('Squaring a matrix of dimension %d took %f seconds
        using %d cores\n',N,elapsed,n);
end
```

Example Code 7.6 MATLAB code to compare how calculation speed depends on the number of cores

In MATLAB and numpy many linear algebra operations are implemented so that they automatically take advantage of all available cores on the machine. Many other functions, for instance the optimization routine `fmincon`, offer options to operate in parallel. Parallel programming is not always straightforward, and so when looking for additional performance it is generally best to first attempt to leverage code that already operates in parallel.

7.3.2 Parallel Computing in MATLAB and Python

We examine the basics of writing parallel code in both MATLAB and Python. Though there is a distinction between a core and a processor, we equate these terms in this section to agree with the typical terminology of parallel computing. In both languages we usually begin by writing code intended for a single core, e.g. code that does some basic start-up tasks and defines how many processors to use when operating in parallel. We then author code intended for parallel execution.

We first consider the MATLAB command, `parpool`, which establishes multiple instances of MATLAB, called a pool of workers, capable of running simultaneous commands. The workers are often called labs in MATLAB. Using `parpool` alone will establish a pool whose size is the number of physical cores available, whereas the command `parpool(P)` will establish a pool of P workers. MATLAB may call `parpool` automatically when using the `spmd` or `parfor` directives. Once a pool exists there is no need to call `parpool` again for future calculations.

Consider Example Code 7.7 as a first MATLAB example. The output of this code should look similar to:

```
Hello there!
Lab 1:
   Hello world, from lab 1
Lab 2:
```

7.3 Parallel Computing 279

```
  Hello world, from lab 2
Lab 3:
  Hello world, from lab 3
Lab 4:
  Hello world, from lab 4
All finished.
```

Example Code 7.7 begins with one processor producing the output "Hello there." The next statement, spmd, stands for single program multiple data, and it indicates that the code inside the following block is to run on all workers in the pool. MATLAB has multiple instances at the spmd command, each simultaneously executing the code inside the smpd block. Each worker is assigned a unique value of the special variable labindex, and each instance prints an identifying message before reaching the end of the smpd block. All but one of the workers has quit executing at the end of the block, and the "original" process continues. A visualization of execution with only two workers is illustrated in Fig. 7.4.

```
fprintf('Hello there!');
spmd
    fprintf('Hello world, from lab %d\n',labindex);
end
fprintf('All finished.\n');
```

Example Code 7.7 Example parallel code in MATLAB

Example Code 7.8 demonstrates that workers automatically inherit a copy of the variables in the workspace of the original MATLAB instance as an spmd block begins. Each lab can then modify its private copy of those variables as well as introduce new variables that are invisible to the other labs. All variables defined in the labs are indexed by labindex and are available to the original MATLAB process at the end of the spmd block. The original MATLAB process does not proceed past the spmd block until all workers have completed their tasks. Figure 7.5 illustrates how the code would execute with two workers.

We now turn to the computational task of determining the prime numbers within a large range, see Example Code 7.9. This code uses spmd to simultaneously check the integers in sub-ranges, storing those found in a master list of results. Each process determines its unique sub-range by examining its value of labindex. We note that an alternative approach would be to use multiple processors to decide if a single integer is prime, a tactic that should be investigated by an interested student.

We let $T_1(N)$ be the time required by an algorithm to complete its task on a single processor with an input of size N. T_1 is called the serial time of the algorithm. We similarly let $T_P(N)$ be the analogous definition with P processors. The speedup from 1 to P processors is

```
                Processor 1                        Processor 2

       execution line              state
       ┌─────────────────────┐  ┌─────────┐
       │ fprintf('Hello      │  │         │
       │    there');         │  │         │
       └─────────────────────┘  └─────────┘

       execution line              state         execution line              state
       ┌─────────────────────┐  ┌─────────┐   ┌─────────────────────┐  ┌─────────┐
       │ fprintf('Hello world,│  │labindex=1│  │ fprintf('Hello world,│  │labindex=2│
  time │    from lab %d',    │  │         │   │    from lab%d',     │  │         │
   │   │    labindex);       │  │         │   │    labindex);       │  │         │
   │   └─────────────────────┘  └─────────┘   └─────────────────────┘  └─────────┘
   │
   │   execution line              state
   │   ┌─────────────────────┐  ┌─────────┐
   ▼   │ fprintf('All        │  │         │
       │    finished.');     │  │         │
       └─────────────────────┘  └─────────┘
```

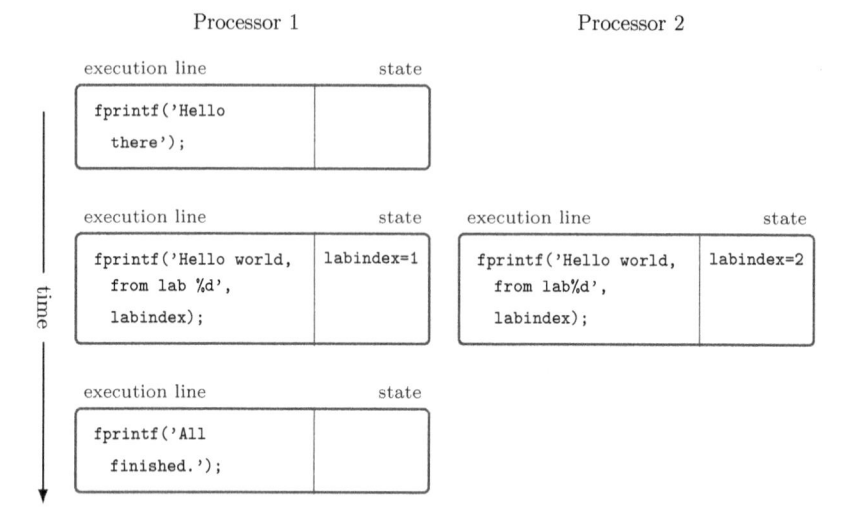

Fig. 7.4 An illustration of Example Code 7.7 executing with two workers.

```
A=3;
fprintf('Initially, A=%d\n',A);
spmd
    B=labindex;
    if mod(labindex,2)==0
        A=4;
    end
    fprintf('A+B=(%d)+(%d)=%d\n',A,B,A+B);
end
for i=1:length(A)
    fprintf('At the end, A{%d}=%d,B{%d}=%d\n',i,A{i},i,B{i});
end
```

Example Code 7.8 A second example of parallel code in MATLAB. All value changes are completely independent across MATLAB instances. Values are automatically available to the original process at the end of an **spmd** block

$$S_p(N) = \frac{T_1(N)}{T_P(N)}.$$

Using P processors would ideally produce an algorithm that is P times faster than the serial time, and an algorithm that achieves this goal exhibits linear speedup. Algorithms that fail to meet this goal exhibit sublinear speedup, whereas algorithms that exceed this goal exhibit super-linear speedup. The latter distinction is rare and typically relies on hardware advantages. We further define efficiency as

$$E_P(N) = S_P(N)/P,$$

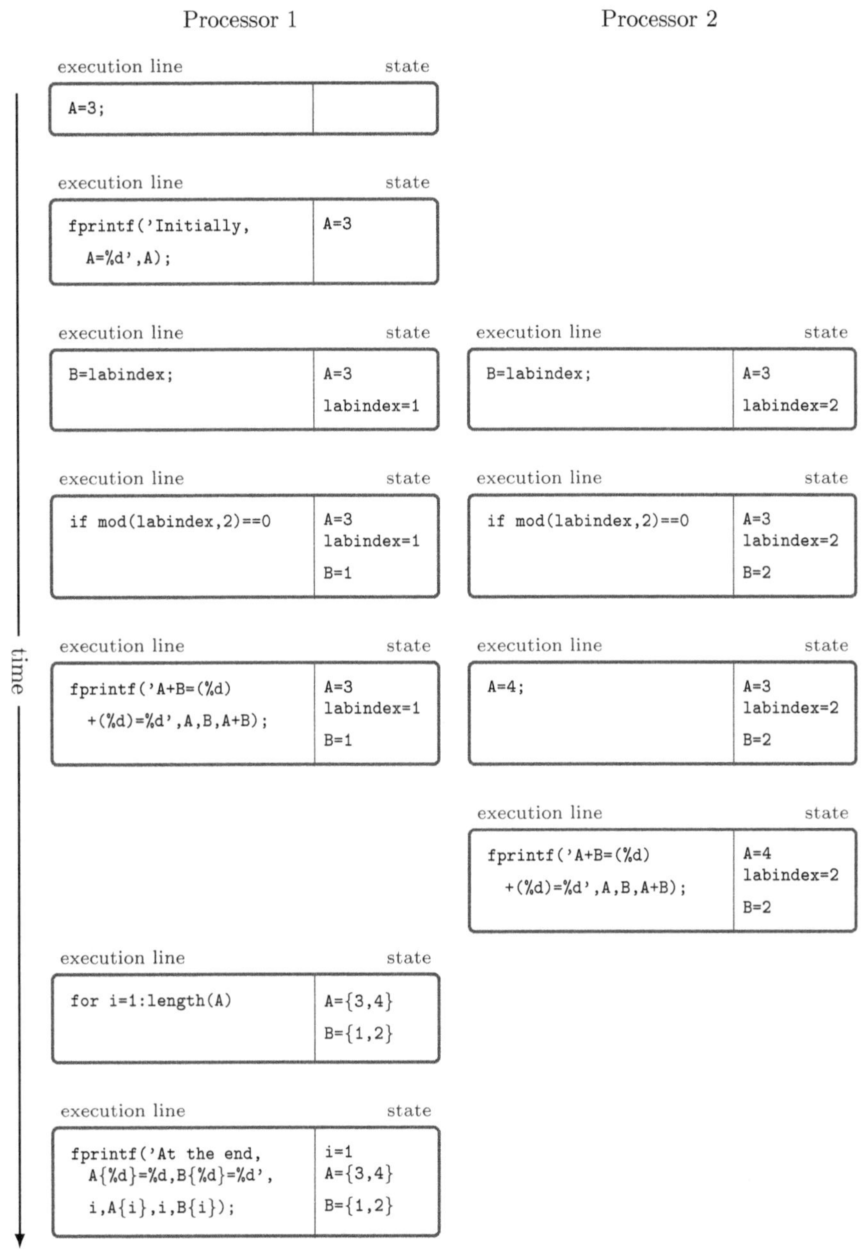

7.3 Parallel Computing

                    Processor 1   Processor 2

execution line	state
A=3;	

execution line	state
fprintf('Initially, A=%d',A);	A=3

execution line	state		execution line	state
B=labindex;	A=3 labindex=1		B=labindex;	A=3 labindex=2

execution line	state		execution line	state
if mod(labindex,2)==0	A=3 labindex=1 B=1		if mod(labindex,2)==0	A=3 labindex=2 B=2

execution line	state		execution line	state
fprintf('A+B=(%d) +(%d)=%d',A,B,A+B);	A=3 labindex=1 B=1		A=4;	A=3 labindex=2 B=2

execution line	state
fprintf('A+B=(%d) +(%d)=%d',A,B,A+B);	A=4 labindex=2 B=2

execution line	state
for i=1:length(A)	A={3,4} B={1,2}

execution line	state
fprintf('At the end, A{%d}=%d,B{%d}=%d', i,A{i},i,B{i});	i=1 A={3,4} B={1,2}

↓ time

Fig. 7.5 An illustration of the code in Fig. 7.8 executing on two workers. The main worker does not begin executing code after the spmd block until all workers have finished executing all the code in the block.

```
function [results,et]=spmd_isprime(N,P)
M=N/P;
tic;
spmd(P)
    %each worker iterates over a distinct range of numbers that
        is M long, checking for primality
    start=tic;
    for i=1:M
        result(i)=isprime((labindex-1)*M+i);
    end
    elapsed=toc(start);
    fprintf('Worker %d worked for %.3f seconds\n',labindex,
        elapsed);
end
for i=1:numel(result)
    results((i-1)*M+1:i*M)=result{i};
end
et=toc();
```

Example Code 7.9 Determining primality of a large range of integers using spmd in MATLAB. Each process determines what to do based on its labindex. Results are collected at the end of the script

which measures how close an algorithm comes to ideal speedup.

We see in Fig. 7.7 that Example Code 7.9 performs reasonably well, although its efficiency diminishes as the number of processors increases. Such performance is normal, and most algorithms do not exhibit perfect efficiency, especially for a large number of processors. The reason for the decreasing performance in this case is clear from the code's output. For instance, running spmd_isprime(3e6,4) on a modern laptop indicates that labs one, two, three, and four ran for about 25, 30, 32, and 35 s, respectively. That is, lab one sits idle for 10 s while lab four continues to work. Moreover, only lab four works for the last 3 s, which means the algorithm proceeds in serial at that point. The mathematical problem lies with the fact that it requires more work to decide the primality of large integers than it does small integers. Therefore lab four is assigned more "hard" work than is lab one, creating a discrepancy in time. This problem is called a load imbalance, and Fig. 7.6 illustrates this concern with two workers.

A scheme to automatically reduce load imbalance would dynamically allocate tasks to help ensure that each worker is given similarly "difficult" work. Small chunks of work are assigned to the workers in such a scheme, and new chunks are assigned as workers complete their current tasks. Example Code 7.10 demonstrates the use of parfor in MATLAB, which dynamically allocates work to parallelize a for loop.

The commands for and parfor similarly run once for every value of the indexed variable. However, parfor can assign the indexed tasks to different processors, and loop iterations may be completed in any order and on any

7.3 Parallel Computing

Processor 1

execution line	state
M=N/P;	N=4e6 P=2

execution line	state
for i=1:M	M=2e6 lbidx=1

execution line	state
Tests primality of 1. This is very fast. result(i)= isprime((lbidx-1)*M+i);	M=2e6 lbidx=1 i=1

execution line	state
Tests 2, very fast. result(i)= isprime((lbidx-1)*M+i);	M=2e6 lbidx=1 i=2 result=[0]

⋮

execution line	state

Tests 2000000,
finishing.
result(i)=
  isprime((lbidx-1)*M+i);
	M=2e6
	lbidx=1
	i=2000000
	result=
	  [0,1,...]

execution line	state

Had to wait for
Processor 2 to finish.
for i=1:numel(result)
	result={
	  [0,1,1,...],
	  [0,0,1,...]}

Processor 2

execution line	state
for i=1:M	M=2e6 lbidx=2

execution line	state
Tests 2000001, slightly slower than testing 1. result(i)= isprime((lbidx-1)*M+i);	M=2e6 lbidx=2 i=1

execution line	state
Tests 2000002. Slower than 2. result(i)= isprime((lbidx-1)*M+i);	M=2e6 lbidx=2 i=2 result=[0]

⋮

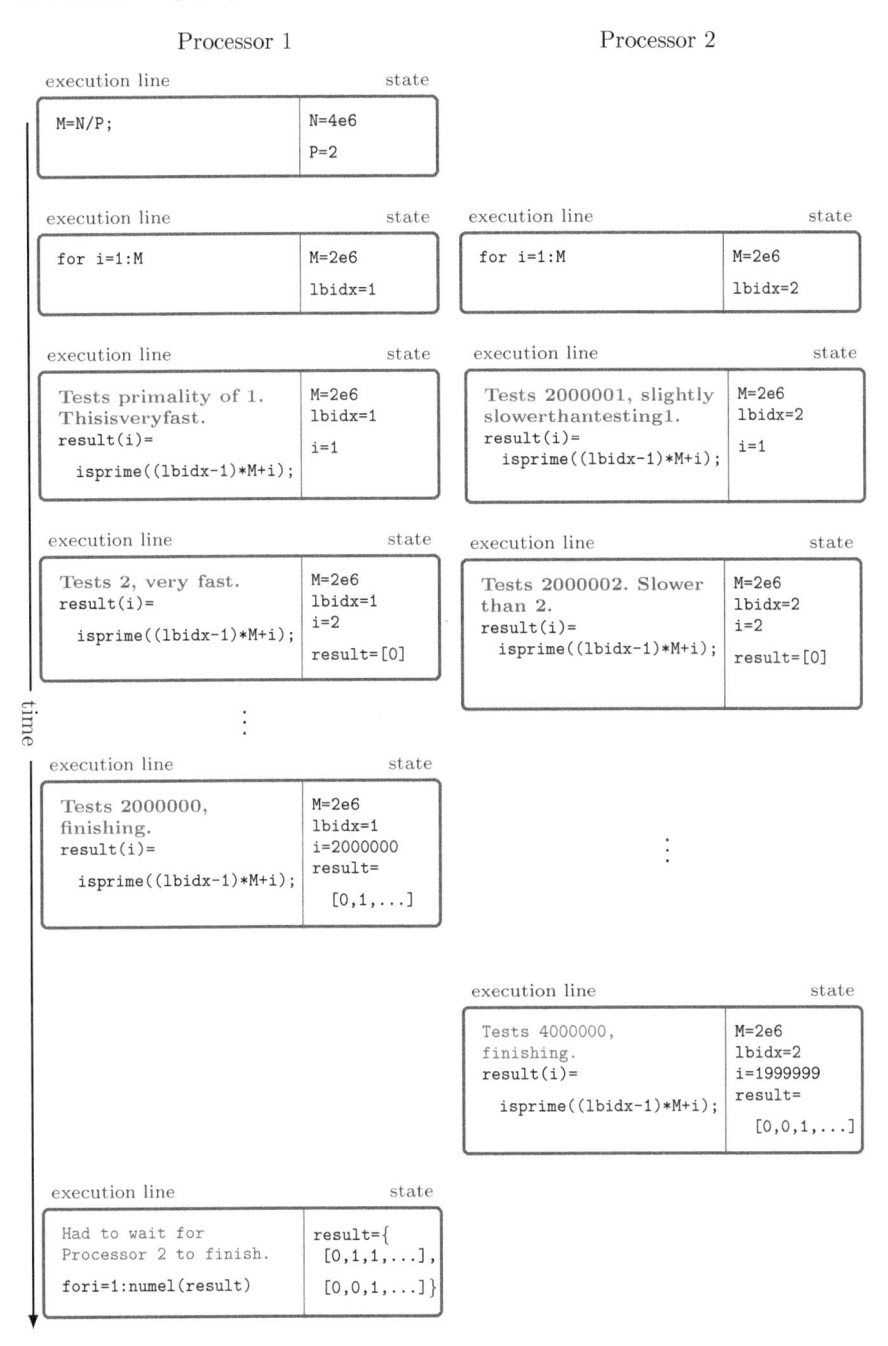

execution line	state

Tests 4000000,
finishing.
result(i)=
  isprime((lbidx-1)*M+i);
	M=2e6
	lbidx=2
	i=1999999
	result=
	  [0,0,1,...]

time ↓

Fig. 7.6 An illustration of Example Code 7.9 executing on two workers with $N = 4{,}000{,}000$.

```
function [results,et]=parfor_isprime(N,P)
%determine the primality of each integer between 1 and N using
    P processors
tic;
parfor (i=1:N, P)
        results(i)=isprime(i);
end
et=toc();
```

Example Code 7.10 A parfor loop in MATLAB uses dynamic work allocation to achieve parallelization

lab. The second argument to parfor is optional and is the number of possible workers. Notice that parfor is not suitable if it is imperative that consecutive iterations be completed in order. Workers have no concept of labindex with parfor, and the loop index is instead used to determine behavior. MATLAB uses dynamic work allocation to assign each lab a few indices, and new indices are assigned as calculations finish. Hence, different labs may execute the loop a different number a times, although they should work for about the same amount of time. Figure 7.7 indicates that parfor gives a performance boost over spmd, though its efficiency still declines as the number of processors increases. Results from parfor are automatically assembled, but there are a number of rules regarding how variables may be indexed. Students should refer to the MATLAB documentation for current guidelines. An illustration of how Example Code 7.10 would run is in Fig. 7.8.

7.3.3 Parallel Computing in Python

We briefly consider Python's parallel tools by considering a long-standing and open problem in mathematics called the Collatz Conjecture. Let a_0 be a positive integer, and define the following sequence,

$$a_k = \begin{cases} \frac{a_{k-1}}{2} & \text{if } a_{k-1} \text{ is even,} \\ 3\,a_{k-1}+1 & \text{otherwise.} \end{cases}$$

The Collatz Conjecture posits that $a_n = 1$ for some n regardless of a_0. The smallest index n such that $a_n = 1$ is called the termination time of the sequence.

Example Code 7.11 defines a Python function to compute the termination time of the sequence beginning with a_0. Calculating a single termination time is not amenable to parallelization since each iteration of the loop depends on the result of the previous iteration. If, however, we are interested in testing the validity of the Collatz Conjecture, then we might decide to compute

7.3 Parallel Computing

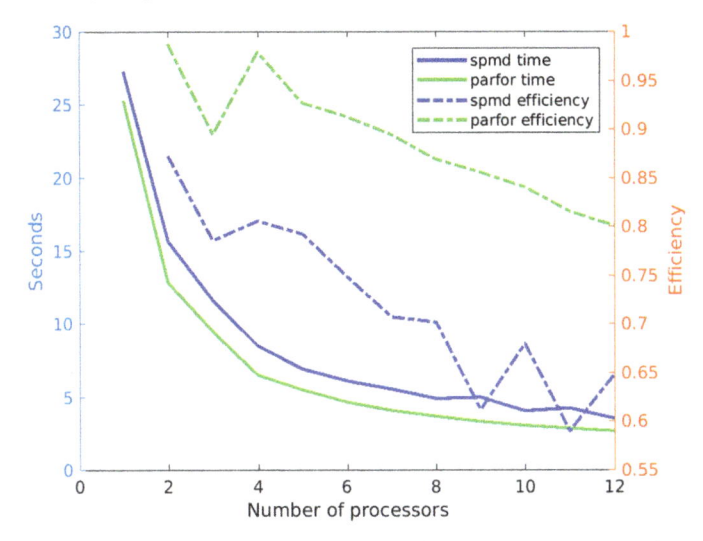

Fig. 7.7 Elapsed time and efficiency for Example Codes 7.9 and 7.10. Both codes determine the primality of each integer from 1 to 1,386,000.

the termination time of all Collatz sequences with initial values between 1 and N. This computational test is parallelizable since we can simultaneously calculate termination times for different starting values.

```
def termination_time(a0):
        itn=0;
        val=a0;
        while val!=1:
                if val%2==0:
                        val=val/2;
                else:
                        val=3*val+1;
                itn+=1;
        return itn
```

Example Code 7.11 Python code for determining the termination time of a Collatz sequence

Example Code 7.12 demonstrates the use of mult.Pool(P) to establish a pool of P workers. The command Pool.map then runs the function termination_time on each of the inputs in the list range(1,N). The calculations run in parallel with the workers in Pool similar to the way parfor works in MATLAB. The output, termination_time, appropriately indexes the results according to i. There are numerous options other than multiprocessing and map, but this tactic is the recommended first try.

Processor 1

execution line	state
i is assigned randomly. results(i)=isprime(i);	i=200 results=[]

executionline	state
Tests 400, fast. results(i)=isprime(i);	i=400 results=[...]

execution line	state
Tests 525, fast. results(i)=isprime(i);	i=525 results=[...]

execution line	state
Determines whether 71 is prime, fast. results(i)=isprime(i);	i=71 results=[...]

⋮

execution line	state
Tests 204165, which happens to be the last integer left to check. results(i)=isprime(i);	i=204165 results=[...]

Processor 2

execution line	state
Tests 399989, slow. results(i)=isprime(i);	i=39989 results=[]

execution line	state
Tests 207191, slow. results(i)=isprime(i);	i=207191 results=[...]

⋮

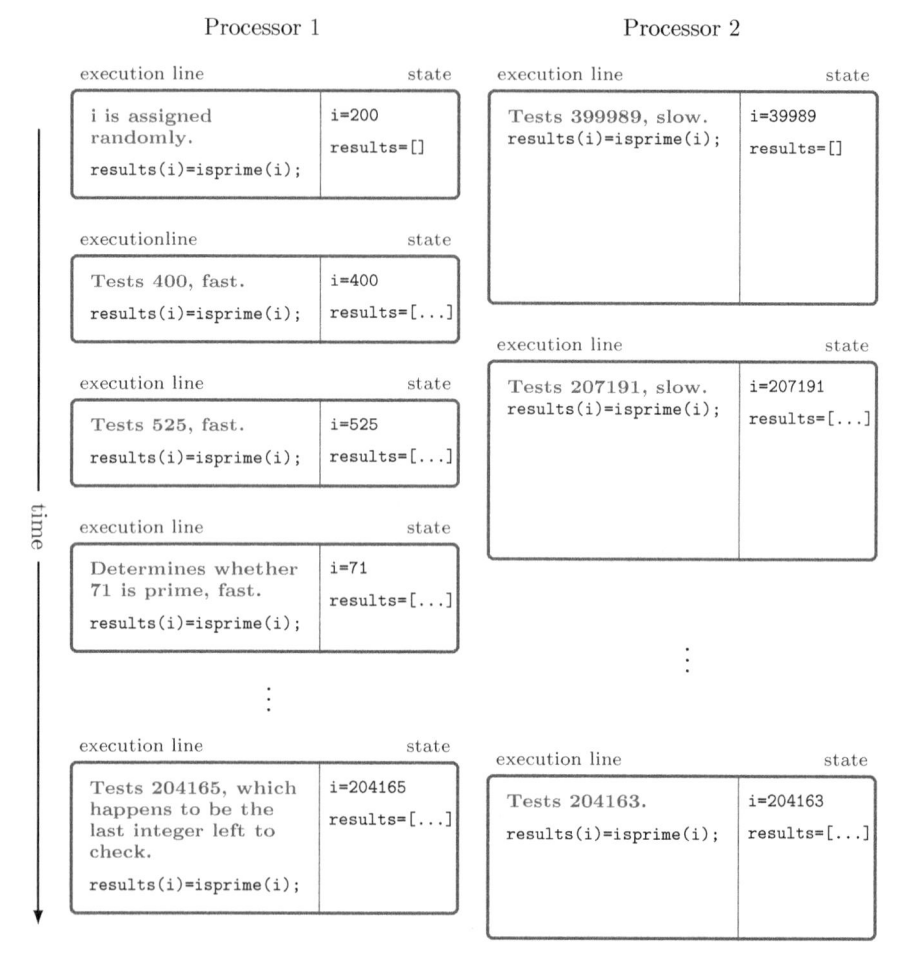

execution line	state
Tests 204163. results(i)=isprime(i);	i=204163 results=[...]

Fig. 7.8 A visualization of how Example Code 7.10 might complete with two workers. The parfor loop assigns new indices to workers as they complete previously assigned tasks, ensuring that each worker is active for roughly the same amount of time.

```
import multiprocessing as mult
P=4;
Pool=mult.Pool(P);
N=1000000;
termination_times=Pool.map(termination_time,range(1,N));
M=max(termination_times);
```

Example Code 7.12 Python code demonstrating the use of a worker pool for calculating the termination time of several Collatz sequences

7.3.4 Pipelining

Not all problems seamlessly lend themselves to parallelization, and we now examine a problem in which it is not immediately clear how to parallelize the serial algorithm. The computational task we consider is to calculate the LU factorization of an $n \times n$ matrix A over p processors. We assume n is large, otherwise serial computation would suffice.

One of the most important decisions to make when parallelizing an algorithm is deciding how to distribute data among processors. There are numerous ways to distribute an $n \times n$ matrix across processors, e.g. each processor could hold several rows of A, several columns of A, or a rectangular "block" of A. One of these partitioning concepts might be particularly advantageous depending on the larger computational context. We assume for the sake of this example that each row of A is stored on a different processor, see Fig. 7.9.

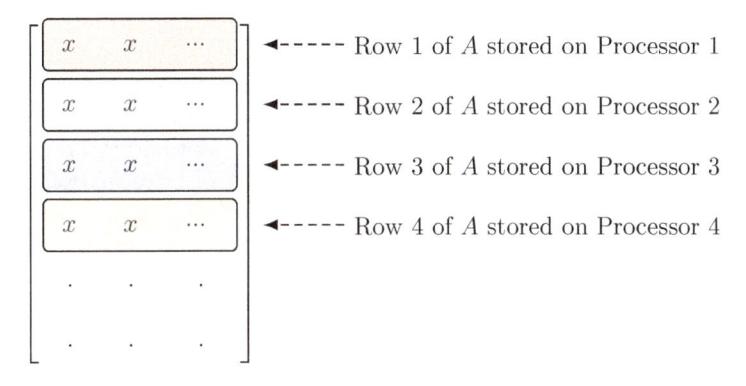

Fig. 7.9 A row partitioning of A. Each row of A is stored on a different processor; we assume that there are enough processors for this to be feasible.

The standard serial algorithm for calculating an LU decomposition proceeds by using row 1 to eliminate the terms in column 1 of rows 2 through n. Row 2 is then used to eliminate the entries in column 2 of rows 3 through n, and so on. Figure 7.10 illustrates this process.

One inherently serial aspect of this algorithm is that we must perform elimination in column 1 before proceeding to column 2, which must be done before column 3, and so on. So we may not parallelize the column eliminations. However, within one column elimination, say column i, we may perform elimination in all rows with index greater than i in parallel by having worker

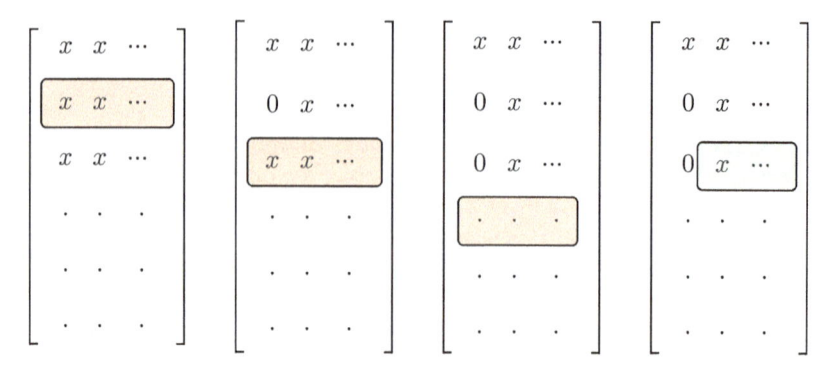

Fig. 7.10 An illustration of an LU decomposition in serial. In this illustration, a row is highlighted if column elimination will be performed on that row. The color indicates which column is being eliminated, red means column one and green means column two. An entry of x in this diagram represents an arbitrary matrix entry, and an entry of 0 indicates that the entry is certainly 0. Two entries marked x need not have the same value.

i distribute a copy of row i to all workers with index greater than i. These workers may then safely perform column elimination in column i. We use the MATLAB command `labSend` to transfer information between workers, which sends data from one worker to another. The destination worker must call a matching `labReceive` to retrieve the data.

We note that processor j is immediately ready to perform elimination in column $i+1$ once it has completed elimination in column i, which further hastens our calculation. Indeed, row i is no longer changed once processor i has performed elimination in column $i-1$, and hence, column i can be used for elimination in column i by all other workers. Example Code 7.13 demonstrates the algorithm, and an illustration of the computational process is in Fig. 7.11. We are taking row i of A and forwarding it from one worker to the next, with each worker forwarding row i to the next worker before performing any necessary work with its copy of row i. An algorithm in this style is said to be pipelined, with each piece of data (a matrix row in this case) moving down a pipeline of operations (row operations). This scheme allows us to have multiple column eliminations, in different columns, happening at the same time.

Extending our algorithm to the case in which there are more rows than processors, which is the practical case since n is assumed to be very large, is not difficult since we can simply store multiple rows per processor. Once processor j receives a copy of row i, the processor forwards row i to the succeeding processor, and then performs elimination in column i on all its

7.3 Parallel Computing

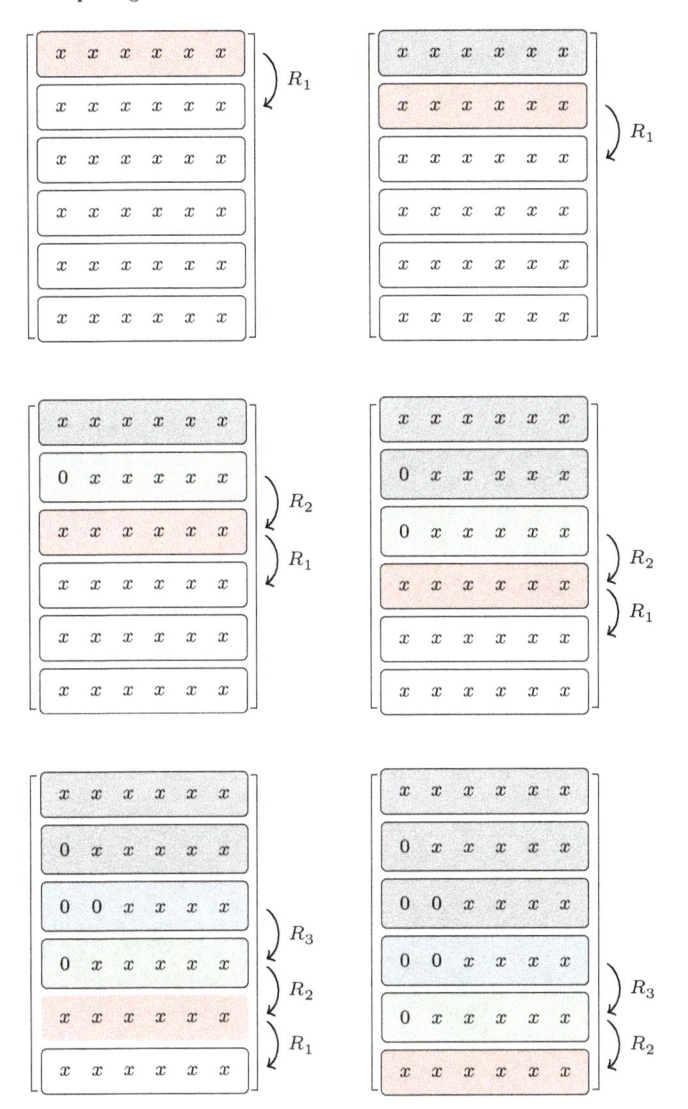

Fig. 7.11 An illustration of Example Code 7.13 progressing through a few steps on a 6×6 matrix with 6 processors. As soon as processor i forwards row j, denoted as R_j, to processor $i+1$, processor i performs elimination in column i. Color indicates what column the processor is currently working on, with red meaning column 1, green column two, and blue column 3. Gray processors are finished working. An entry of x represents an arbitrary matrix entry, and an entry of 0 indicates that the entry is certainly 0. Entries marked x need not have the same value.

```
function [L,U]=simple_parallel_lu(A)

N=size(A,1);
spmd(N)
    %store only row labindex of A
    my_U=A(labindex,:);
    %store a correspongly sized L matrix
    my_L=zeros(size(my_U));
    my_L(1,labindex)=1;
    %now, do elimination with all columns <= labindex
    for i=1:labindex-1
        %Recieve the row with which to do elimination.
        %This says "recieve data from the lab with index
            labindex-1,
        %recieve a message that is labelled as row i.
        R=labReceive(labindex-1,i);
        %Forward this row on to the next worker.
        %This says, "send R to the lab with index labindex+1,
            label the
        %messages as containing row i.
        if labindex<numlabs
            labSend(R,labindex+1,i);
        end
        %Now, do elimination.
        my_L(1,i)=my_U(1,i)/R(1,i);
        my_U=my_U-my_L(1,i)*R;
    end
    %elimination is done, up to column labindex-1, so this row
        is complete.
    %Forward it on.
    if labindex<numlabs
        labSend(my_U,labindex+1,labindex);
    end
end
%Now, assemble all the results
for i=1:N
    L(i,:)=my_L{i};
    U(i,:)=my_U{i};
end
```

Example Code 7.13 Code that performs a parallel LU decomposition of a matrix A, assuming that it is possible to allocate one worker per row of A. Division by zero concerns are ignored

rows. We use a striped partition of A to achieve a good load balance, which is visualized in Fig. 7.12. Rows with small indices go through fewer column eliminations, and the striped partition somewhat evenly distributes the work among processors.

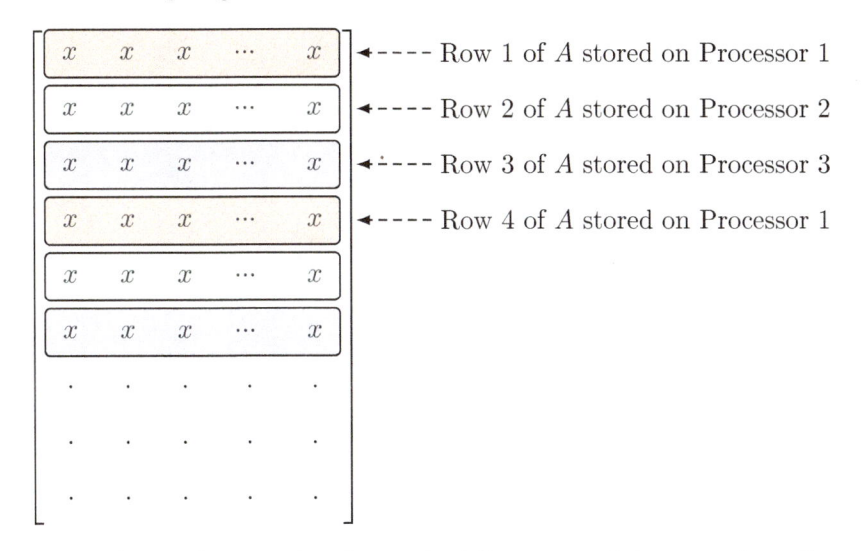

```
┌─────────────────────┐
│ x  x  x  ···  x     │ ←---- Row 1 of A stored on Processor 1
│ x  x  x  ···  x     │ ←---- Row 2 of A stored on Processor 2
│ x  x  x  ···  x     │ ←---- Row 3 of A stored on Processor 3
│ x  x  x  ···  x     │ ←---- Row 4 of A stored on Processor 1
│ x  x  x  ···  x     │
│ x  x  x  ···  x     │
│  ·  ·  ·   ·    ·   │
│  ·  ·  ·   ·    ·   │
│  ·  ·  ·   ·    ·   │
└─────────────────────┘
```

Fig. 7.12 A striped partitioning of a matrix. Adjacent rows of A are stored on adjacent processors rather than on the same processor. This way all processors have a similar number of high-, mid-, and low-index rows.

7.3.5 Ahmdal's Law

Our parallel pursuits are thus far encouraging, and we have improved every algorithm that we have examined. Indeed, we might hope to solve problem P in less than time T by using enough processors. However, this is a dashed hope, and almost all algorithms have some aspect that cannot be parallelized—be that reading input from a disk, assembling some final aspect of the calculation, computing many terms of a sequence, or something else altogether. Let us assume that we are interested in solving problem P on an input of size N. Recall that $T_1(N)$ is the serial time for a problem instance of size N. Let f be the fraction of $T_1(N)$ spent on inherently serial work, and write

$$T_1(N) = f\,T_1(N) + (1-f)\,T_1(N).$$

We do not expect to decrease the time spent on inherently serial work by using p processors in a parallel algorithm, and a linear speed up on the rest of the work is the best we can hope for. So an optimistic estimate of the time required to solve an instance of problem P with size N on p processors is

$$T_p(N) = f\,T_1(N) + \frac{1-f}{p}\,T_1(N).$$

The best possible speedup is thus

$$S_p(N) = \frac{T_1(N)}{fT_1(N) + \frac{1-f}{p}T_1(N)} = \frac{p}{f(p-1)+1},$$

with a best efficiency of

$$E_p(N) = \frac{S_p(N)}{p} = \frac{1}{f(p-1)+1}.$$

We have as the number of processors increases that

$$\lim_{p \to \infty} S_p(N) = \frac{1}{f} \quad \text{and} \quad \lim_{p \to \infty} E_p(N) = 0.$$

These results are called Ahmdal's Law, and they indicate the eventual impracticability of adding processors to a parallel algorithm. At some point there will be no appreciable speedup, even without accounting for additional slowdowns like the overhead of starting multiple processes or communication time. Indeed, it is usually the case in practice that adding too many processors results in an overall increase in processing time due to the aforementioned overhead.

Parallel computing is prominent today despite the disappointing results of Ahmdal's Law, and ever larger supercomputers are being built. Ahmdal's Law might seem to indicate that such machines are without merit. However, the fraction f of serial work is often a function of problem size. On small problems a significant portion of time might be spent reading data from a file, which is inherently serial. So f is likely large for small problems. Although large problems still require time to read data, this burden is regularly dwarfed by the time spent doing parallelizable calculations, thereby decreasing f. If it is the case that $f(N) \to 0$ as $N \to \infty$, where $f(N)$ is the fraction f for a problem of size N, then we have the favorable result that

$$\lim_{N \to \infty} E_p(N) = 1.$$

In conclusion, if you need to solve a specific problem of size N, then adding more processors to your system will not improve performance beyond a certain point. However, more processors will likely be beneficial as the size of your problem grows.

7.3.6 GPU Computing

Programmable graphics processing units (GPUs) became available in the early twenty-first century, originally intended to facilitate visual effects in computer games. It was soon discovered that the characteristics that made these units especially suitable for graphics also made them quite fast at calcu-

7.3 Parallel Computing

lations common to computational science. Vendors then began to make hardware designed specifically for such calculations, calling them general purpose graphical processing units (GPGPUs). A GPGPU provides the opportunity for large-scale parallelism on a personal computer. The 4 to 8 full-feature cores of a modern CPU are staggeringly diminished by the 256 or 512 streaming processors common to current GPGUs. These streaming processors are often not as fast (in terms of instructions per second) as a CPU core, and their performance can degrade dramatically if they need to communicate or if there are many conditional branches, e.g. if statements. That said, GPGPUs can offer sizable speedups on problems for which they are suited.

Many modern laptops and desktops are equipped with graphics cards appropriate for scientific applications, and at the time of this writing, the two primary languages for writing GPU code are CUDA, developed by NVIDIA, and OpenCL, which is an open source language. CUDA is the dominant language of the two. Languages other than CUDA and OpenCL often provide access to GPU computing without having to write specialized code. MATLAB offers useful GPU computing tools in the Parallel Computing Toolbox, which we discuss momentarily, and other languages, like Python and Octave, are developing similar tools.

The examples in this section are based on the MATLAB documentation for the parallel computing toolbox, in particular the section illustrating GPU computing. We begin by defining the Mandelbrot set as the set of points z_0 in the complex plane such that the sequence

$$z_k = z_{k-1}^2 + z_0$$

remains bounded. The Mandelbrot set is a fractal, and we examine using GPU computing to speed up the process of ascertaining which points in a given set belong to the Mandelbrot set.

We start by considering serial code. Example Code 7.14 takes a list of starting points, z0s, and determines if $|z_k| < 2$ for the first 500 iterates for each z_0. We assume z_0 is a likely member of the Mandelbrot set in this case, for if any iterate satisfies $|z_k| \geq 2$, then z_0 is outside the set.

We re-express our algorithm in terms of matrix operations in an attempt to speed up the code, as MATLAB is extremely well tuned to perform matrix operations quickly. In addition, MATLAB will run our matrix operations in parallel on the CPU if possible, thereby parallelizing our code. Using elementwise operations we re-write Example Code 7.14 as Example Code 7.15. Although we expect the rewrite to produce a speedup, the new code in fact slows down, see Fig. 7.13. The issue is that the first algorithm stops as soon as $|z_k| > 2$, whereas the second algorithm computes all iterates even if $|z_k| > 2$ after a few iterations. While the rewritten code does not produce a speedup, we note that expressing algorithms in terms of matrix operations in MATLAB is overwhelming good practice.

```
function ismandel=serial_mandelbrotMember(z0s)

    ismandel=ones(size(z0s));
    for i=1:size(z0s,1)
        for j=1:size(z0s,2)
            z=z0s(i,j);
            for k=1:500
                z=z^2+z0s(i,j);
                if abs(z)>=2
                    ismandel(i,j)=0;
                    break;
                end
            end
        end
    end
end
```

Example Code 7.14 Serial code that loops over each entry z_0 in z0s and tries to determine whether z_0 belongs to the Mandelbrot set

```
function ismandel=serial_vec_mandelbrotMember(z0s)

    z=z0s;
    for k=1:500
        z=z.^2+z0s;
    end
    ismandel=abs(z)<2;
```

Example Code 7.15 A second algorithm for determining which points likely belong to the Mandelbrot set, rewritten from Example Code 7.14 to use MATLAB's matrix operations

MATLAB has already implemented many algorithms to run on the GPU. In particular, almost all matrix operations and many common scientific codes have already been implemented. We must first transfer data to operate to a separate GPU memory, and we use the gpuArray command to achieve this goal. We can perform matrix operations as we normally would, and the operations will be carried out on the GPU with full GPU performance. We use gather to transfer data back to the main memory. Modified Example Code 7.16 produces a sizable speedup over either of the CPU codes. This method of GPU computing, i.e. doing large matrix computations on the GPU, should be the first strategy for speeding up code with a GPU.

Finally, the MATLAB function arrayfun, much like the Python function map, applies a function to multiple inputs in parallel. If the inputs happen to reside on a GPU, then arrayfun will use streaming processors to apply the function to the inputs. To illustrate, Example Code 7.17 takes a single input z0 and iterates to try and determine if it is a member of the Mandelbrot set. We apply isMandelMember to each of the inputs in z0s in parallel in Example Code 7.18. The streaming processors execute the code in parallel because z0s

7.3 Parallel Computing

```
function ismandel=gpu_vec_mandelbrotMember(z0s)

    z0s=gpuArray(z0s);
    z=z0s;
    for k=1:500
        z=z.^2+z0s;
    end
    ismandel=abs(z)<2;
    ismandel=gather(ismandel);
```

Example Code 7.16 Code that takes advantage of matrix operations on the GPU to estimate membership in the Mandelbrot set. We first move the matrix z0s to the GPU and then operate as normal. We conclude by gathering the results back on the CPU

is moved to the GPGPU. The function `isMandelMember` is composed solely of simple arithmetic, with the only `if` statement causing the code to exit early, and so we expect this function to run well on the GPGPU. This approach is the GPGPU equivalent to our first approach in Example Code 7.14, and it retains the favorable computational ability to stop early if possible.

```
function tf=isMandelMember(z0)
    z=z0;
    tf=1;
    for k=1:500
        z=z^2+z0;
        if abs(z)>=2
            tf=0;
            break;
        end
    end
end
```

Example Code 7.17 MATLAB code that estimates whether the initial value z0 is a member of the Mandelbrot set

```
    ismember=arrayfun(@isMandelMember,gpuArray(z0s));
    ismember=gather(ismember);
```

Example Code 7.18 MATLAB code that applies the function `isMandelMember` to each input in the matrix z0s in parallel, using the streaming processors on the GPU

Figure 7.13 shows that GPGPU computing provides substantial time savings. Indeed, using 2,250,000 initial values, the slowest CPU code took roughly 11.9 s to complete. In contrast, the fastest GPU code took roughly 0.2 s to complete, a near 60-fold speedup.

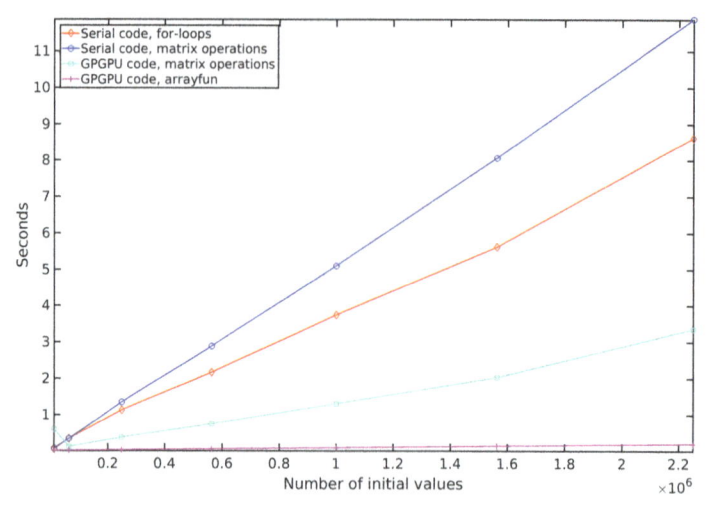

Fig. 7.13 A comparison of different algorithms for the Mandelbrot set problem. GPGPU computing gives excellent speedup on this problem.

7.4 Exercises

1. Create a function with declaration

   ```
   plotMandel(deltaz)
   ```

   that produces a visualization of the Mandelbrot set. The function should first produce a list of all complex numbers in the complex plane on a grid with spacing Δz. That is, all z such that $|\mathrm{Re}(z)| < 2$, $|\mathrm{Im}(z)| < 2$ and $z = (-2+k\Delta z)+(-2+\ell\Delta z)i$ with k,ℓ integers. It should then use any of the parallel examples in this section to estimate which of the grid points belongs to the Mandelbrot set. In order to visualize the set it should place a dot at position (a,b) if the complex number $a+bi$ is estimated to be part of the Mandelbrot set. Investigate the behavior of your code as you use more processors to ensure that you are indeed utilizing your hardware effectively.
2. Use C to implement matrix–vector multiplication following the algorithm in Example Code 7.1. Compile your C code using `mex`. Implement the algorithm in MATLAB, and compare the speed of the two codes. Then, compare to the built-in matrix multiplication of MATLAB. Be sure to use `maxNumCompThreads(1)` in order to turn off any parallelization in the default matrix multiplication. Summarize your results with a well-formatted graph displaying the time required for each implementation on several different problem sizes.
3. Matrix–vector multiplication allows for parallelization. Use a `parfor` loop to parallelize your MATLAB implementation in Exercise 2. Experiment

7.4 Exercises

with different placements of the `parfor` loop. Summarize your results with an appropriate graph.

4. Compare the speed of built-in matrix–vector multiply on the CPU to matrix–vector multiply on the GPU on several different matrix sizes. Report your results in the form of a graph displaying runtime required by each algorithm as a function of matrix size.
5. Implement an escape time algorithm similar to that in Example Code 7.14 in Python. Parallelize the code using `map` from the `multiprocessing` module. Investigate the performance of your parallel code by reporting the speedup observed using 2, 4, 8 processors on several different problem sizes.
6. Implement a naive LU factorization algorithm in C, assuming that no pivoting is required, and that L and U are returned in separate matrices. Your code should return matrices L and U such that $LU = A$, if possible, and should return a `stat` variable indicating the success or failure of the code. The following call should return as expected.

    `[L,U,stat]=mexLU(A);`

7. Extend the code in Example Code 7.13 to allow matrices that have more rows than the number of processors available. Use a striped matrix partitioning. Measure the speed of your code using as many different processors as possible, on $N \times N$ matrices with $N = 10^1, 10^2, 10^3, 10^4$. Produce a graph that plots speedup as a function of number of processors for each N. Interpret these results in the context of Ahmdal's Law.
8. Monte Carlo integration is a natural application for parallelization. Use GPU computing to speed up the `MCint` function from Chap. 6 Exercise 7.
9. The function `simulateBaseballGame`, available at http://www.springer.com, simulates one game of baseball given information about the batters in the batting order. The first input vector, `probSingle`, gives in position i the probability that the player batting i-th in the lineup will hit exactly a single, assuming that this probability does not change with time. The other input vectors are similar. The output is the number of runs scored by this lineup in one simulated game.
Write a new function with declaration

    `[meanRuns,stddevRuns,maxRuns]=`
    `lineupData(probSingle,probDouble,probTriple,`
        `probHr,probWalk,N)`

    that simulates N games played by the lineup and returns relevant statistics about the performance of the lineup.

    The input arguments are:

    `probSingle` - A vector that contains in position i the probability that the player batting i-th will hit exactly a single.

probDouble - A vector that contains in position i the probability that the player batting i-th will hit exactly a double.
probTriple - A vector that contains in position i the probability that the player batting i-th will hit exactly a triple.
probHr   - A vector that contains in position i the probability that the player batting i-th will hit exactly a home-run.
probWalk   - A vector that contains in position i the probability that the player batting i-th will be walked.
N   - The number of games to simulate.

The output arguments are:

meanRuns   - The mean number of runs scored in all games.
stddevRuns - The sample standard deviation of the number of runs scored.
maxRuns   - The maximum number of runs scored in any simulated game.

Use a **parfor** loop to parallelize this code.

10. Given a collection of baseball players there are many possible batting orders in which the players could appear. Write a function with declaration

    [bestOrders,meanRuns,maxRuns,stddevRuns]=
        bestBattingOrder(probSingle,probDouble,
        probTriple,probHR,probWalk,N)

    that collects statistics on all possible batting orders using the players described, and then returns a matrix describing the five batting orders with the highest expected number of runs per game.

    The input arguments are the same as those for Exercise 9.

    The output arguments are:

    bestOrders - A 5×9 matrix describing the five batting orders with the highest expected number of runs per game. For example, if the first row of bestOrders is $[1, 3, 9, 8, 5, 2, 7, 6, 4]$, then the batting order with the highest number of expected runs is to first bat player 1, then player 3, then player 9, and so on.
    meanRuns   - A 5×1 matrix giving the expected number of runs per game for each of the batting orders given in bestOrders.

7.4 Exercises

maxRuns - A 5 × 1 matrix giving the maximum number of runs scored in any simulation by each of the batting orders given in bestOrders.

stddevRuns - A 5 × 1 matrix giving the sample standard deviation in runs scores for each of the batting orders in bestOrders.

There are two different possible levels of parallelization; parallelizing at the level of each batting order, and then simulating each of N games in serial, or fixing a batting order and then simulating N games in parallel. Implement the code both ways, and report your results.

11. Write a MATLAB function named termination_time that determines the termination time of a Collatz sequence with given initial value, similar to the Python code in Example Code 7.12. Make figure similar to Fig. 7.14 displaying your results. Use spmd to parallelize your code, then use parfor, and then use arrayfun to parallelize your code on the GPU. Produce a plot indicating the runtime for each parallel algorithm on several different problem sizes.

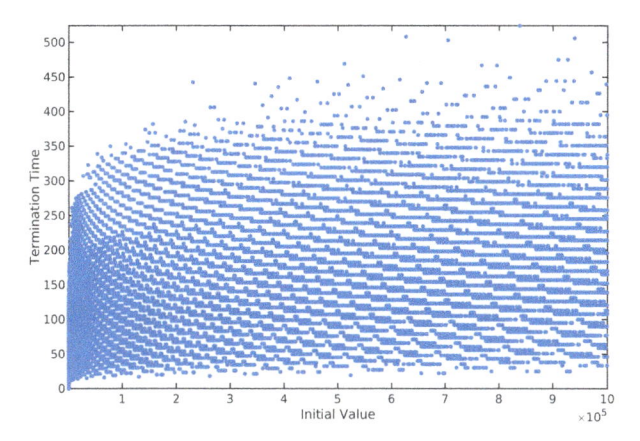

Fig. 7.14 Termination time (number of terms in Collatz sequence) as a function of the initial value of the sequence.

12. Write the termination_time code from Exercise 11 in C, and use mex to compile the code. Compare the speed of the MATLAB implementation to the C implementation when producing figures such as the one in Exercise 11.
13. Use parfor or spmd to speed up the game of War simulation from Chap. 6 Exercise 1.
14. Local optima are a constant source of concern for any optimization algorithm. One simple method for addressing the concern is called a multistart method. A multistart method simply initiates a conventional

optimization routine from several randomly selected starting points, and then selects the most optimal solution obtained from any of the starting points.

Use a technique of your choice to implement a function with declaration

```
[xbest,fbest,x0,itrcnt, stat]=multistartGradDesc
    (f,Df,X0,N,gradTol,xTol,itrBound,s)
```

that performs a parallel multistart gradient descent using N randomly selected started points from a region defined by $X0$.

The input arguments are:

f	- A filename or function handle that evaluates the function to be minimized.
Df	- A filename or function handle that evaluates the gradient of f.
X0	- An $m \times 2$ matrix, where m is the dimension of the input variables to f, that defines the domain from which initial iterates should be selected. For example, if the domain of f is two dimensional, then $X0 = [1, 3; 10, 20]$ indicates that each initial iterate should satisfy $1 \leq x_1 \leq 3$ and $10 \leq x_2 \leq 20$.
N	- The number of initial iterates to use. Each initial iterate should be randomly selected to respect the bounds in X0.
gradTol	- A termination criterion, applied to each search, on $\|Df_x(x_{best})\|$.
xTol	- A termination criterion, applied to each search, on $\|x_k - x_{k-1}\|$.
itrBound	- A maximum number of iterations to perform on any individual search.
s	- A line search parameter encoded as follows:

    s = 1: Use $\alpha = 1$ for all iterations.

    s = 2: Decide α with a Lagrange polynomial with $\alpha_1 = 0$, $\alpha_2 = 1$, and $\alpha_3 = 2$.

    s = 3: Decide α with bisection over the interval $0 \leq \alpha \leq 1$; if $\nabla f(x)$ and $\nabla f(x+d)$ have the same sign, then set $\alpha = 1$.

Each search should terminate if any of the termination tolerances are satisfied or if the iteration bound is exceeded.
The output arguments are:

7.4 Exercises

```
xbest   - The best calculated solution arising from any search.
fxbest  - The function value at xbest.
x0      - The initial iterate that resulted in finding xbest.
itrcnt  - The number of iterations performed on the search that
          resulted in xbest, counting x0 as iteration 0.
stat    - A status variable, encoded as follows:
```

   s = 0: The algorithm succeeded, x_{best} satisfies at least one convergence criterion.

   s = 1: The algorithm failed, in which case x_{best} is the best solution that was encountered.

15. The Sieve of Eratosthenes is a classic method for determining all the prime numbers between 2 and k. This algorithm and its numerous variations are still reasonable methods for finding moderately sized prime numbers today.

    The sieving method works by alternately classifying a previously unknown prime number, and then removing all multiples of the newly identified prime from further consideration. To begin, we start with a list, L that contains all integers 2 through k.

    $$L = \{2, 3, 4, 5, 6, 7, 8, 9, 10, 11, 12, 13, 14, 15, 16, 17, \ldots, k\}$$

    Two is the smallest number in the list, so we take it as prime. We now walk through the list, removing all multiples of two, leaving us with a new list

    $$L = \{2, 3, 5, 7, 9, 11, 13, 15, 17, \ldots, k\}.$$

    The next smallest integer after two in the list is three, which we now know is prime. We mark 3 as prime and remove all multiples of three from the list

    $$L = \{2, 3, 5, 7, 11, 13, 17, \ldots, k\}.$$

    The next smallest entry in the list is five, which we mark as prime, and remove all multiples of five from the list. This process, identifying the smallest remaining unidentified integer p as prime, then removing all multiples of p, is repeated until we have identified and removed all multiples of a prime $p \geq \sqrt{k}$. At that point all remaining entries in L may be safely classified as prime.

    We may think of the sieving method as a sequence of filters that are applied to L. First is the filter that discards all multiples of two. The first element of L to survive the "2-filter" is three, and so we run a filter that discards all multiples of three on the output of the first filter. The efficacy of this algorithm relies on running the filtering steps in order; we cannot apply the filtering steps simultaneously, because the result of the "2-filter" is used to identify the next prime, and subsequently run the "3-filter." We may, however, use a pipelining approach to the problem.

A pipelined version of this algorithm might work as follows: Lab 0 acts as a list-generator and 2-filter. It checks every integer l between 3 and k to see if l is a multiple of two. Every number that is not a multiple of two is sent to the next processor using labSend. Lab 1 first receives 3 from rank 0, and so becomes a 3-filter, passing on only inputs it receives from Lab 0 that are not multiples of 3. Lab 2 becomes a 5-filter, Lab 3 a 7-filter, and so on. See Fig. 7.15.

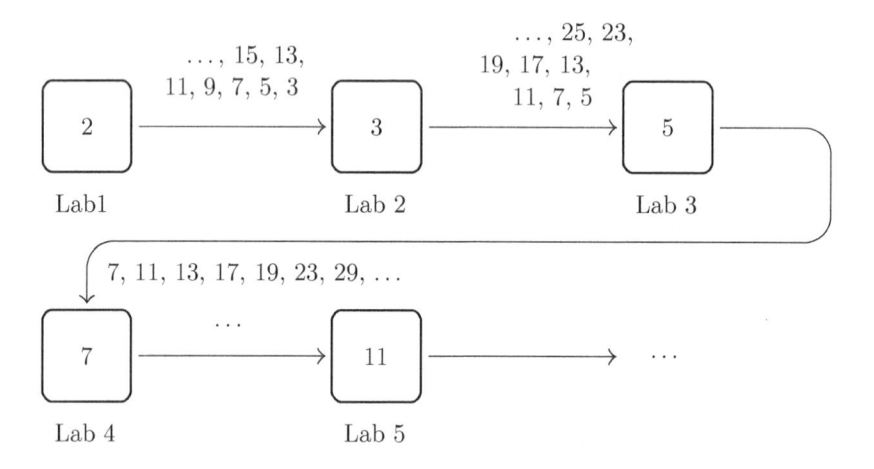

Fig. 7.15 Illustration of the pipelined algorithm for the Sieve of Eratosthenes.

If the last worker first receives integer p and thus becomes a p-filter, that processor can safely classify any integer less than p^2 as prime.

Write a function with declaration

[L,certainty]=findAllPrimes(k,P)

that uses P labs (P may be larger than the number of processors on your computer) to filter all integers $2, \ldots, k$, and classifies the remaining integers as certainly prime or possibly prime.

The input arguments are:

k - The code will filter all integers $2, 3, \ldots, k$ and return those integers that are not divisible by any other integers in the list.
P - The number of labs to start.

7.4 Exercises

The output arguments are:

L — The list of integers that successfully pass all filters, including the filters themselves (so, including 2, 3, etc.).

certainty — A 0-1 list indicating whether the entries of L are certainly prime or possibly prime. If `certainty(i)==1`, then `L(i)` is guaranteed to be prime.

Part II
Computational Modeling

Part II of this text addresses computational modeling, and the forthcoming chapters illustrate how the computational methods of Part I can combine to solve problems in engineering and science. The term "computational modeling" isn't as definitive as related topics like those of "mathematical modeling" or "statistical modeling," so a few words about how we interpret our intent seem warranted. First, computational modeling is not the modeling of computer architectures, data structures, or computing languages, as these topics are more common to computer science. That is not to say that computational modeling disassociates itself from computer science, as the two are certainly intertwined. After all, the applicability and effectiveness of a computational method relies on the computing environment, and hence, computational science depends on the fruits of computer science. Likewise, the scientific aspects of computer science can benefit from computational models to help predict outcomes. For example, we can simulate designs of networks and circuits to decide which are worthy of further pursuit.

Just as computational modeling isn't computer science modeling, neither is it mathematical modeling. Mathematical models are part of our human history, and they arguably intersect every facet of scientific study. This thought has been reflected by some of the greatest minds in science and mathematics:

> We do not master a scientific theory until we have shelled and completely prised free its mathematical kernel. – David Hilbert

> One can understand nature only when one has learned the language and the signs in which it speaks to us; but this language is mathematics and these signs are mathematical figures. – Galileo Galilei

> Profound study of nature is the most fertile source of mathematical discoveries – Joseph Fourier

A mathematical model is an abstraction of a phenomena into a symbolic representation. These symbolic expressions can be studied with mathematical rigor, meaning that we can prove theorems about them, or with experimentation, meaning that we can verify model outcomes with observations to help predict other observations. As an example, the existence and uniqueness theorem introduced in a course on differential equations is a theoretical result that establishes conditions under which we can discuss **the** solution of a differential equation. Knowing that we have a unique solution, we can continue to experiment with model parameters to estimate properties of the solution and how it matches our observations of the real phenomena.

Any delineation between computational modeling and mathematical modeling is nebulous because the relationship is convoluted. Mathematical modeling has traditionally encompassed the process of creating an abstract representation as well as a rigorous study of the abstract model. Mathematical modeling also overlaps the calculation of solutions, either exact or approximated. The related discipline of numerical analysis particularly investigates the mathematics of computing approximate solutions of a mathematical model. The point that we would make here is that it is possible to study mathematical modeling and numerical analysis without leaving the theoretical realm, as there is no requirement that either a mathematical modeler or a numerical analyst do anything other than prove theorems. In this regard there is a tangible distinction with computational modeling.

Our interpretation of computational modeling is that it is the totality of expressing a phenomena for computational studies. As such, computational models are meant to be solved by computational methods. A model's expression can be mathematical, but it can also be algorithmic. Of course algorithms are mathematical entities, so we could argue about whether mathematics or computer science owns a computational model. We shun such tit-for-tat discussions and recognize the reality that computational modeling is an amalgamation of mathematical modeling, algorithmic procedure, the selection of a computing platform, and the intentful actuality of a scientific or engineering impact. So computational modeling cuts across traditional disciplinary boundaries as it harbors aspects of numerous conventional studies.

The requirement of scientific impact is important, and we shouldn't forget that (1) it is possible, if not easy, to model something that we can't compute, and (2) it is also possible to compute something that has no relation to the underlying phenomena. We would generally argue that both are not within the realm of computational modeling since they have strayed beyond the intent of solving a problem in science and engineering. Another way of expressing this sentiment is that computational modeling is a study in computational science, and hence, tangible scientific outcomes are the ultimate goal.

Chapter 8
Modeling with Matrices

Freedom of expression is the matrix, the indispensable condition, of nearly every other form of freedom. – Benjamin Cardozo

The humble linear map, $x \mapsto Ax$, and its associated linear system, $Ax = b$, are so common, so well studied, and so well computed, that their importance can easily slip into the background of computational science. Indeed, most students have likely questioned why they couldn't simply rely on x = A\b as they have studied the numerous ways to solve $Ax = b$ in Chap. 2. The answer is, of course, to gain an appreciation of the intrinsic computational difficulty of solving $Ax = b$, say by either factoring A or with an iterative method. That said, we note that the coverage in this text on solving $Ax = b$ is but a tiny fraction of the expertise couched in the MATLAB command x = A\b. The robust success of this one command speaks for lifetimes of adroit researchers. However, now that this computational power rests with anyone willing to purchase MATLAB or download Octave, one might ask if we can't simply ignore some of the details and let the mantra that "matrices are important" abate as we gain trust in advanced algorithms and software?

The answer to this question in many cases is yes, and you should feel comfortable employing MATLAB commands to undertake important computational tasks. However, linearity is regularly at the core of how we explain phenomena, sometimes due to an observed linear relation and sometimes due to an imposed assumption. A matrix in either case becomes a modeling construct that performs a linear action related to the phenomena. Models for matrix coefficients need not be linear even though their assembled matrices perform linear actions. So models used to generate matrices need care and attention even though their matrix actions can be regularly handled with standard matrix algorithms.

We consider three matrix models in this chapter. The first is the discrete Fourier transform (DFT). The DFT matrix transforms a time signal into

© Springer Nature Switzerland AG 2019
A. Holder, J. Eichholz, *An Introduction to Computational Science*,
International Series in Operations Research & Management Science 278,
https://doi.org/10.1007/978-3-030-15679-4_8

a frequency domain, which can then be processed to, e.g., remove unwanted noise. One of the most important computational aspects of the DFT is that the corresponding matrix action can be efficiently calculated with the fast Fourier transform (FFT). The speed of the FFT, coupled with the applicability of the DFT in engineering and science, makes it one of the most important breakthroughs in computational science.

Our second example is that of designing radiotherapy treatments. In this case the matrix coefficients are a radiobiological model that translates an energy profile external to the anatomy into a biological dose in the anatomy. The intent of the design process is to find an external energetic profile that best treats the cancer. This intent naturally lends itself to an optimization model, which we explore. However, the optimization model inherently depends on the radiobiological model of the matrix coefficients, which is why the model is presented in this chapter.

The third example is that of estimating aeronautic lift from a two- dimensional approximation of a wing's profile. The matrix coefficients of this model grossly explain the air flow around a wing, essentially ignoring the turbulence on the wing's surface. The outcome of solving a resulting linear system to match the freestream air flow, which is distant from the wing, can give an accurate estimate of lift. The effectiveness of this linear model is surprising, and it exemplifies how linearity can be exceptionally useful.

8.1 Signal Processing and the Discrete Fourier Transform

The term signal broadly means an information stream emanating from a source. As examples, a sound wave from a musical instrument, the number of solar flares from the sun, and the voltage from a generator are all signals. The act of processing signals surrounds our modern life and is found in many of the technological devices that we hold dear. The field of signal processing is large, and the applications are vast. Moreover, the field is a curious and pleasing blend of mathematical rigor, engineering savvy, and artistic interpretation. We restrict our development to the classical collection of processing techniques called linear time invariant filters, but even this restriction encourages a wide study.

The action of a filter is mathematically expressed as a convolution operator, which is a binary operation that can be cleverly mapped to a dot product by the discrete Fourier transform (DFT). The DFT has many uses in science, engineering, medicine, and computer science, and it is one of the most desired elements within a computational endeavor because it can be computed with stunning speed by the fast Fourier transform (FFT). The DFT is a linear map and can thus be represented as a matrix vector product, which is already a reasonable computational framework in most settings. However, carefully considering a linear map can lead to improved efficiency and accu-

8.1 Signal Processing and the Discrete Fourier Transform

racy as we have already seen in Sect. 3.3, where we re-expressed the linear map that calculates a spline in terms of its second derivatives to reduce the computational burden and add numerical stability. In this chapter we show how the periodic nature of complex arithmetic allows us to re-express the DFT to reduce the computational burden, the result of which is an FFT.

8.1.1 Linear Time Invariant Filters

We restrict ourselves to discrete time signals, which are typically discrete samples from a continuous signal. So a signal for us is a discrete time sequences $x[n]$ for which n iterates over time steps. As an example, a true continuous signal might be the function $f(t) = \sin(2\pi t) + \cos(4\pi t)$. If we let $\Delta t = 1/8$, then we can sample the continuous signal at multiples of Δt as

$$\ldots, x[-2], x[-1], x[0], x[1], x[2], \ldots$$
$$= \ldots, f(-2/8), f(-1/8), f(0/8), f(1/8), f(2/8), \ldots$$
$$= \ldots, -2.00, -0.71, 1.00, 0.71, 0.00, \ldots.$$

Figure 8.1 illustrates the discrete sample.

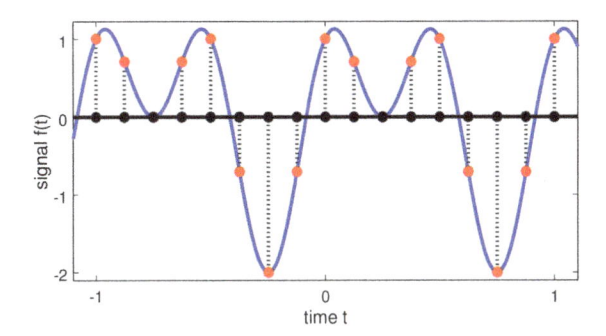

Fig. 8.1 The continuous signal $f(t) = \sin(2\pi t) + \cos(4\pi t)$ is blue. The black dots are steps of $\Delta t = 1/8$, and the sampled discrete time sequence $x[n]$ is the collection of red dots.

Signal processing is the manipulation of a signal to produce another. So if $x[n]$ is the original input signal, then it is processed by an action H to produce an output signal $y[n]$. Processing is expressed mathematically as

$$y[n] = H(x[n]).$$

There are many ways to process a signal, and H depends on the desired action. We assume that H is linear and time invariant, properties that require:

linearity $H(ax_1[n]+bx_2[n]) = aH(x_1[n])+bH(x_2[n])$, where $x_1[n]$ and $x_2[n]$ are signals and a and b are scalars, and

time invariance $H(x[n-d]) = y[n-d]$ for any index shift d.

These requirements might seem restrictive, but they are commonly appropriate. For example, if two sound sequences differ only in their scaling, then you might expect their output sequences to differ similarly. Time invariance forces H to act the same no matter where it starts on the sequence.

The assumed properties of H can be exploited to simplify the calculation of the output signal. The idea is to express the input sequence as a summation with each term having the same functional form. The similarity of the resulting terms means that we only need to know how H acts on the common function, which simplifies and expedites computation.

The function of interest is the Kronecker delta function, which is

$$\delta[n] = \begin{cases} 0, & n \neq 0 \\ 1, & n = 0. \end{cases}$$

The Kronecker delta function is the discrete counterpart of the Dirac delta function, which you may be familiar with from a course in differential equations. The defining relationship between the two is

Discrete: Kronecker Delta	Continuous: Dirac Delta
$\sum_{k=-\infty}^{\infty} x[k]\,\delta[n-k] = x[n]$	$\int_{-\infty}^{\infty} f(t)\,\delta(a-t)\,dt = f(a).$

The intent of both delta functions is to "zero-out" the values of x and f so that only the desired instance remains. If you have prior experience with the Dirac delta, then know that it shares many properties with the Kronecker delta.

From the linearity of H we have

$$H(x[n]) = H\left(\sum_{k=-\infty}^{\infty} x[k]\,\delta[n-k]\right) = \sum_{k=-\infty}^{\infty} x[k]\,H(\delta[n-k]).$$

The terms of the summation on the right have a form common to numerous areas of applied mathematics. Each term is the product of two functions, for which the argument of the first is moving "forward" and the argument of the second is moving "backward." The argument of the second function is also shifted by a value of n. Such summations are so common and important that they have their own name, called convolutions, and notation, denoted by $*$. For our signal notation we have

$$(x * h)[n] = \sum_{k=-\infty}^{\infty} x[k]\,h[n-k] = \sum_{k=-\infty}^{\infty} x[k]\,H(\delta[n-k]) = H(x[n]) = y[n],$$

8.1 Signal Processing and the Discrete Fourier Transform

where $h[n]$ is the unit response of δ under H, i.e. $h[n] = H(\delta[n])$. The conclusion is that the desired output signal $y[n]$ is a convolution of the input signal with the unit response.

The convolution that calculates $y[n]$ can be stated in terms of matrix vector multiplication as long as all sequences are finite. We momentarily restrict ourselves to periodic signals to exemplify the calculation theme. The periodic assumption allows us to restrict our calculations to one period. Consider our previous continuous example of $f(t) = \sin(2\pi t) + \cos(4\pi t)$, which has a period of 1. The input stream is tabulated below for $\Delta t = 1/8$ over the single period of $0 \leq t < 1$.

n	0	1	2	3	4	5	6	7
$f(n\Delta t)$	$f(0/8)$	$f(1/8)$	$f(2/8)$	$f(3/8)$	$f(4/8)$	$f(5/8)$	$f(6/8)$	$f(7/8)$
$x[n]$	1.000	0.707	0.000	0.707	1.000	-0.707	-2.000	-0.707

Assume for the moment that the unit response $h[n]$ is defined by

$$h[0] = h[1] = h[2] = 1 \quad \text{and} \quad h[3] = h[4] = h[5] = h[6] = h[7] = 0,$$

which defines a periodic running sum as we will see. This response is extended to all integers by setting $h[n + 8k] = h[n]$, where $n = 0, 1, \ldots, 7$ and k is any integer. The convolution for one period of the output signal can be calculated as

$$\begin{pmatrix} y[0] \\ y[1] \\ y[2] \\ y[3] \\ y[4] \\ y[5] \\ y[6] \\ y[7] \end{pmatrix} = \begin{bmatrix} h[0] & h[-1] & h[-2] & h[-3] & h[-4] & h[-5] & h[-6] & h[-7] \\ h[1] & h[0] & h[-1] & h[-2] & h[-3] & h[-4] & h[-5] & h[-6] \\ h[2] & h[1] & h[0] & h[-1] & h[-2] & h[-3] & h[-4] & h[-5] \\ h[3] & h[2] & h[1] & h[0] & h[-1] & h[-2] & h[-3] & h[-4] \\ h[4] & h[3] & h[2] & h[1] & h[0] & h[-1] & h[-2] & h[-3] \\ h[5] & h[4] & h[3] & h[2] & h[1] & h[0] & h[-1] & h[-2] \\ h[6] & h[5] & h[4] & h[3] & h[2] & h[1] & h[0] & h[-1] \\ h[7] & h[6] & h[5] & h[4] & h[3] & h[2] & h[1] & h[0] \end{bmatrix} \begin{pmatrix} x[0] \\ x[1] \\ x[2] \\ x[3] \\ x[4] \\ x[5] \\ x[6] \\ x[7] \end{pmatrix}$$

$$= \begin{bmatrix} 1 & 0 & 0 & 0 & 0 & 0 & 1 & 1 \\ 1 & 1 & 0 & 0 & 0 & 0 & 0 & 1 \\ 1 & 1 & 1 & 0 & 0 & 0 & 0 & 0 \\ 0 & 1 & 1 & 1 & 0 & 0 & 0 & 0 \\ 0 & 0 & 1 & 1 & 1 & 0 & 0 & 0 \\ 0 & 0 & 0 & 1 & 1 & 1 & 0 & 0 \\ 0 & 0 & 0 & 0 & 1 & 1 & 1 & 0 \\ 0 & 0 & 0 & 0 & 0 & 1 & 1 & 1 \end{bmatrix} \begin{pmatrix} 1.000 \\ 0.707 \\ 0.000 \\ 0.707 \\ 1.000 \\ -0.707 \\ -2.000 \\ -0.707 \end{pmatrix} = \begin{pmatrix} -1.707 \\ 1.000 \\ 1.707 \\ 1.414 \\ 1.707 \\ 1.000 \\ -1.707 \\ -3.414 \end{pmatrix}.$$

This multiplication is illustrated graphically in Fig. 8.2. Each of the green graphs corresponds with one period of $h[n-k]$. For example, the graph for $h[1-k]$ aligns with the second row of the matrix above, in which the 1st, 2nd, and 8th elements are 1 and the rest are zero. To calculate $y[n]$ we find $h[n-k]$ and multiply this sequence componentwise by $x[k]$. The geometry demonstrates how $h[n-k]$ sweeps to the right as n increases.

If the input and output signals are periodic of length N, then each element of $y[n]$ requires N multiplications as calculated with the convolution definition. Since there are N elements of $y[n]$, the total number of multiplications is N^2. This number is reasonable for small to moderate length sequences, but the calculation becomes burdensome as N climbs, say as Δt decreases. Moreover, speed is paramount in sound and video manipulation, and hence, fast convolution algorithms are important. There are fortunately numerous computational algorithms that dramatically decrease the number of multiplications to increase the speed of signal processing. Such methods require a mapping of the data with the discrete Fourier transform, which is the topic of the next section.

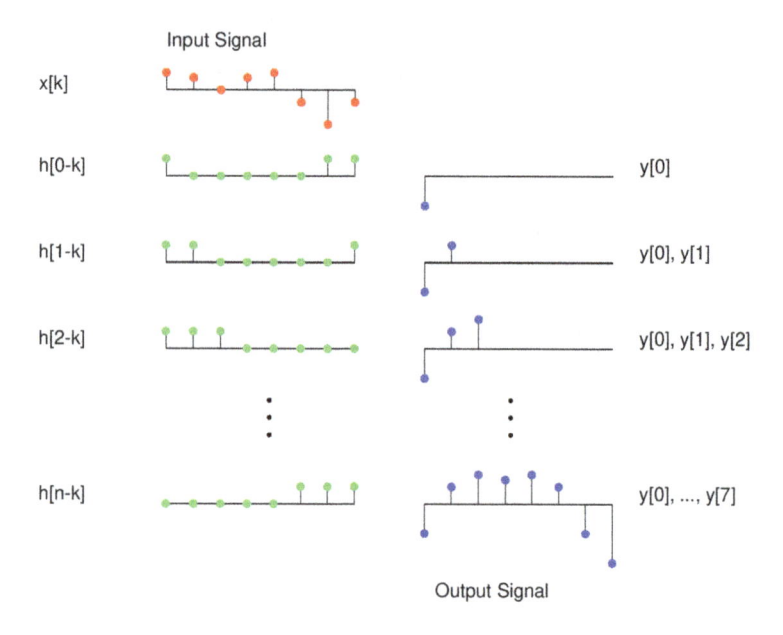

Fig. 8.2 The input signal $x[n]$ is sampled from $f(t) = \sin(2\pi t) + \cos(4\pi t)$ and is shown at the top left in red. The elements of the output sequence $y[n]$ are on the right in blue. Each element of $y[n]$ is the dot product of $x[n]$ with $h[n-k]$, the latter of which is shown in green.

8.1.2 The Discrete Fourier Transform

The DFT is a uniform sampling of the discrete time Fourier series, which is itself a restricted form of the Fourier series. If you have previously studied the Fourier series in calculus or differential equations, then an appropriate mindset is to think of the DFT as a finite discrete counterpart. If you have not seen the Fourier series in another setting, then what is principal to note is that a Fourier series decomposes a function into sines and cosines with varying frequencies. The coefficients of the Fourier series help identify the periodic nature of the function being decomposed. The DFT has the same intent but on a finite signal, which could be a sample from a continuous signal.

The DFT has many uses and properties, but the one we avail is its action on convolution. Let $x[n]$ and $h[n]$ be an input signal and a unit impulse. If $\mathcal{F}$ is the DFT, then the following property shows how the DFT acts on convolution,

$$\mathcal{F}(x * h) = \mathcal{F}(x) \cdot \mathcal{F}(h) = X \cdot H,$$

where $\cdot$ is the componentwise dot-product of the transformed sequences and the capital letters indicate the transformed sequences. If x, h, X, and H are all of length N, then the image of the convolution can be calculated with N multiplications. Since the original convolution requires N^2 multiplications, the DFT maps an N^2 calculation to the much reduced N calculation.

The reduced number of multiplications is only helpful under two conditions. First, calculating X and H must be efficient, for if not, then computing them might negate the intent of mapping the convolution. Second, we must be able to efficiently map $X \cdot H$ back to the value of the convolution, for if this calculation is expensive, then we have again unduly toiled in the hope of gaining the advantage of mapping the convolution. If we assume $\mathcal{F}^{-1}$ exists, then we are interested in the total effort to calculate

$$x * h = \mathcal{F}^{-1}(X \cdot H) = \mathcal{F}^{-1}\left(\mathcal{F}(x) \cdot \mathcal{F}(h)\right).$$

The action of $\mathcal{F}$ is linear, and as developed momentarily, the mapping can be accomplished with matrix vector multiplication. However, multiplying an N vector by an $N \times N$ matrix requires N^2 multiplications. The inverse multiplication would similarly require N^2 multiplications, in addition to the N^3 multiplications that might be required to calculate the inverse matrix. So from a general matrix perspective we seem to have replaced the N^2 calculation of the convolution on the left with the $N^3 + 3N^2 + N$ calculation on the right. The DFT is not looking so good unless its number of multiplications can be reduced.

Fortunately, the structure of the DFT permits its calculation to be completed with far less than N^2 multiplications. Algorithms that efficiently calculate the DFT of a sequence are called fast Fourier transforms (FFTs), and we explain a standard FFT after developing the DFT. The inverse DFT is similarly as efficient, and these algorithms are called inverse fast Fourier transforms (IFFTs). FFTs and their inverses are among the most important computational algorithms in modern scientific computing due to their fast computational ability. Their speed is so desired that researchers commonly exhaust their cleverness to leverage the use of an FFT in their computational efforts.

The DFT of the N element input sequence $x[n]$ is defined by

$$X[k] = \frac{1}{N} \sum_{n=0}^{N-1} x[n] e^{-i2k\pi n/N}$$

$$= \frac{1}{N} \sum_{n=0}^{N-1} x[n] \left(\cos\left(\frac{2k\pi n}{N}\right) - i \sin\left(\frac{2k\pi n}{N}\right) \right), \qquad (8.1)$$

where $i = \sqrt{-1}$. We mention that nearly all materials on signal processing use j instead of i to represent $\sqrt{-1}$. The reason for the change is that i typically represents current in electrical engineering. We maintain the mathematical notation of i. The multiple $1/N$ normalizes the summation to the length of the sequences. Such normalization is not required of the definition, and its use varies among disciplines. We include the normalization factor because the definition then most easily agrees with the traditional Fourier series, of which some readers might be familiar. Expression (8.1) is a consequence of the even and odd properties of sine and cosine and of Euler's formula, which is

$$e^{i\theta} = \cos(\theta) + i \sin(\theta).$$

The common period of $\sin(2k\pi n/N)$ and $\cos(2k\pi n/N)$ is

$$\frac{2\pi}{2k\pi/N} = \frac{N}{k}.$$

Frequency is the reciprocal of period, and hence, the frequency of $X[k]$ is k/N. If n indexes time steps in seconds, then k indexes frequency in Hertz (Hz). The action of the DFT is to map time signals to frequency signals, and we often refer to $x[n]$ as being in the time domain and $X[k]$ as being in the frequency domain.

If we let $w = e^{-i2\pi/N}$, then the DFT in matrix form is

8.1 Signal Processing and the Discrete Fourier Transform

$$\begin{pmatrix} X[0] \\ X[1] \\ X[2] \\ X[3] \\ \vdots \\ X[N-1] \end{pmatrix} = \frac{1}{N} \begin{bmatrix} w^0 & w^0 & w^0 & \cdots & w^0 \\ w^0 & w^1 & w^2 & \cdots & w^{N-1} \\ w^0 & w^2 & w^4 & \cdots & w^{2(N-1)} \\ w^0 & w^3 & w^6 & \cdots & w^{3(N-3)} \\ \vdots & \vdots & \vdots & & \vdots \\ w^0 & w^{N-1} & w^{2(N-1)} & \cdots & w^{(N-1)^2} \end{bmatrix} \begin{pmatrix} x[0] \\ x[1] \\ x[2] \\ x[3] \\ \vdots \\ x[N-1] \end{pmatrix}.$$

Each of the matrix coefficients is w^{kn}, where k is the row index and n is the column index. Since $kn = nk$, the DFT matrix is obviously symmetric. The elements of the matrix are complex, and the transformed signal is thus generally complex even if the input signal is real. The periodic behavior of sine and cosine helps us rewrite the matrix representation so that all powers remain between 0 and $N-1$. For any integer r notice that

$$\begin{aligned} w^{k(n+rN)} = e^{-i2k\pi(n+rN)/N} &= \cos\left(\frac{2k\pi(n+rN)}{N}\right) - i\sin\left(\frac{2k\pi(n+rN)}{N}\right) \\ &= \cos\left(\frac{2k\pi n}{N} + 2kr\pi\right) - i\sin\left(\frac{2k\pi n}{N} + 2kr\pi\right) \\ &= \cos\left(\frac{2k\pi n}{N}\right) - i\sin\left(\frac{2k\pi n}{N}\right) \\ &= e^{-i2k\pi n/N} = w^{kn}. \end{aligned}$$

We conclude that kn can be replaced with $kn(\bmod N)$ in all cases. Suppose $N = 8$ and $k = 3$ to illustrate. Then the third row of the DFT matrix is

$$\begin{aligned} & \left[w^0,\ w^3,\ w^6,\ w^9,\ w^{12},\ w^{15},\ w^{18},\ w^{21}\right] \\ &= \left[w^{0(\bmod N)},\ w^{3(\bmod N)},\ w^{6(\bmod N)},\ \ldots,\ w^{18(\bmod N)},\ w^{21(\bmod N)}\right] \\ &= \left[w^0,\ w^3,\ w^6,\ w^1,\ w^4,\ w^7,\ w^2,\ w^5\right]. \end{aligned}$$

The entire transform with $N = 8$ gives a matrix with powers between 0 and 7. The transformation in this case is

$$\begin{pmatrix} X[0] \\ X[1] \\ X[2] \\ X[3] \\ X[4] \\ X[5] \\ X[6] \\ X[7] \end{pmatrix} = \frac{1}{8} \begin{bmatrix} w^0 & w^0 & w^0 & w^0 & w^0 & w^0 & w^0 & w^0 \\ w^0 & w^1 & w^2 & w^3 & w^4 & w^5 & w^6 & w^7 \\ w^0 & w^2 & w^4 & w^6 & w^0 & w^2 & w^4 & w^6 \\ w^0 & w^3 & w^6 & w^1 & w^4 & w^7 & w^2 & w^5 \\ w^0 & w^4 & w^0 & w^4 & w^0 & w^4 & w^0 & w^4 \\ w^0 & w^5 & w^2 & w^7 & w^4 & w^1 & w^6 & w^3 \\ w^0 & w^6 & w^4 & w^2 & w^0 & w^6 & w^4 & w^2 \\ w^0 & w^7 & w^6 & w^5 & w^4 & w^3 & w^2 & w^1 \end{bmatrix} \begin{pmatrix} x[0] \\ x[1] \\ x[2] \\ x[3] \\ x[4] \\ x[5] \\ x[6] \\ x[7] \end{pmatrix}. \quad (8.2)$$

The top row of the DFT matrix being comprised entirely of $w^0 = 1$ means that $X[0]$ is the average value of $x[n]$. Another property is that the rows from $k = 1$ through $k = N - 1$ have "mirrors" with regard to exponents. For example, the componentwise sums of the exponents of the second and eighth rows are each $0(\bmod 8)$. The same is true for the third and seventh rows, and so on. This relationship means $|X[1]| = |X[7]|$, $|X[2]| = |X[6]|$, and $|X[3]| = |X[5]|$. Sometimes this property is referred to as a "folding in" of the transformed sequence.

The DFT is useful in identifying the fundamental frequencies of a signal. As an example, Exercise 7 in Chap. 3 asks us to find a best fit periodic approximation to a historical record of daily temperatures. One might wonder if the innate periodicity of the data indeed agrees with an annual cycle. If so, then the assumed periodic cycle of 365.25 days in Exercise 7 would be justified. Figure 8.3 shows the 13,499 consecutive daily temperatures of Terre Haute, Indiana from January 1, 1973 through December 31, 2009.

The magnitude of the transformed sequence is plotted in Figs. 8.4 and 8.5. The average daily temperature over the historic record is $X[0] = 53.283$. The sole higher value on the right indicates a fundamental frequency. Its counterpart is somewhat masked on the left due to its proximity to $X[0]$. Figure 8.5 shows $|X[k]|$ for $k = 1, 2, \ldots, 100$, which establishes a fundamental frequency at $k = 37$. Other than $|X[0]|$, $|X[37]|$ and $|X[13463]|$, the latter of which agrees with $|X[37]|$, the transformed signal is nearly zero. From the fundamental frequency we conclude that the data is periodic with an estimated period of

$$\frac{13499}{37} = 364.84 \text{ days.}$$

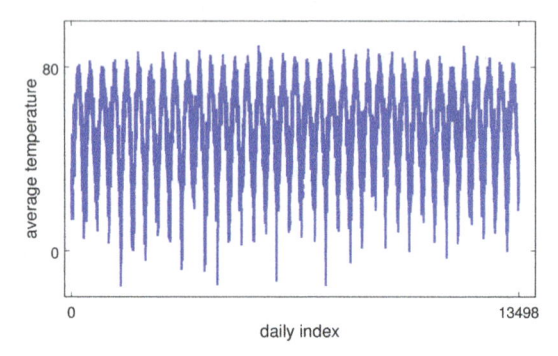

Fig. 8.3 Daily average temperatures in Terre Haute, Indiana from January 1, 1973 through December 31, 2009.

This estimate supports our assumption of an annual cycle. The slight difference between the estimated cycle of 364.84 days and our assumed cycle of

8.1 Signal Processing and the Discrete Fourier Transform

365.25 days is due to the frequencies being decided in discrete steps. Indeed, k would need to be the non-integer $13499/365.25 = 36.985$ to exactly predict our assumed periodic behavior.

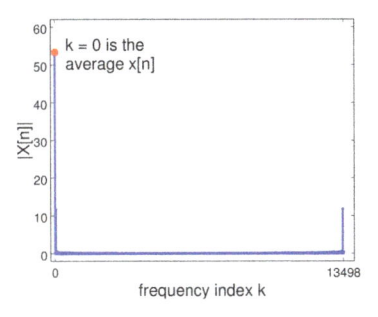

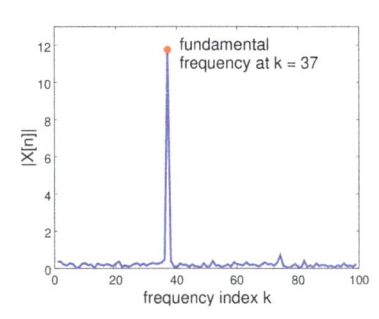

Fig. 8.4 The transformed signal of the daily temperatures in Fig. 8.3.

Fig. 8.5 The transformed signal of the daily temperatures for $k = 1$ through $k = 100$. A fundamental frequency occurs at $k = 37$.

If we let W be the DFT matrix so that the transformation of the length N signal $x[n]$ is

$$\mathcal{F}(x) = \left(\frac{1}{N}\right) Wx = X, \tag{8.3}$$

then an important question is, is W invertible? Moreover, how efficiently can W^{-1} be computed if it exists? Answers to both are encouraging and are stated in the following theorem.

Theorem 17. *Let W be an $N \times N$ DFT matrix as in (8.3). Then,*

- *W is invertible, and if p_{kn} is the exponent so that $W_{kn} = w^{p_{kn}}$, then*

$$W^{-1}_{kn} = \frac{1}{N} w^{-p_{kn}}.$$

- *The matrix $(1/\sqrt{N})W$ is unitary, meaning that*

$$\left(\frac{1}{\sqrt{N}} W\right)^{-1} = \left(\frac{1}{\sqrt{N}} W\right)^H,$$

where the superscript H indicates the conjugate (Hermitian) transpose.

Consider the 8×8 DFT matrix in (8.2) to help interpret Theorem 17. From the first property we have that the inverse matrix exists and is

$$\begin{bmatrix} w^0 & w^0 & w^0 & w^0 & w^0 & w^0 & w^0 & w^0 \\ w^0 & w^1 & w^2 & w^3 & w^4 & w^5 & w^6 & w^7 \\ w^0 & w^2 & w^4 & w^6 & w^0 & w^2 & w^4 & w^6 \\ w^0 & w^3 & w^6 & w^1 & w^4 & w^7 & w^2 & w^5 \\ w^0 & w^4 & w^0 & w^4 & w^0 & w^4 & w^0 & w^4 \\ w^0 & w^5 & w^2 & w^7 & w^4 & w^1 & w^4 & w^3 \\ w^0 & w^6 & w^4 & w^2 & w^0 & w^6 & w^4 & w^2 \\ w^0 & w^7 & w^6 & w^5 & w^4 & w^3 & w^2 & w^1 \end{bmatrix}^{-1} = \frac{1}{8} \begin{bmatrix} w^0 & w^0 & w^0 & w^0 & w^0 & w^0 & w^0 & w^0 \\ w^0 & w^{-1} & w^{-2} & w^{-3} & w^{-4} & w^{-5} & w^{-6} & w^{-7} \\ w^0 & w^{-2} & w^{-4} & w^{-6} & w^0 & w^{-2} & w^{-4} & w^{-6} \\ w^0 & w^{-3} & w^{-6} & w^{-1} & w^{-4} & w^{-7} & w^{-2} & w^{-5} \\ w^0 & w^{-4} & w^0 & w^{-4} & w^0 & w^{-4} & w^0 & w^{-4} \\ w^0 & w^{-5} & w^{-2} & w^{-7} & w^{-4} & w^{-1} & w^{-4} & w^{-3} \\ w^0 & w^{-6} & w^{-4} & w^{-2} & w^0 & w^{-6} & w^{-4} & w^{-2} \\ w^0 & w^{-7} & w^{-6} & w^{-5} & w^{-4} & w^{-3} & w^{-2} & w^{-1} \end{bmatrix}.$$

The form of the above inverse can be verified directly by multiplying the matrix on the right by W, see Exercises 5. The second conclusion is then a by-product of the symmetry of W coupled with the fact that

$$\begin{aligned} w^{-kn} = e^{-i2(-k)\pi n/N} &= \left(\cos\left(\frac{2(-k)\pi n}{N}\right) - i\sin\left(\frac{2(-k)\pi n}{N}\right) \right) \\ &= \left(\cos\left(\frac{2k\pi n}{N}\right) + i\sin\left(\frac{2k\pi n}{N}\right) \right) \\ &= \overline{\left(\cos\left(\frac{2k\pi n}{N}\right) - i\sin\left(\frac{2k\pi n}{N}\right) \right)} \\ &= \overline{e^{-i2k\pi n/N}} = \overline{w^{kn}}. \end{aligned}$$

Hence each w^{-nk} component in the above inverse can be replaced with $\overline{w^{nk}}$, with result being the conjugate transpose.

The importance of Theorem 17 is that the inverse of a DFT matrix comes at no additional computational expense, except for that of a sign change in the imaginary parts of the components. Hence, as we return to our goal of computing

$$x * h = \mathcal{F}^{-1}(X \cdot H) = \mathcal{F}^{-1}\left(\mathcal{F}(x) \cdot \mathcal{F}(y)\right),$$

we find that calculating the matrix for $\mathcal{F}^{-1}$ to be essentially free. What remains is an efficient way to calculate the forward and reverse transforms, which is the topic of the next section.

8.1.3 The Fast Fourier Transform

A fast Fourier transform (FFT) is an efficient algorithm to calculate the DFT. There isn't one FFT algorithm, but rather "the" FFT is more appropriately a family of algorithms from which efficient calculation methods can be selected. We develop a classical FFT that assumes N is a power of 2. This assumption might seem restrictive, but we can often control sampling so that signals have such lengths. There are numerous extensions that permit arbitrary signal length, with some of the fastest being for small values of N.

8.1 Signal Processing and the Discrete Fourier Transform

The sagacious computational effort of an FFT is due to a re-indexing of the defining equation. Assume $N = 2^p$ for some integer p. Then $N/2 = 2^{p-1}$ is an integer, and we have

$$X[k] = \frac{1}{N} \sum_{n=0}^{N-1} x[n]\, w^{kn}$$

$$= \frac{1}{N} \sum_{n=0}^{N/2-1} \left(x[n]\, w^{kn} + x[n+N/2]\, w^{k(n+N/2)} \right)$$

$$= \frac{1}{N} \sum_{n=0}^{N/2-1} \left(x[n] + x[n+N/2]\, w^{kN/2} \right) w^{kn}.$$

The last summation suggests that the original DFT can be rewritten as an associated DFT that is half the size of the original, that is if the parenthetical factor could be expressed independent of $w^{kN/2}$. We can remove $w^{kN/2}$ by noticing that

$$w^{kN/2} = e^{-i2k\pi N/2N} = e^{-ik\pi} = \cos(k\pi) + i\sin(k\pi) = \cos(k\pi) = (-1)^k.$$

Hence,

$$X[k] = \frac{1}{N} \sum_{n=0}^{N/2-1} \left(x[n] + x[n+N/2](-1)^k \right) w^{kn}.$$

Consider the $N = 4$ case as a small example. Then,

$$X[0] = x[0]w^0 + x[1]w^0 + x[2]w^0 + x[3]w^0$$
$$= (x[0] + x[2])w^0 + (x[1] + x[3])w^0,$$

$$X[1] = x[0]w^0 + x[1]w^1 + x[2]w^2 + x[3]w^3$$
$$= (x[0] - x[2])w^0 + (x[1] - x[3])w^1,$$

$$X[2] = x[0]w^0 + x[1]w^2 + x[2]w^0 + x[3]w^2$$
$$= (x[0] + x[2])w^0 + (x[1] + x[3])w^2, \text{ and}$$

$$X[3] = x[0]w^0 + x[1]w^3 + x[2]w^2 + x[3]w^1$$
$$= (x[0] - x[2])w^0 + (x[1] - x[3])w^3.$$

The most important observation is that the 4^2 multiplications of the first expressions, i.e. 4 multiplications for each of the four values of $X[k]$, have been replaced by a total of 8 multiplications in the second expressions. The matrix expression for the $N = 4$ case is

$$\begin{pmatrix} X[0] \\ X[2] \\ X[1] \\ X[3] \end{pmatrix} = \frac{1}{4} \left[\begin{array}{cc|cc} w^0 & w^0 & 0 & 0 \\ w^0 & w^2 & 0 & 0 \\ \hline 0 & 0 & w^0 & w^1 \\ 0 & 0 & w^0 & w^3 \end{array} \right] \begin{pmatrix} x[0] + x[2] \\ x[1] + x[3] \\ x[0] - x[2] \\ x[1] - x[3] \end{pmatrix}.$$

The reduction in the number of multiplications is obvious from the matrix expression because the off diagonal blocks do not need to be considered as the product is calculated. Notice that the output signal on the left-hand side has been reordered to agree with the block structure of the matrix.

The advantage of N being a power of 2 is that the same reduction can be repeated, with each application halving the number of multiplications required to calculate the matrix vector product. Division by 4 results in

$$\begin{aligned} X[k] &= \frac{1}{N} \sum_{n=0}^{N/4-1} \Big(x[n] w^{kn} + x[n+N/4] w^{k(n+N/4)} \\ &\qquad\qquad + x[n+2N/4] w^{k(n+2N/4)} + x[n+3N/4] w^{k(n+3N/4)} \Big) \\ &= \frac{1}{N} \sum_{n=0}^{N/4-1} \Big(x[n] + x[n+N/4] w^{kN/4} \\ &\qquad\qquad + x[n+2N/4] w^{2kN/4} + x[n+3N/4] w^{3kN/4} \Big) w^{kn} \\ &= \frac{1}{N} \sum_{n=0}^{N/4-1} \Big(x[n] + x[n+N/4](-i)^k \\ &\qquad\qquad + x[n+2N/4](-1)^k + x[n+3N/4]\, i^k \Big) w^{kn}. \end{aligned}$$

The resulting matrix expression for $N=8$ is

$$\begin{pmatrix} X[0] \\ X[4] \\ X[1] \\ X[5] \\ X[2] \\ X[6] \\ X[3] \\ X[7] \end{pmatrix} = \frac{1}{8} \left[\begin{array}{cc|cc|cc|cc} w^0 & w^0 & 0 & 0 & 0 & 0 & 0 & 0 \\ w^0 & w^4 & 0 & 0 & 0 & 0 & 0 & 0 \\ \hline 0 & 0 & w^0 & w^1 & 0 & 0 & 0 & 0 \\ 0 & 0 & w^0 & w^3 & 0 & 0 & 0 & 0 \\ \hline 0 & 0 & 0 & 0 & w^0 & w^2 & 0 & 0 \\ 0 & 0 & 0 & 0 & w^0 & w^5 & 0 & 0 \\ \hline 0 & 0 & 0 & 0 & 0 & 0 & w^0 & w^3 \\ 0 & 0 & 0 & 0 & 0 & 0 & w^0 & w^7 \end{array} \right] \begin{pmatrix} x[0] + x[2] + x[4] + x[6] \\ x[1] + x[3] + x[5] + x[7] \\ x[0] - ix[2] - x[4] + ix[6] \\ x[1] - ix[3] - x[5] + ix[7] \\ x[0] - x[2] + x[4] - x[6] \\ x[1] - x[3] + x[5] - x[7] \\ x[0] + ix[2] - x[4] - ix[6] \\ x[1] + ix[3] - x[5] - ix[7] \end{pmatrix}.$$

A careful count of the total number of complex multiplications, some of which are part of the right-hand side vector, shows that an FFT of a signal of length $N = 2^p$ can be accomplished with no more than

8.1 Signal Processing and the Discrete Fourier Transform

$$\frac{N}{2} \log_2 \left(\frac{N}{2}\right) \quad \text{multiplications.}$$

If $N = 2^{10} = 1024$, then the original matrix multiplication of the DFT requires $N^2 = 1{,}048{,}576$ multiplications, whereas the FFT computation requires only 4608—a scant 0.44% of the straightforward matrix multiplication. The 99.56% savings in multiplications is striking!

The inverse FFT uses similar manipulations to achieve the same number of multiplications as its forward counterpart. With regard to the convolution

$$x * h = \mathcal{F}^{-1}(X \cdot H) = \mathcal{F}^{-1}\left(\mathcal{F}(x) \cdot \mathcal{F}(y)\right),$$

the direct calculation on the left requires N^2 multiplications. Each of $\mathcal{F}(x)$ and $\mathcal{F}(y)$ can be calculated with no more than $(N/2)\log_2(N/2)$ multiplications. The dot product $\mathcal{F}(x) \cdot \mathcal{F}(y)$ uses no more than N multiplications, and the inverse transform no more than $(N/2)\log_2(N/2)$. So the maximum number of multiplications on the right-hand side is

$$N + \frac{3N}{2} \log_2 \left(\frac{N}{2}\right).$$

So if $N = 2^{10} = 1094$, we then have

$$N^2 = 1{,}048{,}576 \quad \text{and} \quad N + \frac{3N}{2} \log_2 \left(\frac{N}{2}\right) = 14{,}848.$$

This is a 98.584% savings in the number of multiplications. Moreover, the computational advantage increases as N grows.

State-of-the-art implementations of the FFT draw on advanced computing elements and sophisticated data structures. Thankfully both MATLAB and Octave have trustworthy and efficient algorithms to calculate FFTs and IFFTs. The standard commands are `fft` and `ifft`. The convolution of two periodic sequences x and y of common length can be computed with

```
ifft( fft(x) .* fft(y) ).
```

This calculation is called a circular convolution, a term motivated by the periodic assumption.

We can calculate the aperiodic convolution of two sequences of arbitrary length by padding them with zeros. As an example, suppose we want to convolve

$$x[n] = [1, 3, -1, 3] \quad \text{and} \quad y[n] = [2, -1, 4].$$

We then calculate the circular convolution of

$$\hat{x} * \hat{y} = [1, 3, -1, 3, 0, 0] * [2, -1, 4, 0, 0, 0],$$

where each signal has been padded with enough zeros to make the common length $N_x+N_y-1 = 4+3-1 = 6$. The role of the zeros is to remove the innate periodic calculation of convolution. You should calculate this convolution by hand to see the impact of the added zeros. The command to calculate an aperiodic convolution in MATLAB and Octave is `conv`.

8.1.4 Filtering Signals

We revisit the idea of filtering suggested in Sect. 8.1.1 with our newfound understanding of the DFT and its computationally efficient calculation, the FFT. An ideal low pass filter $h[n]$ eliminates high frequencies, thus $\mathcal{F}(h) = H$ should have the property that

$$H[k] = \begin{cases} 1, & 0 \leq k \leq k_{\max} \\ 0, & k_{\max} < k < N - k_{\max} \\ 1, & N - k_{\max} \leq k \leq N - 1. \end{cases}$$

This way $X[k] \cdot H[k]$ removes frequencies whose indices exceed $k_{\max}$, which is assumed to be less than $N/2$. The previously mentioned folding in property of the DFT means that we only need to zero frequencies to the midpoint of the frequency index, as the values to the right of $N/2$ mirror those to the left. This symmetry is the reason for the high and low values of k being set to 1.

The green signal in Fig. 8.6 is a corrupted sample of $f(t) = \cos(2\pi t) + \sin(4\pi t)$. There are $N = 2^{10} = 1024$ evenly spaced samples for $0 \leq t \leq 1$, and the noise was normally distributed with mean 0 and standard deviation 0.4. The first 20 elements of the FFT of the corrupt sequence are depicted in Fig. 8.7. The fundamental frequencies occur within the first 5 values of k, and we decide to remove all information whose index exceeds $k_{\max} = 10$. The red signal in Fig. 8.6 is the result of the filter, i.e. the red signal is

$$\text{iff(fft(x) .* H)}.$$

The true signal of $f(t)$ is plotted in blue under the approximating red signal, and the near perfect agreement between the red and blue signals indicates that our low pass filter has removed the high frequency noise and decisively identified the true signal.

Filter design is often easier in the frequency domain than in the time domain because the standard dot product naturally models the filter's action. A high pass filter maintains high frequencies but removes others, and a band pass filter keeps frequencies within a band but removes those outside the

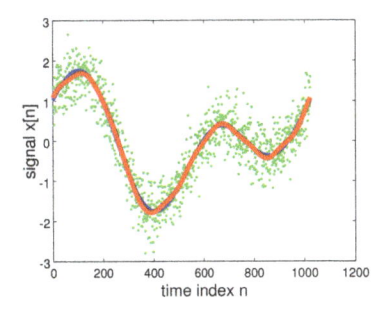

8.1 Signal Processing and the Discrete Fourier Transform 325

Fig. 8.6 The green signal is a randomly corrupted sample of $f(t) = \cos(2\pi t) + \sin(4\pi t)$. The red signal is the result of a low pass filter that only maintains the first 10 frequency indices. The blue curve under the red curve is the true signal of $f(t)$.

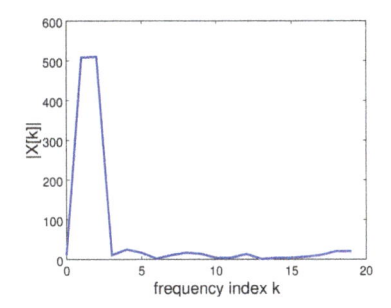

Fig. 8.7 The first 20 elements of the FFT of the corrupted signal. The only meaningful frequency indices are less than 5.

band. High, low, and band pass filters are "ideal" because they act exactly as intended in the frequency domain. However, their inverse FFTs are impossible to implement perfectly in the time domain, and approximating them in the time domain requires technical skill and artistic talent.

8.1.5 Exercises

1. Explain why padding signals with zeros allows us to calculate an aperiodic convolution with a cyclic convolution of the padded signals.
2. Stereo speakers often include a crossover network that splits a signal into different frequencies. The individual signals are then directed to speakers specifically designed for their frequency range. Write a function that splits a sound stream into low, medium, and high frequencies. Use the MATLAB functions `record` and `play` to test your crossover network.
3. One of the earliest applications of Fourier analysis was the study of tides. Indeed, Lord Kelvin built a physical machine to decompose a tidal signal into its constituent frequencies. Tidal data is widely available at numerous weather sites. Collect tidal data at a location of interest and use the `fft` command to identify the main frequency components.
4. The solar cycle is a natural semi-periodic event that helps explain the frequency of solar flares. Identify the fundamental frequencies of a historical record of flare activity. Such data is widely available. What is your best estimate of the period of the solar cycle?

5. Prove the first conclusion of Theorem 17.
6. The DFT extends to two dimensions by acting consecutively on rows and then columns (some fancy indexing is required). Write a program that accepts a gray scale image and then uses a low pass filter to remove high frequency noise. Experiment with different cutoff values and various amounts of random noise. The two-dimensional FFTs in MATLAB and Octave are `fft2` and `ifft2`.

8.2 Radiotherapy

Cancer has three primary treatment modalities, those being surgical, in which tumors are physically removed; pharmaceutical, in which cancer cells are treated with chemotherapeutic drugs; and radiobiological, in which cancerous tissues are targeted by radioactive sources. The radiobiological modality encompasses numerous treatment techniques such as proton therapy, brachytherapy, photodynamic therapy, and external beam therapy. Treatment design for these radiobiological techniques involves customizing treatments to patients to heighten their therapeutic benefits. We consider a linear model that estimates the radiobiological dose received by a patient during a treatment modality called intensity modulated radiotherapy (IMRT).

8.2.1 A Radiobiological Model to Calculate Dose

The use of external radioactive sources to treat cancer originated with Conrad Roentgen's discovery of x-rays in 1895, which were soon thereafter used by the physician Leopold Freund to treat hairy moles. The use of ionizing radiation to treat unwanted and potentially dangerous tissues was thus born. Modern external beam radiotherapy is a highly successful, non-invasive cancer treatment in which beams of radiation emanate from a source external to a patient. Beams are focused on the cancerous region(s) from various angles, delivering anatomical dose as they progress through the anatomy, mostly along their central trajectories. Dose is measured in units of Grays (Gy), with a Gy being a joule per kilogram. The mathematical model we develop is linear because anatomical dose is proportional to the time under which a patient is exposed.

The modern beams of IMRT are created by a linear accelerator, which is housed in a gantry capable of rotating around a patient. Beams are polymorphic in that they are amalgamations of multiple energies. The spectral properties of the beam are typically flattened as a treatment unit undergoes a commissioning protocol to ensure a trusted and uniform energetic profile. A common beam used for numerous cancers has an average energy of 6 megavolts (MV), although most accelerators can adjust this upwards several fold to 18 MV. Increased energies skew the maximal dose rate deeper into the anatomy and can be advantageous in some cases.

A beam's geometry can be modulated to help shield parts of the anatomy. Suppose you were to look at a patient along the beam's trajectory, a visual pathway called the beam's-eye-view. If a beam's cross-sectional field is rectangular, then you would only see the tissue within that rectangle as the beam progressed through the anatomy. The cancerous target, or at least some portion of it, should be within your purview, but other non-cancerous and life-critical tissues might also be within the beam's-eye-view. The non-cancerous organs are called organs-at-risk (OARs) because they are likely to receive radiation in an attempt to treat the cancerous target, and if they receive too much, then serious side effects or even death might be triggered. We can shield OARs from radiation by partially or completely blocking portions of the beam to better contour the regions of the target that are treatable without unduly harming the intersecting OARs.

Shielding was once done with wedges of heavy metal that altered the energy gradient across the beam, with the thick portion of the wedge nearly blocking all radiation and the thin part shielding little. Modern treatment facilities instead use a collimator to shape the beam. A collimator is composed of multiple plates, commonly called leaves, that can open or close to create complex shapes, and non-binary modulation is achieved by alternating the shape while exposing the patient to the beam. For example, a collimator could originally conform to the external shape of the target as an initial dose is delivered, and it could then adjust to shield OARs as additional dose is delivered. Continuing with this process results in a varied and non-uniform dose to the target, which is generally a bad outcome, but the OARs would hopefully be spared a complicating dose.

Target dose heterogeneity is overcome by moving the gantry to another location to deliver dose along a different trajectory. The dose profile can again vary through a sequence of collimator adjustments to limit dose to any intersecting OARs. The ultimate goal is to have the varied target doses from the different gantry locations accrue into a uniform, tumoricidal dose over the target while the threatened non-cancerous tissues are simultaneously spared. This technique is called step-and-shoot, with each step being a gantry movement and each shot being the delivery of radiation through an individual collimator shape. A two-dimensional schematic is seen in Fig. 8.8.

The broad beam is divided into sub-beams by the leaves of the collimator. These sub-beams are called pencils, and our objective is to construct a linear model that maps pencil strengths into anatomical dose. The unit of pencil strength is called fluence, which equates to exposure time, although this re-

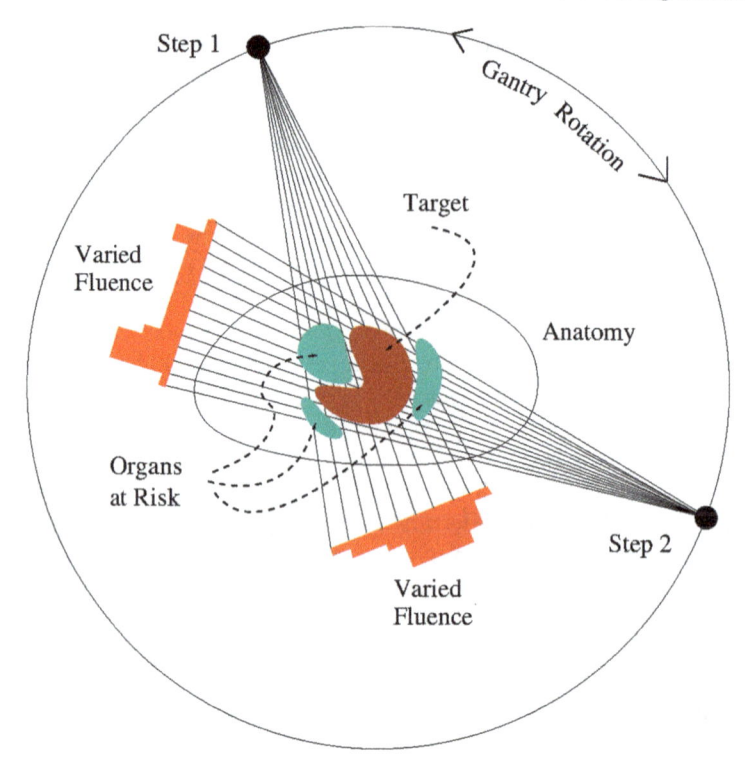

Fig. 8.8 A two-dimensional schematic of an IMRT treatment.

lationship is machine dependent. Consider the varied dose profile of the first gantry location, i.e. Step 1, in Fig. 8.8. The most outside pencils have minute fluences because they pass through OARs as they barely intersect the target. The fluence values increase through the more centralized pencils because they treat the target without as much damage to the OARs. The fluences of Step 2 are similarly interpreted. The high fluence values of Step 2 are associated with pencils covering portions of the target left under treated by Step 1, and combined, the target receives a higher and more uniform dose than could be achieved by either of the individual steps.

The patient in Fig. 8.8 is lying on a "couch" made of a low density material that leaves a beam mostly unaffected. The patient's feet would be thrusting out of this image toward the reader's face. The couch can be raised or lowered, moved in or out, and it can be rotated within the horizontal plane in which it rests. The two-dimensional treatment illustrated is reasonable in some clinical settings since a patient can be treated with a progression of two-dimensional treatments, each representing a three-dimensional slice of the anatomy. This modality is called tomotherapy. However, the general IMRT framework adds the flexibility to treat the patient from most any three-dimensional position on a sphere surrounding the patient, although some limitations exist to ensure that the couch does not intersect the rotation of the gantry. Steps often lie on a few great circles to facilitate timely and accurate treatments.

We use three indices to describe our dose model. Let a index the possible gantry positions. In a two-dimensional model a would index a collection of angles on a great circle circumscribing the patient. If we would rather consider a full three-dimensional model in which couch rotations were permitted, then a would more completely index the locations on the sphere about the patient, i.e. two angles in spherical coordinates. The beam's pencils are indexed by i, and their number and geometry are decided by the collimator. The image of the patient's anatomy is divided into pixels, say, for example, by imposing a mesh over the images. These pixels are indexed by p. The size of the pixels can vary depending on the type of tissue and its importance to the dose calculations. Fine meshes give low volume pixels and increase our computational trust, but they also lead to larger and more computationally cumbersome problems. The cumulative pixel dose is assumed to be amassed at the center of each pixel, a location called a dose-point.

Let $D_{p,a,i}$ be the grays delivered to anatomical point p along pencil i from gantry position a per unit fluence. The realistic modeling assumption that anatomical dose scales with fluence, i.e. that dose scales with the length of exposure, interprets $D_{p,a,i}$ as the rate at which dose is absorbed at anatomical position p per unit fluence of pencil i from gantry position a. Three different modeling rubrics to estimate $D_{p,a,i}$ are common, those being pencil beam models, convolution models, and simulation models. All have been used in commercial systems and have helped treat patients, and all have been approved by the Federal Drug Administration. The simulation models are the gold standard, but these are rarely used in clinical practice because they require lengthy calculations. Convolution models are the workhorses of the industry as they nicely combine moderate efficiency with accuracy, especially with regard to abrupt changes in anatomical density. Pencil beam models provide the most efficient calculations, but they are less accurate if anatomical densities vary. Pencil beam models work reasonably well in many head cases due to the uniformity of tissue. We only consider a pencil beam model since they are the most straightforward to calculate.

Beams attenuate as they progress through the anatomy, meaning that they lose energy as dose is absorbed and as radiation scatters outside the beam's trajectory. Maybe somewhat surprising is the fact that a beam's maximal dose rate occurs well within the patient, after which the beam begins to decay due to the attenuation process. The maximal dose rate is approximately 1.5 cm into the anatomy for a 6 MV beam, but this depth increases for higher energy beams. The dose rate at the surface is approximately 60% of the maximal dose rate, and we linearly scale dose over the first 1.5 cm to achieve this value. A mathematical model that accounts for dose build-up in the shallow depths less than 1.5 cm and then attenuates for depths beyond 1.5 cm is

$$D_{p,a,i} = \begin{cases} \left(P_0 e^{-\mu(d-1.5)}\left(1-e^{-\gamma r}\right) + \frac{r d \alpha_d}{r+1.5}\right) \cdot \text{ISF} \cdot \text{O}, & d \geq 1.5 \\ \left(0.4\frac{d}{1.5} + 0.6\right)\left(P_0 \left(1-e^{-\gamma r}\right) + \frac{1.5\, r \alpha_d}{r+1.5}\right) \cdot \text{ISF} \cdot \text{O}, & 0 \leq d < 1.5. \end{cases}$$

The input to the model is the triple (p, a, i), and from these indices we calculate the depth d of the anatomical position p. Depth is calculated along the central axis of the pencil as shown in Fig. 8.9. The remaining parameters are tailored to the specific machine and/or to the beam geometry and energy. The primary dose contribution for depths of at least 1.5 cm is $P_0 e^{-\mu(d-1.5)}$, where P_0 depends on the treatment machine and μ is an energy parameter that explains the primary contribution's attenuation as depth increases. We use the experimentally validated values of $P_0 = 0.873$ and $\mu = 0.0469$ for a 6 MV energy with square pencils of width 3 mm. So r is assumed to be 1.5 mm in Fig. 8.9, and each orthogonal cross section of the pencil is square.

The percentage of the entire beam's radiation contained within a pencil is $1 - e^{-\gamma r}$, where γ is experimentally dependent on the machine and r is a radial component of the pencil. If pencils are assumed to be conical, then the value of r is the distance illustrated in Fig. 8.9, but if pencils are instead square, which we assume, then the value of r is replaced by $1.122\, r$. Hence, for our model we use $1.122\, r = 1.122 \cdot 1.5 = 1.6830$. The radial component of a pencil is measured at the distance of the gantry's rotation, i.e. a distance of l_{gc} from the gantry, and all pencils are assumed to have the same radial component. The value of γ for the beam under consideration is 5.2586.

The gain from the scatter of nearby pencils is modeled by the term $r\, d\, \alpha_d/(r+1.5)$. The parameter r is the same radial component as just described, and the parameter α_d experimentally satisfies

$$\alpha_d = 0.1299 - 0.0306 \ln(d).$$

The final two multipliers are called the inverse square factor (ISF) and the off axis factor (O). The ISF is the square of the ratio of l_{gc} to l_{gp},

$$\text{ISF} = \left(\frac{l_{gc}}{l_{gp}}\right)^2.$$

The off axis factor, O, adjusts dose with regard to energy lost due to orthogonal scatter. This factor is machine dependent and is calculated by linearly interpolating tabulated experimental data. The distance o in Fig. 8.9 is the minimum distance from point p to the pencil's central trajectory. The value of o is scaled to use the tabulated data defining the off axis factor O, which is assessed as a function of

8.2 Radiotherapy

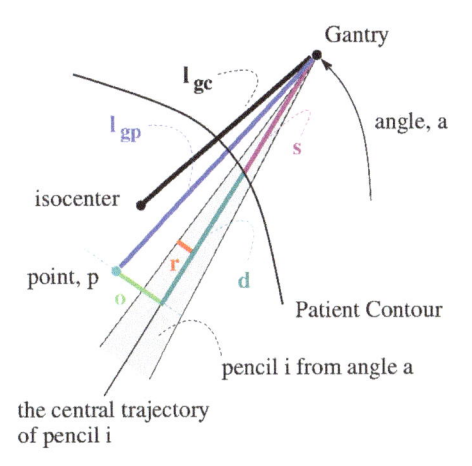

Fig. 8.9 An illustration of the distances used to calculate dose.

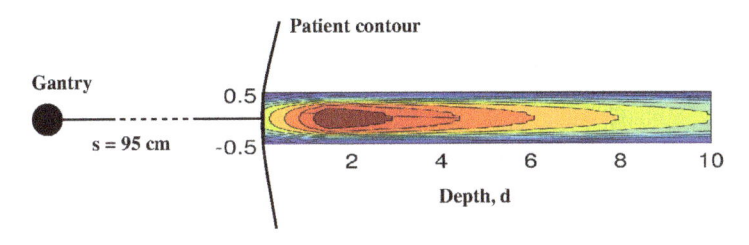

Fig. 8.10 A 1 cm profile of a 3 mm pencil from a 6 MV beam.

$$\psi = \frac{o \cdot l_{gc}}{s + d}, \tag{8.4}$$

see Exercise 1.

Figure 8.10 depicts a beam's profile. The energy is 6 MV, and the pencil is assumed to be square with a 3 mm width. The distance from the gantry to the patient's anatomy is 95 cm, and the distance from the gantry to the isocenter is 1 m. Hence the isocenter lies along the pencil's central trajectory at a depth of 5 cm. The maximal dose rate is visible at a depth of 1.5 cm. Although the pencil is only 3 mm wide at a depth of 5 cm, a path of 1 cm is displayed to illustrate the breadth of scattering. A well-tuned model is accurate to within 5% for most beam geometries as long as tissue density is homogeneous, although accuracy suffers as the radius diminishes.

The dose coefficients $D_{p,a,i}$ are arranged to form a matrix D, where rows are indexed by dose points and columns by pencils. A typical arrangement is

$$D = \begin{array}{c} \text{pencil} \rightarrow \\ \\ \\ \\ \\ \end{array} \begin{array}{c} \overbrace{\begin{array}{cccc} 1 & 2 & \cdots & q \end{array}}^{\text{angle 1}} \quad \cdots \quad \overbrace{\begin{array}{cccc} 1 & 2 & \cdots & q \end{array}}^{\text{angle n}} \\ \left[\begin{array}{cccc|c|cccc} D_{1,1,1} & D_{1,1,2} & \cdots & D_{1,1,q} & \cdots & D_{1,n,1} & D_{1,n,2} & \cdots & D_{1,n,q} \\ D_{2,1,1} & D_{2,1,2} & \cdots & D_{2,1,q} & \cdots & D_{2,n,1} & D_{2,n,2} & \cdots & D_{2,n,q} \\ \vdots & \vdots & & \vdots & & \vdots & \vdots & & \vdots \\ D_{m,1,1} & D_{m,1,2} & \cdots & D_{m,1,q} & \cdots & D_{m,n,1} & D_{m,n,2} & \cdots & D_{m,n,q} \end{array} \right] \begin{array}{c} 1 \\ 2 \\ \vdots \\ m \end{array} \\ \text{dose point} \uparrow \end{array}$$

Here we assume that each beam is divided into q pencils and that the gantry can be positioned at n angles. The patient image is divided into m pixels.

8.2.2 Treatment Design

The dose matrix D defines the linear transformation $x \mapsto Dx$, where x is a vector of fluences, called a fluence map, indexed in the same way as the columns of D. The resulting output Dx is the anatomical dose upon treatment with x. So, the transformation $x \mapsto Dx$ explains how energy from the gantry, encoded by the fluence map x, is deposited into the anatomy. The units of x are fluence, the units of D are Gy per fluence, and the units of Dx are Gy.

Treatment design begins by partitioning the anatomy by tissue type. Let T be the collection of targeted dose points, C be the collection of dose points within the OARs, and N be the remaining dose points of the normal tissue. Further let D_T, D_C, and D_N be the respective rows of D as indicated by the set subscript. The maps $x \mapsto D_T x$, $x \mapsto D_C x$, and $x \mapsto D_N x$ then result in the doses for the target, the OARs, and the remaining normal tissues under treatment x.

The overriding goals of treatment design are to find a fluence map x for which

- the target receives a uniform, tumoricidal dose,
- the dose to the OARs is limited and is as small as possible, and
- normal tissues are not over irradiated, and in particular, they should not contain "hot-spots."

Treatment design depends on a prescription, which is a collection of sought-after restrictions on the dose. For example, a physician might ask for the target to receive 70 Gy while the surrounding OARs receive less than 45 Gy. The target dose further defines the context of a hot-spot, which is typically any 1 cm cube that receives at least 110% of the target dose. For a two-dimensional treatment we interpret the definition of a hot spot as a 1 cm square that receives at least 110% of the target dose.

8.2 Radiotherapy

The mathematical counterparts of the treatment goals are to find a fluence map x satisfying

$$D_T x \approx P_T$$
$$D_C x \leq P_C$$
$$D_N x \leq P_N, \text{ and}$$
$$x \geq 0,$$

where P_T is a prescription vector of tumoricidal doses for the target, P_C is a prescription vector of upper bounds for the OARs, and P_N is a prescription vector of upper bounds for normal tissues. We often interpret our desire to treat the cancer as $D_T x \geq P_T$ to guarantee a tumoricidal dose. Indeed, we often perceive that we want to maximize $D_T x$, although this sentiment requires consideration since optimizing a vector is ambiguous. The stated approximation is the biologically correct desire, as over treating the target has undesirable effects. Normal tissues can be restricted like OARs but with (typically) higher bounds, or they can be somewhat ignored unless a hot spot is detected. The nonnegativity of fluence is a strict requirement as it is impossible to draw radiation out of a patient.

The less complicated beam modulation of early treatment systems readily lent itself to reasonable estimations of appropriate fluence maps. In fact, note that only a single value is needed per beam in the absence of modulation. Early design tools asked a treatment designer to select angles and fluence maps, after which the anatomical dose was calculated. The candidate treatment was then scrutinized to decide if it was satisfactory. If so, then treatment could begin. If not, then the designer would update the original guess and scrutinize the new candidate treatment. This trial-and-error process would repeat until a satisfactory treatment was found. Treatment design regularly took hours per patient, including the tedium of delineating anatomical structures on CAT scans.

The modern ability to modulate the beam has complicated human design because (tens of) thousands of fluences need to be tuned to customize a treatment per patient. Computational models were thus developed to aid planners as they worked to optimize treatments that could leverage the new technology. Some of the trial-and-error process has remained. For instance, planners still navigate the selection of angles, i.e. the gantry positions, upon which a treatment is designed. Once a candidate collection of angles is posited, a computational model optimizes the associated fluence maps. The candidate is then assessed, and if needed, the collection of angles is updated.

Assume D and the associated sub-matrices D_T, D_C, and D_N correspond with a collection of gantry positions selected by a treatment designer. A linear optimization model that expresses the treatment goals is

$$\left.\begin{aligned}
\min \quad & \lambda_T \sum_i \gamma_i + \lambda_C \sum_i \rho_i + \lambda_N \sum_i \sigma_i \\
\text{such that} \quad & \\
& P_T \leq D_T x \leq P'_T + \gamma \\
& D_C x \leq P_C + \rho, \\
& D_N x \leq P_N + \sigma, \\
& x \geq 0 \\
& \gamma \geq 0 \\
& \rho \geq -\eta\, P_C, \\
& \sigma \geq 0.
\end{aligned}\right\} \quad (8.5)$$

The inequalities of the constraints are componentwise. The prescription is the vector of lower bounds on the target, P_T, the vector of upper bounds on the target, P'_T, the vector of upper bounds on the OARs, P_C, the vector of upper bounds on the normal tissue, P_N, and the scalar η, which determines the percentage of unpenalized critical tissue dose. The variables of the problem are the fluence map, x, the overdose to the target and normal tissues, γ and σ, respectively, and the over/under dose to the OARs, ρ. The components of γ, ρ, and σ help assess dose at the corresponding dose point. For example, the p-th element of $D_C x$ is the dose delivered to dose point p. If this dose exceeds $[P_C]_p$, then ρ_p is forced to increase beyond 0, which means that dose point p will receive more dose than is desired.

The objective function is

$$\lambda_T \sum_i \gamma_i + \lambda_C \sum_i \rho_i + \lambda_N \sum_i \sigma_i,$$

and the problems seeks x, γ, ρ, and σ to minimize this total. The second summation accumulates the total deviation from the prescription for the OARs, and since the goal is to minimize the total summation, the model seeks to reduce over irradiating the OARs. The elements of γ and σ are similarly interpreted. The lower bound of $-\eta P_C$, together with the minimization of the objective, expresses an indifference to OAR dose once it falls below $(1-\eta)P_C$, where η is a modeling parameter between 0 and 1.

Each summation in the objective is weighted by a lambda scalar. Deciding these scalars is non-trivial because their relational value is often unclear. Such weights are typically normalized, and it is generally reasonable to assume

$$\lambda_T = \frac{w_T}{|T|}, \quad \lambda_C = \frac{w_C}{|C|}, \quad \text{and} \quad \lambda_N = \frac{w_N}{|N|} \quad \text{with} \quad w_T \geq w_C \geq w_N,$$

where $|T|$, $|C|$, and $|N|$ are the numbers of dose points of each tissue type. The w scalars subsequently express penalties for deviations away from the prescription per dose point.

8.2 Radiotherapy

Linear programs like (8.5) can be solved in MATLAB and Octave or with your own code from Sect. 4.2.1. MATLAB uses the command linprog in the optimization toolbox, and Octave uses the freeware GLPK. The problem has to be formatted into a specific structure to use these optimizers. As an example, to use linprog the optimization problem in (8.5) should be re-expressed so that every inequality becomes a less-than-or-equal constraint with all variables on the left-hand side. Let

$$A = \begin{bmatrix} -D_T & 0 & 0 & 0 \\ D_T & -I & 0 & 0 \\ D_C & 0 & -I & 0 \\ D_N & 0 & 0 & -I \end{bmatrix}, \quad b = \begin{pmatrix} -L_T \\ U_T \\ U_C \\ U_N \end{pmatrix}, \quad \text{and } c = \begin{pmatrix} 0 \\ \lambda_T e \\ \lambda_C e \\ \lambda_N e \end{pmatrix},$$

where e is a vector of ones whose length is decided by the context of its use. For example, the length of the vector e multiplied by λ_T is the number of targeted dose points. The sizes of the zero and identity submatrices of A are also decided by the contexts of their use. If we let

$$z = \left(x^T, \gamma^T, \rho^T, \sigma^T\right)^T,$$

then the constraints of (8.5), less the variable bounds, are the same as

$$Az \leq b.$$

If we further let

$$lb = \left(0^T, 0^T, -\eta U_C^T, 0^T\right)^T,$$

then problem (8.5) is succinctly,

$$\min\{c^T z \,:\, Az \leq b,\, z \geq lb\},$$

which can be solved in MATLAB as

[z, optVal] = linprog(c, A, b, [], [], lb, []).

If we had had equality constraints, say as expressed by the matrix system $Mz = m$, then the first two empty arguments would have been M and m, respectively. The last empty argument could have been used to impose upper bounds on the variables. GLPK has a similar, albeit more flexible, syntax.

We use the linear program in (8.5) to decide a best, equally spaced, 8 beam treatment. The search iterates over 5° rotations of the angles $k\pi/4$, for $k = 0, 1, \ldots, 7$. The pencils are assumed to be 3 mm square, with each beam having 21 pencils. The other parameters of the beam agree with those of the last section:

Energy	P_0	μ	r	γ	α_d
6 MV	0.873	0.0469	1.683	5.2586	$0.1299 - 0.0306 \ln(d)$

The two-dimensional profile of the patient is assumed to be circular with a radius of 10 cm, which approximates a large hat size. The pixelated image places dose points on a 2 mm square grid throughout the anatomy. The isocenter is located at the center of the patient, and the gantry's radius of rotation is 100 cm.

We consider a target that has grown around an OAR at the center of the patient geometry. The geometry crudely approximates how a cancer might wrap itself around parts of the brain stem. The goal for the target is to deliver at least 70 Gy but no more than 95 Gy. The OAR is to receive no more than 45 Gy, and an OAR dose of less than 25 Gy is not considered harmful. The remaining normal tissue has a desired upper bound of no more than 100 Gy. The prescription vectors of (8.5) are thus

$$P_T = 70\,e, \quad P'_T = 95\,e, \quad P_C = 45\,e, \quad P_N = 100\,e, \text{ and } \eta = 25/45,$$

where e is again a vector of ones. The objective weights are experimentally decided to be

$$\lambda_T = \frac{2000}{|T|}, \quad \lambda_C = \frac{1800}{|C|}, \text{ and } \lambda_N = \frac{100}{|N|}.$$

The magnitudes of the numerators could have theoretically been reduced, but the sizes of T, C, and N would have rendered the weights quite small if the numerators had instead been something like 2, 1.8, and 0.1. Such weights should be adjusted per case to achieve reasonable outcomes.

The quality of each treatment is measured by the objective function, which is negative for each of the 9 rotations of the 8 equally spaced beams. The only possible way for the objective to be negative is for the average dose to the OAR to be under its desired upper bound of 45 Gy, i.e. the only possible negative contribution to the objective is from ρ. From this perspective all treatments delivered a tumoricidal dose while sufficiently limiting damage to the OAR. The best rotation has a 5° shift, and the worst has a 25° shift. Figures 8.11 and 8.13 are contour plots of the optimized treatments for the best and worst rotations (see also Figs. 8.12 and 8.13).

Treatment planners do not measure quality solely on a numeric value like that of our objective. Instead, treatments are reviewed by inspecting the dose as overlayed on CAT scans. The combined images allow a planner to inspect the geometry and location of the irradiated tissues. The contour plot in Fig. 8.11 shows high dose values over the target and low dose values over the OAR. The image further shows tumoricidal dose values outside the target, e.g. just above the target on the left side. The color indicates a similarity between the target dose and the high doses within the normal tissue. The contour plot of the worst rotation in Fig. 8.13 similarly depicts a uniform tumoricidal dose over the target, but the color over the target is less red due to the much higher dose below the OAR. The conclusion is that an area of the normal tissue is receiving a much higher dose than is the target, which is generally an unfavorable outcome.

8.2 Radiotherapy

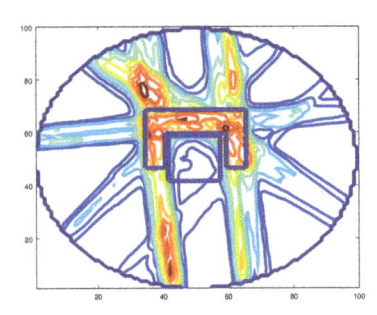

Fig. 8.11 A contour plot of the dose (Gy) for the best optimized treatment overlayed on the delineated anatomical structures.

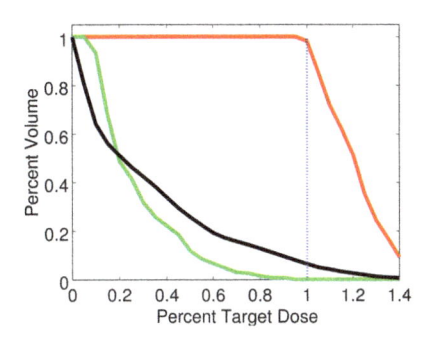

Fig. 8.12 A dose-volume histogram for the best treatment shown in Fig. 8.11. The target curve is red, the OAR is green, and the normal tissue is black.

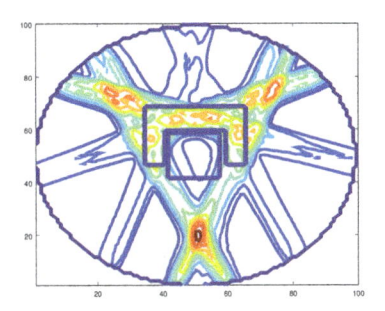

Fig. 8.13 A contour plot of the dose (Gy) for the worst optimized treatment overlaid on the delineated anatomical structures.

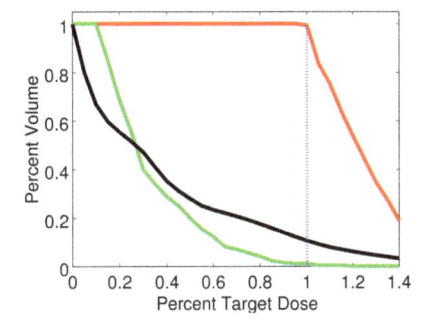

Fig. 8.14 A dose-volume histogram of the worst treatment in Fig. 8.11. The target curve is red, the OAR is green, and the normal tissue is black.

Large three-dimensional targets require much from the planner during assessment. Consider the problem of looking at multiple two-dimensional dose maps in an attempt to accurately interpret the full three-dimensional dose. Such assessments are obviously non-trivial. Modern design systems compensate by building three-dimensional dose images, but the task remains challenging. A routine alternative is to use a dose-volume histogram (DVH). DVHs for the best and worst treatments are shown in Figs. 8.12 and 8.14. These graphs normalize all doses with respect to the target dose, which is 70 Gy in our case. A curve mapping percent dose to percent volume is created for each structure. A perfect curve for the target would remain at 100% volume until it reached 100% dose, after which it would plummet to 0% volume. A perfect curve for any other structure would fall sharply toward 0% volume as percent dose increased. The red curves in Figs. 8.12 and 8.14 favorably indicate near perfect adherence for the target, although the tail after 100% dose means that much of the target received more radiation than was desired. The green curves for the OAR show that only small percentages of the critical structure received elevated doses similar to that of the target, although the worst case has more tissue receiving the tumoricidal dose. The black lines similarly describe the dose to the normal tissues.

Dose-volume histograms are good visual tools that aid initial assessment. If the target curve falls too quickly or an OAR curve does not decline quickly enough, then the treatment can be discarded. A planner could otherwise initiate a careful inspection of the spacial location of the dose, either through a series of two-dimensional images or with an advanced three-dimensional tool. These more detailed investigations are needed because the spacial location of dose is unclear in a DVH.

The linear model in (8.5) is only one of many models that have been suggested for the optimization of IMRT treatments. Some of the alternatives are non-linear, either due to nonlinear terms in the objective or due to combinatorial variables (or both). All the optimization schemes rely on the linear model $x \mapsto Dx$, and it is this linear map that requires the approval of the Food and Drug Administration. Hence great care is required to test and verify the accuracy of the computed dose, Dx, and many computed treatments are verified by delivering them to a phantom prior to treating patients.

8.2.3 Exercises

1. Implement the pencil beam model in Sect. 8.2.1. Assume square 3 mm pencils of a 6 MV beam, and use the parameters as stated. Use the tabulated data OffAxisData, available at http://www.springer.com, to linearly interpolate the off axis factor, O, as a function of ψ, see (8.4). The first column of this file is an experimental sample of ψ, and the second is the corresponding experimental sample of O. Recreate the profile shown in Fig. 8.10 to verify your model.
2. Use your solution to Problem 1 to create a dose matrix D assuming the patient geometry is circular. The isocenter is assumed to be at the center of the circle, and the gantry's radius is assumed to be 100 cm. The mesh of dose points and the number of pencils should be adjustable. Test your model by selecting fluence maps and visualizing contours of the dose.
3. Design a scheme to encode anatomical structures, and use your scheme to divide the dose matrix from Problem 2 into D_T, D_C, and D_N. Design optimal treatments for various anatomical geometries by solving (8.5) with MATLAB or Octave. Be sure to investigate different prescriptions and objective weights. Evaluate your treatments with contour plots and DVHs.
4. Relax the reliance of Problem 3 on a circular patient. The primary difficulty is the need for a line search to calculate s in Fig. 8.9, which is required to calculate d.

    Further relax the assumption of Problem 3 that all beams have the same number of pencils. While the beam is divided into the same number of pencils independent of gantry position, only the pencils that reasonably strike the target should be considered during treatment design. Adjust your model so that only those pencils that intersect the target are used in the creation of D.

5. Several quadratic adaptations of the model in (8.5) have been suggested, with one being

$$\min \lambda_T \|D_T x - P_T\|^2 + \lambda_C \|D_C x - P_C\|_+^2 + \lambda_N \|D_N x\|^2 \text{ such that } x \geq 0.$$

The $+$ subscript on the norm indicates that negative components are set to zero in the calculation. For instance,

$$\|\langle 2, 0, -1, 1, -2 \rangle\|_+ = \|\langle 2, 0, 0, 1, 0 \rangle\| = \sqrt{5}.$$

Repeat Exercises 3 and 4 with this model. Which model do you prefer?

8.3 Aeronautic Lift

The airflow surrounding a moving wing exhibits complicated dynamics, and modeling and analyzing such airflows is an important aeronautic study. Analytic solutions are often impractical, and exact solutions are commonly approximated with numerical counterparts. We present a linear model that can reasonably approximate a wing's lift from its 2D profile. Our development is succinct and assertive in support of illustrating the linearity of the model. A course in aerodynamics or fluid flows would bolster our introductory development.

The modeling premise is to discretize a wing into numerous flat panels, each of which is associated with an individual lift estimate. These estimates then aggregate into a reasonable calculation for the entire wing. The panel method solves a linear system to calculate the individual lift estimates relative to the wing's geometry and other assumptions. An overview of the geometry and notation is illustrated in Fig. 8.15.

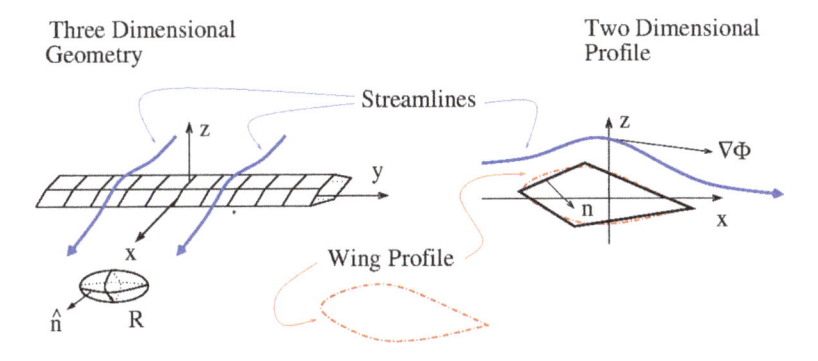

Fig. 8.15 An illustration of the geometry and notation associated with the panel method. The true wing geometry is approximated with a piecewise linear approximation.

8.3.1 Air Flow

Airflow is studied as a potential Φ, which is a function of the spacial coordinates so that the flow's velocity is $\nabla \Phi$. The units of Φ are m^2/s, i.e. velocity times distance, and the resulting units of $\nabla \Phi$ are m/s. A streamline of the flow is a parametric curve $r(t)$ such that

$$\frac{d}{dt} r(t) = \nabla \Phi.$$

Let m be the mass within a small control volume outside the wing, say R. The rate at which mass changes in the volume is

$$\frac{dm}{dt} = \frac{\partial}{\partial t} \int_R \rho \, dV - \int_{\mathrm{bd}(R)} \rho \left(\nabla \Phi \cdot \hat{n} \right) dS,$$

where $\hat{n}$ is an outwardly pointing normal, ρ is the density of the air, and $\mathrm{bd}(R)$ is the boundary surface of the control region R. The product $\rho \left(\nabla \Phi \cdot \hat{n} \right)$ is the rate at which mass enters (or leaves) the control region along the normal $\hat{n}$, and the expression for dm/dt ensures that the rate at which mass changes in the control region is the rate at which mass is exchanged across the region's boundary together with the rate at which the mass is changing (compressing or incompressing) in the volume.

We assume air is homogeneous and incompressible, an assumption that ensures a constant density with regard to space and time. The partial time derivative of the mass within the control region is then zero, i.e.

$$\frac{\partial}{\partial t} \int_R \rho \, dV = 0.$$

We further assume that mass is conserved so that $dm/dt = 0$, which gives

$$\frac{dm}{dt} = \rho \int_{\mathrm{bd}(R)} \left(\nabla \Phi \cdot \hat{n} \right) dS = 0.$$

We conclude that conservation of mass is guaranteed by an airflow with velocity vectors $\nabla \Phi$ satisfying

$$\int_{\mathrm{bd}(R)} \nabla \Phi \cdot \hat{n} \, dS = 0.$$

The divergence theorem from calculus re-expresses this surface integral as a volume integral in terms of the ∇ operator over the control volume, which gives

$$\int_R \nabla \cdot \nabla \Phi \, dV = \int_{\mathrm{bd}(R)} \nabla \Phi \cdot \hat{n} \, dS = 0.$$

The integrand on the left-hand side is the Laplacian of Φ and is often expressed as $\nabla^2 \Phi$. Since the integrals are zero for any control region, the in-

8.3 Aeronautic Lift

tegrands themselves must be zero, and in particular, we arrive at Laplace's equation,

$$0 = \nabla \cdot \nabla \Phi$$
$$= \langle \partial/\partial x, \partial/\partial y, \partial/\partial z \rangle \cdot \langle \partial \Phi/\partial x, \partial \Phi/\partial y, \partial \Phi/\partial z \rangle$$
$$= \frac{\partial^2 \Phi}{\partial x^2} + \frac{\partial^2 \Phi}{\partial y^2} + \frac{\partial^2 \Phi}{\partial z^2}.$$

Laplace's equation follows from the assumption of incompressibility, but an irrotational assumption is needed to identify the flow in terms of a potential. A flow is irrotational if the curl of the velocity is zero,

$$\nabla \times \nabla \Phi = 0.$$

In the language of fluid dynamics, this equation states that there is no vorticity. A convenient interpretation is that air within the control region R does not spin as it progresses along the stream line.

There are two basic solutions to the equations that model an incompressible and irrotational air flow, i.e. there are two standard velocity potentials Φ that satisfy

$$\nabla \cdot \nabla \Phi = 0 \quad \text{and} \quad \nabla \times \nabla \Phi = 0.$$

These two solutions are sources (sinks) and doublets, denoted, respectively, by Φ_s and Φ_d. As a side comment, a doublet is a limiting conjunction of a source and sink, and hence, it is a "double" solution at a single location. We only work with doublets in the xz-plane since these solutions support our introductory calculation of lift for non-trivial wing geometries. The form of a doublet centered at (a, c) with strength μ in the direction of n is

$$\Phi_d^\mu(x, z) = \frac{-\mu\, n \cdot \langle x - a, z - c \rangle}{2\pi \|\langle x - a, z - c \rangle\|^2} = \mu\, \Phi_d(x, z).$$

The superscript μ indicates the doublet's strength, and the unit of μ can be inferred from the definition. Since both the denominator and $n \cdot \langle x - a, z - c \rangle$ are in terms of m², the strength μ shares the m²/s units of the left-hand side. We remove the superscript to express the doublet in the factored form $\mu\, \Phi_d(x, z)$, where $\Phi_d(x, z)$ is unitless.

The geometry of our forthcoming numerical model is conveniently manipulated so that all doublets point in the z-direction. A vertical doublet in the z-direction has a strength vector of $\mu \langle 0, 1 \rangle$, and the expression of Φ_d^μ reduces to

$$\Phi_d^\mu(x, z) = \frac{-\mu\,(z - c)}{2\pi \|\langle x - a, z - c \rangle\|^2}.$$

The velocity vectors in this case are

$$\nabla \Phi_d^\mu(x, z) = \frac{\mu \langle 2(x-a)(z-c),\, (z-c)^2 - (x-a)^2 \rangle}{2\pi \|\langle x-a,\, z-c\rangle\|^4}. \tag{8.6}$$

Assume a doublet at the origin for illustrative purposes, i.e. $a = c = 0$. A useful observation is that the streamlines defined by these velocities satisfy the implicit equation

$$\Psi \,\|\langle x, z\rangle\|^2 = x$$

for a constant Ψ. To substantiate this fact, let $r(t) = \langle x(t), z(t)\rangle$ be a streamline. Then Ψ as a function of t is

$$\Psi(t) = \frac{x(t)}{\|\langle x(t), z(t)\rangle\|^2}.$$

From the chain rule in Calculus III we have

$$\begin{aligned}\frac{d}{dt}\Psi(t) &= \nabla \Psi(x, z) \cdot \frac{d}{dt} r(t) \\ &= \nabla \Psi(x, z) \cdot \nabla \Phi_d^\mu(x, z) \\ &= \left(\frac{\langle z^2 - x^2,\, -2xz\rangle}{\|\langle x, z\rangle\|^4}\right) \cdot \left(\frac{\mu \langle 2xz,\, z^2 - x^2\rangle}{2\pi \|\langle x, z\rangle\|^4}\right) \\ &= 0.\end{aligned}$$

We conclude that Ψ is constant along a streamline, and for this reason, Ψ is called a streamline function.

Streamlines associated with a doublet are easy to identify in polar form because

$$\Psi = \frac{x}{\|\langle x, y\rangle\|^2} = \frac{r\cos(\theta)}{r^2} = \frac{\cos(\theta)}{r} \;\Rightarrow\; r = \Psi \cos(\theta).$$

These are circles of radius $\Psi/2$ centered at $(\Psi/4, 0)$. Figure 8.16 illustrates some of the streamlines associated with a doublet centered at the origin in the z-direction.

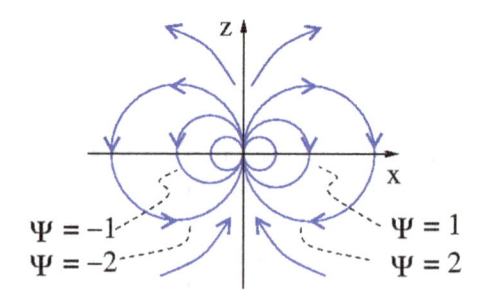

Fig. 8.16 Illustrative streamlines of a doublet centered at the origin pointing in the z-direction.

8.3.2 Flow Around a Wing

Our development of sources and doublets assumes an undisturbed airflow, but a wing disturbs the streamlines as it moves. Fortunately, the more complicated airflows around a wing can be modeled with integrated combinations of sources and doublets. Specifically, the streamline potentials around the wing are

$$\Phi(x,z) = \int_C (\mu\,\Phi_d - \sigma\,\Phi_s)\,dl + \Phi_f,$$

where Φ_f is the free-stream potential away from the wing, C is the curve defining the wing and its wake, the subscripts d and s indicate doublets and sources, and dl is the differential of arc length. The free-stream potential describes the airflow away from the wing's disturbance, and hence, $\nabla\Phi_f$ defines the streamlines of the airflow some distance from the wing.

Doublet solutions suffice to make reasonable approximations, and the model reduces to

$$\Phi(x,z) = \int_C \mu\,\Phi_d\,dl + \Phi_f.$$

The integral term requires a brief study to ensure a correct interpretation. Whereas Φ_d is a function of (x,z), these variables are assigned as arguments of the potential on the left-hand side. The variables of integration are thus not x and z. Instead, the doublet is located at position (a,c), and the integration is across the locations of the doublets. A common interpretation is to continually place doublets along the profile of the wing with continuous strength profile $\mu(a,c)$. The entire potential at position (x,z) is then the cumulative potential of the doublets over the wing's surface together with the free-stream potential.

Divide C into N linear pieces with the wake being one linear segment. Each linear segment corresponds with a rectangular panel of the full three-dimensional geometry. The wing's wake is an important modeling construct and is somewhat an extension of the wing itself. The wake need not be a single panel, but we make this simplifying assumption for our numerical study.

From the approximated wing geometry we have

$$\Phi(x,z) = \int_C \mu\,\Phi_d\,dl + \Phi_f \approx \left(\sum_{j=1}^N \int_{L_j} \mu\,\Phi_d\,dl\right) + \Phi_f,$$

where L_j is the j-th linear segment. If we assume that doublet strengths are constant over a panel, then each panel is assigned a constant strength μ_j and the result is

$$\Phi(x,z) \approx \left(\sum_{j=1}^N \mu_j \int_{L_j} \Phi_d\,dl\right) + \Phi_f, \tag{8.7}$$

The linear model approximating airflow is becoming evident. The integrals over the panels can be calculated based on a unit strength doublet, and since the free-stream potential is assumed to be known, our model of the potential around the wing is affine in the N strengths of the panel doublets.

A boundary condition is needed to finalize the streamline model. A common boundary assumption is

$$n \cdot \nabla \Phi(x, z) = 0$$

as (x, z) traces C, where n is normal to C, see Fig. 8.15. Since $\nabla \Phi$ is the velocity vector of the streamline, this orthogonality statement assumes that the velocity vectors lie along the tangents of the wing's geometry. This assumption coincides with how we might expect an airflow to "hug" a wing's profile as it flows around the wing. With the tacit assumption of continuity to support the interchange of integration and differentiation, the boundary condition for the i-th panel applied to (8.7) is

$$0 = n_i \cdot \nabla \Phi(x, z) \approx \sum_{j=1}^{N} \mu_j \underbrace{\left(n_i \cdot \int_{L_j} \nabla \Phi_d \, dl \right)}_{\text{influence coefficients}} + n_i \cdot \nabla \Phi_f, \qquad (8.8)$$

where n_i is normal to panel i. The factors being multiplied by μ_j are called influence coefficients, and they express how the combined disturbances of the panels reconcile the boundary condition on panel i.

The influence coefficients can be calculated numerically as line integrals by moving the dot product inside the integration. However, the geometry can also be transformed to a common orientation so that all integrals can be evaluated formulaically. The influence coefficients are then calculated by solving a linear system created upon reversing the geometric transformation.

We begin by translating and rotating the geometry so that the j-the panel is horizontal and centered at its midpoint with transformed normal $\langle 0, -1 \rangle$. Assume the panel has length $2d$ and spans the horizontal interval from $-d$ to d. Doublets of unit strength in the (outward) vertical direction are placed continually along the panel at locations $(a, 0)$ as a traverses from $-d$ to d. From (8.6) we have

$$\left[\int_{L_j} \nabla \Phi_d \, dl \right]_{p_j} = \frac{1}{2\pi} \int_{-d}^{d} \frac{\langle 2(x-a)z, \, z^2 - (x-a)^2 \rangle}{\|\langle x-a, z \rangle\|^4} \, da \qquad (8.9)$$

$$= \frac{1}{2\pi} \left\langle \frac{z}{\|\langle x+d, z \rangle\|^2} - \frac{z}{\|\langle x-d, z \rangle\|^2}, \, \frac{x-d}{\|\langle x-d, z \rangle\|^2} - \frac{x+d}{\|\langle x+d, z \rangle\|^2} \right\rangle.$$

The subscript p_j on the left-hand side indicates that the integration is undertaken with regard to the transformed coordinates of panel j. The point of influence (x, z) is called a collection point, and the influence of panel j on the boundary condition over panel i is collapsed onto a single collection

8.3 Aeronautic Lift

point for the i-th panel. Deciding the location of the collection points is part of the modeling process, and we assume influence is collected at a panel's midpoint. As long as $i \neq j$, the coordinates of x and z can be replaced with the transformed midpoint of panel i. The resulting vector is rotated back to the original coordinates, after which the coefficient of μ_j for the i-th panel is calculated by computing the dot product with n_i.

A panel's influence on itself necessitates a special calculation. In this case $x = z = 0$, and the panel integral has a singularity as a passes through the origin. The panel integral with its own collection point is

$$\left[\int_{L_j} \nabla \Phi_d \, dl\right]_{p_j} = \frac{1}{2\pi} \int_{-d}^{d} \frac{\langle 0, -a^2 \rangle}{\|\langle -a, 0 \rangle\|^4} \, da = \frac{1}{2\pi} \int_{-d}^{d} \left\langle 0, \frac{-1}{a^2} \right\rangle da.$$

The improper integral of the second coordinate would have the value

$$\lim_{h \to 0^+} \int_{h}^{d} \frac{-1}{a^2} \, da + \lim_{q \to 0^-} \int_{-d}^{q} \frac{-1}{a^2} \, da = \lim_{h \to 0^+} \left(\frac{1}{d} - \frac{1}{h}\right) + \lim_{q \to 0^-} \left(\frac{1}{q} + \frac{1}{d}\right).$$

However, these limits approach negative infinity independent of how h and q approach 0 (from the right and left, respectively), which is impractical. The physical counter to the singularity is to perturb the collection point vertically so that $z \neq 0$, which is reasonable since the thickness of the wing's material would permit such an adjustment. The influence at the collection point is then calculated as z returns to zero, which is

$$\lim_{z \to 0} \left[\int_{L_j} \nabla \Phi_d \, dl\right]_{p_j}$$

$$= \lim_{z \to 0} \frac{1}{2\pi} \left\langle \frac{z}{\|\langle d, z \rangle\|^2} - \frac{z}{\|\langle d, z \rangle\|^2}, \frac{-d}{\|\langle d, z \rangle\|^2} - \frac{d}{\|\langle d, z \rangle\|^2} \right\rangle$$

$$= \left\langle 0, \frac{-1}{\pi d} \right\rangle.$$

This influence vector is in transformed coordinates, and as in the case with $i \neq j$, the influence coefficient is calculated by rotating the vector back to the original coordinate system and then forming the dot product with n_j. We denote the transformed vector back in standard coordinates by $[\langle 0, -1/(\pi d) \rangle]_S$, making the influence coefficient

$$n_i \cdot \left[\left\langle 0, \frac{-1}{\pi d} \right\rangle\right]_S.$$

The linear system defining the panel doublet strengths μ_j is nearly complete. List the panels counterclockwise from 1 to N as illustrated in Fig. 8.17; the N-th panel being the wake. Let

$$A_{ij} = \begin{cases} n_i \cdot \int_{L_j} \nabla \Phi_d \, dl, & i \neq j \\ n_i \cdot [\langle 0, -1/(\pi d) \rangle]_S, & i = j, \end{cases} \quad (8.10)$$

where i indexes the $N-1$ panels of the wing and j indexes the N collection points (both wing and wake). If $i \neq j$, then A_{ij} can be calculated by transforming the geometry with regard to the j-th panel, evaluating the integral by substituting the transformed coordinates of the i-th panel's collection point into (8.9), and then computing the dot product after returning to the original coordinates. From the approximation in (8.8), the boundary condition (approximately) imposes the linear system:

$$\sum_{j=1}^{N} A_{1j}\mu_j = -n_1 \cdot \nabla \Phi_f \quad \begin{pmatrix} \text{boundary condition over panel 1;} \\ \text{uses the first panel's collection point} \end{pmatrix}$$

$$\sum_{j=1}^{N} A_{2j}\mu_j = -n_2 \cdot \nabla \Phi_f \quad \begin{pmatrix} \text{boundary condition over panel 2;} \\ \text{uses the second panel's collection point} \end{pmatrix}$$

$$\vdots$$

$$\sum_{j=1}^{N} A_{(N-1)j}\mu_j = -n_{N-1} \cdot \nabla \Phi_f \quad \begin{pmatrix} \text{boundary condition over panel } N-1; \\ \text{use the last non-wake collection point.} \end{pmatrix}$$

The free-stream velocities are assumed to be known, and the right-hand sides of these equations are subsequently constant.

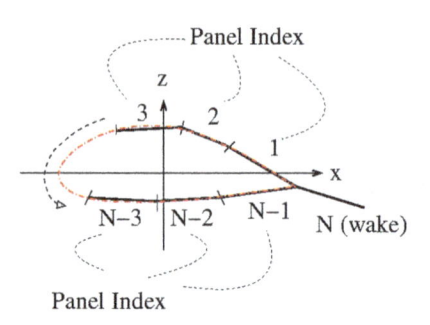

Fig. 8.17 Panels are indexed counterclockwise with the wake being the final panel.

The doublet strengths μ_j are not uniquely defined by the linear system because it has $N-1$ constraints and N variables, but a wake assumption can be added to give a square system. The most common wake assumption is called the Kutta condition, which assumes

$$\mu_N = \mu_1 - \mu_{N-1}.$$

8.3 Aeronautic Lift

The Kutta condition asserts that streamlines progress along the wake as they come off the trailing edge of the wing. The Kutta constraint can be added as an N-th equation, or all occurrences of μ_N can be replaced with $\mu_1 - \mu_{N-1}$ to give a square $(N-1) \times (N-1)$ system. We use the latter approach, and upon letting

$$A = \begin{bmatrix} A_{11} + A_{1N} & A_{12} & \cdots & A_{1(N-1)} - A_{1N} \\ A_{21} + A_{2N} & A_{22} & \cdots & A_{2(N-1)} - A_{2N} \\ \vdots & \vdots & & \vdots \\ A_{(N-1)1} + A_{(N-1)N} & A_{(N-1)2} & \cdots & A_{(N-1)(N-1)} - A_{(N-1)N} \end{bmatrix},$$

we conclude that the linear system

$$A\mu = b,$$

defines the doublet profile, where μ is the column vector with components μ_j and b is the column vector with components $-n_j \cdot \nabla \Phi_f$. The doublet strengths are uniquely decided as long as A is invertible, which is expected.

The difference between consecutive doublet strengths estimates the change in the velocity as the air moves around the wing, and hence, the sum

$$(\mu_2 - \mu_1) + (\mu_3 - \mu_2) + \ldots + (\mu_{N-1} - \mu_{N-2}) = \mu_{N-1} - \mu_1 = -\mu_N$$

approximates the sum of the velocities around the wing. Note that the Kutta condition gives the last equality. The integrated sum of the velocity around the wing in a counterclockwise direction is called the circulation and is denoted by Γ. The approximation above suggests that

$$\Gamma \approx -\mu_N,$$

where the units of circulation match that of μ and are m^2/s. The interpretation is that the strength of the wake doublet estimates the circulation around the wing.

The Kutta-Joukowski Theorem relates circulation to lift (L). The result states that

$$L = \Gamma v \rho,$$

where v is speed and ρ is the density of the air. The unit of lift as expressed on the right-hand side is (m^2/s)(m/s)(kg/m^3) = (1/m)(kg·m/s^2) = N/m. So the two-dimensional model results in the lifting force in Newtons per meter of wing span in the y direction. This lift calculation is called sectional lift. The result of the Kutta-Joukowski theorem as estimated by our model is

$$L \approx -\mu_N \, \|\nabla \Phi_f\| \, \rho. \tag{8.11}$$

8.3.3 Numerical Examples

An example with just two panels is considered before proceeding to a complete numerical study. Consider the two panels on the left of Fig. 8.18. We let $\overrightarrow{AB}$ and $\overrightarrow{BC}$ be the three-dimensional vectors $\langle -3, 0, 3/2 \rangle$ and $\langle -1, 0, -1 \rangle$, i.e. think of the y-axis as being orthogonal to the image. The inward facing normals of the panels in three dimensions are then

$$\frac{\langle 0, -1, 0 \rangle \times \overrightarrow{AB}}{\|\langle 0, -1, 0 \rangle \times \overrightarrow{AB}\|} = \langle -0.4472, 0, -0.8944 \rangle \text{ and}$$

$$\frac{\langle 0, -1, 0 \rangle \times \overrightarrow{BC}}{\|\langle 0, -1, 0 \rangle \times \overrightarrow{BC}\|} = \langle 0.7071, 0, -0.7071 \rangle.$$

The resulting two-dimensional normals are $n_1 = \langle -0.4472, -0.8944 \rangle$ and $n_2 = \langle 0.7071, -0.7071 \rangle$. The (oriented) angles between these normals and their transformed counterpart $\langle 0, -1 \rangle$ are

$$\theta_1 = \cos^{-1}(n_1 \cdot \langle 0, -1 \rangle) \approx 0.4637 \text{ and}$$
$$\theta_2 = -\cos^{-1}(n_2 \cdot \langle 0, -1 \rangle) = -\pi/4 \approx -0.7854.$$

The second angle is negated because our transformations assume counterclockwise rotations (notice that the second panel rotates clockwise through its reference angle to achieve the desired orientation).

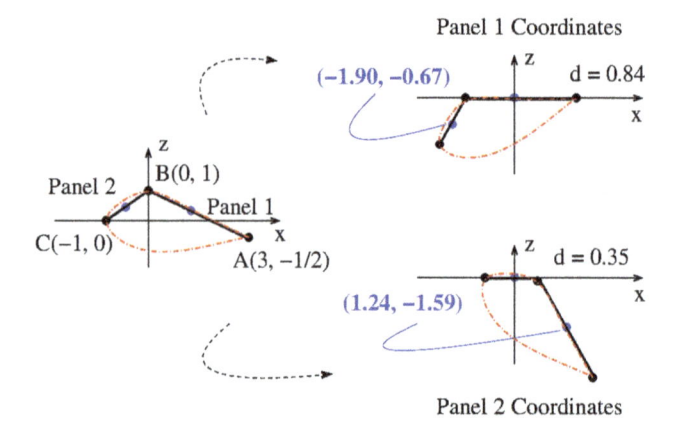

Fig. 8.18 Two panels on the left and coordinate transformations for the panels on the right.

The coordinate transformations are defined in terms of rotational matrices of the form,

8.3 Aeronautic Lift

$$R(\theta) = \begin{bmatrix} \cos(\theta) & -\sin(\theta) \\ \sin(\theta) & \cos(\theta) \end{bmatrix}.$$

Left multiplication by $R(\theta)$ rotates a vector counter clockwise an angle of θ if θ is positive, and clockwise if θ is negative. The midpoints of panels 1 and 2 are $(1.5, 0.25)$ and $(-0.5, 0.5)$, and the transformations from the original coordinate axes to the panel coordinates are, respectively,

$$R(\theta_1)\begin{pmatrix} x - 1.5 \\ z - 0.25 \end{pmatrix} = \begin{bmatrix} 0.8944 & -0.4472 \\ 0.4472 & 0.8944 \end{bmatrix}\begin{pmatrix} x - 1.5 \\ z - 0.25 \end{pmatrix}$$

and

$$R(\theta_2)\begin{pmatrix} x + 0.5 \\ z - 0.5 \end{pmatrix} = \begin{bmatrix} 0.7071 & 0.7071 \\ -0.7071 & 0.7071 \end{bmatrix}\begin{pmatrix} x + 0.5 \\ z - 0.5 \end{pmatrix}.$$

The transformation with θ_1 renders the first panel horizontal and centers it at its midpoint. The second transformation with θ_2 similarly transforms the second panel. The transformed midpoints of the other panels are noted in Fig. 8.18 and are calculated as

$$R(\theta_1)\begin{pmatrix} -0.5 - 1.5 \\ 0.5 - 0.25 \end{pmatrix} = \begin{bmatrix} 0.8944 & -0.4472 \\ 0.4472 & 0.8944 \end{bmatrix}\begin{pmatrix} -2.0 \\ 0.25 \end{pmatrix} = \begin{pmatrix} -1.90 \\ -0.67 \end{pmatrix}$$

and

$$R(\theta_2)\begin{pmatrix} 1.5 + 0.5 \\ 0.25 - 0.5 \end{pmatrix} = \begin{bmatrix} 0.7071 & 0.7071 \\ -0.7071 & 0.7071 \end{bmatrix}\begin{pmatrix} 2.0 \\ -0.25 \end{pmatrix} = \begin{pmatrix} 1.24 \\ -1.59 \end{pmatrix}.$$

The lengths of the first and second panels are $2d_1 = 3.3542$ and $2d_2 = 1.4142$, and hence $d_1 = 1.6771$ and $d_2 = 0.7071$. Substituting the transformed midpoint of the first panel into (8.9) with d_2 results in

$$\left[\int_{L_2} \nabla \Phi_d \, dl\right]_{p_2} = \langle 0.0499, -0.0190 \rangle.$$

Solving

$$R(\theta_2)\left(\int_{L_2} \nabla \Phi_d \, dl\right) = \left[\int_{L_2} \nabla \Phi_d \, dl\right]_{p_2}$$

$$\Leftrightarrow \begin{bmatrix} 0.7071 & 0.7071 \\ -0.7071 & 0.7071 \end{bmatrix}\left(\int_{L_2} \nabla \Phi_d \, dl\right) = \begin{pmatrix} 0.0499 \\ -0.0190 \end{pmatrix},$$

we find

$$\int_{L_2} \nabla \Phi_d \, dl = \langle 0.0487, 0.0219 \rangle,$$

and hence,

$$A_{12} = n_1 \cdot \int_{L_2} \nabla \Phi_d \, dl = \langle -0.4472, -0.8944 \rangle \cdot \langle 0.0487, 0.0219 \rangle = -0.0413.$$

An analogous calculation upon substituting the transformed midpoint of the second panel into (8.9) with d_1 shows that solving

$$R(\theta_1) \left(\int_{L_1} \nabla \Phi_d \, dl \right) = \left[\int_{L_1} \nabla \Phi_d \, dl \right]_{p_1}$$

$$\Leftrightarrow \begin{bmatrix} 0.8944 & -0.4472 \\ 0.4472 & 0.8944 \end{bmatrix} \left(\int_{L_1} \nabla \Phi_d \, dl \right) = \begin{pmatrix} -0.2055 \\ 0.0282 \end{pmatrix},$$

gives

$$A_{21} = n_2 \cdot \int_{L_1} \nabla \Phi_d \, dl = \langle 0.7071, -0.7071 \rangle \cdot \langle -0.1712, 0.1171 \rangle = -0.2039.$$

The diagonal elements are computed similarly, but in these cases we instead use the special limiting values in (8.10) to solve for vectors v_1 and v_2 that satisfy

$$R(\theta_1) v_1 = \lim_{z \to 0} \left[\int_{L_1} \nabla \Phi_d \, dl \right]_{p_1} = \begin{pmatrix} 0 \\ -1/(\pi d_1) \end{pmatrix}$$

$$\Leftrightarrow \begin{bmatrix} 0.8944 & -0.4472 \\ 0.4472 & 0.8944 \end{bmatrix} v_1 = \begin{pmatrix} 0 \\ -0.1898 \end{pmatrix}$$

and

$$R(\theta_2) v_2 = \lim_{z \to 0} \left[\int_{L_1} \nabla \Phi_d \, dl \right]_{p_1} = \begin{pmatrix} 0 \\ -1/(\pi d_2) \end{pmatrix}$$

$$\Leftrightarrow \begin{bmatrix} 0.7071 & 0.7071 \\ -0.7071 & 0.7071 \end{bmatrix} v_2 = \begin{pmatrix} 0 \\ -0.4502 \end{pmatrix}.$$

The solutions are $v_1 = \langle -0.0849, -0.1698 \rangle$ and $v_2 = \langle 0.3183, -0.3183 \rangle$, and the resulting diagonal elements are

$$A_{11} = n_1 \cdot v_1 = \langle -0.4472, -0.8944 \rangle \cdot \langle -0.0849, -0.1698 \rangle = 0.1898$$

and

$$A_{22} = n_2 \cdot v_2 = \langle 0.7071, -0.7071 \rangle \cdot \langle 0.3183, -0.3183 \rangle = 0.4502.$$

The coefficient matrix for this two panel example is thus

$$A = \begin{bmatrix} 0.1898 & -0.0413 \\ -0.2038 & 0.4502 \end{bmatrix}.$$

8.3 Aeronautic Lift

The next example is more complete. Consider the piecewise linear approximation to a wing's profile in Fig. 8.19. The wake model is defined by the point $(3.85, -1.44)$, which is calculated as

$$\langle 3, -1 \rangle - \frac{1}{2}\left(\frac{\overrightarrow{AB}}{\|\overrightarrow{AB}\|} + \frac{\overrightarrow{AC}}{\|\overrightarrow{AC}\|}\right) = \langle 3.85, -1.44 \rangle,$$

where the coordinates of points A, B, and C are, respectively, $(3,-1)$, $(2,0)$, and $(0,-0.5)$. The midpoint of each panel is the collection point, and after calculating the matrix coefficients as previously done in the small example, we have the following 7×8 matrix

$$\frac{1}{100} \times \begin{bmatrix} 45.02 & -17.21 & -2.98 & -0.19 & -0.05 & 2.42 & -27.01 & -12.96 \\ -7.97 & 28.47 & -10.72 & -0.37 & -0.14 & 0.55 & -9.82 & -1.68 \\ -1.31 & -6.29 & 20.81 & -1.91 & -0.74 & -13.09 & 2.52 & -0.49 \\ -0.60 & -1.98 & -67.91 & 142.35 & -50.02 & -22.36 & 0.52 & -0.17 \\ -0.16 & -0.83 & -21.61 & -50.02 & 142.35 & -68.46 & -1.27 & 0.10 \\ 0.95 & 0.85 & -13.48 & -0.64 & -1.78 & 21.21 & -7.11 & 0.60 \\ -3.22 & -12.54 & 1.88 & 0.07 & -0.17 & -6.95 & 20.93 & 4.00 \end{bmatrix}.$$

Adding the last column to the first and subtracting it from the second-to-last enforces the Kutta condition and permits us to remove the last column. The resulting matrix is

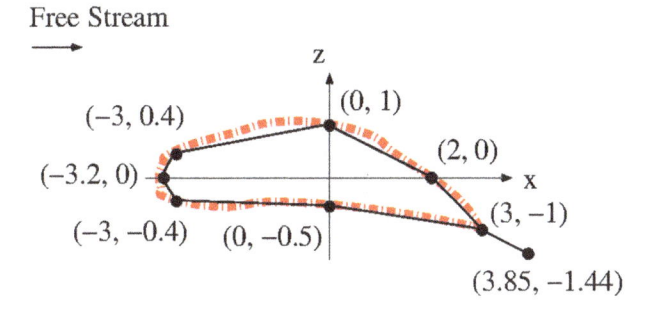

Fig. 8.19 An example linear discretization of a wing profile.

$$A = \frac{1}{100} \times \begin{bmatrix} 32.06 & -17.21 & -2.98 & -0.19 & -0.05 & 2.42 & -14.05 \\ -9.65 & 28.47 & -10.72 & -0.37 & -0.14 & 0.55 & -8.14 \\ -1.80 & -6.29 & 20.81 & -1.91 & -0.74 & -13.09 & 3.01 \\ -0.77 & -1.98 & -67.91 & 142.35 & -50.02 & -22.36 & 0.68 \\ -0.06 & -0.83 & -21.61 & -50.02 & 142.35 & -68.46 & -1.37 \\ 1.56 & 0.85 & -13.48 & -0.64 & -1.78 & 21.21 & -7.71 \\ 0.78 & -12.54 & 1.88 & 0.07 & -0.17 & -6.95 & 16.93 \end{bmatrix}.$$

The free stream decides the right-hand side vector, which is assumed to be in the horizontal direction with a magnitude of 76 m/s (about 180 mph—the approximate take off speed of an Airbus A320), see Fig. 8.19. So,

$$b = 76 \times \begin{pmatrix} n_1 \cdot \langle 1,0 \rangle \\ n_2 \cdot \langle 1,0 \rangle \\ n_3 \cdot \langle 1,0 \rangle \\ n_4 \cdot \langle 1,0 \rangle \\ n_5 \cdot \langle 1,0 \rangle \\ n_6 \cdot \langle 1,0 \rangle \\ n_7 \cdot \langle 1,0 \rangle \end{pmatrix} = \begin{pmatrix} -53.74 \\ -33.99 \\ 14.90 \\ 67.98 \\ 67.98 \\ 2.53 \\ 12.49 \end{pmatrix}.$$

The doublet profile is calculated by solving $A\mu = b$, which results in

$$\mu = \langle -277.15, -189.29, 76.05, 151.51, 159.00, 99.58, -19.71 \rangle.$$

The default, unit-strength doublet that is being scaled by μ_i points in the outward direction of panel i, i.e. away from the inside of the wing. So doublets with positive strengths indicate flows away from the wing, and those with negative strengths indicate flows toward the wing, see Fig. 8.20. Notice that the doublet directions and magnitudes agree with our intuition of how the "pressures" around the wing should be. Indeed, Bernoulli's equation provides pressure estimates, a development we forgo.

The doublet strengths give the following estimate of circulation,

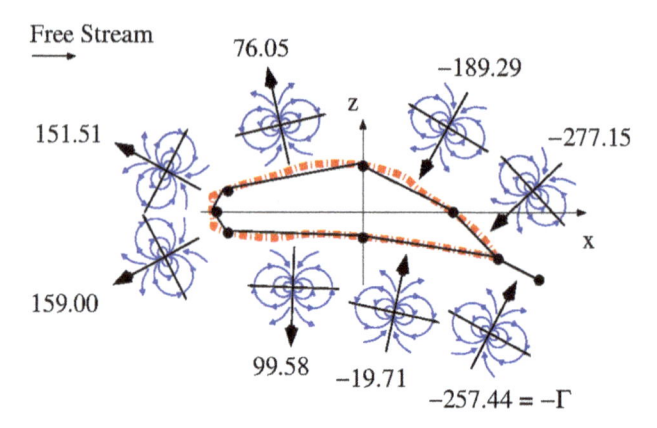

Fig. 8.20 The direction and magnitude of the solution doublets. The wake doublet estimates the negative of the circulation $-\Gamma$.

$$\Gamma \approx = \mu_7 - \mu_1 = -\mu_8 = 257.44$$

8.3 Aeronautic Lift

where μ_8 is the doublet strength of the wake as assumed by the Kutta condition. The extension of our indexing to the wake results in its normal facing downward, and hence, the fact that $\mu_8 = -257.44$ is negative indicates an upward facing doublet. This wake doublet coincides with the definition of circulation, which measures the flow around the wing in a counterclockwise rotation. The sectional lift of our model from the Kutta-Joukowski Theorem as noted in (8.11) is

$$L = -\mu_i \, \|\nabla \Phi_f\| \, \rho = 257.44 * 76 * 1.225 = 23{,}967.57 \text{ N/m},$$

where the air density is calculated at sea level at about 15 °C. An Airbus A320 can have a mass up to 70,000 kg, and if such a plane were fitted with a wing as modeled above, then we would estimate the need of a wing span of at least

$$\frac{70{,}000 * 9.8}{23{,}967.57} = 28.62 \text{ m}$$

to create sufficient lift to leave the ground. The actual wingspan of an A320 is 31 m, which includes the fuselage. So while the model has many approximations, the result is surprisingly reasonable. Increasingly accurate models are, of course, possible if more than just doublets are used.

8.3.4 Exercises

1. Build a computational model that accepts as its inputs:
   - a sequence of points that defines a collection of panels as an estimation of a wing's two-dimensional profile, and
   - a free steam vector.

   The code should produce doublet strengths, an estimate of circulation, and a lift calculation.
2. Use the program from Exercise 1 to study how the wake model affects lift for the 7-panel approximation depicted in 8.19. Design a collection of numerical experiments to illustrate how the wake's length, angle, and position of collection point alter lift estimates.
3. Use the program from Exercise 1 to study how the positions of the collection points alter lift estimates for a specified wing geometry. What locations would you recommend to make the lift estimates as large as possible?
4. Use the program from Exercise 1 to design a wing for a human-powered craft.

Chapter 9
Modeling with Ordinary Differential Equations

> Newton has shown us that a law is only a necessary relation between the present state of the world and its immediate subsequent state. All the other laws since discovered are nothing else; they are in sum, differential equations. – Henri Poincaré

Ordinary differential equations have long been studied due to their ability to model applications across numerous branches of science and engineering. They are of such importance that they have spawned much of the mathematics that we know today. Our task here is to model and solve problems to illustrate how computational modeling can aid scientific and engineering pursuits. We consider the three problems of modeling a Couette flow, an insulin bolus, and a plug-flow reactor.

9.1 Couette Flows

A Couette flow is created by sandwiching a fluid between two surfaces. The bottom surface is stationary, but the top surface moves horizontally at a rate U. We consider the rectangular element of the fluid depicted in Fig. 9.1. The shear stress, τ, is a smooth function of the element's distance from the bottom plate, denoted by y. If we assume that the shear stress at the bottom of the element is τ, then the first order approximation of the stress at the top of the element is $\tau + (d\tau/dy)\Delta y$. Balancing the forces along the top and bottom of the element, we have

$$0 = \left(\tau + \frac{d\tau}{dy}\Delta y\right)\Delta x - \tau \Delta x = \frac{d\tau}{dy}\Delta y \Delta x.$$

The implied outcome of this equality is the differential equation

$$\frac{d\tau}{dy} = 0. \tag{9.1}$$

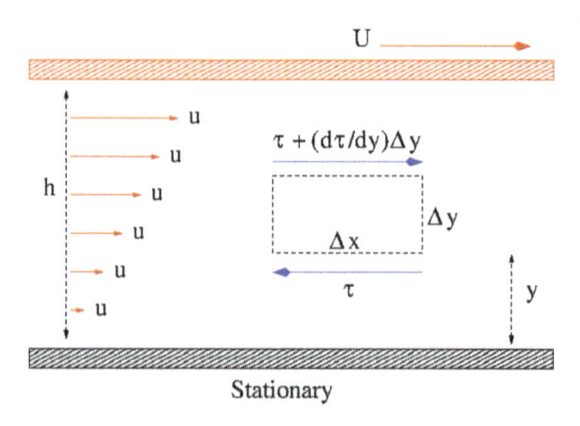

Fig. 9.1 A diagram of a Couette flow. The bottom black plate is stationary, whereas the top red plate is moving at a rate of U. The horizontal velocity profile of the intermediate fluid is u.

We let $u(y)$ be the horizontal velocity of the fluid at height y. The rate at which this horizontal velocity changes with respect to height is called the shear rate, and a fluid is classified by its relationship between its shear stress and its shear rate. If shear stress is proportional to shear rate, then the fluid is Newtonian, and hence, a Newtonian fluid is defined by the relationship

$$\tau = \mu \frac{du}{dy}, \tag{9.2}$$

where the constant of proportionality, μ, is the fluid's viscosity. We have upon combining (9.1) and (9.2) that

$$\frac{d}{dy}\left(\mu \frac{du}{dy}\right) = 0,$$

which implies

$$\mu \frac{du}{dy} = c, \tag{9.3}$$

where c is a constant decided by the boundary conditions $u(0) = 0$ and $u(h) = U$.

Although our goal is to study Couette flows computationally, we shouldn't devalue the benefit of calculating an exact analytic solution to aid our understanding of the standard cases. In particular, we can solve (9.3) with the technique of separation of variables if μ is constant, for which

$$u = \int \frac{du}{dy}\,dy = \int \frac{c}{\mu}\,dy = \frac{c}{\mu}y + k.$$

9.1 Couette Flows

We have $k = 0$ and $c = U\mu/h$ from the boundary conditions $u(0) = 0$ and $u(h) = U$, resulting in the specific solution

$$u = \frac{U}{h}y. \tag{9.4}$$

This linear relationship between horizontal velocity and height is the signature of a Couette flow of a Newtonian fluid, and we say that a Couette flow behaves Newtonian if $u(y)$ is linear, even if the fluid is non-Newtonian.

There are several types of non-Newtonian fluids, each having a different functional relationship among shear stress, shear rate, and time. Three common examples in which viscosity is time independent are found in Table 9.1. Non-Newtonian fluids for which the viscosity depends on time are divided into those whose viscosities reduce as time increases, called thixotropic fluids, and those whose viscosities increase with time, called rheopectic fluids. Examples of the former and latter are spinal fluid and whipping cream.

Fluid type	Defining equation	Example(s)
Bingham plastic	$\tau = \tau_0 + \mu(du/dy)$	Toothpaste, mayonnaise
Dilatant fluid	$\tau = \mu(du/dy)^n$, $n > 1$	Viscous coupling fluid
Power law fluid	$\tau = \mu(du/dy)^n$, $n < 1$	Paint, blood, syrup

Table 9.1 Common examples of non-Newtonian fluids. The term τ_0 is called a yield stress.

Non-Newtonian fluids such as toothpaste, peanut butter, and drilling slurries don't flow until a yield stress is surpassed, and we consider a fluid with a non-constant yield stress for which

$$\tau_0 = (15 - u/4000) \text{ kg/m·s}^2.$$

Suppose further that the fluid is rheopectic and that the time dependent viscosity is

$$\mu = \frac{t}{40(t+1)} \text{ kg/m·s}.$$

The resulting shear stress is

$$\tau = \left(15 - \frac{u}{4000}\right) + \frac{t}{40(t+1)}\frac{du}{dy} \text{ kg/m·s}^2. \tag{9.5}$$

We have from Eq. (9.1) that

$$\frac{d}{dy}\left(\left(15 - \frac{u}{4000}\right) + \frac{t}{40(t+1)}\frac{du}{dy}\right) = 0,$$

and hence,
$$\frac{du}{dy} = \left(c + \frac{u}{4000} - 15\right)\left(\frac{40(t+1)}{t}\right), \quad (9.6)$$

where c is again a constant decided by the boundary conditions $u(0) = 0$ and $u(h) = U$. We can again use the technique of separation of variables to solve the equation analytically by evaluating the following integrals,

$$\int \left(c + \frac{u}{4000} - 15\right)^{-1} du = \int \left(\frac{40(t+1)}{t}\right) dy. \quad (9.7)$$

Solving (9.7) results in the following time dependent specific solution assuming that $U = 0.3\,\text{m/s}$ and $h = 0.05\,\text{m}$,

$$u_t(y) = 60{,}000 - \frac{0.3\left(2 \times 10^5 \times e^{(5 \times 10^{-4}(t+1)/t)} - 1.9999 \times 10^5\right)}{e^{(5 \times 10^{-4}(t+1)/t)} - 1}$$

$$+ 4000\, e^{\left(40yt + 40y + 4000t \ln\left(7.5 \times 10^{-5}/(e^{5 \times 10^{-4}(t+1)/t} - 1)\right)\right)/4000t}.$$

The solution $u_t(y)$ permits us to ask if the fluid exhibits the linear relationship in (9.2) as t increases. If so, then the rheopectic fluid actually becomes Newtonian over time. We substitute $u_t(y)$ into (9.5) and evaluate the limit of $\tau/(du_t/dy)$ as $t \to \infty$, which gives

$$\lim_{t \to \infty} \frac{\tau}{du_t/dy} = \frac{0.3787}{e^{(0.01y - 1.8974)}}. \quad (9.8)$$

Since this limit is not constant in y, we know that the fluid does not become Newtonian as time increases.

We transition to a computational experiment to answer further questions about whether or not the characteristics of the Couette flow become Newtonian even though the fluid itself does not, i.e. do the horizontal velocities become approximately linear in height as time proceeds? One benefit of a numerical model is that it facilitates queries for a host of different situations, many of which might challenge the most sophisticated of integration prowess.

We use ode45 in MATLAB to calculate solutions of our model, and we begin by solving the Couette flow model for the Newtonian fluid in (9.3) with $h = 0.05\,\text{m}$, $U = 0.3\,\text{m/s}$, and $\mu = 0.5\,\text{kg/m·s}$, which is approximately the viscosity of SAE 30 weight oil. The solver requires a function whose return value is du/dy, which can be included in a function file to solve the problem. For example, creating a file called CouetteNewtonianFluid.m with the code in Example Code 9.1 allows us to solve the problem with the command

```
[y,u] = CouetteNewtonianFluid(0.3, 0.05, 0.5);
```

9.1 Couette Flows

The return arguments y and u are vectors so that $u(y_i) \approx u_i$. See the documentation for more information about ode45 and its arguments. Note that the last two arguments of our call to ode45 are passed directly to the function calculating the derivative. The solution is shown in Fig. 9.2. The outcome $u(y)$ is linear as expected.

```
function [u,y] = CouetteNewtonianFluid(U,h,mu);
  [y,u] = ode45(@calcDer, [0 h], [0], [], U*mu/h, mu);
  plot(u,y)
  ylabel('height (y)', 'fontsize', 24)
  xlabel('velocity (u)', 'fontsize', 24)

function uPrime = calcDer(y, u, c, mu)
  uPrime = c/mu;
```

Example Code 9.1 A MATLAB function to solve (9.3)

We return to the question of whether or not the non-Newtonian fluid with the shear stress in (9.5) develops Newtonian characteristics. Specifically, we investigate if u becomes linear in y. We again use ode45 in MATLAB, but the function defining du/dy changes. Unlike the Newtonian case, the constant c needs to be calculated for each t so that the speed at the top of the fluid matches the speed of the top plate. We calculate c with Newton's method by letting $u_t(h|c)$ be the calculated velocity at the top of the fluid if $d\tau/dy = c$. The derivative $du_t(h|c)/dc$ is estimated by the forward difference,

$$\frac{du_t(h|c)}{dc} = \frac{u_t(h|c+\delta) - u_t(h|c)}{\delta},$$

The syntax is in Example Code 9.2. The command

$$\texttt{CouetteNonNewtonianFluid(0.3,0.2,timeStp)},$$

with timeStp = [10^(-4)*[1:3:15], 0.5], solves the Couette model with $U = 0.3$ m/s and $h = 0.2$ m. The output is shown in Fig. 9.3. The solutions become increasingly linear as t lifts from zero, and we conclude that the horizontal velocities mimic those of a Newtonian fluid within a tenth of a second.

We can use our computational model to calculate the ratio of shear stress to shear rate. This ratio is constant for a Newtonian fluid, and since our non-Newtonian Couette flow quickly adopts Newtonian characteristics, we might expect $\tau/(du_t/dy)$ to approach a constant. Figure 9.4 depicts $\log_{10}(\tau/(du_t/dy))$ for the same values of t as in Fig. 9.3. A logarithmic scale is used since otherwise the magnitude of $\tau/(du_t/dy)$ is so great at $t = 10^{-4}$ that all other ratios appear constant by comparison. The graphs approach being constant as t increases from zero, further indicating convergence toward Newtonian behavior. Figure 9.5 shows $\tau/(du_t/dy)$ for $t = 1$ and $t = 10$, which

```
function CouetteNonNewtonianFluid(U,h,Time)
    c = 150;
    delta = 10^(-8);
    tol = 0.01;
    figure;
    hold on;
    for timeIndx = 1:length(Time)
      err = inf;
      while err > tol
        [y1,u1] = ode45(@calcDer, [0 h], [0], [], c, Time(timeIndx));
        [y2,u2] = ode45(@calcDer, [0 h], [0], [], c+delta, Time(timeIndx));
        U1 = u1(length(u1));
        U2 = u2(length(u2));
        err = min(err, abs(U-U1));
        dudc = (U2-U1)/delta;
        c = c-(U1-U)/dudc;
      end
      plot(u1,y1)
    end
    ylabel('height (y)', 'fontsize', 24);
    xlabel('velocity (u)', 'fontsize', 24);

function uPrime = calcDer(y, u, c, t)
    uPrime = (c + u/4000 - 15)*40*((t+1)/t);
```

Example Code 9.2 A MATLAB function to solve (9.6)

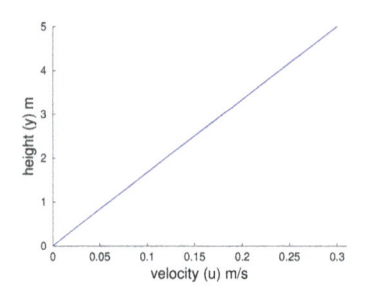

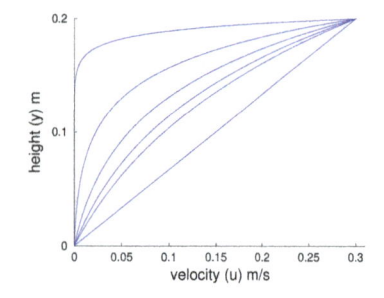

Fig. 9.2 The solution to (9.3) with $\mu = 0.5\,\text{kg/m·s}$, $h = 0.05\,\text{m}$, and $U = 0.3\,\text{m/s}$.

Fig. 9.3 Solutions to (9.6). As time increases from 10^{-4} to 0.0432, with $h = 0.2\,\text{m}$ and $U = 0.3\,\text{m/s}$. Notice that u adopts a Newtonian behavior.

confirms the fact that this ratio is not identically constant for larger values of t. However, $\tau/(du_t/dy)$ varies within $0.036\,\text{kg/m·s}$ after 1 s and within $0.019\,\text{kg/m·s}$ after 10 s, which are sufficiently small deviations to support the characteristics of a Newtonian flow.

9.1 Couette Flows   361

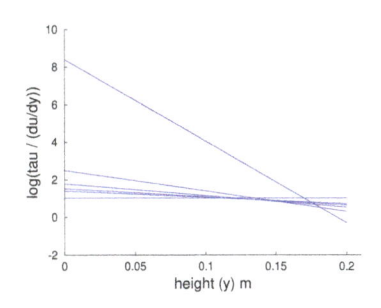

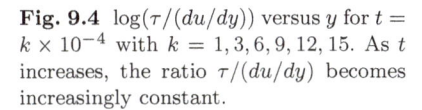

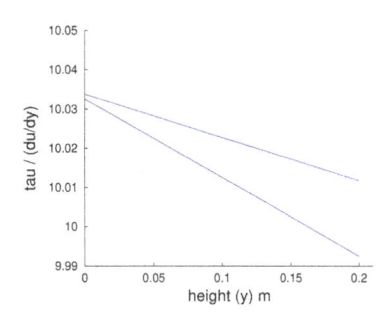

Fig. 9.4 $\log(\tau/(du/dy))$ versus y for $t = k \times 10^{-4}$ with $k = 1, 3, 6, 9, 12, 15$. As t increases, the ratio $\tau/(du/dy)$ becomes increasingly constant.

Fig. 9.5 $\tau/(du/dy)$ versus y for $t = 1$ and $t = 10$. As t increases, the variation in $\tau/(du/dy)$ decreases.

Our juxtaposition of an analytical study and a computational study shows their inherit differences. Analytic solutions precisely express a solution's functional form and can perfectly answer some queries, e.g. does this non-Newtonian fluid become Newtonian as $t \to \infty$? One could argue if this question is "experimental," as we can never observe t at infinity. Analytic solutions are also limited by our ability to integrate and by the fact that we are always dealing with a model of reality. Indeed, even the best of computer algebra systems can only integrate a small portion of the functions that can be integrated, and hence, analytic analysis is restricted to models that lend themselves to integration.

A computational study by comparison only approximates values of the true solution. An approximate solution does not require an integral's analytic evaluation, and as such, a computational study can answer questions about a wider variety of models. Computational solutions often resonate within scientific and engineering disciplines because they promote models beyond those that have been simplified for analytic study. The resulting benefit is that computational solutions regularly coincide with actual questions of interest. For instance, our study here shows that a non-Newtonian fluid essentially adopts the characteristics of a Newtonian fluid after a few seconds, even though the analytic solution never exhibits Newtonian characteristics.

9.1.1 Exercises

1. Create and solve Couette flow models for some Newtonian fluids of your choice. There are several online resources that can help decide fluid characteristics. Use `ode45` in MATLAB to solve the models and assume $c = U\mu/h$.
2. Adapt your code from Exercise 1 to search for c to ensure the boundary conditions $u(0) = 0$ and $U(h) = U$. Experiment with different root finding techniques and report your findings.
3. Build a computational model for a Couette flow of a thixotropic fluid with a shear stress of

$$\tau = \left(\frac{1}{(u+1)e^t}\right) \frac{du}{dy} \text{ kg/m·s}^2.$$

   Assuming $h = 0.1$ m, investigate how u behaves if $0 < t < 2$ (s) and $0 < U < 2$ (m/s). For each of your selected values of U generate (1) a graph of $u_t(y)$ as t varies, and (2) a graph of $\tau/(du/dy)$ as t varies. What conclusions do you reach about the flow characteristics?
4. A laminar flow of a fluid moving down an incline due to gravity satisfies

$$\tau = \gamma(h-y)\sin(\theta) \text{ kg/m·s}^2,$$

   where γ is the specific weight of the fluid and θ is the angle of incline. As with a Couette flow, h is the thickness of the fluid, y is the depth of the element, u is the flow rate at y, and τ is the shear stress at y. The value of y is measured perpendicular to the inclined surface. We assume $u(0) = 0$.

   a. Calculate u for SAE 30 weight oil, which is Newtonian. For this fluid we know that $\gamma = 8630$ kg/m²·s² and that $\mu = 0.5$ kg/m·s. Assume that $h = 0.05$ m. Vary the angle of inclination between 0 and $\pi/6$ to study the variation in $u(y)$.
   b. Calculate u for the thixotropic fluid in Problem 3. Assume that $h = 0.05$ m and that $\gamma = 9220$ kg/m²·s². Graph $u(y)$ as θ varies over the interval $[5°, 10°]$ and as time varies over $(0, 0.1]$. Do the flow characteristics match those of the Newtonian fluid in part (a)?
   c. Calculate u for the rheopectic fluid in (9.5). Assume that $\gamma = 7850$ kg/m²·s², which is about the specific weight of ethyl alcohol, and that $h = 0.05$ m. Vary the angle of incline so that $5° \leq \theta \leq 6°$, and assume that $10 \leq t \leq 11$ to ensure the viscosity is somewhat stable. For each of your selected values of θ plot $u_t(y)$ as t varies between 10 and 11 s.

9.2 Pharmacokinetics: Insulin Injections

Differential equations regularly express a principle of conservation, a fact that we promote as we study models that predict insulin in blood plasma. These models divide the anatomy into compartments and then trace the flow of insulin through the anatomical divisions. We require the amount of insulin to be conserved so that no insulin is either gained or lost.

Models of anatomical drug distribution fall within the discipline of pharmacokinetics. The compartment framework is, however, much broader, having applications in science, social science, engineering, and economics. Compartment models naturally suit situations in which a substance spreads among sub-divisions that define separate compartments. Flow rates between these compartments are known or assumed.

Consider the system of three compartments in Fig. 9.6. Compartment i has volume of solvent V_i, and the flow rate between compartments i and j is R_{ij}. External flows in and out of each compartment are denoted with a 0 subscript. The substance of interest, called the solute, is assumed to be uniformly dissolved within each compartment. So if the system blends alcohol in petroleum, then the solute alcohol is assumed to be uniformly blended with the petroleum in each compartment.

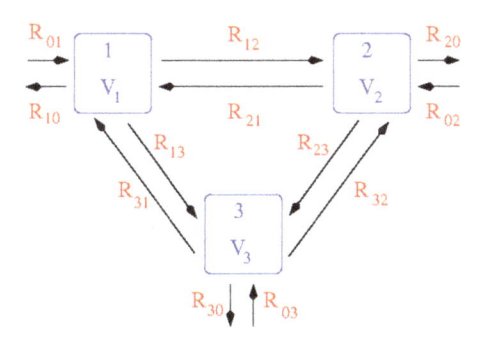

Fig. 9.6 A diagram for a three compartment model. The flow rate between compartments i and j is R_{ij}, with flow in and out of the system being indicated with a 0 subscript. The volume of compartment i is V_i.

Let the time dependent amount of solute in compartment i be x_i, from which the concentration of the dissolvent in compartment i is x_i/V_i. The conservation principle requires the rate, dx_i/dt, at which the dissolvent in compartment i is changing at time t, to be

Rate In	Rate Out

$$\left(R_{0i}C_{0i} + \sum_{\substack{j\,:\,j\neq i,\\ j>0}} R_{ji}\left(\frac{x_j}{V_j}\right) \right) - \left(R_{i0}\left(\frac{x_i}{V_i}\right) + \sum_{\substack{j\,:\,j\neq i,\\ j>0}} R_{ji}\left(\frac{x_i}{V_i}\right) \right),$$

where C_{0i} is the concentration of the dissolvent entering compartment i (possibly different for each i).

The conservation equations are the same as the following matrix system for the model depicted in Fig. 9.6,

$$\frac{d}{dt}\begin{pmatrix} x_1 \\ x_2 \\ x_3 \end{pmatrix} = \begin{bmatrix} -\frac{R_{10}+R_{12}+R_{13}}{V_1} & \frac{R_{21}}{V_2} & \frac{R_{31}}{V_3} \\ \frac{R_{12}}{V_1} & -\frac{R_{20}+R_{21}+R_{23}}{V_2} & \frac{R_{32}}{V_3} \\ \frac{R_{13}}{V_1} & \frac{R_{23}}{V_2} & -\frac{R_{30}+R_{31}+R_{32}}{V_3} \end{bmatrix}\begin{pmatrix} x_1 \\ x_2 \\ x_3 \end{pmatrix} + \begin{pmatrix} R_{01}C_{01} \\ R_{02}C_{02} \\ R_{03}C_{03} \end{pmatrix}.$$

The diagonal elements express the flows out of the compartments, and the off-diagonal elements express the flows into the compartments from the other compartments. The system's structure extends analogously to larger models. Transfers in and out of a compartment can be nullified by setting the appropriate flow rates to zero, and hence, the system need not permit all possible exchanges.

Conservation of the dissolved species can be verified by noticing that

$$\frac{d}{dt}(x_1 + x_2 + x_3)$$
$$= (R_{01}C_{01} + R_{02}C_{02} + R_{03}C_{03}) - \left(\frac{R_{10}}{V_1}x_1 + \frac{R_{20}}{V_2}x_2 + \frac{R_{30}}{V_3}x_3\right).$$

The first parenthetical sum is the rate at which the system gains solvent, and the second is the rate at which the system losses solvent. If the system is externally disassociated so that $R_{0i} = R_{j0} = 0$ for all i and j, then

$$\frac{d}{dt}(x_1 + x_2 + x_3) = 0.$$

The conclusion is that no solvent is lost or gained as the solvent progresses through the system itself.

The flow rates need not be constant and can depend on time and/or the compartment concentrations. A compartment's solvent volume is static only if the flow rates into and out of the compartment balance. If the flow rates and volumes are constant, then the system is linear with constant coefficients, which is a particularly well studied class of problems. Nonlinear models arise if, for example, the flow rates depend on the solute levels.

9.2 Pharmacokinetics: Insulin Injections

The solute in the pharmacokinetics examples of this section is insulin, and the compartments are anatomical tissues. The systems of differential equations that predict insulin concentrations for type-1 diabetics are listed in Table 9.2. The underlying medical study had patients ingest a meal, after which a bolus, i.e. an extra amount of insulin, was given to normalize the plasma concentration at 12 mU/ℓ/kg. Patients had only water for the next 720 min. Insulin was continuously supplied by an insulin pump at an average rate of 860 mU per hour, or about 14.333 mU per minute. Insulin delivered by the pump entered the anatomy through subcutaneous tissue. The two sources of insulin can be combined into a single compartment or separated into individual compartments.

We use the following notation to denote the insulin in the various compartments:

Q_1 Insulin (mU) in a subcutaneous compartment not accessible to plasma

Q_2 Insulin (mU) in a subcutaneous compartment accessible to plasma

Q_{1a} One of two possible divisions of Q_1

Q_{1b} A second possible division of Q_1

Q_{2a} One of two possible divisions of Q_2

Q_{2b} A second possible division of Q_2

Q_3 Insulin (mU) in plasma

The first four models use only compartments Q_1, Q_2, and Q_3, and hence, these models simplify the flow of insulin as it moves through the three main compartments of the subcutaneous tissue inaccessible to plasma, the subcutaneous tissue accessible to plasma, and the plasma. The first model is linear with constant coefficients, but Models 2 through 4 are nonlinear.

The conservation principle is established by summing the derivatives of the model. For instance, Models 1 and 5 maintain

$$\text{Model 1:} \quad \frac{d}{dt}(Q_1 + Q_2 + Q_3) = u - k_e Q_3, \text{ and}$$

$$\text{Model 5:} \quad \frac{d}{dt}(Q_{1a} + Q_{1b} + Q_{2a} + Q_{2b} + Q_3) = u_i + u_b - k_e Q_3.$$

The parameter u in Model 1 is the rate at which insulin enters the system from the insulin pump. This parameter is split into u_i and u_b in Model 5 under the assumption that $u = u_i + u_b$. This division permits insulin to enter the system along two different channels. The rate at which insulin leaves either system is $k_e Q_3$, and the total rate at which either system gains or loses insulin is the difference between the rate at which insulin enters the system and the rate at which insulin leaves the system. So the amount of insulin would be constant if the systems were externally disassociated.

Compartment diagrams like those in Figs. 9.7 and 9.8 depict the flows of a model, in these cases for Models 1 and 3, respectively. The flow parameters u, k_{a1}, and k_e of Model 1 are static, and the first model is thus linear with constant coefficients. The transition rates of Model 3 vary and depend on the time dependent level of insulin in the leaving compartment. The result is a nonlinear model. Models 2 and 4 have similar diagrams. Models 5 through 10 split the Q_1 and Q_2 compartments, and hence, their diagrams are more involved.

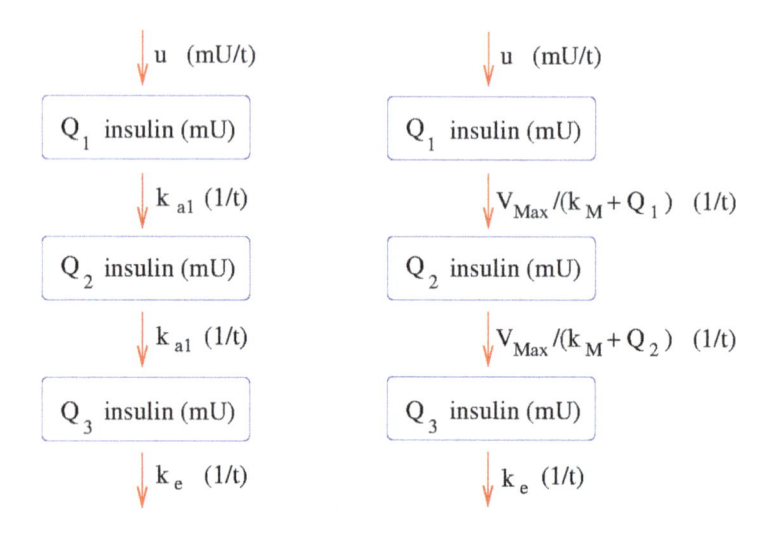

Fig. 9.7 A compartment diagram for Model 1. Units are parenthetical.

Fig. 9.8 A compartment diagram for Model 3. Units are parenthetical.

Plasma insulin is experimentally measured, and each model is assessed on how well Q_3 matches the clinical observations. The outcomes are normalized by considering a 1 kg mass with a volume of V deciliters of insulin. The plasma concentration in each model is Q_3/V, and all but Model 7 assume a constant volume of 4.2 dl of plasma per kg of mass. A statistical report of the experimental observations is shown in Table 9.3.

The rate at which insulin leaves plasma is $k_e = 0.35$ min^{-1}. The remaining model parameters are listed in Tables 9.4 and 9.5. These values are calibrated to approximate the observed plasma concentrations. In all instances we assume the initial value of Q_1 or Q_{1b}, depending on the model, is 5950 mU, which represents the initial bolus. The initial value of plasma insulin, i.e. $Q_3(0)$, is 50.4 mU. The compartment between the initial bolus and the plasma is assumed to be the average of $(5950 + 50.4)/2 = 3000.2$ mU in Mod-

9.2 Pharmacokinetics: Insulin Injections

Model 1
$Q'_1 = u - k_{a1}Q_1$
$Q'_2 = k_{a1}Q_1 - k_{a1}Q_2$
$Q'_3 = k_{a1}Q_2 - k_eQ_3$

Model 2
$Q'_1 = u - (k_{a1} - aQ_1)Q_1$
$Q'_2 = (k_{a1} - aQ_1)Q_1 - (k_{a1} - aQ_2)Q_2$
$Q'_3 = (k_{a1} - aQ_2)Q_2 - k_eQ_3$

Model 3
$Q'_1 = u - V_{\text{Max}}Q_1/(k_M + Q_1)$
$Q'_2 = V_{\text{Max}}Q_1/(k_M + Q_1)$
$\quad - V_{\text{Max}}Q_2/(k_M + Q_2)$
$Q'_3 = V_{\text{Max}}Q_2/(k_M + Q_2) - k_eQ_3$

Model 4
$Q'_1 = u - V_{\text{Max}}Q_1/(k_M + Q_3)$
$Q'_2 = V_{\text{Max}}Q_1/(k_M + Q_3)$
$\quad - V_{\text{Max}}Q_2/(k_M + Q_3)$
$Q'_3 = V_{\text{Max}}Q_2/(k_M + Q_3) - k_eQ_3$

Model 5
$Q'_{1a} = u_i - k_{a1}Q_{1a}$
$Q'_{1b} = u_b - k_{a2}Q_{1b}$
$Q'_{2a} = k_{a1}Q_{1a} - k_{a1}Q_{2a}$
$Q'_{2b} = k_{a2}Q_{1b} - k_{a2}Q_{2b}$
$Q'_3 = k_{a1}Q_{2a} + k_{a2}Q_{2b} - k_eQ_3$

Model 6
$Q'_{1a} = ku - k_{a1}Q_{1a}$
$Q'_{1b} = (1-k)u - k_{a2}Q_{1b}$
$Q'_{2a} = k_{a1}Q_{1a} - k_{a1}Q_{2a}$
$Q'_{2b} = k_{a2}Q_{1b} - k_{a2}Q_{2b}$
$Q'_3 = k_{a1}Q_{2a} + k_{a2}Q_{2b} - k_eQ_3$

Model 7
$Q'_1 = u - k_{a1}Q_1$
$Q'_2 = k_{a1}Q_1 - k_{a1}Q_2$
$Q'_3 = k_{a1}Q_2 - k_eQ_3$
$X' = V_{\text{Avg}} - k_V X$
$V = V_0(1 + V_{\text{Max}}X/(k_M + X))$

Model 8
$Q'_{1a} = ku - k_{a1}Q_{1a}$
$Q'_{1b} = (1-k)u - k_{a2}Q_{1b}$
$Q'_2 = k_{a1}Q_{1a} - k_{a1}Q_2$
$Q'_3 = k_{a1}Q_2 + k_{a2}Q_{1b} - k_eQ_3$

Model 9
$Q'_{1a} = ku - k_{a1}Q_{1a}$
$Q'_{1b} = (1-k)u - k_{a2}Q_{1b}$
$Q'_2 = k_{a1}Q_{1a} - k_{a1}Q_2$
$Q'_3 = k_{a1}Q_2 + k_{a2}Q_{1b} - k_eQ_3$
$u = u_i + Bu_b$

Model 10
$Q'_{1a} = ku - k_{a1}Q_{1a} - LD_a$
$Q'_{1b} = (1-k)u - k_{a2}Q_{1b} - LD_b$
$Q'_2 = k_{a1}Q_{1a} - k_{a1}Q_2$
$Q'_3 = k_{a1}Q_2 + k_{a2}Q_{1b} - k_eQ_3$
$LD_a = V_{\text{Max}}Q_{1a}/(k_M + Q_{1a})$
$LD_b = V_{\text{Max}}Q_{1a}/(k_M + Q_{1b})$

Model 11
$Q'_3 = -k_eQ_3 + u_b t^{s-1} s T_{50b}^s / (T_{50b} + t^s)^2 + \int_0^t u(\tau) \frac{(t-\tau)^{s_i-1} s_i T_{50i}^{s_i}}{(T_{50i}^{s_i} + (t-\tau)^{s_i})^2} d\tau$

$T_{50b} = au_b + b$
$T_{50i} = b_i$

Table 9.2 Eleven pharmacokinetics models used to study insulin flow.

Time	0	1	2	3	4	5	6	7	8	9	10	11
mU/dl	12	35.33	42	38	33.67	30	27.33	21.33	20.67	20	18.33	14.67
95%	±1.33	±4	±4	±4.67	±3.67	±3.33	±4	±4	±3.33	±3.67	±3	±2.33

Time	12	13	14	15	16	17	18	19	20	21	22	23
mu/dl	14.67	15.33	15.33	14	13.33	12.67	12	10.33	9.67	10	9.67	9.33
95%	±2	±2.67	±2.33	±2.33	±3	±3.33	±2.67	±2.33	±2.67	±2.33	±2.33	±2.67

Table 9.3 Average concentrations of plasma insulin (mU/dl) over the experimental population. Each time interval is 30 min. The row labeled 95% identifies the 95% confidence interval.

Model	k_{a1} $\times 10^{-4}$	k_{a2} $\times 10^{-4}$	k_M $\times 10^{3}$	k_v $\times 10^{-2}$	k $\times 10^{-2}$	a $\times 10^{-9}$	u_i	u_b	V_{Max}	V_{Avg}	B $\times 10^{-2}$
1	166										
2	183					148					
3			66						1,140		
4			66						1,140		
5	189	158					14.333	0			
6	101	180			61						
7	166		66	1					1,140	0.01	
8	251	124			71						
9	251	124			71		14.333	10			15
10	251	124	66		71				1,140		
Units (Mod 7)	$\min^{-1}$ $\min^{-1}$	$\min^{-1}$	mU mu/l		$\min^{-1}$	$\min^{-1}$	mU/l	mU/l	mU/min unitless	mU/min^2	

Table 9.4 Parameters for Models 1 through 10.

Parameter	a	b	b_i	s	s_i	u_b
Value	2.44	53.45	79.19	2.01	2.86	12

Table 9.5 Parameters for Model 11.

els 1 through 7 at $t = 0$. Models 5 and 6 use the initial value of 3000.2 for Q_{2b}, whereas the initial values of Q_{1a} and Q_{2a} are both 0. The initial value of Q_2 is assumed to be zero in Models 8, 9, and 10.

Solvers like `ode45` in MATLAB and `lsode` in Octave are well suited to solving the suggested models. The syntax is similar to that of Sect. 9.1, the only difference being that an m-file for a system returns a vector of rates. An

9.2 Pharmacokinetics: Insulin Injections

example function for the first model is shown in Example Code 9.3, which can then be used to solve the model in Octave with

```
lsode("CompMod1", [5950; 3000.2; 50.4], (t = 0:0.1:720'));
```

Similar commands for ode45 in MATLAB also solve the problem. A graph of the solution against the empirical observations is in Fig. 9.9. The solution raises some doubt about the appropriateness of the first model. Whereas the solution favorably matches the value of the highest insulin level, this prediction is a bit premature. Model predictions agree with observations after 600 min as the amount of insulin steadies to a near constant of 10 mU. However, the model underpredicts the amount of insulin as it decreases from its peak value. The solution being outside the 95% confidence interval over much of the experiment is dubious, and we are left to wonder if there is a problem with the model or its parameters. That said, problems in the life sciences are regularly difficult, if for no other reason than humans vary significantly.

```
function Qdot = CompMod1(Q,t)
    u = 860/(60);
    ka1 = 0.0166;
    ke = 0.35;
    Qdot(1) = u - ka1*Q(1);
    Qdot(2) = ka1*Q(1) - ka1*Q(2);
    Qdot(3) = ka1*Q(2) - ke*Q(3);
end
```

Example Code 9.3 An example function file that works with Octave's lsode function for the first model.

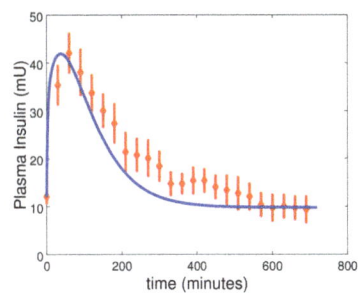

Fig. 9.9 A solution to the first model is shown in blue. The empirical measurements with their 95% confidence intervals are depicted in red.

The eleventh model is unique against the others, as it attempts to directly estimate the plasma insulin without the compartment framework. This model is an integro-differential equation since the rate at which Q_3 changes depends on a time dependent integral. The function defining Q'_3 needs to estimate the stated integral, and any of the quadrature rules of Sect. 6.2 can be used. One option is to approximate the integral stochastically, although this technique is ill-advised since slight cumulative effects between iterations could easily change sign. The result would be, for example, that small anticipated gains might appear as losses.

9.2.1 Exercises

1. Create compartment diagrams like those in Figs. 9.7 and 9.8 for the first ten models in Table 9.2.
2. Provide an interpretation of the integral in the eleventh model. In particular, explain the integral as a replacement for the compartments prior to Q_3 in the other models.
3. Show that each of the first 10 models in Table 9.2 is conservative. What is different about Model 9, and what does this difference mean?
4. Solve each of the eleven models in Table 9.2 with the parameters and initial conditions stated in this section. Plot each solution against the observed data as in Fig. 9.9.
5. Develop a metric to assess how well a model matches the observed data. Use this metric to select a model based on the parameters and initial conditions of this section.
6. The concept of sensitivity relates a model's outcomes with its parametric inputs. A model is sensitive to a parameter if small amounts of perturbation in its value lead to large deviations in the outcome. A model is less sensitive to a parameter if large changes in its value lead to small changes in the outcome. Conduct a sensitivity study of your own design for the eleven models. Which parameters should be decided most carefully?
7. Design and conduct a search to identify a best set of parameters for each model relative to your metric from Exercise 5. Which model(s) would you promote if you were allowed to optimize the parameters to best agree with the observational data?
8. Parameter values and initial conditions are arguably random, and the models are thus stochastic equations. Impose distributions on the parameters and initial conditions and generate a sample of solutions for each model by solving it as the parameters and initial conditions are sampled. Use your sample to estimate the expected outcome of the model. Give a 95% confidence interval on the expected outcome.

    Adapt your metric from Exercise 5 so that it assess how well your stochastic solutions match the observational data. Which stochastic model would you promote?
9. The probability distributions assumed in Exercise 8 depend on parameters. For instance, a normal variable depends on its mean and variance. Design a search similar to Exercise 7 to identify quality parameters for your distributions. Which model would you favor once parameters are tuned?

9.3 Chemical Reactions

Chemical systems surround us; indeed, we are essentially chemical machines living in a chemical environment. The study of complex and interwoven chemical systems underlies modern disciplines such as systems biology and atmospheric chemistry, as well as traditional mainstays such as chemical engineering. These disciplines have prompted significant advances in disparate arenas such as healthcare and global environmental policy, illustrating how chemical applications are thus ubiquitous and salient.

Chemical reactions are often studied as systems of differential equations, although we warn that such systems are commonly difficult to solve and are routinely stiff. Hence, differential equations arising from chemical systems should be approached knowing the advantages of different numeric solvers, and in particular, if a standard RK4(5) algorithm gives questionable results or requires a suspicious amount of time, then a stiff solver should be considered. If stiff solvers also prove insufficient, then one may need to alter settings, consider model reformulations, or customize the solver.

Conservation of mass is the motivating premise of the modeling process. Suppose the four chemical species A, B, C, and D combine as

$$A + 2B \underset{k_-}{\overset{k_+}{\rightleftharpoons}} C + D.$$

The notation implies a forward reaction that combines one A with two Bs to form one C and one D. The forward reaction occurs at the kinetic rate of k_+. A backward reaction combines one C with one D to form one A and two Bs, and the kinetic rate in this case is k_-.

The rate constants are interpreted through the principle of mass action, which effectually asserts that interactions among chemical species occur at a rate that is proportional to the product of the concentrations. The standard notation for concentration is $[\cdot]$, where we assume the units of moles per liter (mol/ℓ). So the concentration of A is $[A]$ and the concentration of B is $[B]$. Since species A is created upon the interaction of C and D, the principle of mass action asserts that the rate at which $[A]$ increases is $k_-[C][D]$. The rate at which $[A]$ decreases is likewise $k_+[A][B][B] = k_+[A][B]^2$. The resulting nonlinear differential equation is

$$\frac{d[A]}{dt} = k_-[C][D] - k_+[A][B]^2.$$

The rate coefficients must have different units for this equation to make sense. The backward reaction is second-order since two species combine, and the forward reaction is third-order since three (non-unique) species combine. We infer, assuming that time is measured in seconds, that the unit of k_- is ℓ/s·mol and the unit of k_+ is ℓ^2/s·mol^2.

Using the same modeling process for $[B]$, $[C]$, and $[D]$ results in the system,

```
function dC = calcDer(t,C,kf,kb)
    dC(1) = kb*C(3)*C(4) - kf*C(1)*C(2)^2;
    dC(2) = kb*C(3)*C(4) - kf*C(1)*C(2)^2;
    dC(3) = kf*C(1)*C(2)^2 - kb*C(3)*C(4);
    dC(4) = kf*C(1)*C(2)^2 - kb*C(3)*C(4);
end
```

Example Code 9.4 An example function file that works with Octave's ode45 command for the chemical system in (9.9).

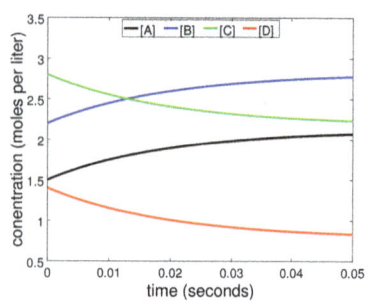

Fig. 9.10 An example solution for the chemical system in (9.9). The initial concentrations for $[A]$, $[B]$, $[C]$, and $[D]$ were, respectively, 1.5, 2.2, 2.8, and 1.4 mol/ℓ.

$$\left.\begin{aligned}\frac{d[A]}{dt} &= k_-[C][D] - k_+[A][B]^2 \\ \frac{d[B]}{dt} &= k_-[C][D] - k_+[A][B]^2 \\ \frac{d[C]}{dt} &= k_+[A][B]^2 - k_-[C][D] \\ \frac{d[D]}{dt} &= k_+[A][B]^2 - k_-[C][D].\end{aligned}\right\} \quad (9.9)$$

The right-hand sides of the first and last paired equations are the same because each pair describes either the forward or backward reaction. If the initial concentrations of $[A]$ and $[B]$ agree, then the model mandates that $[A]$ and $[B]$ be identical throughout the process. Similarly, $[C]$ and $[D]$ are identical if they share a common initial condition. The concentrations of each pair differ as the initial conditions vary. An example solution with each species having a unique initial concentration is illustrated in Fig. 9.10. The solutions were calculated with Octave's ode45 command with

ode45(@calcDer, 0:maxTime/100:maxTime, initCond, kf, kb).

The calcDer function is in Example Code 9.4, and kf and kb are, respectively, $k_+ = 1$ and $k_- = 10$, i.e. the backward reaction rate is 10 times that of the forward reaction. The initial conditions are

$$[A]|_{t=0} = 1.5 \text{ mol}/\ell, \quad [B]|_{t=0} = 2.2 \text{ mol}/\ell,$$
$$[C]|_{t=0} = 2.8 \text{ mol}/\ell, \quad \text{and} \quad [D]|_{t=0} = 1.4 \text{ mol}/\ell.$$

9.3 Chemical Reactions

We now consider an example associated with reactor design by specifically considering the catalytic reaction of an ideal, plug-flow reactor, which is a chemical reaction occurring in a flow through a pipe. The chemical process is the oxidation of CO in the presence of H_2, and the two reactions are

$$CO + \frac{1}{2}O_2 \to CO_2 \quad \text{and} \quad H_2 + \frac{1}{2}O_2 \to H_2O. \tag{9.10}$$

Figure 9.11 illustrates a plug-flow reactor. Reactions occur as materials flow down the pipe, and reactant concentrations change with the flow. Consider an element of the pipe as illustrated in Fig. 9.11. The conservation principle is that of material balance of the reactants as they relate to the amount of catalyst. The flow rate through any differential volume within the pipe is v, in units of volume per time, and the material balance equation for any species A related to the catalyst is

$$v\, d[A] = r_A\, dW. \tag{9.11}$$

The grams of catalyst within the differential volume is W, and the parameter r_A is a catalytic rate in units of moles per time per grams-catalyst. Since the units of $[A]$ are moles per volume, both sides of the equation are in terms of moles per time.

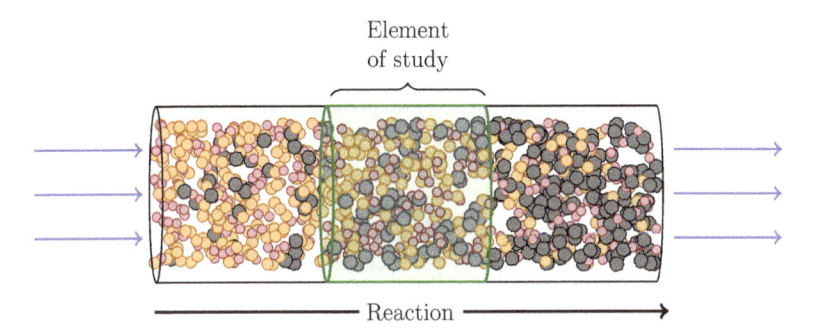

Fig. 9.11 An illustration of a plug-flow reactor. As material flows down the pipe the concentration of reactants, products, and available catalyst change.

The experimentally approximated rates for our reaction are

$$r_{CO} = -k_1\, p_{CO}\, p_{O_2}^{1/2} \quad \text{and} \quad r_{H_2} = -k_2 \frac{p_{H_2}\, p_{O_2}^{1/2}}{(1 + K_{CO}\, p_{CO})^2},$$

where p_A is the partial pressure of species A in atmospheres. The constants k_1 and k_2 are in units of moles per time per grams-catalyst per atmosphere$^{1.5}$, and K_{CO} is in units of inverse atmospheres. We assume a constant velocity and set $\tau = W/v$ so that $d\tau = dW/v$. The differential equations arising from (9.11) are

$$\frac{d[CO]}{dW/v} = \frac{d[CO]}{d\tau} = -k_1\, p_{CO}\, p_{O_2}^{1/2} \quad \text{and}$$
$$\frac{d[H_2]}{dW/v} = \frac{d[H_2]}{d\tau} = -k_2\, \frac{p_{H_2}\, p_{O_2}^{1/2}}{(1+K_{CO}\, p_{CO})^2}. \quad (9.12)$$

System (9.12) appropriately models the reactions, but the equations are inappropriate for numerical solvers. The problem is that the derivatives are expressed in terms of the partial pressures of CO and H_2 and not in terms of their concentrations. We rectify this conundrum by asserting the ideal gas law, which is

$$PV = nRT.$$

Here P is pressure, V is volume, n is the number of moles, R is a gas constant, and T is degrees Kelvin. Noticing that n/V is concentration, we rewrite (9.12) as

$$\frac{dp_{CO}}{d\tau} = -k_1\, RT\, p_{CO}\, p_{O_2}^{1/2} \quad \text{and}$$
$$\frac{dp_{H_2}}{d\tau} = -k_2\, RT\, \frac{p_{H_2}\, p_{O_2}^{1/2}}{(1+K_{CO}\, p_{CO})^2}. \quad (9.13)$$

This nonlinear system of differential equations is ready for numerical computation once parameters are deduced. The partial pressure of O_2 is required, but this can be inferred from the partial pressures of CO and H_2 through the stoichiometry associated with the reactor. The following parameters correspond with an ideal plug-flow reactor at 1 atmosphere and 100 °C and with a feed containing 1% mol CO and 30% mol H_2:

$$k_1 RT = 4743.72075\ \ell/\text{g·min·atm}^{0.5},$$
$$k_2 RT = 59.67907\ \ell/\text{g·min·atm}^{0.5},$$
$$K_{CO} = 1000\ \text{atm}^{-1}, \text{ and} \quad (9.14)$$
$$p_{O_2} = 10^{-4} + 0.5(p_{CO} - 10^{-5}) + 0.5(p_{H_2} - 0.2866).$$

The evaluation of p_{O_2} depends on τ since p_{CO} and p_{H_2} are functions of τ, and this evaluation of p_{O_2} is designed to keep the concentration of CO to 10 ppm as the flow leaves the reactor.

The system shows signs of being stiff, and ode23 outperforms ode45. We use odeset in MATLAB to further improve solution quality. A MATLAB function to calculate a solution is in Example Code 9.5, and a solution is depicted in Fig. 9.12.

9.3 Chemical Reactions

```
function [tau, parPressure] = plugFlowReactor()
  opts = odeset('AbsTol', 10^(-12), 'RelTol', 10^(-6), 'MaxStep', 10^(-4));
  [tau, partPressure] = ode23(@calcRate, [0, 0.028], [0.01, 0.3], opts);
end

function dy = calcRate(t,y)
  pO2 = 10^(-4) + 0.5*(y(1)-10^(-5)) + 0.5*(y(2)-0.2866);
  kRT = [4743.72075, 59.67907];
  KCO = 1000;
  dy = [-kRT(1)*y(1)*nthroot(max(pO2,0),2); ...
        -kRT(2)*y(2)*nthroot(max(pO2,0),2)/(1+KCO*y(1))^2];
end
```

Example Code 9.5 A MATLAB script to calculate a solution to system (9.13) with parameter settings in (9.14)

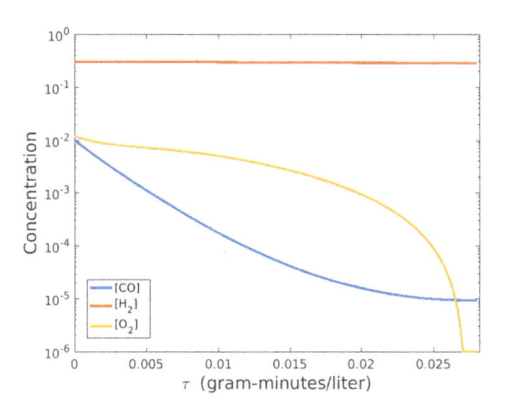

Fig. 9.12 The concentrations of the reactants of the ideal plug-flow reactor modeled by (9.13) with parameter settings in (9.14).

9.3.1 Exercises

1. The decomposition of acetaldehyde, which is the chemical species CH_3CHO, is the multiple reaction system,

$$CH_3CHO \xrightarrow{k_1} CH_3 + CHO$$
$$CH_3 + CH_3CHO \xrightarrow{k_2} CH_4 + CH_3CO$$
$$CH_3CO \xrightarrow{k_3} CH_3 + CO$$
$$CH_3 + CH_3 \xrightarrow{k_4} C_2H_6.$$

a. Write the system of seven differential equations corresponding with this system.
b. Solve your system from part (a) for the kinetic coefficients,

$$k_1 = 28.7, \quad k_2 = 4.2, \quad k_3 = 11.4, \quad \text{and} \quad k_4 = 0.001.$$

Assume $[CH_3CHO]|_{t=0} = 0.1$ and that all other initial concentrations are zero.

2. The chemical species CH_3 and CH_3CO in Exercise 1 are free radicals and are often assumed to be transitional constructs satisfying the pseudo-steady-state approximations

$$\frac{d[CH_3]}{dt} = 0 \quad \text{and} \quad \frac{d[CH_3CO]}{dt} = 0.$$

Use your equations from Exercise 1a to show that the sum of these assumed equations permits us to express $[CH_3]$ in terms of $[CH_3CHO]$. Then replace $[CH_3]$ in your equation for $d[CH_3CHO]/dt$ in Exercise 1a to conclude that

$$\frac{d[CH_3CHO]}{dt} \approx k\,[CH_3CHO]^{3/2}$$

under the additional assumption that $[CH_3CHO]$ is significantly less than $[CH_3CHO]^{3/2}$. What is k in terms of k_1, k_2, k_3, and k_4?

3. Compare the approximate solution of $[CH_3CHO]$ from Exercise 2 to the solution from Exercise 1. First plot all concentrations from Exercise 1 and argue that all chemical species have largely reached equilibrium after 0.2 s. Use the concentration of $[CH_3CHO]$ at $t = 0.2$ as the initial condition for the approximate equation in Exercise 2. Plot the solutions of $[CH_3CHO]$ from both the system in Exercise 1 and the approximate equation in Exercise 2 for $0.2 \leq t \leq 1$. At what time and by what amount do these solutions differ the most? You may want to approximate both solutions with a cubic spline to better answer this question.

Chapter 10
Modeling with Delay Differential Equations

> As has been asked by many students in many classrooms, "Why study this subject?" Why study differential equations with time delays when so much is known about equations without delays, and they are so much easier? The answer is because so many of the processes, both natural and manmade, in biology, medicine, chemistry, physics, engineering, economics, etc., involve time delays. Like it or not, time delays occur so often, in almost every situation, that to ignore them is to ignore reality. – Yang Kuang

Although modeling phenomena with differential equations has a long and successful history over a wide range of applications, some situations lend themselves to adaptations that more seamlessly capture the entity being modeled. This chapter studies one such adaptation called a delay differential equation (DDE). A DDE is an ordinary differential equation that permits dependencies on historical information, and initial conditions are replaced with legacy assumptions that detail the solution's previously observed behavior. A DDE is a welcome framework for processes that naturally depend on historical trajectories and not just initial values.

The general form of a delay differential equation is

$$y'(t) = f(t, y(t - \tau_1), y(t - \tau_2), \ldots,$$
$$y(t - \tau_{p-1}), y'(t - \tau_p), y'(t - \tau_{p+1}), \ldots, y'(t - \tau_q)).$$

The parameters $\tau_1, \tau_2, \ldots, \tau_q$ are time delays, which are often called lags. If y' is independent of previous evaluations of y', then the DDE is called retarded. Otherwise y' depends on previous evaluations of y', and the DDE is called neutral. Note that an ODE fits within the DDE framework since $y' = f(t, y)$ is a DDE with a single zero lag.

The models and exercises of this chapter include the effectiveness of a catalytic converter, epidemiology studies, temperature oscillations due to the El-Niño–La-Niña cycle, the charge through a partial element circuit, and cloud computing. Some of these models can be altered to become traditional ODEs, but others cannot.

10.1 Is a Delay Model Necessary or Appropriate?

Some delay differential equations can be re-cast as ordinary differential equations, and such problems can be solved directly with ODE solvers. However, the DDE paradigm can still aid modeling and interpretation. For instance, the efficacy of a filter, such as a catalytic converter, depends on its cumulative use. If we allow $P(t)$ to be the number of particles removed by a filter at time t, then a reasonable model is

$$\frac{dP}{dt} = -\alpha P - \beta \int_{\tau}^{t} P\, dx, \tag{10.1}$$

where τ defines the start of the filter's use. Setting $y(t) = \int_{\tau}^{t} P\, dx$, we find that the following system re-states Eq. (10.1),

$$\frac{dP}{dt} = -\alpha P - \beta y \quad \text{and} \quad \frac{dy}{dt} = P. \tag{10.2}$$

The initial conditions for this system are $y(0) = \int_{\tau}^{0} P(t)\, dt$ and $P(0) = P_0$, where we assume that $\tau \leq 0$.

The integral form of the history defining $y(0)$ leads to two observations. First, the Fundamental Theorem of Calculus allows us to reformulate the DDE as a system of ODEs. Hence, we can solve this DDE with a standard ODE solver. Second, the cumulative dependence represented by the integral removes the importance of the individual values of P over its historical range as long as the integral remains unchanged. To illustrate, notice that both $P(t) = |t|$ and $P(t) = 3t^2$ both satisfy, for $\tau = -1$,

$$\int_{\tau}^{0} P(t)\, dt = 1 \quad \text{and} \quad P(0) = 0,$$

These different histories give the same solution since their cumulative historical effects are identical. With regard to the filter, identifying when particles have been removed is unimportant, and a solution only requires the aggregate total of removed particles.

While the filter model can be stated and solved as a system of ODEs, the phenomena is less naturally expressed in this format. The value of the DDE model in this case is its conceptual agreement with the entity being modeled. Indeed, the DDE model has a simpler interpretation and defense.

10.2 Epidemiology Models

For example, system (10.2) with the initial conditions $P(0) = 1$ and $y(0) = 1$ hides the fact that y is the aggregate total of the particles that have been removed. However, the DDE in (10.1) explicitly and naturally describes the model in notation that agrees with the behavior of the filter. Models that seamlessly express the underlying processes are generally preferred, and this can be one of the advantages of the DDE framework.

We note that system (10.2) could have been obtained by differentiating (10.1) to produce

$$P'' = -\alpha P' - \beta P.$$

Setting $Q = P'$ gives

$$Q' = -\alpha Q - \beta P \quad \text{and} \quad P' = Q,$$

which is the same as (10.2) after an appropriate change of variable.

10.2 Epidemiology Models

A stalwart introductory model in epidemiology is the SIR model. Let $S(t)$, $I(t)$, and $R(t)$ trichotomize a population, where

- $S(t)$ is the number of people who are susceptible to a disease,
- $I(t)$ is the number of people who are infected with a disease, and
- $R(t)$ is the number of people who have recovered from a disease.

A succinct traditional SIR model is

$$\frac{dS}{dt}(t) = -\alpha S(t) I(t)$$

$$\frac{dI}{dt}(t) = \alpha S(t) I(t) - \beta I(t) \qquad (10.3)$$

$$\frac{dR}{dt}(t) = \beta I(t).$$

The $S(t)I(t)$ terms measure the interactions between the susceptible and infected populations, and the parameter α notes the percentage of these interactions that become infected (per person per time). The parameter β is the percentage of the infected population that recovers (per time). The SIR model conserves the size of the entire population because

$$\frac{d}{dt}(S(t) + I(t) + R(t)) = (-\alpha S(t)I(t) + \alpha S(t)I(t) - \beta I(t) + \beta I(t)) = 0.$$

Hence, the total population remains the same size as t varies.

One limitation of the SIR model is that those who recover are no longer susceptible. However, some who recover remain susceptible in many situations. We adapt the SIR model to consider this case,

$$\frac{dS}{dt}(t) = -\alpha S(t)I(t) + \kappa \beta I(t)$$

$$\frac{dI}{dt}(t) = \alpha S(t)I(t) - \beta I(t) \qquad (10.4)$$

$$\frac{dR}{dt}(t) = (1-\kappa)\beta I(t),$$

where κ is the proportion of those who remain susceptible after recovery. Notice again the conservation of the entire population, which is ensured by

$$\frac{d}{dt}(S(t) + I(t) + R(t)) = 0.$$

Solutions to (10.3) and (10.4) are shown in Figs. 10.1 and 10.2 with $\alpha = 0.08$, $\beta = 0.8$, $\kappa = 0.3$, $S(0) = 60$, $I(0) = 2$, and $R(0) = 38$.

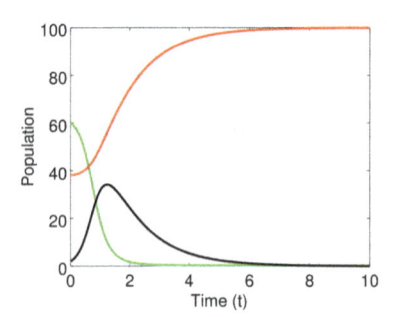

Fig. 10.1 Solutions to the traditional SIR model: green is the susceptible population, black the infected population, and red the recovered population.

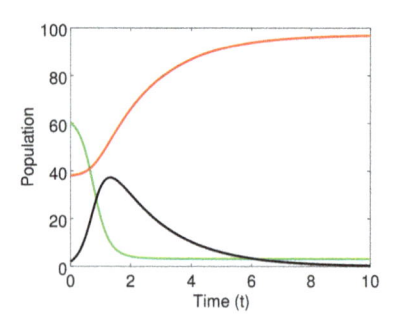

Fig. 10.2 Solutions to the adapted SIR model in (10.4): green is the susceptible population, black the infected population, and red the recovered population.

The solutions in Figs. 10.1 and 10.2 appear similar, but their tails are different. In the traditional model all individuals (essentially) leave the susceptible state by $t = 3$, but in the adapted model a small portion of the population persists being susceptible.

10.2 Epidemiology Models

A common SIR adaptation is

$$\frac{dS}{dt}(t) = \rho - \alpha S(t) \frac{I(t)}{N(t)} - \delta S(t),$$

$$\frac{dI}{dt}(t) = \alpha S(t) \frac{I(t)}{N(t)} - (\beta + \delta + \varepsilon)I(t),$$

$$\frac{dR}{dt}(t) = \beta I(t) - \delta R(t),$$

$$N(t) = S(t) + I(t) + R(t),$$

where we assume positive initial populations. The interaction terms $S(t)I(t)$ in the original SIR model are now normalized by the size of the total population. The parameters α and β have interpretations similar to their prior use, with the unit of α being changed to per time due to the normalization. The new parameters are:

ρ is the recruitment rate (person per time),

δ is the natural mortality rate (per time), and

ε is the disease induced death rate (per time).

The extended model does not conserve the population size, but solutions are bounded. This follows because

$$\frac{d}{dt}(S(t) + I(t) + R(t)) = \frac{dN}{dt}(t) = \rho - \delta N(t) - \varepsilon I(t) \leq \rho - \delta N(t). \quad (10.5)$$

The last inequality holds because $I(t)$ cannot change sign due to the equilibrium solution $I(t) = 0$, which would force $dI/dt = 0$. Hence, the most extreme case satisfies

$$\frac{dN}{dt}(t) = \rho - \delta N(t).$$

Using the technique of separation of variables, we find that

$$\int \frac{1}{\rho - \delta N} \frac{dN}{dt} dt = \int 1\, dt = t + k.$$

Hence,

$$\ln|\rho - \delta N| = -\delta t + k,$$

and solving for N, we have

$$N(t) = \frac{\rho - Ce^{-\delta t}}{\delta},$$

where C is a constant decided by the size of the initial population. As $t \to \infty$ the right-hand side converges to ρ/δ, and hence, the size of the population cannot grow without bound even in the most extreme case.

An intuitive delayed extension of the SIR model is motivated by the study of a vaccination policy. Individuals are not immediately immune after receiving an inoculation, and hence, the transport from the susceptible population to the recovered population via vaccination has a natural and important delay. Allowing u to be the constant percentage of the susceptible population that receives a vaccination, we study

$$\left.\begin{aligned}\frac{dS}{dt}(t) &= \rho - \alpha\, S(t)\, \frac{I(t)}{N(t)} - \delta\, S(t) - u\, S(t-\tau), \\ \frac{dI}{dt}(t) &= \alpha\, S(t)\, \frac{I(t)}{N(t)} - (\beta + \delta + \epsilon) I(t), \\ \frac{dR}{dt}(t) &= \beta\, I(t) - \delta\, R(t) + u\, S(t-\tau), \\ N(t) &= S(t) + I(t) + R(t),\end{aligned}\right\} \qquad (10.6)$$

where τ is the lag between being vaccinated and being immune. This delayed SIR model assumes that u percent of the susceptible population receives a vaccination at time $t - \tau$ and that these individuals enter the recovered population at time t. The delayed model importantly satisfies (10.5), and so, the total population is bounded.

Vaccination policies to help curb the affect of the H1N1 influenza strain in Morocco were studied with model (10.6). Parameters tailored to Morocco's demographics were:

$$\tau = 10, \quad \alpha = 0.3095, \quad \beta = 0.2, \quad \rho = 1174.17,$$
$$\epsilon = 0.0063, \text{ and } d = 3.9139 \times 10^{-5}.$$

Constant histories terminating with the initial conditions $S(0) = 30 \times 10^6$, $I(0) = 30$, and $R(0) = 28$ were assumed.

Figures 10.3 and 10.4 illustrate the affect of doubling u from 1 to 2% of the susceptible population. In particular, Fig. 10.4 suggests that the time during which the population is threatened reduces to about 60 days with a modest doubling of the vaccination rate, which is about half the time of the smaller u. Solutions were created with the Octave command,

`ode45d(@calcDP, [0,180], history, [lag], history, options),`

where

`lag = 10,`
`history = [30*10^(6); 30; 28], and`
`options = odeset('NormControl', 'on', 'MaxStep', 1).`

10.3 The El-Niño–La-Niña Oscillation

Fig. 10.3 The susceptible (green) and recovered (red) populations for $u = 0.01$ (dashed) and $u = 0.02$ (solid).

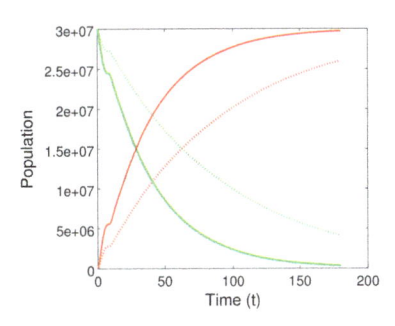

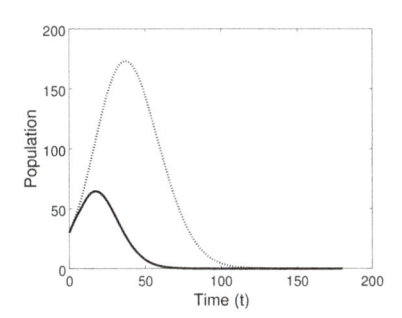

Fig. 10.4 The infected populations for $u = 0.01$ (dashed) and $u = 0.02$ (solid).

The function @calcDP with $u = 0.01$ is in Example Code 10.1.

```
function [dP] = calcDP(t, P, Pdel)
    u = 0.01;
    Gamma = 1174.17;
    beta = 0.3095;
    d = 3.9139*10^(-5);
    eps = 0.0063;
    gamma = 0.2;
    dP = zeros(3,1);
    N = P(1)+P(2)+P(3);
    dP(1) = Gamma - beta*P(1)*P(2)/N - u*Pdel(1,1);
    dP(2) = beta*P(1)*P(2)/N - (gamma+d+eps)*P(2);
    dP(3) = gamma*P(2) - d*P(3) + u*Pdel(1,1);
end
```

Example Code 10.1 A function to calculate the derivatives of the DDE in (10.6). The calculations of dP(1), dP(2), and dP(3) are, respectively, for S, I, and R

We note that the delayed SIR model motivated by the study of vaccinations does not have an intuitive re-expression in terms of an integral, i.e. it does not lend itself to being re-expressed in terms of a cumulative population. Hence, the vaccination model is distinct from the original filter model of this chapter. The vaccination delays are natural, important, and they accurately reflect the intent of the modeler. Since MATLAB and Octave both have solvers for delay equations, such models are welcome, and indeed encouraged, in similar situations.

10.3 The El-Niño–La-Niña Oscillation

Delay differential equations have been suggested as climatology models, and the following two have been used to describe the El-Niño–La-Niña oscillation,

$$\frac{dT}{dt} = -\alpha T(t - \tau) + T, \ \alpha > 0, \ \tau > 0, \tag{10.7}$$

and
$$\frac{dT}{dt} = -\alpha T(t-\tau) + T - T^3, \ \alpha > 0, \ \tau > 0. \tag{10.8}$$

The variable T is the ocean's surface temperature, and the dependence on $T(t-\tau)$ expresses a delayed decrease in the temperature's rate of change. This decrease is proportionate to the temperature observed τ units of time earlier. The remaining terms express the autonomous relationship between the current temperature and the rate at which the temperature is changing. The first model assumes the rate at which the current temperature is changing is the difference of the current temperature and a scaled prior temperature. The second model further assumes the rate of change in the temperature decreases with the cube of the current temperature.

We emphasize the importance of the history by considering solutions to systems (10.7) and (10.8) as h is one of $h(t) = 1$, $h(t) = e^t$, or $h(t) = 1 + 2\sin(t)$. Solutions with $\alpha = 1.2$ and $\tau = 10$ are depicted in Figs. 10.5 and 10.6. Observe the disparity in the solutions even though each satisfies $T(0) = h(0) = 1$. The solutions were calculated in MATLAB with ddesd. For example, the solution to system (10.7) with history $h(t) = e^t$ was calculated with

```
ddesd(@(t,T,Tdel)(-1.2*Tdel+T),[10],@(t)exp(t),[0,3]).
```

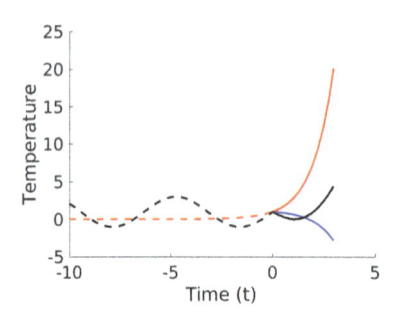

Fig. 10.5 Solutions to (10.7). The blue curve assumes the history $h(t) = 1$, the red curve $h(t) = e^t$, and the black curve $h(t) = 1 + 2\sin(t)$. Histories are dashed.

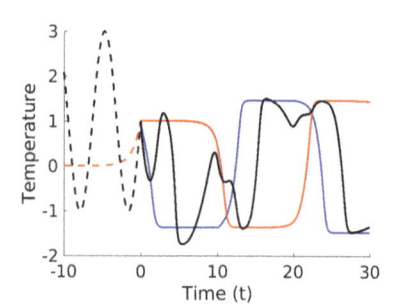

Fig. 10.6 Solutions to (10.8). The blue curve assumes the history $h(t) = 1$, the red curve $h(t) = e^t$, and the black curve $h(t) = 1 + 2\sin(t)$. Histories are dashed.

Legacy sea surface temperatures (SST data) exists for the El-Niño-La-Niña oscillation,[1] and we can use this data to accurately reflect the true historical record. We consider monthly measurements of the north pacific from 1948 through 2013. While the historical records imbue a pseudo-periodic oscillation, the trend doesn't appear sufficiently rhythmic to impose a periodic

---

[1] See *Indices of El Nino Evolution*, K. Trenberth and D. Stepaniak, J. Climate, 14, 1697–1701; with latest update at www.esrl.noaa.gov/psd/data/climateindices/List/#TNI.

10.4 Exercises

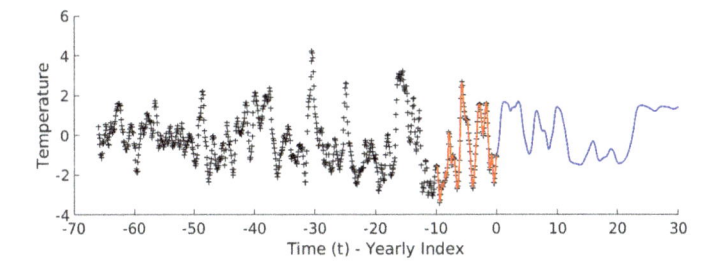

Fig. 10.7 Monthly historical data from 1948 through 2013 is plotted with the 'plus symbol'. A 10-year history is shown in red as a natural cubic spline of the measured data. System (10.8) is solved with this 10-year history, and the predicted temperatures for the next 30 years are shown in blue.

model, say with regression. Instead, we use a natural cubic spline to infer a smooth functional form and use this as our history. The results with $\alpha = 1.2$ and $\tau = 10$ for system (10.8) are shown in Fig. 10.7.

The lag of $\tau = 10$ means that we only use 10 of the 66 years of historical data. We could use the historical data to vet the model's efficacy. For example, we could crop the data so that it ended in year 2003, and then using $\tau = 10$ we could predict temperatures with one or both of (10.7) and/or (10.8). Our predictions could then be assessed against the true historical record. Moreover, we could search for a lag that would minimize the error of our predictive model. Such questions are considered in the exercises.

10.4 Exercises

1. Let $\hat{I}(t)$ be the rate at which people are infected at time t, and assume that people are sick for τ units of time. A reasonable assumption of the infected population in model (10.3) is that

$$I(t) = \int_{t-\tau}^{t} \hat{I}(y) \, dy.$$

Alter the SIR model in (10.3) to include this delayed construct. The new model should define dS/dt, dI/dt, dR/dt, and $\hat{I}(t)$, the latter of which is a rate. Does the adaptation lead to a model that necessitates or benefits from the introduction of a delay?

2. A two-delay model of the El-Niño–La-Niña cycle is

$$\frac{dT}{dt} = -\alpha \tanh(\kappa T(t - \tau_1)) + \beta \tanh(\kappa T(t - \tau_2)) + \gamma \cos(2\pi t). \quad (10.9)$$

Solve this model for the following parameter settings:

Instance	α	β	γ	κ	τ_1	τ_2
1	1	0	1	100	0.01	0
2	1	0	1	100	0.15	0
3	1	0	1	100	0.995	0
4	1	1	1	10	0.9	0.1
5	1.2	0.8	1	10	0.6	0.6

Solve each with the constant histories $T(t) = 1$ and $T(t) = 0$, for $t \leq 0$. Work with other histories to experiment with a delay solver and to sample the model's predictions.

3. Repeat Exercise 2 but use the sea surface data referenced in Fig. 10.7.
4. The following linear, neutral DDE has been used to study partial element circuits,
$$y'(t) = Ly(t) + My(t-\tau) + Ny'(t-\tau), \; t \geq 0,$$
where

$$L = 100 \begin{bmatrix} -7 & 1 & 2 \\ 3 & -9 & 0 \\ 1 & 2 & -6 \end{bmatrix}, \; M = 100 \begin{bmatrix} 1 & 0 & -3 \\ -1/2 & -1/2 & -1 \\ -1/2 & -3/2 & 0 \end{bmatrix},$$

$$\text{and } N = \frac{1}{72} \begin{bmatrix} -1 & 5 & 2 \\ 4 & 0 & 3 \\ -2 & 4 & 1 \end{bmatrix}.$$

Solve this problem with various values of τ and a history of
$$h(t) = (\sin(t), \sin(2t), \sin(3t)),$$
for $t \leq 0$. Explain the behavior of the circuit as the delay varies.

5. Use the SST historical SST data associated with Fig. 10.7. Tune model (10.8) for optimal historical efficacy by designing a search over reasonable values of α and τ. Construct a history that accurately reflects the legacy data for each (α, τ) pair as the history sweeps across its historical reference. Assess your model's predictions against the historical data that postdates the history. For instance, you could set $\alpha = 1.3$ and assume a 10-year history starting in 1950. Your model could then predict 10 years past the assumed end of your history, i.e. you could predict temperatures for the decade from 1960 through 1970 based on a history from 1950 through 1960. The model's error could be the maximum deviation between the historical record and the model's predictions over the 1960 through 1970 time frame. You should investigate assessment metrics based on the 1-, 2-, and inf-norms. Which settings would you defend as reasonable?

6. Repeat Exercise 5 with model (10.9).

10.4 Exercises

7. Design a model of your own creation to study wide-scale computations in a cloud computing environment. The model should be based on the following two assumptions:

   - Each processor can divide its capability into tasks such as managing its queue, e.g. by deciding which tasks to send to other processors, completing tasks on its queue, etc.
   - There is a time delay between processors. So if one processor requests information from another or sends a task to another, then there is a delay between when the information is sent and received.

   Draft a report that explains your model, its implementation details, and the computed results. You should investigate how performance depends on the time lags and possibly on the network's topology. Interested readers should be able to reproduce your experiments based solely on your report.

Chapter 11
Partial Differential Equations

Science is a differential equation. Religion is a boundary condition. – Alan Turing

We have the only cookbook in the world that has partial differential equations in it. – Nathan Myhrvold

The field of partial differential equations is arguably the workhorse of applied mathematics. While the field is steeped with a rich and fruitful history supporting volumes of research, our modest goal is to present a couple of the standard models and to show how to solve them with introductory methods. We encourage those who are interested to pursue further coursework in numerical analysis and computational science.

11.1 The Heat Equation

The heat equation explains how heat flows through a (uniform) medium, and in this section we model heat flow in a two-dimensional plate, though extensions to higher dimensions are common. Let $u(x, y, t)$ indicate the temperature of the plate at position (x, y) at time t, and consider a differential element of the plate centered at (x, y) with width Δx and depth Δy, see Fig. 11.1. The total heat contained within the element at time t is modeled as

$$u(x, y, t)\Delta x \Delta y.$$

This expression illustrates that temperature is an average kinetic energy over an area, or more generally a volume, and thus multiplying temperature by area gives a measure of thermal energy. The model assumes that $u(x, y, t)$ is the average temperature over the area of the element,

$$[x - \Delta x/2, x + \Delta x/2] \times [y - \Delta y/2, y + \Delta y/2].$$

Let $\mathbf{F}_x(x, y, t)$ be the rate, in units of heat per time per length, at which heat flows from left to right across a vertical boundary of fixed length centered at point (x, y) at time t. Let $\mathbf{F}_y(x, y, t)$ be the corresponding rate at which heat flows from bottom to top. The rates, in units of heat per time, at which heat flows across vertical and horizontal boundaries of corresponding lengths Δy and Δx are then $\mathbf{F}_x(x, y, t)\Delta y$ and $\mathbf{F}_y(x, y, t)\Delta x$.

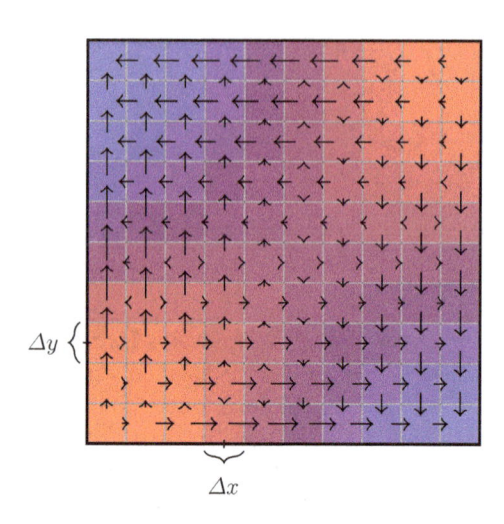

Fig. 11.1 A uniform two-dimensional slab. In each differential unit temperature is indicated by color, with red indicating high heat. Each arrow describes flux in the indicated direction, with longer arrows indicating higher flux.

If we assume that heat energy is conserved, and not used, for example, in thermal expansion, then the rate, in units of heat per time, at which heat within a differential element changes is the rate at which heat flows across the boundaries. In particular,

$$\frac{\partial (u \Delta x \Delta y)}{\partial t}(x, y, t) \approx \Delta y \mathbf{F}_x\left(x - \frac{\Delta x}{2}, y, t\right) - \Delta y \mathbf{F}_x\left(x + \frac{\Delta x}{2}, y, t\right)$$
$$+ \Delta x \mathbf{F}_y\left(x, y - \frac{\Delta y}{2}, t\right) - \Delta x \mathbf{F}_y\left(x, y + \frac{\Delta y}{2}, t\right). \tag{11.1}$$

The negative signs occur due to flow rates being signed, that is, they indicate flow from left to right or bottom to top.

Dividing by $\Delta x \Delta y$ in (11.1) and taking the limit at $\Delta x \to 0$ and $\Delta y \to 0$, we find that

$$\frac{\partial u}{\partial t} + \frac{\partial \mathbf{F}_x}{\partial x} + \frac{\partial \mathbf{F}_y}{\partial y} = 0.$$

11.1 The Heat Equation

If we let $\mathbf{F} = \langle \mathbf{F}_x, \mathbf{F}_y \rangle$ and ∇ be the operator $\nabla = \langle \partial/\partial x, \partial/\partial y \rangle$, then we arrive at the continuity equation,

$$\frac{\partial u}{\partial t} + \nabla \cdot \mathbf{F} = 0. \tag{11.2}$$

The vector valued function $\mathbf{F}$ is called the heat flux, though the word flux is more generally used to describe any flow. The continuity equation is a fundamental equation in physics applicable whenever we are describing the time evolution of a quantity that is conserved in all space regions. For example, the equation is fundamental in electrodynamics, fluid dynamics, and thermodynamics.

We use Fourier's law of heat conduction to complete a description of heat flow. The law is

$$\mathbf{F} = -\alpha \nabla u, \tag{11.3}$$

which asserts that heat flows from high to low temperatures at a rate (heat per length per time) proportional to the spatial temperature variation. Inserting $\mathbf{F}$ from (11.3) into the continuity equation gives the heat equation

$$\frac{\partial u}{\partial t} - \alpha \nabla \cdot \nabla u = 0. \tag{11.4}$$

The function $\nabla \cdot \nabla u$ is called the Laplacian of u and is often denoted Δu, which is not to be confused with a difference in u values. It is simply expressed as

$$\Delta u = \nabla \cdot \nabla u = \begin{pmatrix} \frac{\partial}{\partial x} \\ \frac{\partial}{\partial y} \end{pmatrix} \cdot \begin{pmatrix} \frac{\partial u}{\partial x} \\ \frac{\partial u}{\partial y} \end{pmatrix} = \frac{\partial^2 u}{\partial x^2} + \frac{\partial^2 u}{\partial y^2}.$$

In practice the coefficient α is a material property, often written as

$$\alpha = \frac{\kappa}{\rho c},$$

where ρ is the material density, c is the heat capacity of the material, and κ is the thermal conductivity. Moreover, the heat equation requires initial and boundary conditions to define a unique solution. Initial conditions describe the original state of the plate and are prescribed with a function $f(x,y)$ so that

$$u(x, y, 0) = f(x, y).$$

Boundary conditions have different forms depending on the physical scenario. For instance, if we have arranged things so that we can precisely control the temperature of the plate on its boundary, then we have a Dirichlet boundary condition of the form,

$$u(x, y, t) = g(x, y, t), \text{ if } (x, y) \text{ is on the boundary of the plate.}$$

In this case we assume the function $g(x, y, t)$ is known. Alternatively, the plate could be insulated, and we could then assume that no heat flows through the boundary. In this case we have a Neumann boundary condition,

$$\nabla u(x, y, t) \cdot n(x, y) = 0, \text{ if } (x, y) \text{ is on the boundary of the plate}.$$

The vector $n(x, y)$ is the unit vector orthogonal to the boundary at point (x, y) in the outward facing direction. Recall that the flux vector $\mathbf{F}(x, y, t)$ is related to the rate at which heat flows at (x, y). It can be shown that $\mathbf{F}(x, y, t) \cdot w$ is the rate at which heat flows in direction w at point (x, y). Recalling that $\mathbf{F} = -\alpha \nabla u$, we have that the Neumann boundary condition is

$$\mathbf{F}(x, y, t) \cdot n(x, y) = -\alpha \left(\nabla u(x, y, t) \cdot n(x, y) \right) = 0$$

for all boundary points (x, y). So the condition requires that no heat flows across the boundary of the plate.

11.2 Explicit Solutions by Finite Differences

Solving a partial differential equation is generally more complex than solving an ordinary differential equation. Part of the complication with numerical techniques is the requirement to discretize an n-dimensional region while accounting for the various boundary conditions. We consider a simple solution method that estimates a derivative with a finite difference. Substituting these differences into the heat equation provides an iterative calculation that steps forward in time, and hence, this technique is akin to Euler's method of Chap. 5. The algorithm only requires direct calculations from the problem statement and is thus explicit. Straightforward finite difference algorithms like the one below are often used as an initial solution technique, although more sophisticated procedures are commonly employed if this straightforward approach proves insufficient.

Consider the temperature of a well-insulated rod. If we think of the rod as one-dimensional and occupying space $0 \leq x \leq L$, then we need to solve the one-dimensional heat equation

$$\frac{\partial u}{\partial t} = \frac{k}{c\rho} \frac{\partial^2 u}{\partial x^2}.$$

The forward difference approximation of $\partial u / \partial t$ is

$$\frac{\partial u}{\partial t}(x, t) \approx \frac{u(x, t + \Delta t) - u(x, t)}{\Delta t},$$

and the central difference approximation of $\partial^2 u / \partial x^2$ is

11.2 Explicit Solutions by Finite Differences

$$\frac{\partial^2 u}{\partial x^2}(x,t) \approx \frac{1}{\Delta x}\left(\frac{u(x+\Delta x,t)-u(x,t)}{\Delta x}-\frac{u(x,t)-u(x-\Delta x,t)}{\Delta x}\right)$$

$$= \frac{u(x+\Delta x,t)-2u(x,t)+u(x-\Delta x,t)}{\Delta x^2}.$$

Substituting these approximations into

$$\frac{\partial u}{\partial t} = \frac{k}{c\rho}\frac{\partial^2 u}{\partial x^2},$$

we find

$$u(x, t + \Delta t) \approx$$
$$u(x,t) + \left(\frac{k\Delta t}{c\rho\Delta x^2}\right)(u(x+\Delta x,t)-2u(x,t)+u(x-\Delta x,t)). \qquad (11.5)$$

In a finite difference scheme we define a finite number of space- and time-nodes, and we attempt to estimate u at each of the space-node/time-node pairs by using a finite difference equation such as (11.5). Let us define evenly spaced nodes $x_1, x_2, \ldots, x_N$ and $t_1, t_2, \ldots, t_M$ satisfying:

$$x_1 = 0, \qquad\qquad t_0 = 0,$$
$$x_i = x_{i-1} + \Delta x,\ i > 1, \qquad t_k = t_{j-1} + \Delta t,\ j > 0,$$
$$x_N = L,\ \text{and} \qquad\qquad t_M = T.$$

Define $u_{i,j}$ to be our finite-difference approximation to $u(x_i, t_j)$. Referencing Eq. (11.5), we see that it is natural to require

$$u_{i,j+1} = u_{i,j} + \left(\frac{k\,\Delta t}{\rho c\,\Delta x^2}\right)(u_{i+1,j} - 2\,u_{i,j} + u_{i-1,j}). \qquad (11.6)$$

Calculating $u_{i,j}$ according to (11.6) gives an approximation of u so that $0 \le x \le L$ and $0 \le t \le T$ (Fig. 11.2).

Since the expression on the right-hand side of (11.6) depends on the temperature at t and not $t + \Delta t$, this equation estimates the temperature at $t + \Delta t$ from previously (estimated) temperatures. As an example, consider a well-insulated steel rod of length 2 cm, with thermal conductivity $k = 0.13$, heat capacity $c = 0.11$, and density $\rho = 7.8$. Let $\Delta x = 0.05$ and $\Delta t = 0.005$, and assume the initial heat profile increases linearly with x as

$$u(x, 0) = 2x + 1. \qquad (11.7)$$

Further assume the temperature on the left-hand endpoint is held constant at $1\,°\mathrm{C}$ and that the temperature on the right-hand endpoint decreases exponentially so that

$$u(0,t) = 1 \text{ and } u(2,t) = 4e^{-t} + 1. \tag{11.8}$$

In light of the initial condition (11.7), the boundary conditions in (11.8), and the finite difference equation (11.6), the full finite difference scheme is

$$u_{i,0} = 2x_i + 1, \ u_{1,j} = 1, \ u_{N,j} = 4e^{-t_j} + 1, \text{ and}$$

$$u_{i,j+1} = u_{i,j} + \left(\frac{k\Delta t}{\rho c \Delta x^2}\right)(u_{i+1,j} - 2u_{i,j} + u_{i-1,j}) \text{ at all interior nodes.}$$

There are stability concerns with the scheme above, but these can be controlled in the one-dimensional case by ensuring that Δx and Δt satisfy

$$\Delta t < \frac{1}{2}\frac{c\rho \Delta x^2}{k}.$$

The calculation of dt in Example Code 11.1 guarantees that Δt is chosen to be $0.45 \cdot c\rho \Delta x^2 / k$, which will satisfy the stability condition. If we alter this calculation to

```
dt = (0.51)*c*p*dx^2/k;
```

then the computed solution has the irrational behavior shown in Fig. 11.3.

Two items of note arise from this inequality. First, choosing Δt in such a way that the stability condition is violated will quickly lead to nonsensical solutions. It is often startling how very similar choices of Δt, one slightly respecting the stability condition and one violating it, will lead to wildly varying solutions. Second, the stability condition mandates that Δt decrease more substantially than Δx. If we wish to decrease Δx by a factor of 10 in order to achieve a more accurate solution, we will be required to decrease Δt by a factor of 100 in order to maintain stability. For this reason the scheme presented here is not used in applications where high accuracy is a concern.

We now consider a two-dimensional example that combines Dirichlet and Neumann boundary conditions. The geometry is that of a circular washer whose outer radius is 2 cm and whose inner radius is 1 cm. The washer is made of steel with $k = 0.13$, $c = 0.11$, and $\rho = 7.8$. The inner region is held at a constant temperature of $4\,°C$, and the washer's initial temperature is $0\,°C$ on its interior. We assume that the temperature along the outer boundary satisfies

$$\frac{\partial u}{\partial n} = \nabla u \cdot n = -0.3,$$

where n is an outward facing unit normal to the circle $x^2 + y^2 = 4$. This condition states that heat is being lost to the surrounding environment, and since the inner core is warmer than the disk, heat should flow from the inner radius, through the disk, and then into the surrounding environment. The problem is mixed since both Dirichlet and Neumann boundary conditions are used.

The two-dimensional heat equation is

11.2 Explicit Solutions by Finite Differences

```
k = 0.13;
c = 0.11;
p = 7.8;
dx = 0.05;
dt = (0.45)*c*p*dx^2/k;
EndTime = 10;
RodLength = 2;
NumPlots = 10;

PosX = unique([0:dx:RodLength,
RodLength]);
Time = unique([0:dt:EndTime, EndTime]);

Sol = zeros( length(PosX),
length(Time));
Sol(:,1) = 2*PosX + 1;
Sol(1,:) = 1;
Sol(length(PosX),:) = 4*exp(-Time)+1;

r = (k*dt)/(c*p*dx^2);
indx = 2:length(PosX)-1;
for t=2:length(Time)
  Sol(indx,t) = (1-2*r)*Sol(indx,t-1) ...
      + r*(Sol(indx-1,t-1) +
Sol(indx+1,t-1));
end
```

Example Code 11.1 Code to solve the one-dimensional heat equation with the finite difference iteration in (11.6) and the initial conditions in (11.7) and Dirichlet boundary conditions in (11.8)

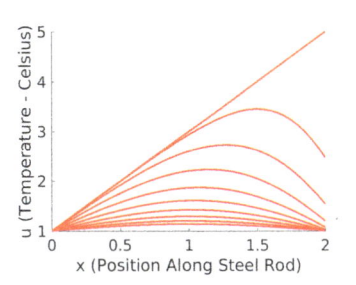

Fig. 11.2 Solutions for the one-dimensional heat equation using the finite difference iteration in (11.6) with the Dirichlet boundary conditions in (11.7) and (11.8).

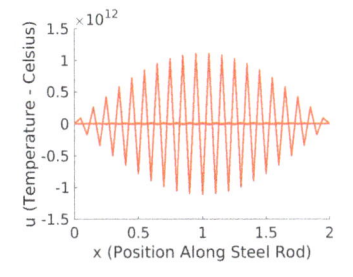

Fig. 11.3 Incorrect results with $k\,\Delta t/c\,\rho\,\Delta x^2$ being too large.

$$\frac{\partial u}{\partial t} = \frac{k}{c\rho}\left(\frac{\partial^2 u}{\partial x^2} + \frac{\partial^2 u}{\partial y^2}\right).$$

Employing a forward difference in the time derivative and a centered difference for both space derivatives, similar to the motivation for (11.6), we arrive at the scheme

$$u_{i,j,k+1} = u_{i,j,k} + \left(\frac{k\Delta t}{c\rho\Delta x^2}\right)(u_{i+1,j,k} - 2u_{i,j,k} + u_{i-1,j,k})$$
$$+ \left(\frac{k\Delta t}{c\rho\Delta y^2}\right)(u_{i,j+1,k} - 2u_{i,j,k} + u_{i,j-1,k}). \quad (11.9)$$

In this method we use $u_{i,j,k}$ as the approximation to $u(x_i, y_j, t_k)$. Similar to the finite difference scheme for the one-dimensional heat equation, the space and time steps must meet a certain stability condition. In this case we must have

$$\Delta t < \frac{c\rho \Delta x^2 \Delta y^2}{2k\left(\Delta x^2 + \Delta y^2\right)}.$$

The inner boundary condition of $u(x, y, t) = 4$ if $x^2 + y^2 = 1$ is approximated by $u_{i,j,k} = 4$ if $|x_i^2 + y_j^2 - 1| \approx 0$. The code that generated Figs. 11.4, 11.5, 11.6, and 11.7 set $u_{i,j,k} = 4$ if $|x_i^2 + y_j^2 - 1| < \min\{\Delta x, \Delta y\}/1.5$. The outer boundary condition requires that $\nabla u = -0.3\,n$, which requires more care to implement. Consider a point (x, y) on the boundary of the domain. The average of the forward and backward differences of the spatial partials gives the following approximations:

$$\frac{\partial u}{\partial x}(x, y, t) \approx \frac{u(x + \Delta x, y, t) - u(x - \Delta x, y, t)}{2\Delta x} = -0.3\,n_1 \quad \text{and}$$

$$\frac{\partial u}{\partial y}(x, y, t) \approx \frac{u(x, y + \Delta y, t) - u(x, y - \Delta y, t)}{2\Delta y} = -0.3\,n_2. \qquad (11.10)$$

We further have for our problem that

$$n_1 = \frac{x - x_0}{\sqrt{(x - x_0)^2 + (y - y_0)^2}} \quad \text{and} \quad n_2 = \frac{y - y_0}{\sqrt{(x - x_0)^2 + (y - y_0)^2}}, \qquad (11.11)$$

where (x_0, y_0) is the center of the washer.

If n_1 and n_2 are both positive, then $(x + \Delta x, y + \Delta y)$ is not part of the washer. However, for the sake of implementing the boundary conditions we introduce a "faux" temperature at $(x + \Delta x, y + \Delta y)$, as this point should not be far outside the washer. Then, combining (11.10) and (11.11) we require

$$u_{i+1,j,k} = u_{i-1,j,k} - \frac{0.6\,(x - x_0)\,\Delta x}{\sqrt{(x - x_0)^2 + (y - y_0)^2}} \quad \text{and}$$

$$u_{i,j+1,k} = u_{i,j-1,k} - \frac{0.6\,(y - y_0)\,\Delta y}{\sqrt{(x - x_0)^2 + (y - y_0)^2}}.$$

These approximations are used in (11.9) along the northeast portion of the outer boundary since they depend only on temperature evaluations within the geometry. If n_1 and n_2 had instead been negative with (x, y) on the southwest portion of the outer border, then we would have used

$$u_{i-1,j,k} = u_{i+1,j,k} + \frac{0.6\,(x - x_0)\,\Delta x}{\sqrt{(x - x_0)^2 + (y - y_0)^2}} \quad \text{and}$$

$$u_{i,j-1,k} = u_{i,j+1,k} + \frac{0.6\,(y - y_0)\,\Delta y}{\sqrt{(x - x_0)^2 + (y - y_0)^2}}.$$

The cases in which n_1 and n_2 differ in sign use similar approximations.

11.3 The Wave Equation

If we set $\Delta x = \Delta y = 0.1$, and $\Delta t = 0.0037125$, which guarantees stability, then calculating the temperature distribution through the first 2 s results in the heat distributions shown in Figs. 11.4, 11.5, 11.6, and 11.7. Initially the only nonzero temperature is along the inner boundary, but the temperature approaches uniformity as time progresses. After 2 s the temperature is approximately constant at 4 °C.

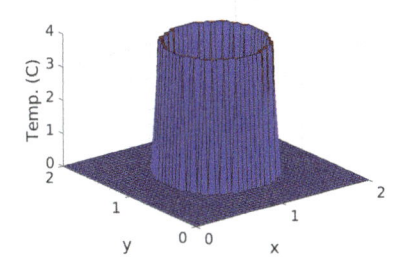

Fig. 11.4 Heat distribution at $t = 0$ s.

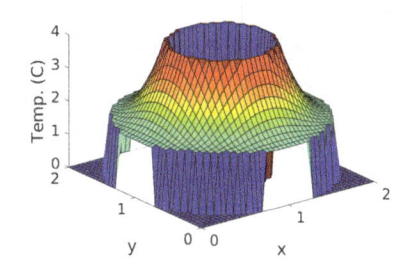

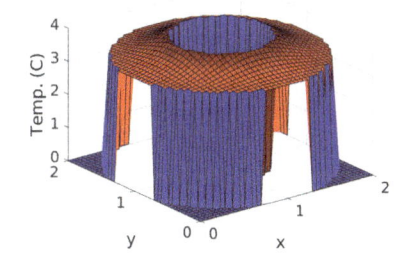

Fig. 11.5 Heat distribution at $t = 2/3$ s.

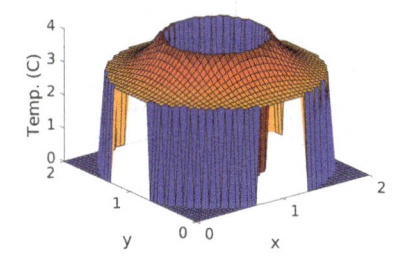

Fig. 11.6 Heat distribution at $t = 4/3$ s.

Fig. 11.7 Heat distribution at $t = 2$ s.

11.3 The Wave Equation

Another classical partial differential equation is the wave equation, which like the heat equation is stated in terms of the Laplacian,

$$\frac{\partial^2 u}{\partial t^2} = \alpha \left(\nabla \cdot \nabla u \right).$$

As before, ∇ is the partial differentiation operator with respect to the spatial coordinates. This equation is germane to numerous applications, with the most common being the original interpretation of the movement of a vibrating wire.

Consider a taught horizontal wire of length L, and suppose the wire is displaced vertically. Let $u(x,t)$ be the vertical displacement of the wire at position x at time t. Moreover, assume the only force exerted on the wire is that of tension. A depiction of the wire is in Fig. 11.8.

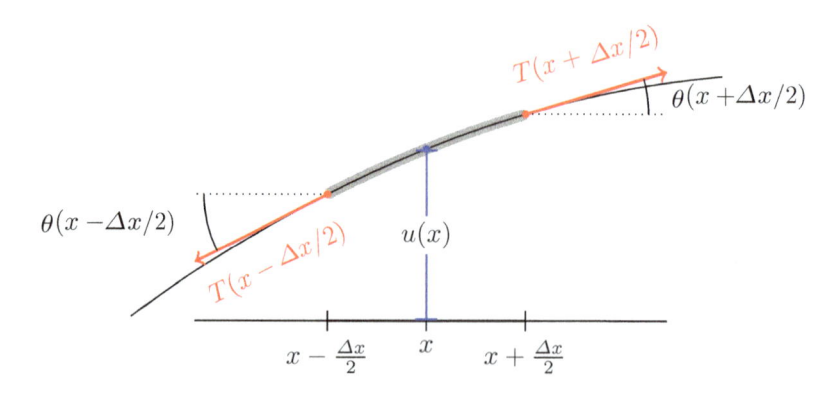

Fig. 11.8 The depiction of a small amount of stretched wire is highlighted in gray. We assume that tension is the only force acting on the wire; the magnitude of the tension force at point x is $T(x)$.

We regard a differential element of wire occupying space near position x as a point-mass. At each end of the section of wire tension force is acting in the direction tangent to the wire, say with magnitude $T(x)$. Applying Newton's second law,
$$ma = F_{net},$$
to the differential element of wire produces
$$\rho \Delta x \frac{\partial^2 u}{\partial t^2}(x,t) = F_{net}, \tag{11.12}$$
where ρ is the linear density of the wire at position x and F_{net} is the net vertical force.

Let us assume that the wire only moves vertically, a reasonable assumption if the displacement is small and the wire is taught. In this case the horizontal components of the tension forces balance, which implies
$$\cos\left(\theta\left(x - \frac{\Delta x}{2}\right)\right) T\left(x - \frac{\Delta x}{2}\right) = \cos\left(\theta\left(x + \frac{\Delta x}{2}\right)\right) T\left(x + \frac{\Delta x}{2}\right).$$

This equality holds at all points x, and as such, the magnitude of the horizontal component of the tension vector is common at all points on the string. We let H be this common value, and hence,

11.3 The Wave Equation

$$H = \cos(\theta(x))T(x) \text{ for any } x. \tag{11.13}$$

The net vertical force acting on the wire is

$$F_{net} = \sin\left(\theta\left(x + \frac{\Delta x}{2}\right)\right)T\left(x + \frac{\Delta x}{2}\right) - \sin\left(\theta\left(x - \frac{\Delta x}{2}\right)\right)T\left(x - \frac{\Delta x}{2}\right).$$

From (11.13) we find that at any point x,

$$T(x)\sin(\theta(x)) = H\tan(\theta(x)),$$

and so

$$F_{net} = H\left[\tan\left(\theta\left(x + \frac{\Delta x}{2}\right)\right) - \tan\left(\theta\left(x - \frac{\Delta x}{2}, t\right)\right)\right].$$

The value $\tan(\theta(x))$ is the slope of the tension vector at point x. Because the tension force is acting tangent to the displacement of the wire, we have that

$$\tan(\theta(x)) = \frac{\partial u}{\partial x}(x, t).$$

We now have that

$$F_{net} = H\left[\frac{\partial u}{\partial x}\left(x + \frac{\Delta x}{2}, t\right) - \frac{\partial u}{\partial x}\left(x - \frac{\Delta x}{2}, t\right)\right].$$

Substituting this expression of F_{net} into (11.12) and solving for $\frac{\partial^2 u}{\partial t^2}(x, t)$ gives

$$\frac{\partial^2 u}{\partial t^2}(x, t) = \frac{H}{\Delta x \rho}\left[\frac{\partial u}{\partial x}\left(x + \frac{\Delta x}{2}, t\right) - \frac{\partial u}{\partial x}\left(x - \frac{\Delta x}{2}, t\right)\right].$$

The wave equation is the result upon letting $\Delta x \to 0$,

$$\frac{\partial^2 u}{\partial t^2}(x, t) = \frac{H}{\rho}\frac{\partial^2 u}{\partial x^2}(x, t). \tag{11.14}$$

The wave equation requires boundary and state conditions to identify a specific solution. The most common boundary conditions are the Dirichlet conditions,

$$u(0, t) = \alpha, \quad \text{and} \quad u(L, t) = \beta.$$

These conditions state that the ends of the wire stay in fixed positions as time progresses. We must specify the initial state of the wire in addition to the boundary conditions, including both initial displacement and velocity. The initial conditions are given in the form

$$u(x,0) = f(x), \quad \text{and} \quad \frac{\partial u}{\partial t}(x,0) = g(x).$$

One simple method for approximating a solution to the wave equation numerically is premised on inserting central difference approximations into the second partials, which produces

$$\frac{u(x, t+\Delta t) - 2u(x,t) + u(x, t - \Delta t)}{\Delta t^2}$$

$$\approx \frac{H}{\rho} \frac{u(x+\Delta x, t) - 2u(x,t) + u(x - \Delta x, t)}{\Delta x^2}.$$

This motivates us to require that our approximation satisfies

$$u_{i,j+1} = 2u_{i,j} - u_{i,j-1} + \left(\frac{H \Delta t^2}{\rho \Delta x^2}\right)(u_{i+1,j} - 2u_{i,j} + u_{i-1,j}), \quad (11.15)$$

where $u_{i,j}$ approximates $u(x_i, t_j)$.

Suppose we have the initial conditions

$$u(x,0) = 0 \quad \text{and} \quad \frac{\partial u}{\partial t}(x,0) = e^{-10(x-L/2)^2} - e^{-10(L/2)^2}, \quad (11.16)$$

along with the boundary condition $u(0,t) = u(L,t) = 0$. The derivative initial condition implies that

$$u(x, \Delta t) \approx u(x, 0) + \frac{\partial u}{\partial t}(x, 0)\, \Delta t = \left(e^{-10(x-L/2)^2} - e^{-10(L/2)^2}\right) \Delta t.$$

We therefore require that our approximation satisfies

$$u_{i,2} = u_{i,1} + \left(e^{-10(x_i - L/2)^2} - e^{-10(L/2)^2}\right) \Delta t.$$

Hence we have (estimates) for displacements at $t = 0$ and $t = \Delta t$, which combine with the boundary condition to specify the displacement values required by the iterative step in (11.15). If we have $H/\rho = 0.1$, then several of the computed solutions from Example Code 11.2 with $L = 2$, $\Delta x = 0.01$ and $\Delta t = 0.028$ are in Fig. 11.9. This scheme for approximating a solution to the wave equation requires that

$$\Delta t < \Delta x \sqrt{\frac{\rho}{H}}$$

in order to remain stable.

The wave equation extends to higher dimensions, and some of the exercises seek such sequels. Other adaptations include damping and forcing terms.

11.4 Exercises

```
L = 2;
T = 30;
alpha = 0.1;
dt = 0.028;
dx = 0.01;
PosX = unique([0:dx:L,L]);
Time = unique([0:dt:T,T]);
Sol = zeros(length(PosX), length(Time));
Sol(:,2) = (exp(-10*(PosX-(L/2)).^2) ...
   - exp(-10*(L/2).^2))*dt;
r = alpha*dt^2 / dx^2;
indx = 2:length(PosX)-1;
for j = 3:length(Time)
  Sol(indx,j) = 2*Sol(indx,j-1) - ...
    Sol(indx,j-2) + r*(Sol(indx+1,j-1) - ...
    2*Sol(indx,j-1) + Sol(indx-1,j-1));
end
```

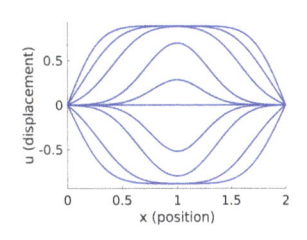

Fig. 11.9 Ten solutions equally spaced with respect to time over the interval $0 \le t \le 10$.

Example Code 11.2 MATLAB/Octave code to solve an instance of the wave equation.

11.4 Exercises

1. Solve the heat equation over the region depicted below. Assume the Dirichlet condition that $u(x, y, t) = 60\,°C$ if (x, y) is on either the left or the right side of the central square. Assume the Neumann condition

$$\frac{\partial u}{\partial n}(x, y, t) = \nabla u(x, y, t) \cdot n = -1,$$

provided that (x, y) is a point on the hemispherical top or bottom, where n is an outward facing normal. The initial temperature is $u(x, y, 0) = 60$. Consider thermal parameters for several materials of your choice, e.g. copper, nickel, tin, etc., and assess the uniformity of the object's temperature after 2 s. Which material would you select for the most uniform heat profile?

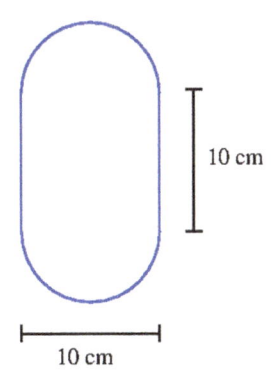

2. The result of adding a damping term to (11.14) is

$$\frac{\partial^2 u}{\partial t^2} = \frac{H}{\rho}\frac{\partial^2 u}{\partial x^2} - \kappa \frac{\partial u}{\partial t},$$

where damping is assumed proportional to velocity. Develop and implement an explicit solution procedure and experiment with values of κ to estimate the smallest value for which displacement remains nonnegative. Use the initial conditions in (11.16).

3. Solve the following wave equation over the unit disc,

$$\frac{\partial^2 u}{\partial t^2} = 2\left(\nabla \cdot \nabla u\right), \ u(x,y,0) = \frac{x^2 + y^2}{10}, \ \frac{\partial u}{\partial t}(x,y,0) = 0,$$

with the boundary condition that $u(x,y,t) = 0.1$ if $x^2 + y^2 = 1$.

4. Extend the previous problem to solve the wave equation over the unit disc for various p-norms,

$$\frac{\partial^2 u}{\partial t^2} = 2\left(\nabla \cdot \nabla u\right), \ u(x,y,0) = \frac{\|(x,y)\|_p}{10}, \ \frac{\partial u}{\partial t}(x,y,0) = 0,$$

with the boundary condition that $u(x,y,t) = 0.1$ if $\|(x,y)\|_p = 1$. Solve this problem for $p = 1$, $p = 3$, and $p = \infty$.

5. Finite difference solvers require substantial space and time for fine-mesh discretizations, and MATLAB and Octave can become unbefitting in such cases. Extend MATLAB, Octave, or Python by writing a finite difference solver in C\C++ to solve the two-dimensional heat equation, say as either a mex-file or by employing SWIG (see Chap. 7). Your extension should return a set of states that can be used to create a movie illustrating the flow of heat. Use your extension to solve Exercise 1.

6. Repeat Exercise 5 for the two-dimensional wave equation and solve Exercise 3.

Chapter 12
Modeling with Optimization and Simulation

> For since the fabric of the universe is most perfect, and is the work of the most wise Creator, nothing whatsoever takes place in the universe in which some relation of maximum or minimum does not appear. – Leonhard Euler
>
> True optimization is the revolutionary contribution of modern research to decision processes. – George Dantzig
>
> If you optimize everything, you will always be unhappy. – Donald Knuth

Earlier chapters have already developed several examples demonstrating how an optimal property can characterize a computational study. For instance, optimization was used in the method of least squares in Sects. 3.1 and 3.2, the development of principal component analysis in Sect. 3.4, the examples and techniques in Chap. 4, and the design of radiotherapy treatments in Sect. 8.2. This chapter introduces two models associated with optimization and simulation, the latter of which is regularly employed within a search for optimality, see Sects. 4.3.1 (simulated annealing) and 4.3.2 (genetic algorithms) as examples. The first model of this chapter optimizes the selection of stocks for a portfolio, where stock prices are stochastic and simulated. The second model uses simulation to predict abrupt changes in a material property. We specifically study the Ising model to computationally illustrate magnetic phase transitions. Both models are famous and have had profound impacts.

12.1 Stock Pricing and Portfolio Selection

We consider the problem of selecting stocks to create a portfolio. The goal is to simulate future stock prices and then use them to construct an efficient portfolio, meaning that the portfolio is optimally balanced between risk and return. Specifically, a portfolio is efficient if every portfolio with less risk has lower return and if every portfolio with higher return has higher risk. The surface of such portfolios is called the efficient frontier or the Pareto surface, and we present a quadratic programming problem whose solution is efficient.

12.1.1 Stock Pricing

We begin with a brief discussion of the stochastic differential equation used to model a stock's price as a function of time. The standard, stochastic differential equation defining the stock's price $P(t)$ is

$$dP = \mu P\, dt + \sigma P\, dz, \qquad (12.1)$$

where μ is a drift parameter and σ is a volatility parameter. This differential equation expresses the marginal change in a stock's price as the sum of the deterministic term $\mu P\, dt$ and the random term $\sigma P\, dz$. We first note that if there is no volatility, that is if $\sigma = 0$, then the differential equation reduces to

$$dP = \mu P\, dt,$$

and the solution is $P(t) = P(0)\, e^{\mu t}$. So the model assumes, quite naturally, that an investment's growth is that of continually compounded interest with a rate of μ if there is no volatility.

The stochastic term $\sigma P\, dz$ has an intuitive interpretation based on the assumption that z is a Weiner process, which means that $z(t)$ is a time dependent random variable. Hence $z(t)$ is distributed with respect to a probability distribution for each t. The assumption of a Weiner process means that z has the following properties.

Conditions that define $z(t)$ as a Weiner process

- $z(0) = 0$.
- $z(t)$ and $z(\tau)$ are independent random variables for any $t \neq \tau$.
- $z(t)$ is sufficiently continuous,[1] and
- $z(t) - z(\tau)$ is normally distributed with mean 0 and variance $t - \tau$.

---

[1] z need not be continuous in the usual manner, but rather, the probability that z is continuous almost everywhere needs to be 1.

12.1 Stock Pricing and Portfolio Selection

The last condition shows that

$$z(t + \Delta t) - z(t) = \sqrt{\Delta t}\, \phi,$$

where ϕ is a standard normal variable. We have the following differential expression as Δt decreases to zero,

$$dz = \sqrt{dt}\, \phi.$$

We focus our discussion on the stochastic modeling assumption by assuming for the moment that there is no deterministic drift. In this case $\mu = 0$ and (12.1) reduces to

$$dP = \sigma P\, dz = \sigma P \sqrt{dt}\, \phi.$$

Hence, for sufficiently small Δt, we have

$$\frac{P(t + \Delta t) - P(t)}{P(t)} \approx \sigma \sqrt{\Delta t}\, \phi.$$

This relation expresses our stochastic modeling assumption, which is that the relative change in a stock's price is normally distributed with variance $\sigma^2 \Delta t$. So the relative change in a stock's price without drift is equally likely to increase or decrease due to the symmetry of the normal distribution. Moreover, the variance of the relative price increases linearly with Δt, and the standard deviation is the volatility at $\Delta t = 1$.

Other probabilistic claims are possible. For example, suppose that $P(t) = \$100$, the volatility is 0.15, and that there is no drift. If $\Delta t = 1/10$, then the probability of the relative change of the stock's price being between $-0.15/\sqrt{10}$ and $0.15/\sqrt{10}$ is approximately 68.2%, i.e. the probability that a normal variable is within one standard deviation of its mean is approximately 68.2%. For this stock we have a 68.2% likelihood of the future price satisfying

$$95.26 \approx 100 - \frac{0.15}{\sqrt{10}} 100 \leq P(t + 1/10) \leq 100 + \frac{0.15}{\sqrt{10}} 100 \approx 104.74.$$

If we increase Δt to 1, then the same calculation gives a 68.2% likelihood of satisfying

$$85 \leq P(t + 1) \leq 115.$$

A discrete, random approximation of a stock's prices can be generated by repeatedly sampling from a normal distribution and then updating the price with

$$P(t + \Delta t) \approx P(t) + \mu P(t) \Delta t + \sigma P(t) \sqrt{\Delta t}\, \phi, \qquad (12.2)$$

which is the discrete counterpart to (12.1). The iterates from this calculation form a random walk, also commonly called a random path or a sample path. Consider the example above with $P(0) = \$100$, $\mu = 0.03$ (annual return), and $\sigma = 0.15$ (annual volatility). Assume there are 260 work days per year so that $\Delta t = 1/260$. Further assume that we observe the following 10 sequential, independent random samples of ϕ,

t (day)	1	2	3	4	5	6	7	8	9	10
ϕ	1.21	0.24	−1.95	−1.02	−2.15	1.37	−0.30	0.38	1.37	1.25

The sample path for the stock's price as it changes per day is then

$P(0)$	$P(1)$	$P(2)$	$P(3)$	$P(4)$	$P(5)$	$P(6)$	$P(7)$	$P(8)$	$P(9)$	$P(10)$
100.00	101.14	101.38	99.55	98.61	96.65	97.90	97.63	97.99	99.25	100.42

Several sample paths for the stock are shown in Figs. 12.1 and 12.2. Notice that the general trends differ in these two simulations even though the parameters are identical. Figure 12.1 suggests large possible returns, whereas Fig. 12.2 is more pessimistic and suggests possible losses. The difference is of course due to the random sample of ϕ.

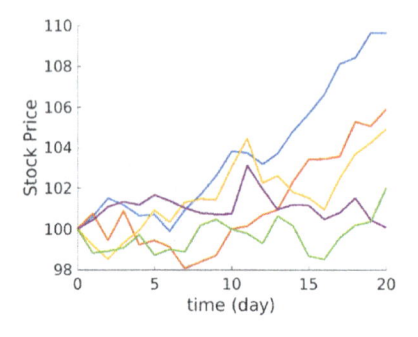

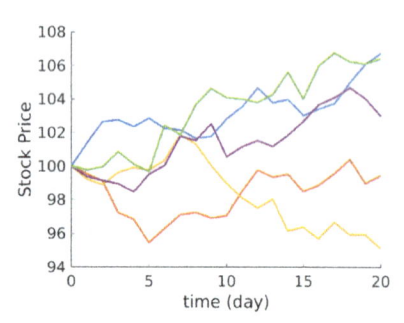

Fig. 12.1 Five sample paths of a stock's price with $P(0) = 100$, $\mu = 0.03$ and $\sigma = 0.15$.

Fig. 12.2 Five sample paths of a stock's price with $P(0) = 100$, $\mu = 0.03$ and $\sigma = 0.15$.

Itō's Lemma and Solving Stochastic Differential Equations

Solving (12.1) mathematically requires a foray into stochastic calculus, which extends the study of differentials and integrals into the realm of stochastic processes. Itō's Lemma shows how to calculate stochastic differentials, and we use it to solve (12.1).

Lemma 1 (Itō's Lemma). *Let $f(P, t)$ be twice differentiable and P be a solution to (12.1). Then*

$$df = \left(\frac{\partial f}{\partial t} + \mu P \frac{\partial f}{\partial P} + \frac{\sigma^2 P^2}{2} \frac{\partial^2 f}{\partial P^2} \right) dt + \sigma P \frac{\partial f}{\partial P} dz.$$

Itō's Lemma doesn't explicitly solve (12.1), but rather, the lemma states that if we already have a solution $P(t)$ together with a twice differentiable function

12.1 Stock Pricing and Portfolio Selection

f, then the differential df can be calculated with the stated expression. Itō's Lemma establishes a necessary condition of any solution P, and while the result might appear odd, such necessary statements are common and useful.

Extracting practical merit from a necessary condition often requires a clever application, and the standard tactic in this case is to select

$$f(P,t) = \ln(P).$$

We then have from Itō's Lemma that

$$d\ln(P) = \left(\frac{\partial \ln(P)}{\partial t} + \mu P \frac{\partial \ln(P)}{\partial P} + \frac{\sigma^2 P^2}{2} \frac{\partial^2 \ln(P)}{\partial P^2}\right) dt + \sigma P \frac{\partial \ln(P)}{\partial P} dz$$

$$= \left(0 + \mu P \left(\frac{1}{P}\right) + \frac{\sigma^2 P^2}{2} \left(\frac{-1}{P^2}\right)\right) dt + \sigma P \left(\frac{1}{P}\right) dz$$

$$= \left(\mu - \frac{\sigma^2}{2}\right) dt + \sigma dz.$$

Using the assumption that z is a Weiner process, we have for any future time τ that

$$\ln\left(\frac{P(\tau)}{P(0)}\right) = \ln(P(\tau)) - \ln(P(0)) = \int_0^\tau d\ln(P)$$

$$= \int_0^\tau \left(\mu - \frac{\sigma^2}{2}\right) dt + \int_0^\tau \sigma\, dz$$

$$= \left(\mu - \frac{\sigma^2}{2}\right)\tau + \sigma\left(z(\tau) - z(0)\right)$$

$$= \left(\mu - \frac{\sigma^2}{2}\right)\tau + \sigma\sqrt{\tau}\,\phi.$$

We conclude upon replacing τ with t that $\ln(P(t)/P(0))$ is a normally distributed random variable with mean $(\mu - \sigma^2/2)\,t$ and variance $\sigma^2 t$. Notice that if $\mu - \sigma^2/2 > 0$, then we expect the price to increase. However, if the volatility increases so that $\mu - \sigma^2/2 < 0$, then we expect the price to decrease even if the drift parameter is positive. This captures the reality that if a stock's volatility is sufficiently large, then we might expect to lose money. Solving for P results in the anticipated exponential form of the solution, which is

$$P(t) = P(0)e^{\left(\mu - \frac{\sigma^2}{2}\right)t + \sigma\sqrt{t}\,\phi}. \tag{12.3}$$

If volatility is zero, then again notice that the outcome is a continuous compound interest model with rate μ. However, the expected price incurs a degradation of $\sigma^2/2$ in the presence of volatility, and the expectation can be negative if σ is sufficiently large, even if the average trend μ is positive. Of course there is also the possibility of huge gains if σ is large.

12.1.2 Portfolio Selection

One of the most steeped tenets of investing is that of diversification, and investing in a range of assets generally distributes an investor's exposure to risk over the broad economy. The sentiment is that a diversified portfolio is more likely to approximate the overriding economy's (hopefully) upward trend by imitating the economy's aggregate return. So while some investments might be doing well, others might be doing poorly. The goal is to have the aggregate trend with the market as the performances of individual assets meander over time.

The interplay between risk and return suggests a reciprocal relationship. Diversified, low-risk portfolios should have a reduced anticipated return to counter their diminished risk. Less diversified, high-risk portfolios have the potential of high returns, but they simultaneously have the downside of increased risk of loss. Low-risk portfolios tend to have less variability, and hence, they are more certain about their return. High-risk portfolios typically have higher variability and less certainty about their returns. So a high-risk portfolio increases the chance of both an outstanding return and a catastrophic loss. Selecting a collection of investments to balance the trade-off between risk and return, called portfolio selection, is innately an optimization problem, and we consider one of the classical models in this section.

Suppose we have V dollars to invest in n stocks, indexed by i. Let $x_i(t)$ be the number of shares invested in stock i at time t so that the value of the portfolio at time t is

$$V(t) = \sum_{i=1}^{n} P_i(t) x_i(t),$$

where $P_i(t)$ is the price of stock i at time t. Let $w_i(t)$ be the percentage, often called the weight, invested in stock i,

$$w_i(t) = \frac{P_i(t) x_i(t)}{V(t)}.$$

The per invested dollar return over a time interval of length Δt for stock i is

$$R_i(t) = \frac{(P_i(t + \Delta t) - P_i(t)) x_i(t)}{P_i(t) x_i(t)} = \frac{P_i(t + \Delta t) - P_i(t)}{P_i(t)}, \qquad (12.4)$$

and the per invested dollar return of the entire portfolio is

12.1 Stock Pricing and Portfolio Selection

$$R(t) = \frac{V(t+\Delta t) - V(t)}{V(t)}$$

$$= \frac{1}{V(t)} \sum_{i=1}^{n} (P_i(t+\Delta t) - P_i(t)) x_i(t)$$

$$= \sum_{i=1}^{n} \left(\frac{P_i(t)x_i(t)}{V(t)}\right) \left(\frac{P_i(t+\Delta t) - P_i(t)}{P_i(t)}\right)$$

$$= \sum_{i=1}^{n} w_i(t) R_i(t). \tag{12.5}$$

Recall that the stochastic differential equation in (12.1) is our assumed model of a stock's price. Hence, for each stock we have

$$\frac{dP_i}{P_i} = \mu_i \, dt + \sigma_i \, dz = \mu_i \, dt + \sigma_i \sqrt{dt} \, \phi,$$

which we approximate with a forward difference as

$$R_i(t) = \frac{P_i(t+\Delta t) - P_i(t)}{P_i(t)} = \mu_i \Delta t + \sigma_i \sqrt{\Delta t} \, \phi. \tag{12.6}$$

This approximation expresses the per invested dollar return on stock i as a normal random variable with mean $\mu_i \Delta t$ and variance $\sigma_i^2 \Delta t$. Substituting the mean into (12.5) shows that the expected return is

$$\sum_{i=1}^{n} w_i \mu_i \Delta t.$$

If we let r be the vector with $r_i = \mu_i \Delta t$, then the expected return at time t is $r^T w$, where w is the corresponding weight vector.

The risk of investing in any single stock is often measured in terms of the variance of the stock's return. We assume from (12.6) that the risk of making an investment of $x_i(t)$ dollars at time t in stock i is

$$\left(\frac{P_i(t)x_i(t)}{V(t)} \sigma_i \sqrt{\Delta t}\right)^2 = w_i^2(t) \sigma_i^2 \, \Delta t.$$

This risk model multiplies the square of the portfolio's weight in stock i by the stock's variance, and it imposes the philosophy that low risk portfolios can only include large holdings of stocks provided that they have low volatilities.

A portfolio's risk is not typically the sum of the individual stocks' variances because stock prices often correlate. For example, investments in construction and concrete might trend similarly since they support each other, whereas investments in bread and tortillas might trend oppositely since they compete against one another. We model such trends with a correlation factor, denoted by ρ_{ij}, which measures how the i-th and j-th prices trend in tandem, see Sect. 3.2.1. The correlation factor is always between -1 and 1, with 1 indicating a perfect correlation, i.e. the stocks trend up and down identically, and a -1 indicating a perfect negative correlation, i.e. the stocks always trend opposite to one another. The correlation between a stock and itself is always $\rho_{ii} = 1$.

We assign the risk of co-investing in stocks i and j to be C_{ij}, and the weighted risk of co-investing in stocks i and j is

$$\left(\frac{P_i(t)x_i(t)}{V(t)}\sigma_i\sqrt{\Delta t}\right)\left(\frac{P_j(t)x_j(t)}{V(t)}\sigma_j\sqrt{\Delta t}\right)\rho_{ij} = w_i(t)w_j(t)\left(\sigma_i\sigma_j\Delta t\rho_{ij}\right)$$
$$= w_i(t)w_j(t)C_{ij}.$$

Observe that if $i = j$, then we have the case of investing in a single stock since $\rho_{ii} = 1$. The covariance matrix is the $n \times n$ matrix whose i, j element is C_{ij}, and a portfolio's risk is

$$\sum_{ij} w_i w_j C_{ij} = w^T C w,$$

which is a convenient quadratic form representing risk. This quadratic is convex, see Exercise 3.

The overriding goal of portfolio design is to select stocks that give the highest possible expected return with the least risk. This sentiment suggests consideration of the trade-off between risk and return, and the objective of portfolio design is to minimize a weighted difference of these competing goals, which is

$$(1-\alpha)w^T C w - \alpha r^T w,$$

where $0 \le \alpha \le 1$. The parameter α quantifies an investor's risk tolerance. A portfolio is designed solely on the principle of minimizing risk if $\alpha = 0$ and is alternatively completely focused on maximizing return if $\alpha = 1$.

The convex optimization problem used to design a portfolio for any α is

$$\min (1-\alpha)w^T C w - \alpha r^T w$$
such that
$$\sum_{i=1}^{N} w_i = 1 \qquad (12.7)$$
$$w_i \ge 0 \ \forall \ i.$$

Any solution to this problem is efficient for $0 < \alpha < 1$, meaning that it is impossible to increase the expected return without increasing risk and that it is impossible to decrease risk without decreasing the expected return. We demonstrate this efficiency by letting w^* be an optimal solution for a selected

12.1 Stock Pricing and Portfolio Selection

value of α. Let $\tilde{w}$ be an alternative collection of feasible portfolio weights with less risk. Then, $(w^*)^T C w^* > \tilde{w}^T C \tilde{w}$ and

$$(1-\alpha)(w^*)^T C w^* - \alpha r^T w^* \leq (1-\alpha) \tilde{w}^T C \tilde{w} - \alpha r^T \tilde{w}.$$

We conclude that

$$0 < (1-\alpha)\left((w^*)^T C w^* - \tilde{w}^T C \tilde{w}\right) \leq \alpha \left(r^T w^* - r^T \tilde{w}\right),$$

and hence, $r^T \tilde{w} < r^T w^*$. A similar argument shows that there is no portfolio with a greater return with at most the same amount of risk.

The collection of efficient solutions is called the Pareto set, and for this model it is parameterized by α over the interval $[0, 1]$. The Pareto set can be sampled by repeatedly solving (12.7) as α varies. The parameterized curve that plots risk versus expected return as α ranges over $[0, 1]$ is called the efficient frontier, and this graphical tool allows an investor to quickly assess the trade-off between risk and return.

An Example Calculation

We consider a daily 6-month history of the 30 stocks comprising the Dow Jones Industrial Average starting in July 2014 and ending in December 2014. Each stock has a history of 128 trading days over this period. The daily returns $R_i(t)$, for $i = 1, 2, \ldots, 127$, are estimated with the forward difference calculation in (12.4). The sample means and covariances are

$$r_i = \bar{\mu}_i = \frac{1}{127}\sum_{k=1}^{127} R_i(t_k) \quad \text{and} \quad C_{ij} = \frac{1}{126}\sum_{k=1}^{127}(R_i(t_k) - \bar{\mu}_i)(R_j(t_k) - \bar{\mu}_j).$$

A divisor of $127 - 1 = 126$ is used in the covariance calculation because $\bar{\mu}_i$ is an estimate of the true expectation, and adjusting the divisor to one less than the number of samples provides an unbiased estimate of the variance. This adjustment is called Bessel's correction.

If P is a matrix whose rows index the day and whose columns index the stock, then the MATLAB commands

```
R = (P(2:end,:)-P(1:end-1,:))./P(1:end-1,:);
r = mean(R); and
C = cov(R);
```

calculate a matrix of daily returns (R), a vector of sample returns for each stock (r), and a covariance matrix (C).

The convex quadratic optimization problem in (12.7) can be solved with `quadprog` in MATLAB, which is part of the optimization toolbox. A command like

```
[w, optVal] = quadprog((1-alpha)*2*C, -alpha*r, [], [], ...
              ones(1,n), [1], zeros(n,1), ones(n,1));
```

solves (12.7) for a desired α (n stocks are assumed). The multiple of 2 on the covariance matrix is required because `quadprog(H,f)` minimizes

$$\frac{1}{2} x^T H x + f^T x.$$

The optimal weights are returned as `w` and the optimal value as `optVal`. Octave has a similar quadratic optimizer called `qp`.

Additional constraints are possible. For example, suppose we want to ensure that no more than 20% of the portfolio is invested in a single stock. The adapted quadratic optimization problem is

$$\min\,(1-\alpha)\,w^T C w - \alpha\, r^T w \quad \text{such that} \quad \sum_{i=1}^{n} w_i = 1,\ 0 \leq w_i \leq 0.2\ \forall\, i.$$

A straightforward adjustment to `quadprog` is

```
[w, optVal] = quadprog((1-alpha)*2*C, -alpha*r, [], [], ...
              ones(1,n), [1], zeros(n,1), 0.2*ones(n,1));
```

The efficient frontier is parameterized by adjusting α between 0 to 1. Having more small values of α tends to produce better graphs, e.g.

$$\text{alpha} = [0:0.01:1].\text{\textasciicircum}2;$$

is used to index `alpha` for this example. For each α we calculate the risk $w^T C w$ and the expected return $r^T w$ and plot these against each other. The efficient frontiers with and without the 20% restriction on the maximum weight are shown in Fig. 12.3.

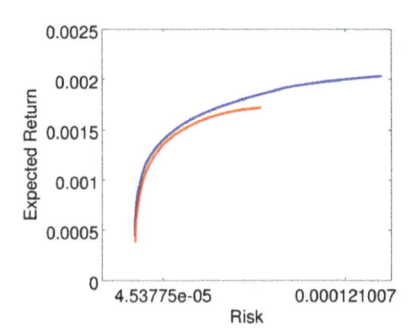

Fig. 12.3 Efficient frontiers for portfolios selected from the Dow Jones Industrial Average. The blue curve has no restriction on the weights while the red curve limits each weight to no more than 0.2.

We consider two illustrative portfolios, those being a risk accepting portfolio with $\alpha = 0.9$ and a risk adverse portfolio with $\alpha = 0.1$. Portfolios are

12.1 Stock Pricing and Portfolio Selection

designed with and without restrictions on the weights, with one being called the unrestricted portfolio and the other being called the restricted portfolio. Only 7 of the 30 stocks were selected over all cases. Historical, statistical, and portfolio information is listed in Table 12.1.

	Stock symbol	HD	INTC	NKE	PG	UNH	V	WMT
	Price on 7/1/2014	$81.20	$30.98	$78.00	$79.28	$81.89	$53.56	$75.28
	Price on 12/31/2014	$104.97	$36.29	$96.15	$91.09	$101.09	$65.55	$85.88
	$\bar{\mu} \times 10^{-3}$	2.0310	1.3809	1.7548	1.1224	1.7349	1.6818	1.0811
	$\bar{\sigma} \times 10^{-4}$	1.3613	2.7445	2.2253	0.5750	1.5156	1.8618	0.8757
$\alpha = 0.9$	Weight (unrest.)	1.00	0.00	0.00	0.00	0.00	0.00	0
	Weight (rest.)	0.20	0.20	0.20	0.00	0.20	0.20	0
$\alpha = 0.1$	Weight (unrest.)	0.45	0.03	0.08	0.23	0.11	0.10	0
	Weight (rest.)	0.20	0.06	0.13	0.20	0.20	0.14	0.07

Table 12.1 Historical information about the 6 stocks selected in an optimal portfolio from the Dow Jones Industrial Average. Optimal weights for a risk accepting and risk adverse portfolio in a restricted and unrestricted design are listed in the last four rows.

We simulate stock prices according to (12.3). The future horizon is 20 days (about 4 weeks), and the simulations use 10 sample paths for each stock. Illustrative sample paths are depicted in Figs. 12.4, 12.5, 12.6, 12.7, and 12.8. Since we have actual price data for the 20 days past our history, we can evaluate our simulation against the true price. The actual price data is shown in red on all graphs, and in each case the simulations reasonably capture the true trajectory. Indeed, if the actual data had not been highlighted, then it would have been unreasonable to postulate which path was the actual data. The cyan path is the first simulation in each case, and we use this path to motivate some of the exercises.

The unrestricted, risky portfolio invested in the single stock HD. The true price of HD fell from $104.97 to $104.42 over the horizon, but the first simulation in Fig. 12.4 instead forecasted an increase to $106.54. Averaging over the 10 sample paths resulted in an expected price of $110.43 at the end of the horizon. The true return over the horizon was −0.0053, whereas the returns of the first and average simulations were, respectively, 0.0147 and 0.0494. So this portfolio would have lost money even though the computed expectation was to make money.

The restricted, risk adverse portfolio varied its investment over the seven stocks HD, INTC, NKE, PG, UNH, V, and WMT. The true and simulated prices at the end of the horizon are in Table 12.2. The first simulation of each price predicted a return of 0.0304, and the expected return over all ten simulations was 0.0305. However, the actual return would have been a loss with a return of −0.0243. The expected return over the ten sample paths matched the actual return in sign for only UNH. So six of the seven price predictions incorrectly forecasted a loss or gain.

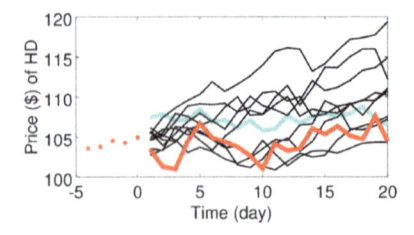

414　12 Modeling with Optimization and Simulation

Fig. 12.4 Ten simulated price trajectories for stock HD for the 20 trading days following Dec. 31, 2014. True prices are shown in red, with the last 5 trading days of 2014 shown to the left of day 0. The first simulated path is shown in cyan.

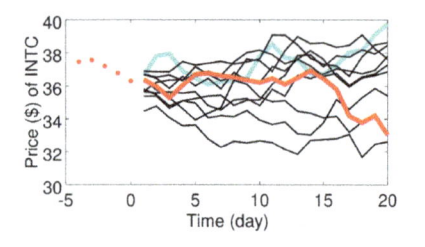

Fig. 12.5 Ten simulated price trajectories for stock INTC for the 20 trading days following Dec. 31, 2014. True prices are shown in red, with the last 5 trading days of 2014 shown to the left of day 0. The first simulated path is shown in cyan.

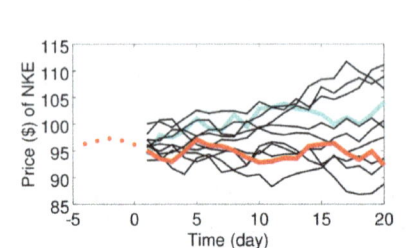

Fig. 12.6 Ten simulated price trajectories for stock NKE for the 20 trading days following Dec. 31, 2014. True prices are shown in red, with the last 5 trading days of 2014 shown to the left of day 0. The first simulated path is shown in cyan.

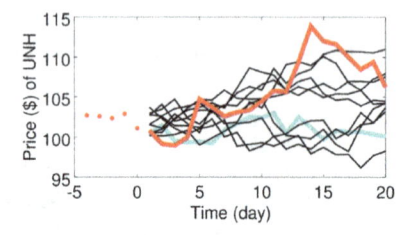

Fig. 12.7 Ten simulated price trajectories for stock UNH for the 20 trading days following Dec. 31, 2014. True prices are shown in red, with the last 5 trading days of 2014 shown to the left of day 0. The first simulated path is shown in cyan.

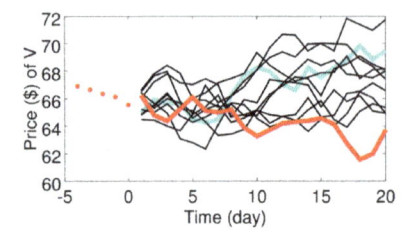

Fig. 12.8 Ten simulated price trajectories for stock V for the 20 trading days following Dec. 31, 2014. True prices are shown in red, with the last 5 trading days of 2014 shown to the left of day 0. The first simulated path is shown in cyan.

12.1 Stock Pricing and Portfolio Selection

	Stock	HD	INTC	NKE	PG	UNH	V	WMT
$t=0$	Actual	$104.97	$36.29	$96.15	$101.09	$65.55	$65.55	$85.88
$t=20$	Actual	$104.42	$33.04	$92.25	$106.25	$63.73	$63.73	$84.98
	First Sim.	$106.54	$39.77	$104.13	$93.37	$100.06	$69.45	$87.09
	Average Sim.	$110.43	$36.76	$100.30	$91.77	$104.47	$67.47	$89.26
Return	Actual	−0.0053	−0.0984	−0.0423	0.0486	−0.0286	−0.0286	−0.0106
	First Sim.	0.0147	0.0875	0.0766	0.0244	−0.0103	0.0562	0.0139
	Average Sim.	0.0494	0.0129	0.0414	0.0075	0.0322	0.0284	0.0378

Table 12.2 Assessment of the restricted, risk adverse portfolio. The unrestricted portfolio invested in all stocks except PG, and the restricted portfolio invested in all six stocks. Optimal weight are listed in Table 12.1.

The weights in Table 12.1 illustrate how portfolio design depends on an investor's willingness to accept risk. Disregarding risk tends to clump a portfolio's resources onto as few stocks as possible in the hopes of investing in those with big potential upsides. The portfolios of more cautious investors tend to include more stocks in an attempt to distribute risk while seeking an aggregate return. This analysis suggests an appropriateness of the model, as its outcomes largely agree with standard advice and practice.

12.1.3 Exercises

1. The efficient frontier satisfies the law of diminishing returns, meaning that we have to accept increasing amounts of risk to increase our expected return. The efficient frontier can be re-cast in terms of proportion by normalizing risk and return by their largest possible values. In this case the α value at which the slope of the curve is 1 is called the point of diminishing returns. Increasing α beyond the point of diminishing returns necessitates a proportionately greater increase in risk than in return. Similarly, decreasing α from the point of diminishing returns results in a portfolio that can proportionately increase the expected return with a lesser increase in risk.

    Acquire the last 6 months of price history for the Dow Jones Industrial Average and calculate an optimal portfolio in which α is selected to be at the point of diminishing returns.
2. Solution (12.3) nicely solves our stochastic model of a stock's price. However, the solution can lead to confusion if forecasting prices. Select a 20-day history of a stock of your choice and calculate $\bar{\mu}$ and $\bar{\sigma}$ over this history. Assume a forecasting horizon of 20 days. First, simulate the stock's price with (12.2). Second, create another simulated price trajectory over the forecasting horizon with (12.3) by

- generating 20 random samples from a standard normal, say $\phi_1, \phi_2, \ldots, \phi_{20}$, and
- calculating

$$P(t) = P(0)e^{\left(\bar{\mu} - \frac{\bar{\sigma}^2}{2}\right) + \bar{\sigma}\sqrt{t}\phi_t},$$

for $t = 1, 2, \ldots, 20$.

Recalculate $\bar{\mu}$ and $\bar{\sigma}$ for each of the two price simulations, i.e. assume the history is replaced with the forecasted prices of each simulation. Explain why the variance estimate from the prices generated with (12.3) is unrealistic. How can (12.3) be used to more accurately calculate a price trajectory?

3. The covariance matrix was shown in Sect. 3.4 to have the factored form

$$C = \frac{1}{m-1}(R - e\bar{\mu})^T (R - e\bar{\mu}), \tag{12.8}$$

where R is the $m \times n$ matrix of returns with R_{ij} being the i-th return for stock j, and $\bar{\mu}$ is the n element row vector of sample means for the returns of the n stocks. Show that the portfolio optimization problem in (12.7) is convex assuming the columns of $(R - e\bar{\mu})$ are linearly independent.

4. Uncorrelated stock prices were assumed in this section's sample paths. Each simulation was conducted with that stock's unique volatility and a collection of independent samples from a standard normal variable. However, stock prices are often correlated. For example, the sample covariances of this section for the Dow Jones Industrial Average were all positive, and hence, these 30 stocks had a tendency to move up or down in unison (this shouldn't be a surprise).

Historical correlations can be incorporated by coupling the random samples. A multivariate normal is a vector in which each coordinate is a (univariate) normal variable, but instead of the distribution being defined by a single mean and a single variance, the random vector is defined by a vector of means and a matrix of covariances. The notation $w \sim \mathcal{N}(\mu, \Sigma)$ commonly denotes that the vector w is distributed so that component w_i is distributed normally with mean μ_i and so that the covariance matrix is Σ. Our development has noted the covariance matrix as C instead of the more standard Σ used in statistics.

If $\phi = (\phi_1, \phi_2, \ldots, \phi_m)^T$ is a sample vector with each ϕ_i being drawn from a standard normal with mean 0 and standard deviation 1, then we can transform the vector to a multivariate sample with mean μ and covariance C by calculating

$$\mu + A^T \phi,$$

12.1 Stock Pricing and Portfolio Selection

where A is any matrix such that $AA^T = C$. Three common choices for A are:

- A can be the lower triangular matrix of the Cholesky factorization,
- $A = U\sqrt{D}$, where $A = UDU^T$ is the orthogonal eigenvector eigenvalue decomposition, or
- $A = (1/\sqrt{m-1})(R - e\bar{\mu})^T$ from (12.8).

Acquire the last 6 months of price history for the Dow Jones Industrial Average and simulate 20 days of new prices in two ways, one assuming returns are uncorrelated and one including the observed correlations. Experiment with different choices of A.

5. Assume the 20-day forecasts of the last exercise are your new histories. Design optimal portfolios and construct the efficient frontiers in both cases. What differences do you find?

6. **Group Project:** Each member of the group is to select at least 10 stocks as possibilities to include in the group's portfolio. Collect the prior year's price history and use this to estimate annual drift and volatility parameters. Run the following experiment:

   a. Decide as a group how to select a portfolio for the next month based on your group's risk tolerance.
   b. Repeatedly simulate future stock prices to assess the performance of your portfolio.
   c. Simulate stock prices for the next month, but only once for each stock. Assume these are the real stock prices observed during the next month.
   d. Assume we can alter the portfolio after each month. Use the previous year's prices, the last month of which has been simulated in the previous step, to re-estimate drift and volatility parameters. Use these updated values to construct a new portfolio.
   e. Repeat steps (b) through (d) for 4 months from the original investing date.
   f. Experiment with your model to see how risk exposure might alter the outcomes.

   Draft a report that explains the models, the implementation details, and the results so that the work is reproducible based solely on your report.

7. **Group Project:** Repeat (or include in) the experiment of the previous exercise the reality of a 5% transaction cost for any adjustment made at the 1 month epochs.

12.2 Magnetic Phase Transitions

Material properties are of substantial interest in science and engineering, and quantifying macro properties from atomic interactions is a regular study of statistical mechanics. The observable macro properties of interest are typically energy and magnetism, and in this section we demonstrate how a simulation of microscopic states can infer such observables. In particular, we develop a two-dimensional Ising model that is a lattice simulation of a simplified physical system.

Ising models are capable of predicting phase transitions induced by changes in temperature. For example, a nickel ferromagnet loses its magnetic attraction once it is heated to about 354 °C, a temperature called the Curie point. A simple experiment is to attach a piece of metal to a suspended nickel magnet and then heat it. The metal will fall once the magnet is sufficiently hot. Predicting this loss of magnetism and studying the material properties near the phase transition have long been approached mathematically by either approximately solving a precise physical model or by exactly solving an approximate physical model. Ising took the latter approach and exactly solved an approximate, one-dimensional model in 1925.

Ernst Ising's original one-dimensional solution did not exhibit a phase transition, but continued work on two-dimensional models successfully forecasted known transitions. The (future) Nobel laureate Lars Onsager exactly solved a simplified two-dimensional model in 1944, and his solution was later extended to the general, two-dimensional case. The search for an exact solution to the general three-dimensional model continued, but in May of 2000 the theoretical computer scientist Sorin Istrail announced that computationally solving the three-dimensional model exactly was NP-Complete, essentially showing that three- dimensional models were intractable. Istrail's argument establishing this fact illustrated the difficulty that physicists had had as they approached their search for exact solutions. The finding was grim enough that the report breaking the news quoted Dante's *The Divine Comedy*:

> I turned like one who cannot wait to see the thing he dreads, and who, in sudden fright, runs while he looks, his curiosity competing with his terror – and at my back I saw a figure that came running toward us across the ridge, a Demon huge and black.

While the computational burden associated with three-dimensional models is daunting, it does not perfectly exclude the existence of exact solutions. After all, the famous question of whether or not P = NP could suggest a reasonable calculation scheme if the equality held. However, the preponderance of current scientific and mathematical thought would indicate that finding an exact solution is unlikely.

The theoretical hindrance of calculating an exact solution does not dismiss the Ising model's utility or importance, but it does redirect our computational effort toward tactics that lead to tractable science. The underlying models

12.2 Magnetic Phase Transitions

are stochastic, and the advent of computation suggests that we consider a sampling procedure to infer estimates of observables. We assume this sampling mentality and show that the optimal nature of Monte Carlo sampling agrees with the physical models.

12.2.1 The Gibbs Distribution of Statistical Mechanics

Statistical Mechanics is the study of the stochastic physical properties of a mechanical system based on its random micro-components. The field includes studies in statistical thermodynamics, with a central theme being the reconciliation of the competing sentiments that a system in equilibrium has both minimum energy and maximum entropy. The competing nature between energy and entropy is explained probabilistically in terms of thermodynamic temperature. The occurrence of a system's state is a random variable, say Σ, and the likelihood of state $\sigma^{\hat{i}}$ is decided by the Gibbs distribution of

$$P(\Sigma = \sigma^{\hat{i}}) = \frac{e^{-E(\sigma^{\hat{i}})/kT}}{\sum_i e^{-E(\sigma^i)/kT}}, \qquad (12.9)$$

where $E(\sigma^i)$ is the energy associated with state σ^i. The summation in the denominator is called a partition function, and we assume a finite sample space so that the summation is over all possible outcomes σ^i. We use a superscript to index different states because we will later index elements of a state with subscripts. The thermodynamic temperature is T, and the constant k is a free parameter that defines the temperature scale. If T is in Kelvin, then k is the Boltzmann constant.

Any finite system whose states adhere to the Gibbs distribution maximizes entropy for a specified energetic average and temperature. The entropy associated with a discrete probability distribution p is

$$-\sum_i p_i \ln(p_i),$$

where $p_i > 0$ and $\sum_i p_i = e^T p = 1$. If state σ^i occurs with probability p_i, then the average energy of the system is

$$\langle E \rangle = \sum_i E(\sigma^i) p_i = E^T p,$$

where the components of E are $E(\sigma^i)$. The notational change of writing the expected value as $\langle \cdot \rangle$ instead of the more common $E(\cdot)$ of Chap. 3 is routine in statistical mechanics. The use of the angled brackets somewhat conflicts with other mathematical uses like that of vector notation and of an inner product.

However, since E naturally denotes the energy function, the angled brackets sidestep an awkward notation like $E(E)$, which would ambiguously mean both the expected value of energy and the energy of the expected value. We use angle brackets here to avoid such conflicts and to agree with the standard development in statistical mechanics.

The distribution that maximizes entropy with an average energy of $\hat{E}$ solves the optimization problem

$$\left.\begin{aligned}\max_p \ &-\sum_i p_i \ln(p_i) \\ \text{such that} \ & \\ E^T p &= \hat{E} \\ e^T p &= 1,\end{aligned}\right\} \qquad (12.10)$$

where p_i is assumed to be positive for each i. Recognizing that the objective is the same as $-\min \sum_i p_i \ln(p_i)$, we can negate entropy and minimize instead of maximize. The resulting minimization problem is convex, and the Lagrangian is

$$\mathcal{L}(p, \lambda, \omega) = \sum_i p_i \ln(p_i) - \lambda(E^T p - \hat{E}) - \omega(e^T p - 1). \qquad (12.11)$$

The convexity of the optimization problem guarantees the sufficiency of the first order optimality conditions, see Theorem 14 in Chap. 4, and solving

$$\nabla \mathcal{L}(p, \lambda, \omega) = 0,$$

we find that

$$p_i = \frac{e^{-\lambda E(\sigma_i)}}{\sum_i e^{-\lambda E(\sigma_i)}},$$

where λ satisfies

$$\hat{E} = \sum_i E(\sigma_i) \left(\frac{e^{-\lambda E(\sigma_i)}}{\sum_i e^{-\lambda E(\sigma_i)}} \right).$$

John Gibbs showed in 1902 that the probabilities with $\lambda = 1/kT$ resulted in a thermodynamically sound model that promoted the macro calculations of energy and magnetism from microscopic atomic interactions. So with energies $E(\sigma^i)$ we know that entropy is maximized if the probability of being in state i is (12.9). This leaves us with the consideration of modeling energy.

There are numerous energetic models, but the one used in the Ising model of the next section is a Hamiltonian. The exact model is presented shortly, but the point here is that we can seek a collection of low energy states for which a system is in equilibrium once an energy formulation is selected. The Gibbs distribution then assigns probabilities to these states so that entropy is maximized for the low energy configurations. The trade-off between minimum

12.2 Magnetic Phase Transitions

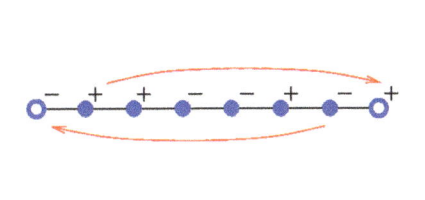

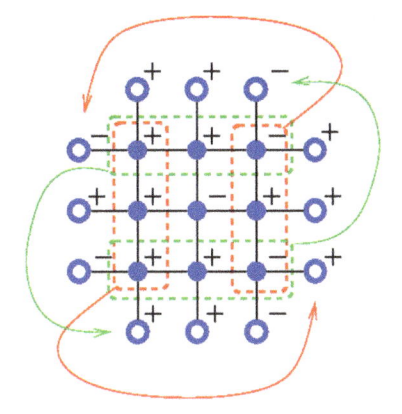

Fig. 12.9 An example spin configuration for a one-dimensional Ising lattice with periodic boundary conditions.

Fig. 12.10 An example spin configuration for a two-dimensional Ising lattice with periodic boundary conditions.

energy and maximum entropy is the thermodynamic temperature T, which alters the relationship between energy and entropy.

12.2.2 Simulation and the Ising Model

The Ising model simplifies the inter-atomic interactions of a material to that of a discrete, finite lattice in which a position's state is only affected by its nearest neighbors. The state at each position must be one of 1 or -1, indicating a positive or negative 'spin' at each location. One- and two-dimensional examples are illustrated in Figs. 12.9 and 12.10. A potential downside of the model is that the boundary positions are different from the others because they lack a full complement of neighbors. A periodic boundary condition is most often used to overcome this concern. For example, the states 'above' the top row of the two-dimensional model are assumed to be those of the bottom row, and the states to the 'right' of the last column are assumed to be those of the first column. We assume the periodic boundary condition. As a point of note, this boundary condition is the same as working over a discrete ring in one dimension and a discrete torus in two dimensions.

Lattice states are encoded as vectors of positive and negative ones. No concern arises due to the dimensionality of the lattice as long as we are careful with our indexing. After all, a matrix is simply a concatenation of its rows or columns, a fact demonstrated by the following matrix vector pair of Fig. 12.10.

$$\begin{bmatrix} 1 & 1 & -1 \\ 1 & -1 & 1 \\ 1 & 1 & -1 \end{bmatrix} \Leftrightarrow \sigma = (1, 1, 1, 1, -1, 1, -1, 1, -1). \tag{12.12}$$

The elements of the matrix are listed in σ as consecutive columns. Both MATLAB and Octave store matrices as such. So if A is the matrix on the left, then A(4) and A(1,2) are both $A_{12} = 1$, and A(9) and A(3,3) are both $A_{33} = -1$.

The energy associated with any state is modeled as the Hamiltonian

$$E(\sigma) = -J \sum_{(i,j) \in \mathcal{A}} \sigma_i \sigma_j - H \sum_i \sigma_i,$$

where $\mathcal{A}$ is the collection of nearest neighbors. The example σ in (12.12) for Fig. 12.10 has

$$\mathcal{A} = \{ (1,4), (2,5), (3,6), (4,7), (5,8), (6,9), (7,1), (8,2), (9,3), \\ (1,2), (4,5), (7,8), (2,3), (5,6), (8,9), (3,1), (6,4), (9,7)\},$$

where the left-most column of nodes is indexed from top to bottom as 1, 2, and 3, the middle column from top to bottom as 4, 5, and 6, and the right-most column from top to bottom as 7, 8, and 9. The first summation of the Hamiltonian models the inter-atomic interactions, and the second models the energetic contributions from an external magnetic field. Ising models are most regularly studied in the absence of an external field, and we assume $H = 0$ in our calculations.

Ferromagnetic materials have positive J values, and antiferromagnetic materials have negative J values. Common magnets like those made from iron and nickel are ferromagnetic, and we assume $J > 0$. Physical experiments show that these magnets undergo a phase transition, and the historical intent of an Ising model is to ask if the model can predict such transitions even with its simplifying assumptions. If so, then we gain trust in our ability to use a simple model to infer material properties of ferromagnets.

The general properties of the Ising model for ferromagnetic materials can be studied under the assumption that $J = k = 1$, an assumption that sets a thermodynamic temperature scale. The value of J could be experimentally customized for a particular material and temperature scale to better agree with the material, but such scalings don't alter the general study because adjusting J and k essentially rescales temperature. Moreover, we lack a theory that shows how to calculate J for a specific material from first principles.

The Gibbs probabilities with $J = k = 1$ are

$$P(\hat{\sigma}) = \frac{e^{-E(\hat{\sigma})/T}}{\sum_\sigma e^{-E(\sigma)/T}} \quad \text{with} \quad E(\sigma) = -\sum_{(i,j) \in \mathcal{A}} \sigma_i \sigma_j.$$

12.2 Magnetic Phase Transitions

States with low energy have neighboring spins that agree, but high energy states have reversed neighboring spins. So a ferromagnet tends to have all its spins agree, and it is the accumulation of these micro spins that creates the magnetic pull we experience. Notice that higher energy states have lower probabilities than do their lower energy counterparts, which means low energy states are probabilistically preferred.

Calculating the Gibbs probabilities is a computational concern. Suppose the two-dimensional lattice size is $n \times n = N$. Then there are 2^N possible states, a number well beyond enumeration in a computational setting for even modest lattice sizes. For example, the number of states exceeds the number of particles in the universe for $n = 10$. Symmetry reductions can reduce the calculation, but it is evident that calculating macroscopic averages directly from the Gibbs probabilities is out of the realm of possibility for reasonable lattice sizes.

Instead of restricting the model to tiny lattices for which exact probabilities can be calculated, we use larger models and approximate the desired quantities by sampling the state space. Our sampling procedure should be biased toward low energy states because physical systems in equilibrium seek minimum energies. This consideration suggests that the probability of including a state in our sample should be high (low) if the energy is low (high). The Gibbs distribution already has this property, and we assume that the probability of sampling state σ^i is proportional to $e^{-E(\sigma^i)/T}$. The sample average of the system's energy then estimates the true average, that is

$$\bar{E} = \frac{1}{N} \sum_{i=1}^{N} E(\sigma_i) \approx \langle E \rangle,$$

where the summation is over the sampled states.

The sentiment of drawing a physically meaningful collection of energetic states that are biased toward low energies coincides with the intent of heuristically solving an optimization problem with simulated annealing. Indeed, the standard acceptance probabilities of simulated annealing in (4.23) agree with the Gibbs probabilities once f is replaced with the Hamiltonian E. The computational paradigm is slightly altered here because the goal is to minimize energy for a fixed temperature so that the search is conducted over low energy states for that temperature. The uphill possibilities of simulated annealing permit the observation of non-minimized energetic states, although the probability of doing so decreases as their energies increase.

The basic iteration of transitioning from one state to the next is a simultaneous application of simulated annealing across a portion of the lattice to decide if a subset of spins should reverse. Let σ^k be the current state, and assume that $\hat{\sigma}$ differs from σ^k in exactly one of the selected spins. The flip is accepted according to the probability

$$P(\sigma^k, \hat{\sigma}, T) = \begin{cases} 1, & E(\hat{\sigma}) < E(\sigma^k) \\ e^{(E(\sigma^k) - E(\hat{\sigma}))/T}, & E(\hat{\sigma}) \geq E(\sigma^k). \end{cases} \quad (12.13)$$

If $P(\sigma^k, \hat{\sigma}, T) \geq y$, with y being a sample of $Y \sim \mathcal{U}(0,1)$, then the sole reversal in $\hat{\sigma}$ is accepted. The process repeats for each of the spins considered for reversal. These decisions are based on the current state σ^k, and once the decisions are complete, the new state σ^{k+1} is created from σ^k by reversing all the accepted reversals. Sampling procedures that select according to probabilities like that of (12.13) are a type of Monte Carlo sampling.

The definition of energy expedites the calculation of the acceptance probability because the decision to flip is solely based on the interacting spins. The energy summation can be expressed as

$$E(\sigma) = -\sum_{(i,j) \in \mathcal{A}} \sigma_i \sigma_j = -\sigma_{\hat{i}}\left(\sigma_{\hat{i}_1} + \sigma_{\hat{i}_2} + \sigma_{\hat{i}_3} + \sigma_{\hat{i}_4}\right) - \sum_{(i,j) \in \mathcal{A}'} \sigma_i \sigma_j,$$

where $\hat{i}_1, \hat{i}_2, \hat{i}_3$, and $\hat{i}_4$ are the nearest neighbors of $\hat{i}$ and $\mathcal{A}'$ is the subset of $\mathcal{A}$ so that neither i nor j is $\hat{i}$. If the only difference between σ^k and $\hat{\sigma}$ is a reversed spin at location $\hat{i}$, then

$$\begin{aligned} E(\sigma^k) - E(\hat{\sigma}) &= \left(-\sigma_{\hat{i}}^k + \hat{\sigma}_{\hat{i}}\right)\left(\sigma_{\hat{i}_1}^k + \sigma_{\hat{i}_2}^k + \sigma_{\hat{i}_3}^k + \sigma_{\hat{i}_4}^k\right) \\ &= \left(-\sigma_{\hat{i}}^k - \sigma_{\hat{i}}^k\right)\left(\sigma_{\hat{i}_1}^k + \sigma_{\hat{i}_2}^k + \sigma_{\hat{i}_3}^k + \sigma_{\hat{i}_4}^k\right) \\ &= -2\sigma_{\hat{i}}^k\left(\sigma_{\hat{i}_1}^k + \sigma_{\hat{i}_2}^k + \sigma_{\hat{i}_3}^k + \sigma_{\hat{i}_4}^k\right). \end{aligned}$$

So the acceptance probabilities only require information about the neighboring spins and not those of the entire lattice.

The computational scheme is to iterate over a range of different temperatures, slightly increasing T per step. For each T we generate a sample of low energy states by repeating the Monte Carlo iteration. The expected energy $\langle E_T \rangle$ is then estimated by its sample average $\bar{E}_T$, where the subscript denotes the dependence on temperature. The process is commonly initiated with a random state, and the first several iterations are discarded since they don't necessarily represent spin configurations that agree with a system in equilibrium. Such initializations are less of a concern for high temperatures, a fact that is depicted in Figs. 12.11 and 12.12 for a lattice of size 30×30. At each iteration 60% of the 900 spins are candidates for reversals. The first iterations are essentially used to ensure a selection of low energy states so that the system is in equilibrium. Sampling for the first temperature is started after the initial search has placed the system near equilibrium. As long as the temperature is adjusted in small increments, subsequent samples for other temperatures do not need initialization and can start from the last sampled state.

The macro quantities of the system are called observables, and the system's expected energy is one such observable. Three others are common,

12.2 Magnetic Phase Transitions

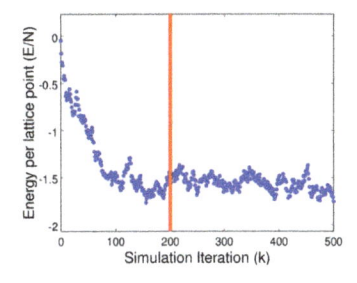

Fig. 12.11 The first 200 iterations of simulated annealing are discarded to let the system equilibrate from its random initiation. The lattice is 30 × 30 and 60% of the spins are candidates for reversal. The temperature is 2.

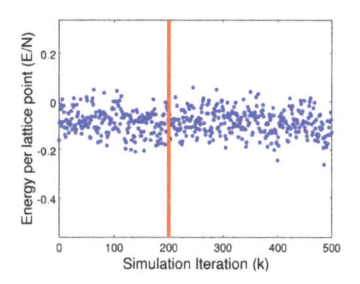

Fig. 12.12 The first 200 iterations of simulated annealing are discarded to ensure reasonable equilibration even though the system randomly starts near equilibrium. The lattice is 30 × 30 and 60% of the spins are candidates for reversal. The temperature is 4.

those being a system's expected magnetization, its heat capacity, and its susceptibility. The magnetization of any state is proportional to the sum of the spins, meaning that for some scalar η the magnetization for state σ is

$$M(\sigma) = \eta \sum_i \sigma_i = \eta \, e^T \sigma.$$

Similar to energy, we estimate $\langle M_T \rangle$ with its sample average $\bar{M}_T$, where the subscript again denotes the dependence on temperature. We assume $\eta = 1$ for our analysis, which is appropriate because we are only concerned with the behavior of the system's expected magnetism instead of pinpointing its exact value for different temperatures.

Heat capacity and susceptibility measure variance in the system's energy and magnetism. The definitions are

Heat Capacity   Susceptibility

$$C_T = \frac{\langle E_T^2 \rangle - \langle E_T \rangle^2}{k\,T^2} \quad \text{and} \quad \chi_T = \frac{\langle M_T^2 \rangle - \langle M_T \rangle^2}{k\,T}.$$

All expected values are replaced with sample averages in our calculations, and as already noted, k is assumed to be 1. These measures are important because high variations suggest phase transitions. Intuitively, material properties are likely to jump between different configurations as a material approaches a phase transition such as the sudden loss of magnetism. So near a Curie temperature we would expect the states to waffle between those with spins mostly in the same direction, which would have high magnetism, and those with spins that largely cancel each other, which would have less magnetism.

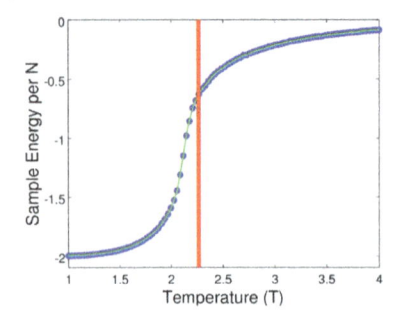

Fig. 12.13 Simulated results estimating $\langle E_T \rangle / N$. The red line is the theoretical Curie temperature as $N \to \infty$.

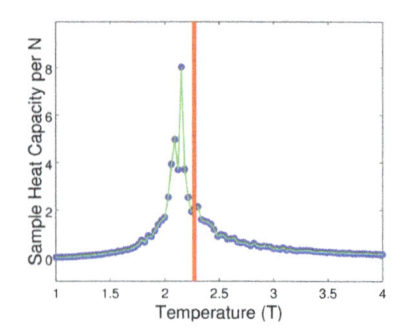

Fig. 12.14 Simulated results estimating C_T/N. The red line is the theoretical Curie temperature as $N \to \infty$.

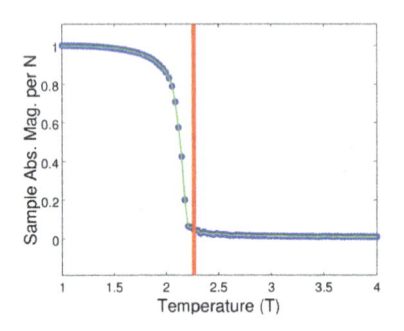

Fig. 12.15 Simulated results estimating $\langle |M_T| \rangle / N$. The red line is the theoretical Curie temperature as $N \to \infty$.

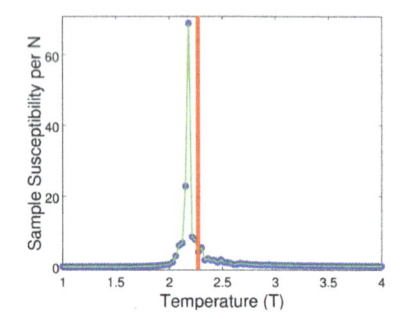

Fig. 12.16 Simulated results estimating χ_T/N, with $\langle M_T \rangle$ replaced with $\langle |M_T| \rangle$. The red line is the theoretical Curie temperature as $N \to \infty$.

Figures 12.13, 12.14, 12.15, and 12.16 illustrate the results of a simulation with a $100 \times 100 = N$ lattice. Temperatures ranged from $T = 1$ to $T = 4$ in steps of 0.03, and the first 1000 iterations were discarded for the initial temperature of $T = 1$. For each temperature a sample of 1000 states was selected, and 60% of the spins were possible reversals at each iteration. All graphs depict the average observable per spin. We are interested in the existence of a magnetic force and not its direction, and for this reason Fig. 12.15 is $\langle |M_T| \rangle$ instead of $\langle M_T \rangle$. Likewise, the susceptibility in Fig. 12.16 replaces $\langle M_T \rangle$ with $\langle |M_T| \rangle$.

The simulation indicates a material event near $T = 2.3$. Low energies associate with low temperatures, but energies climb rapidly toward 0 as T passes from 2.0 to 2.5. Moreover, the heat capacity in Fig. 12.14 shows that our sample witnessed significant variation in this temperature range. Magnetism drops sharply around $T = 2.3$, after which magnetism is essentially zero.

12.2 Magnetic Phase Transitions

This corresponds with the physical experiment mentioned at the beginning of this section, and once the temperature is sufficiently high, the ferromagnetic force quickly dissipates. We conclude that the model predicts a Curie temperature near $T = 2.3$. Figure 12.16 further shows a discernible increase in the statistical variation of the magnetic force as T approaches the apparent Curie temperature. It is possible to prove that the true Curie temperature for the model as $N \to \infty$ is about $T_c = 2.269$. A red vertical line appears on each figure at this theoretical value, and the take away message is that the simulated results reasonably predict the theoretical value.

The functional forms of the observables near a Curie temperature help explain the material properties as the material undergoes a phase transition, and these forms are of noted interest. An analytic study of the Ising model and the sampling procedure gives the following approximations for large $N = n \times n$ and for T close to T_c,

$$|E_T| \approx |T - T_c|^{-1} \approx N,$$
$$|M_T| \approx |T - T_c|^{\beta} \approx N^{-\beta},$$
$$C_T \approx |T - T_c|^{\alpha} \approx N^{\alpha}, \text{ and}$$
$$\chi_T \approx |T - T_c|^{-\gamma} \approx N^{\gamma}.$$

The exponents α, β, and γ are called the critical exponents of the model. There is another exponent related to energy, denoted by ν, but it is known to be 1 in the two-dimensional case.

The exponents can be approximated with linear regression. For example, we use the statistical model

$$|M_T| = \omega\, N^{-\beta} \quad \Leftrightarrow \quad \ln(|M_T|) = \ln(\omega) - \beta \ln(N)$$

to predict β. The model is linear with the regressor being $\ln(N)$ and the response being $\ln(|M_T|)$. To collect data for the regression we simulate solutions for various lattice dimensions N. For each N we locate the temperature with the highest heat capacity. The values of $|M_T|$, C_T, and χ_T are then recorded at this temperature for the regression. Graphical results of these regressions are shown in Figs. 12.17, 12.18 and 12.19, and the calculation details are listed in Table 12.3. The theoretical values of the exponents are derived from the explicit solution to the two-dimensional model. The predictions from the regressions are a bit crude compared to the theoretical benchmarks, although improved predictions from more thorough and long-running simulations are common in the literature

We have shown that approximate solutions of the Ising model, which itself is a simplified physical analogue, can predict phase transitions. The computational tactic of sampling a simplified model supports the calculation of solutions that aid our physical understanding and hasten our ability to experiment.

Simulation details

n	Temp. range	Initiation iterates	Sample iterates
$4, 5, 6, \ldots, 20$	$2.15 \leq T \leq 2.30$	500	5000

Results

Property	Exponent	Predicted value	Theoretical value
Heat capacity	α	0.73765	0.500
Magnetism	β	0.20269	0.125
Susceptibility	γ	1.6969	1.750

Table 12.3 Details of the simulations used to estimate the critical exponents with regression.

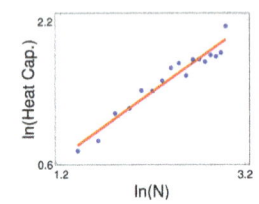

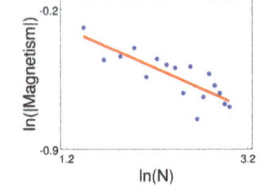

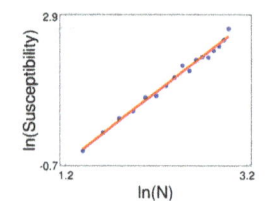

Fig. 12.17 Predicting α with regression.

Fig. 12.18 Predicting β with regression.

Fig. 12.19 Predicting γ with regression.

12.2.3 Exercises

1. Show that problem (12.10) is convex.
2. Solve (12.11) to show that the Gibbs distribution maximizes entropy with an average energy constraint.
3. Prove that the uniform distribution maximizes entropy if the average energy constraint is removed from (12.10).
4. Build a computational model for the one-dimensional Ising model with periodic boundary conditions. Assume $k = J = 1$ and $H = 0$. Make a computational argument that the one-dimensional model does not exhibit a phase transition.
5. The partition function, i.e. the denominator of (12.9), can be factored. In the one-dimensional case with $k = J = 1$ and $H = 0$ we have

$$\sum_i e^{-E(\sigma^i)/T} = \sum_i e^{-(1/T) \sum_{j=1}^{N} \sigma_j^i \sigma_{j+1}^i}$$

$$= \sum_i e^{-\sigma_1^i \sigma_2^i / T} \cdot e^{-\sigma_2^i \sigma_3^i / T} \cdot \ldots \cdot e^{-\sigma_N^i \sigma_1^i / T},$$

12.2 Magnetic Phase Transitions

where the one-dimensional lattice has N spins and the periodic boundary condition assumes $\sigma_{N+1}^i = \sigma_1^i$. The products in the exponents can only be 1 or -1 depending on the state, and counting these options shows that the the partition function can be expressed in terms of matrix multiplication. In particular, if we let

$$V = \begin{bmatrix} e^{-1/T} & e^{1/T} \\ e^{1/T} & e^{-1/T} \end{bmatrix},$$

then

$$\sum_i e^{-E(\sigma^i)/T} = \text{Tr}(V^N), \tag{12.14}$$

where $\text{Tr}(\cdot)$ is the sum of the matrix's diagonal elements (called the trace of the matrix). One of the properties of the trace is that it is also the sum of the eigenvalues of the matrix.

Prove (12.14), and then show that both $2\cosh(1/T)$ and $2\sinh(1/T)$ are the eigenvalues of V (hint: write $\cosh(1/T)$ and $\sinh(1/T)$ in terms of exponentials to guess eigenvectors). Show that the eigenvalues of V^N are $\cosh^N(1/T)$ and $\sinh^N(1/T)$ so that

$$\sum_i e^{-E(\sigma^i)/T} = \cosh^N(1/T) + \sinh^N(1/T).$$

The free energy at any location in the continuum, i.e. as $N \to \infty$, is

$$f(T) = -T \lim_{N \to \infty} \frac{1}{N} \ln \left(\sum_i e^{-E(\sigma^i)/T} \right).$$

This limit is often called the thermodynamic limit. Evaluate the thermodynamic limit to express f explicitly.

The observable $\langle E_T \rangle$ in the thermodynamic limit is

$$\langle E_T \rangle = -T^2 \frac{d}{dT} \frac{f(T)}{T}.$$

Calculate this theoretic observable and compare it against the computed estimate from Exercise 4.

6. Build a computational model for the two-dimensional Ising model with periodic boundary conditions. Assume $k = J = 1$ and $H = 0$. Verify your model by benchmarking it against the results of this section.
7. Adjust the models of Exercises 4 and 6 so that H need not be zero. Computationally investigate the behavior of these models for nonzero values of H. In particular, how does the Curie point of the two-dimensional model depend on H?

8. Two-dimensional geometries other than that of a square lattice are possible. Build Ising models for the interaction structures below.

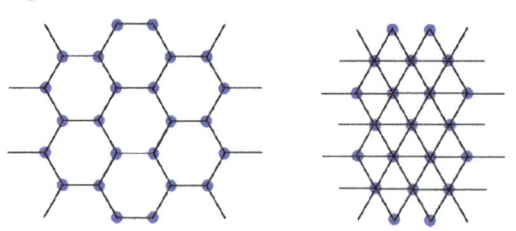

   Use the same Hamiltonian of this section but adjust the set of nearest neighbors to the adjacent spins depicted in the new lattices. Assume $J = k = 1$. Conduct a thorough study with and without H being zero.
9. Repeat the regression analysis used to estimate the critical exponents, but include the calculation of 95% confidence intervals. Assume the standard two-dimensional model with $J = k = 1$ and $H = 0$.
10. It is popular to view the simulated states as a movie, with each frame being a different sample. Each pixel is color coded to distinguish the positive spins from the negative spins. MATLAB can make such movies using commands like `movie` and `getframe`. Make a movie of the states as the temperature increases through the Curie point. Use the standard model with $J = k = 1$ and $H = 0$, although feel free to see how other models behave.

Chapter 13
Regression Modeling

A big computer, a complex algorithm, and a long time does not equate to science.
– Robert Gentleman

Statisticians, like artists, have the bad habit of falling in love with their models. – George Box

The best thing about being a statistician is that you get to play in everyone's backyard. – John Tukey

We learned in Chap. 3 how to calculate and assess model parameters for a linear regression model. However, we did not consider which variables a model should include, and it is this question that we approach here. The process of identifying a collection of independent variables to gain an accurate statistical analysis of a response variable is commonly, and simply, referred to as "model building" in the world of statistics. The act of model building combines intuition and computational skill and is an artful application of mathematics. Numerous appropriate models are often inferred from the same data, and indeed, disparate models are regularly disputed among experts. What is important is to be able to identify a model's benefits and weaknesses so that its strengths can be leveraged and its faults avoided.

Consider the following data as an illustrative example,

y	15.6	21.7	3.1	14.7	6.9	16.1	7.5	16.1	20.2	13.3	15.0	23.3
x_1	2.0	3.6	0.9	2.4	0.1	2.6	0.7	1.0	2.4	1.9	2.3	3.9
x_2	4.7	5.7	1.6	4.8	4.4	6.0	2.6	5.7	4.5	5.0	4.8	6.9
x_3	2.9	4.9	1.4	3.3	1.6	3.7	1.6	3.0	3.1	3.4	3.3	4.8

Suppose we use

$$\hat{y}(x_1, x_2, x_3) = a_0 + a_1 x_1 + a_2 x_2 + a_3 x_3$$

to assess how well X_1, X_2, and X_3 can describe Y. The resulting least-squares model is
$$\hat{y}(x_1, x_2, x_3) = -1.01 + 2.36x_1 + 1.76x_2 + 0.8x_3.$$
The R^2 value is relatively high at 0.87, and if this is the only evaluative concern, then we might accept this model since it describes 87% of the variation in y. However, problems arise if we want to draw inferences from the parameters themselves. For instance, the model imbues the possible interpretation that a unit increase in an observation of X_1 correlates with an increase of 2.36 units in the expected value of Y. The p-value of the null hypothesis $H_0 : \alpha_1 = 0$ is 0.37, and hence, we fail to reject the null hypothesis by common standards. The conclusion is that any perceived correlation between X_1 and Y is dubious as part of the model's explanation of Y. The qualification that the conclusion is relative to "the model's explanation" is paramount, as we are **not** saying that X_1 is potentially uncorrelated to Y in the absence of X_2 and X_3. Indeed, if we model $\hat{y}(x_1) = a_0 + a_1 x_1$, then the least squares solution is $\hat{y}(x_1) = 5.36 + 4.59x_1$. The R^2 value is 0.75 and the p-value of a_1 is 0.00. So we have near perfect confidence that X_1 and Y are correlated and that the sample variance of x_1 explains 75% of the sample variance in y. So the concern of the bigger model is clearly about the ensemble of independent variables and not about their individual relationships with the response.

The p-values for all four parameters are

Parameter	a_0	a_1	a_2	a_3
p-value	0.73	0.37	0.22	0.82

These high p-values do not provide confidence in any of the model's parameters, but again, the concerns are relative to the collection of independent variables. The underlying issue is multicollinearity, which means that some of the independent variables are mimicked by linear combinations of the others. The sample data was designed to illustrate this concern, as it imposed the relationship
$$x_3 \approx 0.6x_1 + 0.4x_2. \tag{13.1}$$
The variable x_3 can then be removed from the model,
$$\begin{aligned}\hat{y}(x_1, x_2, x_3) &= -1.01 + 2.36x_1 + 1.76x_2 + 0.8x_3 \\ &\approx -1.01 + 2.36x_1 + 1.76x_2 + 0.8(0.6x_1 + 0.4x_2) \\ &= -1.01 + 2.84x_1 + 2.00x_2 + 0x_3.\end{aligned} \tag{13.2}$$

So the response could have been reasonably described by the sample data of x_1 and x_2 without x_3. The failure to reject the null hypothesis $H_0 : \alpha_3 = 0$ is obvious because the data supported the possibility that a_3 could have been zero if the sample data had made the approximation in (13.1) exact. This observation confirms the suspicion surrounding the original a_3 parameter raised by the high p-value of 0.82. We could have solved (13.1) for either x_1

13 Regression Modeling

or x_2 to raise doubt about a_1 or a_2, which would have confirmed the high p-values for a_1 and a_2.

The optimization perspective of multicolinearity is that near optimal solutions to the underlying least-squares problem can be had by projecting the optimal solution to "zero-out" some coefficients. For instance, the sum of the squared residuals for the optimal solution $(-1.01, 2.36, 1.76, 0.8)^T$ is 51.7, but the sum of squared residuals for the approximate solution of $(-1.01, 2.84, 2.00, 0.0)^T$ in (13.2) only increases this value by 4% to 53.87. If we let A be the matrix $[e\,|\,x_1\,|\,x_2\,|\,x_3]$, then the least-squares problem defining the optimal coefficients is

$$\min_a \|Aa - y\|^2.$$

The least-squares problem associated with the projected solution with $a_3 = 0$ is the constrained problem

$$\min_a \|Aa - y\|^2 \text{ such that } a_3 = 0.$$

This problem can be re-stated as an unconstrained problem by removing a_3 from the model, which is the same as removing the last column of A. The solution to this problem is not perfectly $(-1.01, 2.84, 2.00, 0.0)^T$ but is rather $(-0.77, 2.88, 2.01, 0)^T$. The difference is that the former is an approximate optimal solution constructed from the approximation in (13.1) whereas the latter is the actual optimal solution.

A verification index factor, or VIF, measures a regressor's multicolinearity against the other regressors. The calculation for the k-th regressor is

$$\text{VIF}(X_k) = \frac{1}{1 - R_k^2},$$

where R_k^2 is the percentage of variation of X_k described by the other regressors. So R_1^2, R_2^2, and R_3^2 are the R^2 values of the following models,

$$\hat{x}_1 = a_0 + a_1 x_2 + a_2 x_3 \quad (R^2 \text{ is } R_1^2)$$
$$\hat{x}_2 = a_0 + a_1 x_1 + a_2 x_3 \quad (R^2 \text{ is } R_2^2)$$
$$\hat{x}_3 = a_0 + a_1 x_1 + a_2 x_2 \quad (R^2 \text{ is } R_3^2).$$

The resulting VIFs are $\text{VIF}(X_1) = 13.51$, $\text{VIF}(X_2) = 6.11$, and $\text{VIF}(X_3) = 26.48$. The rule-of-thumb is that a VIF greater than 10 suggests a multicolinearity, and by this standard, there is a concern. We don't necessarily want to remove all regressors with high VIFs because a high value only indicates a dependence and does not indicate the collection of regressors on which the dependence relies.

Seven models are possible in our example, and their coefficients, p-values, and R^2 values are

Included variables	Coefficient (p-value)	R^2
X_1	4.59 (0.00)	0.75
X_2	3.56 (0.00)	0.71
X_3	4.93 (0.00)	0.84
X_1, X_2	2.88 (0.01), 2.01 (0.02)	0.87
X_1, X_3	0.25 (0.90), 4.70 (0.04)	0.84
X_2, X_3	0.95 (0.37), 3.91 (0.01)	0.86
X_1, X_2, X_3	2.36 (0.37), 1.76 (0.22), 0.80 (0.82)	0.87.

Which model would you suggest? The sample data of x_3 provides the best single variable model, and the combined data of x_1 and x_2 provides the best two variable model. However, note that the descriptive power of the best two variable model only improves by 3% over the best single variable model, i.e. R^2 increases from 0.84 to 0.87. Is the simplicity of a single variable relation sufficient, or does the increase of 3% warrant the complication of a second variable? These are some of the choices of model building.

Another complication arises as a modeler considers variable transformations. For instance, recall that some of the linear models in Chap. 3 provided nonlinear relationships among the variables after transforming the appropriate data. The Faber-Jackson model in (3.21) illustrated how to deduce exponential relationships with logarithmic transformations, and the periodic data in (3.7) was analyzed after transforming the data with sines and cosines. The basic idea is that Y and X might not be linearly related whereas the transformed variables $f(Y)$ and $g(X)$ could be for an appropriate choice of f and g. So instead of considering

$$Y = \alpha_0 + \alpha_1 X + \varepsilon, \quad \text{we might consider} \quad f(Y) = \alpha_1 + \alpha_1 g(X) + \varepsilon.$$

One way to conjecture a variable transformation is to plot the response variable against each of the potential regressors. A transformation isn't recommended if the trend appears linear, but if not, then you can juxtapose the graphical evidence against your expertise to infer an apt transformation. For instance, if the response variable appears to have a quadratic relationship with a regressor, then adding a squared regressor could make sense. The same is true of exponential, logarithmic, and periodic relationships. While it can be tedious to visually inspect the graphical evidence, this is a common tactic if variable transformations are being considered.

13.1 Stepwise Regression

The techniques of stepwise regression direct the navigation of variable selection once a collection of candidate regressors is identified. An obvious contender to a stepwise approach is to consider all possible subsets of regressors, which would require the calculation of $2^q - 1$ regression models, where q is the number of possible regressors. This is clearly impractical for even modest values of q.

The two most common stepwise techniques are forward and backward regression. These algorithms are greedy, and they respectively build up or tear down a collection of regressors toward the goal of finding a reasonable model. As with most greedy approximations to a combinatorial problem, results can be far from optimal. Adaptations and alternatives are regularly considered. Remember, there is no perfect model. We instead seek models for a purpose, and we should be able to defend our results relative to this purpose.

The concept of a greedy design requires a metric to assess the improvement or degradation caused by a model's loss or gain of regressors. The F-statistic is used as this measurement. Suppose the model $\hat{y}$ has q inputs and that we want to compare $\hat{y}$ to a second model with an additional j inputs. If the second model is $\hat{\hat{y}}$, then

$$F = \frac{\frac{1}{j}\left(\sum_{i=1}^{n}(\hat{y}_i - y_i)^2 - \sum_{i=1}^{n}(\hat{\hat{y}}_i - y_i)^2\right)}{\frac{1}{n-q-j-1}\sum_{i}^{n}(\hat{\hat{y}}_i - y_i)^2},$$

where n is the size of the sample. The null hypothesis is

$$H_0 : \alpha_{q+1} = \alpha_{q+2} = \ldots = \alpha_{q+j} = 0.$$

The F-statistic is nonnegative because the sum of squared residuals for $\hat{\hat{y}}$ is no greater than the sum of squared residuals for $\hat{y}$, a fact that follows because one possible model for $\hat{\hat{y}}$ is $\hat{y}$ with the extended parameters being zero. The hypotheses test uses the F-distribution, and the likelihood of making a Type 1 error reduces as F increases. We reject the null hypothesis and conclude that at least one of the new coefficients is significant if F is sufficiently large. In this case we are claiming with an appropriate confidence that at least one of the added variables should be included in the model—although which is unclear. The p-value of the test can be calculated in MATLAB and Octave with `1 - fcdf(F,j,n-q-j-1)`.

Forward regression builds a collection of regressors by iteratively adding them to an existing model. Suppose the response variable is Y and that the possible regressors are $X_1, X_2, \ldots, X_q$. The first step builds the q models

$$\hat{y} = a_0^i + a_1^i x_i,$$

where the superscript i indicates the candidate regressor. The null hypothesis is $H_0 : \alpha_1^i = 0$, and rejecting this hypothesis suggests that X_i should be included as a regressor. The variable with the highest F value is consequently selected since it gives the greatest confidence in rejecting the null hypothesis. We comment that the F-test reduces to the standard t-test in these initial models, so we are adding the most significant variable. Assume X_{i_1} is the selected variable. The next step considers the $q - 1$ models

$$\hat{y} = a_0^i + a_1^i x_{i_1} + a_2^i x_i,$$

where $i \neq i_1$. The model with the highest F value is again selected, creating a new model of the form

$$\hat{y} = a_0^{i_2} + a_1^{i_2} x_{i_1} + a_2^{i_2} x_{i_2}.$$

The process repeats until a stopping criteria is satisfied, typically something like the p-value of the F-tests are all greater than 0.1 or 0.05. This termination criterion leads to a failure to reject the null hypotheses, $H_0 : \alpha_k^i = 0$ for all possible i in iteration k. The result is a model

$$\hat{y} = a_0^{i_p} + a_1^{i_p} x_{i_1} + a_2^{i_p} x_{i_2} + \ldots + a_p^{i_p} x_{i_p},$$

for which adding the next variable would call to question its significance.

Backward regression deconstructs an existing model by selecting a variable to remove. The process reverses that of forward regression, with the initial model containing all possible regressors,

$$\hat{y} = a_0 + a_1 x_1 + a_2 x_2 + \cdots + a_q x_q.$$

The first step considers q models, each with some x_i removed,

$$\hat{y} = a_0 + a_1 x_1 + a_2 x_3 + \ldots + a_{i-1} x_{i-1} + a_{i+1} x_{i+1} + \cdots + a_q x_q.$$

The null hypothesis for the F-test is $H_0 : \alpha_i = 0$, so failure to reject suggests the removal of X_i as a regressor. The model with the lowest F value is subsequently removed. The process repeats until a suitable stopping criteria is achieved, e.g. the p-values of the F-tests are all below 0.05 or 0.1.

Neither forward nor backward regression guarantees a "best" model at termination, after all, both algorithms search at most $q^2 - q + 1$ of the possible $2^q - 1$ models. Forward regression tends to terminate with lean models with fewer regressors, whereas backward regression tends to terminate with larger models with numerous regressors. Forward and backward iterations can be combined into adaptations, and termination criteria can be altered or replaced. As an example, a combined algorithm could add two regressors and then remove one, and the F-test tolerances for adding and removing variables could be different. In this scheme all two-element subsets of the ex-

13.2 Qualitative Inputs and Indicator Variables

Employee Num	1	2	3	4	5	6	7
Car	Yes	Yes	Yes	No	Yes	No	Yes
PubTran	Yes	No	Yes	Yes	Yes	Yes	No
Educ	Bach	HS	HS	Bach	HS	HS	HS
Reliability	0.9141	0.8902	0.8729	0.8963	0.8707	0.8566	0.9017
Employee Num	8	9	10	11	12	13	14
Car	Yes	No	Yes	No	No	Yes	No
PubTran	Yes	No	Yes	Yes	No	No	Yes
Educ	HS	HS	Bach	HS	Adv	HS	HS
Reliability	0.8788	0.8833	0.9173	0.8509	0.9556	0.8911	0.8604

Table 13.1 Employee data from an engineering firm seeking new hires.

cluded regressors are searched in the forward step, and all included regressors are candidates for removal in the backward step.

Automated methods such as forward and backward regression are good for initial model design, but they are often not the conclusion of the modeling process. Model design should be driven by intent, intuition, and expertise, and a modeler should be free to investigate. Indeed, discovering new relationships is without a doubt part of the joy. Lastly, any model should be vetted against its design goals and against its statistical metrics.

13.2 Qualitative Inputs and Indicator Variables

Suppose you work for a growing engineering firm with 14 employees. The company seeks some data analysis to help guide its future hires, and in particular, it seeks a model of reliability. You are presented with the employee records in Table 13.1, and the company has charged you with identifying the characteristics that identify reliable employees.

The data for car ownership, access to public transportation, and education is qualitative, and these regressors need to be encoded prior to a regression analysis. Let X_1 and X_2 be the random variables for car ownership and access to public transportation. These variables have binary sample spaces, and we assign "Yes" a value of 1 and "No" a value of 0. An employee's education status has three states, and we encode these states as binary inputs to discern their individual effects. If X_3 is the random variable with three outcomes, then we express X_3 in terms of the binary variables X_3' and X_3'' as indicated by

Education	X_3	X_3'	X_3''
High school	0	0	0
Bachelor's	1	1	0
Advanced	2	0	1.

The encoded data is in Table 13.2

The model of reliability in terms of car ownership, access to public transportation, and education is

$$\hat{y} = 0.87 + 0.02x_1 - 0.02x_2 + 0.04x_3' + 0.08x_3''.$$

The R^2 value is 0.96, and all p-values for the parametric t-tests are below 0.01. The conclusion is that this model explains the variance of reliability well and that all parameters are significant. The largest VIF is 1.37 for X_2, and hence, there is no concern of multicolinearity.

The model indicates a gain in reliability with car ownership and with increased educational status. An employee's reliability score increases by 0.02, on average, if she or he has a car. The reliability score of an employee with a bachelor's degree is higher by 0.04 over the reliability score of an employee with a high school degree. This difference increases to 0.08 if the employee has an advanced degree. The potentially surprising outcome is that access to public transportation tends to decrease reliability. So, while public transportation is often the preferred method of commuting in a large metropolitan area, it might tend, for example, to delay an employee's arrival. Your advice would be to seek individuals who own their own cars, who have as much education as possible, and who live without easy access to public transportation. Management can decide if it wants to support public transportation due to

Employee Num	1	2	3	4	5	6	7
Car (X_1)	1	1	1	0	1	0	1
PubTran (X_2)	1	0	1	1	1	1	0
Bach (X_3')	1	0	0	1	0	0	0
Adv (X_3'')	0	0	0	0	0	0	0
Educ	Bach	HS	HS	Bach	HS	HS	HS
Reliability (Y)	0.9141	0.8902	0.8729	0.8963	0.8707	0.8566	0.9017
Employee Num	8	9	10	11	12	13	14
Car (X_1)	1	0	1	0	0	1	0
PubTran (X_2)	1	0	1	1	0	0	1
Bach (X_3')	0	0	1	0	0	0	0
Adv (X_3'')	0	0	0	0	1	0	0
Educ	HS	HS	Bach	HS	Adv	HS	HS
Reliability (Y)	0.8788	0.8833	0.9173	0.8509	0.9556	0.8911	0.8604

Table 13.2 The encoded data from Table 13.1.

13.2 Qualitative Inputs and Indicator Variables

other strategic goals, but the data does not support doing so if the goal is to improve reliability.

Another modeling adaptation can help us query if the parameters of our model rely on the states of the others. For instance, the increased reliability of 0.2 for car ownership is an aggregate over the entire population, but this gain might be different for those who have access to public transportation and those who do not. We add interaction terms to asses such possibilities. The interaction term for car ownership and access to public transportation is the product $X_1 X_2$, which can be added as a new regressor by multiplying the sample data. The resulting model has the form

$$\hat{y} = a_0 + a_1 x_1 + a_2 x_2 + a'_3 x'_3 + a''_3 x''_3 + a_{1,2} x_1 x_2.$$

The least squares solution has $a_{1,2} = -0.002$, and the associated p-value is 0.77. This interaction term is not significant, and we conclude that the gain in reliability due to car ownership is not dependent on access to public transportation.

Undertaking the same study for the other interaction terms results in

Interaction	$x_1 x'_3$	$x_1 x''_3$	$x_2 x'_3$	$x_2 x''_3$
Parameter	-0.002	0.000	0.021	0.000
p-value	0.841	NaN	0.000	NaN

The NaN (not a number) outcomes are due to the parameter estimates being zero to numerical tolerance. The significance of $x_2 x'_3$ advocates that we should include this regressor. The resulting model is

$$\hat{y} = 0.87 + 0.2 x_1 - 0.02 x_2 + 0.02 x'_3 + 0.08 x''_3 + 0.02 x_2 x'_3.$$

The R^2 value is again 0.96 and all parameters are significant—the largest p-value is 0.0026. Our analysis and recommendation changes with regard to the deleterious impact of public transportation. If an employee has a bachelor's degree, then $x'_3 = 1$ and the x_2 terms add to zero. So, there is no negative impact on reliability for having access to public transportation if the employee has a bachelor's degree. Similarly, if the employee has access to public transportation, then $x_2 = 1$ and the x'_3 terms combine to $0.04 x'_3$. Otherwise the employee doesn't have access to public transportation, $x_2 = 0$, and the combined x'_3 term is $0.02 x'_3$. So unlike our earlier analysis that claimed an increase in reliability of 0.04 due to a bachelor's degree, we now see that this increase is realized for those who have access to public transportation. Otherwise, employees with bachelor's degrees who lack access to public transportation realize an increase in reliability of only 0.02.

We turn to another example to further illustrate the role of indicator and interaction variables. Consider the (fictitious) data related to cancer risk in

Table 13.3. The gender variable X_1 is encoded so that 1 means male and 0 means female. The model for cancer risk without an interaction term is

$$\hat{y} = 0.02 + 0.04x_1 + 0.10x_2.$$

The R^2 value is 0.97, and all parametric p-values are less than 0.0008. The model is reasonable with regard to the expected metrics, and it imposes different gender trends,

$$\text{Male } \hat{y} = 0.06 + 0.10x_2$$
$$\text{Female } \hat{y} = 0.02 + 0.10x_2.$$

Risk (Y)	0.0915	0.0899	0.0570	0.0645	0.0504	0.0873	0.0846	0.0959
Gender (X_1)	M	M	W	W	W	M	M	M
% Diet (fat) (X_2)	0.3114	0.3075	0.3650	0.4960	0.3091	0.3036	0.2999	0.3684
Risk (Y)	0.1115	0.0525	0.1043	0.0563	0.0457	0.0795	0.0563	0.0572
Gender (X_1)	M	W	M	W	W	M	W	W
% Diet (fat) (X_2)	0.4784	0.3253	0.4042	0.3767	0.2141	0.2460	0.3499	0.4390

Table 13.3 Fictitious data related to the risk of cancer.

The conclusion is that males have a higher risk of cancer, but that the risk of cancer increases at the same rate of 0.1 per percentage increase in dietary fat for both genders.

The analysis is different if we include the interaction term $x_1 x_2$. The resulting model is

$$\hat{y} = 0.03 + 0.01x_1 + 0.06x_2 + 0.08x_1 x_2.$$

The R^2 value increases to 0.99, and the maximum parametric p-value is 0.015. So this model is also reasonable, if not slightly improved over the original. The gender models in this case are,

$$\text{Male : } \hat{y} = 0.04 + 0.14x_2 \text{ and}$$
$$\text{Female : } \hat{y} = 0.03 + 0.06x_2.$$

The conclusion is different from that of the original. Not only do females have lower risk than males for the same fat intake, but their risk increases at a rate less than half that of their male counterparts. So female risk of cancer is much more tolerable of dietary fat.

The examples of this section illustrate how to encode qualitative variables for use with regression. In general, if a variable X has a state space of size m, then it can be decomposed into $m-1$ binary indicators. The first element of the state space is encoded by all indicators being zero, the second element by the first indicator being one, and so on. Binary indicator variables are called 'dummy' variables in many statistical presentations, but we have refrained from this term—indeed, why should such an important variable deserve a demeaning name? Qualitative variables should be scrutinized just like their quantitative sisters, and their importance to a response should be assessed against the other regressors. Lastly, note that analysis and recommendations can change among reasonable models, so explore options before deciding on an analysis.

13.3 Exercises

1. The published data at

    ww2.amstat.org/publications/jse/v16n3/kuiper.xls

    lists sale prices of used 2005 GM cars, along with other characteristics such as mileage, model, and engine specifications.[1] Build a regression model with the sale price being the response. You should consider variable transformations and interaction variables. What conclusions can you draw from your model?
2. Many publicly available datasets detail aviation safety, e.g. the Bureau of Transportation Statistics provides historical data for the number of accidents, the number of fatalities, flight hours, etc. Posit and study several models with the response being the number of fatalities. Consider variable transformations by plotting the number of fatalities against each of the possible regressors.

    Two of the posited models are to be created with forward and backward regression algorithms of your own design. Consider other models as guided by your investigation. Which model would you recommend, and what conclusions are you willing to draw?
3. Repeat Exercise 2 but include technological and regulatory advances as possible qualitative regressors. For instance, de-icing rules changed in 1992, security standards changed in 2001, and engines have largely transitioned to turbines from cylinders. Include interaction variables in your study. Decide if the technological and/or regulatory changes indicate statistical decreases in the number of fatalities.

---

[1] See Shonda Kuiper (2008) *Introduction to Multiple Regression: How Much Is Your Car Worth?*, Journal of Statistics Education, 16:3, DOI: 10.1080/10691898.2008.11889579.

4. Adjust Exercises 2 and 3 to investigate automotive accidents and fatalities. Specifically investigate if cell phones and the legislation surrounding them have had a statistical impact on fatalities.
5. Assume the iterative models of a forward regression differ by a single regressor. Then the normal equations solved to decide which regressor is added are positive definite and can be solved with the method of conjugate directions. A convenient initial iterate is the current solution adjoined with a zero for the coefficient of the potentially new regressor. Expedient convergence is sought, and initiating the method of conjugate directions with this solution hopefully reduces the number of iterates needed to converge.

Design a computational study that compares how a forward regression depends on the manner in which the normal equations are solved. Include the method of conjugate directions as initiated above in the comparison. Other options are a Cholesky factorization with a forward and backward solve, and an LU factorization with a forward and backward solve (as well as others).

6. A potential downside to the method of conjugate directions in a forward regression is that each step converges with a tolerance, and small errors at each step can accrue through the iterative process. Experiment with different tolerances, and compare the concluding parameters. Assess the resulting parameters against those from a 'clean' calculation with the terminal collection of regressors.
7. Explain why initiating the method of conjugate directions as explained in Exercise 5 doesn't directly extend to backward regression. Is there a similar adaptation that you can conceive?
8. Most professional sports leagues provide statistics about their teams and players. Select a sport and build a model to predict a notable outcome. For instance, which player will win the NBA's Most Valuable Player award, or which NHL team will win the Stanley Cup?

You should use a technique called logistic regression. Suppose Y is the binary response, and that 1 indicates the notable event. Assume $X_1, X_2, \ldots, X_p$ are the regressors. Long-standing practice, buttressed by years of application, advocates the assumption that

$$P(Y = 1 | X_1 = x_1, X_2 = x_2, \ldots, X_p = x_p)$$
$$= \frac{e^{a_0+a_1x_1+\ldots+a_px_p}}{1+e^{a_0+a_1x_1+\ldots+a_px_p}} = \frac{1}{1+e^{-(a_0+a_1x_1+\ldots+a_px_p)}} = p(x).$$

13.3 Exercises

The probability model is sigmoidal, and it reaches the value of a half when $a_0 + a_1 x_1 + \ldots + a_p x_p = 0$. In some situations we might predict the binary outcome $y = 1$ if $p(x) > 0.5$, i.e. if $a_0 + a_1 x_1 + \ldots + a_p x_p > 0$. Likewise, we might predict the outcome $y = 0$ if $a_0 + a_1 x_1 + \ldots + a_p x_p < 0$.

The a coefficients are decided by maximizing a likelihood function that helps us assess an accurate probability model. In particular, the a coefficients solve

$$\max_a \prod_{i=1}^{n} p(x_i)^{y_i} (1 - p(x_i))^{1-y_i},$$

where we assume n data points. Notice that we can instead maximize the logarithm of the likelihood since the logarithm is monotonic, which favorably converts the product into a summation. So the a coefficients also solve

$$\max_a \sum_{i=1}^{n} y_i \ln(p(x_i)) + (1 - y_i) \ln(1 - p(x_i)). \tag{13.3}$$

This optimization problem lends itself to several of the optimization algorithms in Chap. 4, but be mindful that we are optimizing with respect to the a coefficients. An initial guess of an optimal solution can be calculated by adjusting the binary sample of Y so that values of 1 become something like $1 - \varepsilon$ and the values of 0 become ε, where ε is some suitably small, positive value. We can then construct the linear model with the transformed sample of Y, denoted by y', as

$$\ln\left(\frac{y'_i}{1 - y'_i}\right) = a_0 + a_1 x_1 + \ldots + a_p x_p.$$

Notice that this equality returns the functional form of the probability model if we solve for y'. We can then use the resulting coefficients to initiate an optimization algorithm such as gradient ascent or Newton's method to solve (13.3).

Consider for illustrative purposes a salesforce working to identify shared characteristics among the clientele who make purchases. They draw a 50-element sample of past customers and collect information about salary, family size, daily temperature, and purchase (0 means no purchase and 1 means a purchase). The results of a linear regression and a logistic regression are depicted in Figs. 13.1, 13.2 and 13.3.

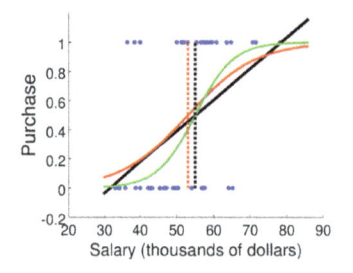

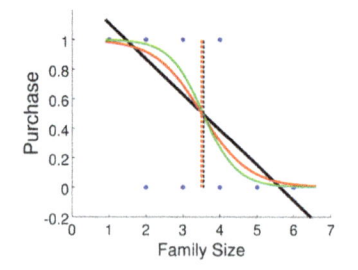

Fig. 13.1 Hypothetical salary versus purchase outcome data.

Fig. 13.2 Hypothetical family size versus purchase outcome data.

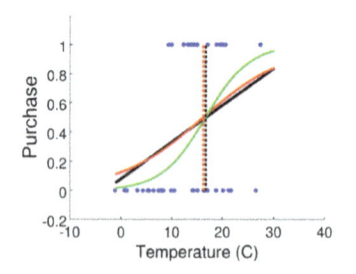

Fig. 13.3 Hypothetical temperature versus purchase outcome data.

In each of Figs. 13.1, 13.2 and 13.3 the 50-element sample is plotted as blue dots, and the black line is a simple linear model. The green sigmoidal function is an initial approximation to the maximum likelihood problem with $\varepsilon = 0.01$, and the red sigmoidal function is the solution to the maximum likelihood problem. The black vertical line depicts where the simple regression model achieves a value of a half, and the red vertical line similarly denotes where the logistic regression model reaches the value of a half.

A common analysis separates the observations into true positives (TP), true negatives (TN), false positive (FP), and false negatives (FN). This classification is relative to the four quadrants defined by one of the vertical lines such that: TPs are in the upper right, TNs are in the lower left, FPs are in the lower right, and FN in the upper left. Tallies of the sample of each type are regularly tabulated as in Table 13.4.

13.3 Exercises

	Predicted (Salary)			Predicted (Fam. Size)			Predicted (Temp.)		
	Purchase		Percentage	Purchase		Percentage	Purchase		Percentage
Observed	0	1	Correct	0	1	Correct	0	1	Correct
Purchase 0	23	6	79.31	24	5	82.76	21	8	72.41
Purchase 1	8	13	61.90	6	15	71.43	12	9	42.86
Pred. %			72.00			78.00			60.00

Table 13.4 Summary of predictions from the linear model based on each of three different regressors.

Predictability is defined as $(TP+TN)/(TP+TN+FP+FN)$, and the fact that the vertical lines differ between the simple and logistic models illustrates how predictability depends on the model of $P(Y=1)$. These tallies are possible with multiple regressors.

The probability model for this exercise is more important than its use to calculate predictability. Your goal is to assign each entity a probability of achieving the notable event. A reasonable task is to find a model that accurately identifies the true "winner" as an entity with a high probability. For instance, can you design a model that historically predicts the MVP of the NBA as one of the top 5 players with the highest probability of doing so?

Appendix A
Matrix Algebra and Calculus

The mathematical and computational medium on which computational science largely resides is the combination of calculus and matrix algebra, and indeed, it would not be egregious to claim that the field of computational science is primarily applied matrix algebra and its associated numerical analysis. Our goal in this appendix is to bridge the gap between a typical Calculus III course and the necessary rudiments of matrix algebra needed to initiate the study of computational science herein. We do not stray significantly into the realm of linear algebra, and we bypass discussions on such topics as (abstract) vector spaces. We instead motivate the use of matrices for analytical pursuits, which is a narrow development that supports a succinct presentation sufficient for our needs. Students are encouraged to continue with coursework in linear algebra and its associated numerical analysis, as continued study will deepen perspectives on nearly all aspects of computational science.

A.1 Matrix Algebra Motivated with Polynomial Approximation

A stalwart of Calculus III is the study of real valued functions of many variables. For example, consider the following function of three variables,

$$f(x, y, z) = x^2 + yz + xy.$$

The gradient of f is the vector

$$\nabla f(x, y, z) = \left\langle \frac{\partial f}{\partial x}, \frac{\partial f}{\partial y}, \frac{\partial f}{\partial z} \right\rangle = \langle 2x + y,\ x + z,\ y \rangle,$$

and the first order Taylor approximation of f at the fixed point (x_0, y_0, z_0) is

$$f(x_0, y_0, z_0) + \left(\frac{\partial f}{\partial x}(x_0, y_0, z_0)\right)(x - x_0)$$

$$+ \left(\frac{\partial f}{\partial y}(x_0, y_0, z_0)\right)(x - y_0) + \left(\frac{\partial f}{\partial z}(x_0, y_0, z_0)\right)(x - z_0)$$

$$= f(x_0, y_0, z_0) + \nabla f(x_0, y_0, z_0) \cdot \langle x - x_0, y - y_0, z - z_0 \rangle$$

$$= f(x_0, y_0, z_0) + \langle 2x_0 + y_0, x_0 + z_0, y_0 \rangle \cdot \langle x - x_0, y - y_0, z - z_0 \rangle,$$

where the last two expressions use the dot product. The last expression nicely aligns with the single variable case in which the first order approximation, i.e. the tangent line, of a function $g(x)$ is

$$g(x_0) + g'(x_0)(x - x_0).$$

The comparison between the single and multiple variable cases suggests that the dot product suffices as a direct replacement for the scalar product in the single variable setting. However, such a replacement isn't obvious if we continue with the second order approximation, and it is this fact from which we launch our study of matrix algebra.

The second order Taylor approximation of $g(x)$ about x_0 is

$$g(x_0) + g'(x_0)(x - x_0) + \frac{1}{2}g''(x_0)(x - x_0)^2,$$

and the question we answer is how to express the quadratic term $g''(x_0)(x - x_0)^2$ in the multivariable case so that it seamlessly reduces to the single variable setting. The natural counterpart of $g''(x)$ for $f(x, y, z)$ is called the Hessian, which is matrix. A matrix is simply a rectangular array of elements from a field, which in our case is always the collection of real numbers $\mathbb{R}$. The size of a matrix is denoted by the number of rows and the number of columns. For example, the following matrix is 2×3,

$$A = \begin{bmatrix} 1 & -1 & 0 \\ 2 & 0 & -1 \end{bmatrix}. \tag{A.1}$$

The elements of a matrix are indexed by subscripts in order of row and then column. So $A_{21} = 2$, and in general A_{ij} is the value in the i-th row and the j-th column. A ":" is used as a wild card and iterates over all possible values. The colon is used to extract row and column vectors, and we have for the matrix above that

$$A_{2:} = (2, 0, -1) \quad \text{and} \quad A_{:3} = \begin{pmatrix} 0 \\ -1 \end{pmatrix}.$$

A.1 Matrix Algebra Motivated with Polynomial Approximation

The Hessian of $f(x, y, z)$ arranges the second partials of f into a 3×3 matrix, and for our example, we have

$$D^2 f(x, y, z) = \begin{bmatrix} \partial^2 f/\partial x^2 & \partial^2 f/\partial y \partial x & \partial^2 f/\partial z \partial x \\ \partial^2 f/\partial x \partial y & \partial^2 f/\partial y^2 & \partial^2 f/\partial z \partial y \\ \partial^2 f/\partial x \partial z & \partial^2 f/\partial y \partial z & \partial^2 f/\partial z^2 \end{bmatrix} = \begin{bmatrix} 2 & 1 & 0 \\ 1 & 0 & 1 \\ 0 & 1 & 0 \end{bmatrix}.$$

Note that the second partials are functions and are thus evaluated at (x, y, z). In this case the functions happen to be constants. For functions with sufficient continuity properties, which we always assume, the Hessian is symmetric, meaning that the matrix remains unchanged if the i, j-th and the j, i-th elements are swapped. The transpose of a matrix is the matrix resulting from such swaps, and the transpose of the 2×3 matrix A in (A.1) is the 3×2 matrix

$$A^T = \begin{bmatrix} 1 & 2 \\ -1 & 0 \\ 0 & -1 \end{bmatrix},$$

where the superscript T indicates the transpose. The Hessian has the property that

$$(D^2 f(x, y, z))^T = D^2 f(x, y, z),$$

which defines the symmetric property.

The Hessian of $f(x, y, z) = x^2 + yz + xy$ is comprised of constants because f is a quadratic. The degree of each term of f is the sum of the exponents of each variable, and hence, the product xy is a degree two term since each exponent is 1. The degree of the multi-variable polynomial f is the maximum degree of the terms. If f had had the term xyz^2, then f would have been at least of degree 4 and the Hessian would have had non-constant entries. A quadratic is selected as a motivational example for a couple of reasons. First, the vast majority of applications in computational science use either the first or second order approximation, and their study is crucial. Second, just as in the single variable case, the matrix algebra extension to the multi-variable setting should result in the second order approximation being the original quadratic function. For instance, for any parameters a, b, and c we have

$$\begin{aligned} g(x) &= ax^2 + bx + c \\ &= g(x_0) + g'(x_0)(x - x_0) + (1/2)g''(x_0)(x - x_0)^2, \end{aligned}$$

independent of the value of x_0. The matrix algebra extension to the multi-variable case must result in this same consistency for a quadratic function of several variables such as f.

The multi-variable extension of $g''(x_0)(x - x_0)^2$ requires an algebra to define something like

$$D^2 f(x_0, y_0, z_0) \langle x - x_0, y - y_0, z - z_0 \rangle^2,$$

but this expression doesn't make sense. For example, if all multiplications became dot products, then we would have

$$f(x_0, y_0, z_0) + \nabla f(x_0, y_0, z_0) \cdot \langle x - x_0, y - y_0, z - z_0 \rangle +$$

$$\tfrac{1}{2} D^2 f(x_0, y_0, z_0) \cdot \langle x - x_0, y - y_0, z - z_0 \rangle \cdot \langle x - x_0, y - y_0, z - z_0 \rangle. \quad (A.2)$$

The first two terms correctly identify the first order approximation, but the third term is algebraically ill-formed. The dot product of the last two vectors is well defined and results in a scalar, but this fact then leads to a dot product between the Hessian matrix and a scalar, which is nonsense since the dot product is only defined for two vectors of equal length. There is a scalar product between a matrix and a scalar that multiplies every element of the matrix by the scalar, but this product is also insufficient since we would then be adding a matrix to scalars to compute the value of the second order approximation.

Expression (A.2) is ill-posed no matter how we re-order the proposed dot products. Hence, we need to consider a new multiplication between matrices and vectors. Although not immediately obvious, the natural definition is commonly called row–column multiplication. In this multiplication each column of the matrix requires a unique entry of the vector. So, if a matrix has 3 columns, then the vector must have three components. We move from writing vectors as sequences sandwiched within $\langle \ldots \rangle$ and write them as column vectors to accommodate this multiplication. So,

$$\langle 3, -2, -3 \rangle \quad \text{is rewritten as} \quad \begin{pmatrix} 3 \\ -2 \\ -3 \end{pmatrix}.$$

All the known operations with vectors from Calculus remain valid. For example, a scalar multiple of a vector multiplies each element of the vector by the scalar, and the dot product multiplies like components and sums them.

The row–column multiplication takes the form

$$\begin{bmatrix} 1 & -1 & 0 \\ 2 & 0 & -1 \end{bmatrix} \begin{pmatrix} 3 \\ -2 \\ -3 \end{pmatrix} = 3 \begin{pmatrix} 1 \\ 2 \end{pmatrix} - 2 \begin{pmatrix} -1 \\ 0 \end{pmatrix} - 3 \begin{pmatrix} 0 \\ -1 \end{pmatrix}$$

$$= \begin{pmatrix} 3 \\ 6 \end{pmatrix} + \begin{pmatrix} 2 \\ 0 \end{pmatrix} + \begin{pmatrix} 0 \\ 3 \end{pmatrix}$$

$$= \begin{pmatrix} 5 \\ 9 \end{pmatrix}.$$

The matrix on the left is 2×3 and the vector on the right is 3×1, and the product is defined because the number of columns of the matrix agrees with

A.1 Matrix Algebra Motivated with Polynomial Approximation

the number of rows, i.e. the number of elements, in the vector. The product wouldn't make sense otherwise. For instance, the proposed product

$$\begin{bmatrix} 1 & -1 & 0 \\ 2 & 0 & -1 \end{bmatrix} \begin{pmatrix} 3 \\ -2 \end{pmatrix} = 3 \begin{pmatrix} 1 \\ 2 \end{pmatrix} - 2 \begin{pmatrix} -1 \\ 0 \end{pmatrix} + ? \begin{pmatrix} 0 \\ -1 \end{pmatrix}$$

is ill-defined because we have no value for "?".

The row–column multiplication between a matrix and a vector extends to define the product of matrices. The matrix–matrix product considers the matrix on the right as a collection of column vectors. So, for example,

$$\begin{bmatrix} 1 & -1 & 0 \\ 2 & 0 & -1 \end{bmatrix} \begin{bmatrix} 2 & 0 \\ 1 & 3 \\ -2 & -3 \end{bmatrix} = \begin{bmatrix} \begin{bmatrix} 1 & -1 & 0 \\ 2 & 0 & -1 \end{bmatrix} \begin{pmatrix} 2 \\ 1 \\ -2 \end{pmatrix}, \begin{bmatrix} 1 & -1 & 0 \\ 2 & 0 & -1 \end{bmatrix} \begin{pmatrix} 0 \\ 3 \\ -3 \end{pmatrix} \end{bmatrix}$$

$$= \begin{bmatrix} -1 & -3 \\ 4 & 3 \end{bmatrix}.$$

Since vectors are matrices with a single column, we also have, for example,

$$\begin{pmatrix} 3 \\ 0 \\ -1 \end{pmatrix}^T \begin{bmatrix} 2 & 0 \\ 1 & 3 \\ -2 & -3 \end{bmatrix} = (3, 0, -1) \begin{bmatrix} 2 & 0 \\ 1 & 3 \\ -2 & -3 \end{bmatrix}$$

$$= \begin{bmatrix} (3, 0, -1) \begin{pmatrix} 2 \\ 1 \\ -2 \end{pmatrix}, (3, 0, -1) \begin{pmatrix} 0 \\ 3 \\ -3 \end{pmatrix} \end{bmatrix}$$

$$= (5, 3).$$

The general definition of the product of two matrices, say the $m \times n$ matrix A and the $n \times p$ matrix B, is the $m \times p$ matrix AB in which

$$[AB]_{ij} = \sum_{k=1}^{n} A_{ik} B_{kj}$$
$$= A_{i:} B_{:j}$$
$$= (A_{i:})^T \cdot B_{:j},$$

where i ranges from 1 to m and j ranges from 1 to p. Importantly, each component of the product AB is a dot product, and hence, matrix–matrix multiplication is calculated by evaluating several dot products.

The product among matrices and vectors allows us to re-address our need to explain the bogus multi-variable expression

$$D^2 f(x_0, y_0, z_0) \langle x - x_0, y - y_0, z - z_0 \rangle^2.$$

A quick dimensional study suggests the correct interpretation. The Hessian is a 3×3 matrix, and the vector is 3×1. The result needs to be a 1×1 scalar, and the only arrangement that results in a scalar is

$$\begin{pmatrix} x - x_0 \\ y - y_0 \\ z - z_0 \end{pmatrix}^T D^2 f(x_0, y_0, z_0) \begin{pmatrix} x - x_0 \\ y - y_0 \\ z - z_0 \end{pmatrix}$$

$$= (x - x_0, y - y_0, z - z_0) \, D^2 f(x_0, y_0, z_0) \begin{pmatrix} x - x_0 \\ y - y_0 \\ z - z_0 \end{pmatrix}.$$

For our example with $f(x, y, z) = x^2 + yz + xy$, we have

$$f(x, y, z) = x^2 + yz + xy$$

$$= f(x_0, y_0, z_0) + (\nabla f(x_0, y_0, z_0))^T \begin{pmatrix} x - x_0 \\ y - y_0 \\ z - z_0 \end{pmatrix}$$

$$+ \frac{1}{2} (x - x_0, y - y_0, z - z_0) \, D^2 f(x_0, y_0, z_0) \begin{pmatrix} x - x_0 \\ y - y_0 \\ z - z_0 \end{pmatrix}$$

$$= x_0^2 + y_0 z_0 + x_0 y_0 + (2x_0 + y_0, x_0 + z_0, y_0) \begin{pmatrix} x - x_0 \\ y - y_0 \\ z - z_0 \end{pmatrix}$$

$$+ \frac{1}{2} (x - x_0, y - y_0, z - z_0) \begin{bmatrix} 2 & 1 & 0 \\ 1 & 0 & 1 \\ 0 & 1 & 0 \end{bmatrix} \begin{pmatrix} x - x_0 \\ y - y_0 \\ z - z_0 \end{pmatrix}.$$

After a bit of tedious multiplication, these equalities can be verified independent of the choice of x_0, y_0, and z_0. Note that the first order term expresses the dot product from calculus as the product of a row vector with a column vector. We have in general that if v and w are vectors of the same length, then

$$\left. \begin{array}{c} v \cdot w \\ \langle \ldots \rangle \text{ form} \end{array} \right\} \text{ is equal to } \left\{ \begin{array}{c} v^T w \\ \text{column vector form.} \end{array} \right.$$

Some restatements from calculus are

$$\|v\| = \sqrt{v^T v}, \quad \cos(\theta) = \frac{v^T w}{\|v\| \|w\|}, \quad \text{and} \quad \text{proj}_v(w) = \frac{v^T w}{v^T v} v,$$

where θ is the angle between v and w. The vectors v and w are orthogonal/perpendicular if and only if $v^T w = 0$.

A.2 Properties of Matrix–Matrix Multiplication

Listing numerous variable arguments in a function's description can be cumbersome, and to streamline notation we let a variable such as x or y represent a vector of variables, with x_i and y_i being the i-th variables. We could let $w = (x, y, z)^T$ so that $w_1 = x$, $w_2 = y$, and $w_3 = z$, which would then allow us to re-write $f(x, y, z)$ as $f(w)$. We don't distinguish between x and $\vec{x}$ and instead simply use x to represent either depending on the context of use. With this notation we have the following.

First Order Approximation The first order approximation of a smooth function f about x_0 is

$$f(x_0) + \nabla f(x_0)^T (x - x_0).$$

Second Order Approximation The second order approximation of a twice smooth function f about x_0 is

$$f(x_0) + \nabla f(x_0)^T (x - x_0) + (1/2)(x - x_0)^T D^2 f(x_0)(x - x_0).$$

Importantly, if x is a scalar variable, then these matrix–vector expressions reduce to the typical results from single variable calculus. For instance, if $f(x) = x^3 + e^x$, then

$$f'(x) = \nabla f(x) = 3x^2 + e^x \text{ and}$$
$$f''(x) = D^2 f(x) = 6x + e^x.$$

Hence, the second order approximation of f about $x_0 = 0$ is

$$\begin{aligned} f(0) + f'(0)x + (1/2)f''(0)x^2 &= f(0) + \nabla f(0)^T x + (1/2)\, x^T D^2 f(0)\, x \\ &= 1 + (1)(x) + (1/2)\, x\, (1)\, x \\ &= 1 + x + x^2/2. \end{aligned}$$

A.2 Properties of Matrix–Matrix Multiplication

Matrix multiplication requires special care as compared with either the scalar or dot products. Most importantly, the matrix product is not commutative, and the order of multiplication is important. If A is $m \times n$ and B is $n \times p$, with $m \neq p$, then AB exists whereas BA is not defined. Even for square matrices the order matters as demonstrated by

$$\begin{bmatrix} 1 & 2 \\ 1 & 1 \end{bmatrix} \begin{bmatrix} 1 & 0 \\ 0 & 0 \end{bmatrix} = \begin{bmatrix} 1 & 0 \\ 1 & 0 \end{bmatrix} \text{ and } \begin{bmatrix} 1 & 0 \\ 0 & 0 \end{bmatrix} \begin{bmatrix} 1 & 2 \\ 1 & 1 \end{bmatrix} = \begin{bmatrix} 1 & 2 \\ 0 & 0 \end{bmatrix}.$$

Matrix addition is defined for matrices of the same size and is commutative. An example is

$$\begin{bmatrix} 2 & -3 \\ 0 & 2 \\ 0 & -2 \end{bmatrix} + \begin{bmatrix} 1 & 0 \\ -1 & -2 \\ 3 & -1 \end{bmatrix} = \begin{bmatrix} 1 & 0 \\ -1 & -2 \\ 3 & -1 \end{bmatrix} + \begin{bmatrix} 2 & -3 \\ 0 & 2 \\ 0 & -2 \end{bmatrix} = \begin{bmatrix} 3 & -3 \\ -1 & 0 \\ 3 & -3 \end{bmatrix}.$$

An illustration of the matrix scalar product is

$$5 \begin{bmatrix} 1 & 0 \\ -1 & -2 \\ 3 & -1 \end{bmatrix} = \begin{bmatrix} 5 & 0 \\ -5 & -10 \\ 15 & -5 \end{bmatrix}.$$

A matrix is diagonal if it is square with the only nonzero elements lying on the main diagonal, which starts at the upper left and ends at the lower right. A lower triangular matrix is a square matrix with elements above the main diagonal being zero, and an upper triangular matrix is a square matrix with every element below the main diagonal being zero. Examples of a diagonal, lower triangular, and upper triangular matrix are, respectively,

$$D = \begin{bmatrix} 2 & 0 & 0 \\ 0 & -2 & 0 \\ 0 & 0 & 3 \end{bmatrix}, \quad L = \begin{bmatrix} 2 & 0 & 0 \\ 1 & -2 & 0 \\ 4 & 0 & 3 \end{bmatrix}, \quad \text{and} \quad U = \begin{bmatrix} 2 & 4 & -2 \\ 0 & -2 & 1 \\ 0 & 0 & 3 \end{bmatrix}.$$

At times these terms are extended to non-square matrices, where the context is clear.

The matrix product is associative, and scalar multiplication is both associative and commutative. Hence, for matrices A, B, and C and scalars α and β we have, assuming appropriate matrix dimensions in each statement, that

1. $ABC = (AB)C = A(BC)$
2. $A(B + C) = A(C + B) = AB + AC = AC + AB$
3. $\alpha A = A\alpha$
4. $\alpha(A + B) = \alpha(B + A) = \alpha A + \alpha B = \alpha B + \alpha A$
5. $(\alpha + \beta)A = (\beta + \alpha)A = \alpha A + \beta A = \beta A + \alpha A$
6. $\alpha \beta A = \alpha(\beta A) = (\alpha \beta)A = (\beta \alpha A) = \beta(\alpha A)$.

In addition, working straight from the definition of the matrix product we can show that the transpose of a product satisfies

$$(AB)^T = B^T A^T.$$

A reasonable question of any product is what is the identity, and for matrices the answer is a square diagonal matrix with ones on the main diagonal. This matrix, of whatever dimension makes sense in the context of its use, is denoted by I. As with any multiplicative identity, multiplication by the identity leaves the matrix unchanged. So,

$$\begin{bmatrix} 1 & 0 & 0 \\ 0 & 1 & 0 \\ 0 & 0 & 1 \end{bmatrix} \begin{bmatrix} 2 & -1 \\ 1 & 1 \\ -2 & 1 \end{bmatrix} = \begin{bmatrix} 2 & -1 \\ 1 & 1 \\ -2 & 1 \end{bmatrix} \quad \text{and} \quad \begin{bmatrix} 2, -1, 3 \end{bmatrix} \begin{bmatrix} 1 & 0 & 0 \\ 0 & 1 & 0 \\ 0 & 0 & 1 \end{bmatrix} = \begin{bmatrix} 2, -1, 3 \end{bmatrix}.$$

A.3 Solving Systems, Eigenvalues, and Differential Equations   455

In both instances the 3×3 identity matrix is appropriate, but notice that these products do not commute due to a conflict in dimensions. For square matrices the identity always commutes. So, provided that A is $n \times n$, we have for the $n \times n$ identity matrix I that

$$AI = IA = A.$$

The additive identity is the matrix of zeros and is denoted simply by 0, the size is again decided by the context of its use. The 0 matrix need not be square, and, for example, we have

$$\begin{bmatrix} 0 & 0 & 0 \\ 0 & 0 & 0 \end{bmatrix} + \begin{bmatrix} 1 & 0 & -1 \\ 1 & 1 & -1 \end{bmatrix} = \begin{bmatrix} 1 & 0 & -1 \\ 1 & 1 & -1 \end{bmatrix}.$$

The vector of all ones is useful and is denoted by e. A sum can then be expressed as a multiplication,

$$\sum_{i=1}^{n} v_i = e^T v,$$

where v is the n-vector whose i-th component is v_i. This notation is adopted, albeit somewhat awkwardly, so that e_i is the vector of zeros except for a one in the i-th position. A particular row, column, or element of a matrix can then be identified through multiplication as illustrated by

$$\begin{bmatrix} 2 & 1 & -1 \\ -1 & 3 & -5 \end{bmatrix} e_2 = \begin{pmatrix} 1 \\ 3 \end{pmatrix}, \quad e_1^T \begin{bmatrix} 2 & 1 & -1 \\ -1 & 3 & -5 \end{bmatrix} = (2, 1, -1),$$

and

$$e_2^T \begin{bmatrix} 2 & 1 & -1 \\ -1 & 3 & -5 \end{bmatrix} e_3 = -5.$$

A.3 Solving Systems, Eigenvalues, and Differential Equations

While the reasons to study linear systems are vast, we use a simple differential equation to motivate the fact that they arise naturally. The motivation being from differential equations is particularly germane to the study of computational science because many computational studies in science and engineering use (variants of) differential equations.

We let x be a vector of n unknown functions of t so that

$$x = \begin{pmatrix} f_1(t) \\ f_2(t) \\ \vdots \\ f_n(t) \end{pmatrix} \quad \text{and} \quad \frac{dx}{dt} = \begin{pmatrix} f_1'(t) \\ f_2'(t) \\ \vdots \\ f_n'(t) \end{pmatrix}.$$

The system of differential equations we consider is

$$\frac{dx}{dt} = Ax - b, \tag{A.3}$$

where A is an $n \times n$ matrix and b is an n-vector, both with constant entries. Suppose that v is a constant vector such that $Av = b$. Then,

$$\frac{d}{dt}v = 0 \quad \text{and} \quad Av - b = 0, \quad \text{which implies that} \quad x = v \quad \text{is a solution.}$$

Constant solutions to the linear system $Ax = b$ are equilibrium solutions to the differential equation.

A.3.1 The Nature of Solutions to Linear Systems

We now have that constant solutions to (A.3) are algebraic solutions to the linear equation $Ax = b$, and we now return to our study of linear algebra. Our algebraic development below does not depend on A being square, and hence, we relax the assumption from our motivating differential equation and let A be $m \times n$ unless otherwise specified. The work of this section is, for the most part, supported by mathematical argument, but toward the end we increasingly assert results without rigor. Since our goal is to motivate and express the requisite elements of matrix algebra to initiate a study in computational science, such assertions appropriately help contain our presentation. Solving linear systems computationally is a main topic of Chap. 2.

We let x_p be a particular solution, meaning that $Ax_p = b$ (x_p would be an equilibrium solution to (A.3)). A homogeneous solution x_h satisfies $Ax_h = 0$, and for any homogeneous solution we have

$$A(x_p + x_h) = Ax_p + Ax_h = b + 0 = b.$$

Hence, each particular solution can be adjoined to any homogeneous solution to generate a new solution. In particular, if x_h is nontrivial, meaning that it isn't the zero vector, then for any scalar α we have $A(\alpha x_h) = \alpha A x_h = 0$. So a single non-trivial homogeneous solution generates an infinite number of homogeneous solutions, all of which can be adjoined to a particular solution.

If $\hat{x}_p$ is a particular solution different than x_p, then

$$A(\hat{x}_p - x_p) = A\hat{x}_p - Ax_p = b - b = 0,$$

and we see that $\hat{x}_p - x_p$ is a homogeneous solution. Since

$$\hat{x}_p = x_p + (\hat{x}_p - x_p),$$

A.3 Solving Systems, Eigenvalues, and Differential Equations

we conclude that any solution to $Ax = b$ is the sum of our original particular solution and a homogeneous solution. This fact leads to an important characterization of the solutions of $Ax = b$.

Theorem 18. *Suppose x_p is a particular solution to $Ax = b$. Then, all solutions are of the form $x_p + x_h$, where x_h satisfies $Ax_h = 0$.*

Theorem 18 and its development have some immediate consequences. First, if $Ax = b$ has a particular solution and a non-trivial homogeneous solution, then there are an infinite number of solutions. If $Ax = b$ has a particular solution but only the trivial homogeneous solution, then the particular solution is the only solution. This invites the question of whether a linear system $Ax = b$ is guaranteed to have a solution. Unfortunately, $Ax = b$ need not always have a solution, e.g.

$$\begin{bmatrix} 1 & -1 \\ -1 & 1 \end{bmatrix} \begin{pmatrix} x_1 \\ x_2 \end{pmatrix} = \begin{pmatrix} 1 \\ 1 \end{pmatrix}$$

has no solution since it forces the quantity $x_1 - x_2$ to be both 1 and -1. Systems that have at least one solution are consistent, whereas systems that have no solution are inconsistent. The following result states the possibilities about the number of solutions to a linear system.

Theorem 19. *The linear system $Ax = b$ must satisfy exactly one of the following:*

- *there is no solution,*
- *there is a single solution and no non-trivial homogeneous solution, or*
- *there are an infinite number of solutions, and subject to knowing a single particular solution x_p, each is uniquely of the form*

$$x_p + x_h,$$

where x_h is a homogeneous solution.

Our next foray characterizes the three situations of Theorem 19 for any particular linear system. Recall that

$$Ax = [A_{:1}|A_{:2}|\ldots|A_{:n}] \begin{pmatrix} x_1 \\ x_2 \\ \vdots \\ x_n \end{pmatrix} = A_{:1}x_1 + A_{:2}x_2 + \ldots + A_{:n}x_n,$$

where the middle expression displays the matrix A as n column vectors. The expression on the right-hand side is a linear combination, and every matrix vector multiplication is a linear combination of the columns of A. Similarly, the row vector resulting from a vector–matrix multiplication is a linear combination of the rows of A.

A collection of vectors is linearly independent if the only possible way to express zero as a linear combination of those vectors is to have each vector be multiplied by zero. In matrix–vector form this is the same as stating that the columns of the matrix A are linearly independent if and only if $x = 0$ is the unique solution to $Ax = 0$. Importantly notice that $x = 0$ is always a solution to $Ax = 0$, but that the columns of the matrix being linearly independent forces this to be the only solution. If the columns of the matrix are not linearly independent, then they are linearly dependent.

The three possibilities of Theorem 19 are intertwined with the concept of linear dependence. If the columns of a matrix are linearly independent, then the system $Ax = b$ must have either no solution or a single solution. This outcome is based entirely on the property of A and is independent of b. If the columns of A are linearly dependent, then the system $Ax = b$ must have either no solution or an infinite number of solutions, an implication that is again decided entirely by A. The role of b is in deciding which of the two possibilities is correct in both instances.

The vector v is a linear combination of the vectors in $\{v_1, v_2, \ldots, v_n\}$ if there are scalars $c_1, c_2, \ldots, c_n$ so that

$$v = c_1 v_1 + c_2 v_2 + \ldots + c_n v_n.$$

In this case we say that v is in the span of $\{v_1, v_2, \ldots, v_n\}$. In terms of the system $Ax = b$, there is a solution if and only if the vector b is in the span of the columns of A. Combining this fact with the consequences of whether or not the columns of A are linearly independent leads to the following characterizations.

- $Ax = b$ has no solution if and only if b is not in the span of the columns of A.
- $Ax = b$ has a unique solution if and only if b is in the span of the columns of A and the columns of A are linearly independent.
- $Ax = b$ has an infinite number of solutions if and only if b is in the span of the columns of A and the columns of A are linearly dependent.

The computational methods of Chap. 2 discuss how to distinguish between these outcomes as a system is solved. However, the terminology of linear (in)dependence and span are important and should be lucid as one studies computational science.

A concept related to linear independence is that of rank, which measures the maximum number of columns of a matrix that form a linearly independent set. If A is $m \times n$ and the rank$(A) = n$, a condition called full column rank, then the columns of A are linearly independent. If rank(A) is instead less than n, then the columns of A form a linearly dependent set. A couple of results about rank are:

- If A is $m \times n$, then rank$(A) \leq \min\{m, n\}$.
- The system $Ax = b$ is consistent if and only if rank$(A) = $ rank$([A|b])$.

A.3 Solving Systems, Eigenvalues, and Differential Equations

The characterizations above can be re-stated in terms of rank.

- $Ax = b$ has no solution if and only if $\text{rank}(A) < \text{rank}([A|b])$.
- $Ax = b$ has a unique solution if and only if $\text{rank}(A) = \text{rank}([A|b])$ and A has full column rank.
- $Ax = b$ has an infinite number of solutions if and only if $\text{rank}(A) = \text{rank}([A|b])$ and A does not have full column rank.

If $\text{rank}(A) = m$, i.e. the number of rows of A, then the matrix has full row rank. A linear system with a full row rank matrix must be consistent for any right-hand side, a fact for which we forego a formal argument. So, if A has full row rank, then $Ax = b$ must have either one solution or an infinite number of solutions for any b.

The case of a square $n \times n$ matrix is of particular interest. In this case the concepts of full column and row rank coincide with $\text{rank}(A) = n$, and the matrix simply has full rank. If A is a square full rank matrix, then $Ax = b$ must have a unique solution for any b. Hence there is a unique solution to

$$Ax = e_i \text{ for each } i = 1, 2, \ldots, n.$$

Let v^i be the solution for each i. Then,

$$A\left[v^1|v^2|\ldots|v^n\right] = \left[Av^1|Av^2|\ldots|Av^n\right] = [e_1|e_2|\ldots|e_n] = I.$$

The matrix $\left[v^1|v^2|\ldots|v^n\right]$ is an inverse since multiplying by it results in the identity, and much like $3 \times (1/3) = 3 \times 3^{-1} = 1$, we use the superscript -1 to denote the inverse. So,

$$A^{-1} = \left[v^1|v^2|\ldots|v^n\right] \text{ and } AA^{-1} = I.$$

If A has an inverse, then A is invertible or nonsingular. Otherwise, A is non-invertible or singular. The juxtaposition of these terms stems from the divide between algebra, in which invertible and non-invertible are common, and analysis, in which singular and nonsingular are common. The terms are used interchangeably, and readers should be deft with their use.

Matrix inverses are important mathematical entities and are used to develop algorithmic procedures. They have many, many properties, of which we state a few. In all of the statements below, the matrices A and B are assumed to be $n \times n$.

- A is invertible if and only if $\text{rank}(A) = n$.
- A is invertible if and only if the columns of A are linearly independent.
- A is invertible if and only if $Ax = b$ has a unique solution for each b.
- If A is invertible, then $AA^{-1} = A^{-1}A = I$.
- If A is invertible, then A^T is invertible and $(A^T)^{-1} = (A^{-1})^T$.
- If A and B are invertible, then AB is invertible and $(AB)^{-1} = B^{-1}A^{-1}$.

The fourth property is particularly useful in notating the solution to $Ax = b$ provided that A is nonsingular. By pre-multiplying on the left by A^{-1} we have
$$x = Ix = A^{-1}Ax = A^{-1}b.$$
So the unique solution to $Ax = b$ is $x = A^{-1}b$ as long as A is invertible. This quick, easily denoted outcome is brilliant mathematics, but it is, unfortunately, often horrid in the realm of computational science due to numerical instabilities and speed. The matrix inverse is a superb mathematical tool and is regularly used to denote solutions, but as Chap. 2 highlights, it is better to solve $Ax = b$ without explicitly calculating A^{-1}.

A.3.2 Eigenvalues and Eigenvectors

We return to our original differential equation (A.3),
$$\frac{dx}{dt} = Ax - b,$$
with A being $n \times n$. The differential equation is homogeneous if $b = 0$, in which case the equilibrium solutions satisfy the homogeneous linear system $Ax = 0$. Just as in the algebraic setting of the previous section, solutions to the differential equation are the sum of a particular and homogeneous solution. To verify that this is indeed true, let x_p and x_h be functions of t such that
$$\frac{dx_p}{dt} = Ax_p - b \text{ and } \frac{dx_h}{dt} = Ax_h.$$
Then,
$$\frac{d}{dt}(x_p + x_h) = \frac{dx_p}{dt} + \frac{dx_h}{dt} = (Ax_p - b) + Ax_h = A(x_p + x_h) - b.$$
We conclude that $x_p + x_h$ is a solution.

The difference between the algebraic problem of solving $Ax = b$ and the analytic problem of solving the differential equation $dx/dt = Ax - b$ is that the former is looking for constant real vectors that simultaneously satisfy each of the n algebraic equations and that the latter is looking for real-valued functions of t that simultaneously satisfy the n differential equations. Even with this difference, the problems mirror each other in many ways, and in particular, we need to study the homogeneous solutions to the differential equation since all solutions are known once we have a single particular solution and a general expression for the homogeneous solution.

The mathematical result that leads to identifying the homogeneous solutions is that

A.3 Solving Systems, Eigenvalues, and Differential Equations

$$\frac{d}{dt} e^{\lambda t} = \lambda e^{\lambda t}.$$

So, if A is the 1×1 matrix $[\lambda]$, then $x_h(t) = e^{\lambda t}$ is a homogeneous solution. Moreover, for any scalar α we have

$$\frac{d}{dt} \alpha e^{\lambda t} = \lambda \alpha e^{\lambda t},$$

and hence, $\alpha e^{\lambda t}$ is also a homogeneous solution. Indeed, it can be argued that all homogenous solutions have this form, and hence, all solutions to

$$\frac{dx}{dt} = 3x - 1,$$

look like $x_p + \alpha e^{3t}$, where x_p is any particular solution yet to be found. The straightforward observation that $(d/dt)1/3 = 3(1/3) - 1 = 0$ shows that $x_p = 1/3$ works, making all solutions of the form $1/3 + \alpha e^{3t}$.

The fact that the 1×1 case extends nearly identically to the $n \times n$ case is striking. Observe that if v is a vector of constants, then

$$\frac{d}{dt} e^{\lambda t} v = \lambda e^{\lambda t} v \stackrel{?}{=} A(e^{\lambda t} v),$$

where the last equality is a puzzle to solve. If we can satisfy the proposed equality, then we have identified homogeneous solutions. Since $e^{\lambda t}$ is positive for any λ and t, the suggested equality is valid for a nonzero solution if and only if

$$Av = \lambda v, \text{ with } v \neq 0.$$

Any λ for which there exists a nonzero vector v satisfying this equation is called an eigenvalue, and any nonzero v for which the equality holds is an eigenvector for that particular λ.

The equation $Av = \lambda v$ is worthy of a protracted study, and even with modern mathematics and computation, efficiently and accurately calculating solutions to this equation remains an important area of study, especially for large matrices. Calculating eigenvalues is regularly introduced in numerical analysis or numerical linear algebra, and students are encouraged to continue with such coursework. Our computational needs do not require a broad knowledge of how to calculate eigenvalues, but much of our development in the first part of the text rests on an understanding of a few key elements that routinely appear in applied mathematics. Here is a list of facts of which any student of computational science should be familiar.

- An $n \times n$ matrix with real entries has n, possibly complex and possibly non-unique, eigenvalues.
- The eigenvalues of a diagonal, upper triangular, or lower triangular (square) matrix are the entries along the main diagonal.

- The collection of eigenvectors for each distinct eigenvalue (together with the zero vector) form an eigenspace.
- A real, symmetric matrix has real eigenvalues.
- The homogeneous equation $Ax = 0$ has a nontrivial solution if and only if 0 is an eigenvalue.
- A is invertible if and only if 0 is not an eigenvalue.
- If A is invertible, then λ is an eigenvalue of A if and only if $1/\lambda$ is an eigenvalue of A^{-1}.
- A and A^T share the same eigenvalues.
- For a symmetric matrix A, the product $v^T A v$ is positive for every $v \neq 0$ if and only if the eigenvalues of A are positive.

The algebraic importance of the eigenspaces is that they are invariant under multiplication of the matrix. For example, if v is an eigenvector for λ, then Av is an eigenvector for the same λ, which follows since

$$Av = \lambda v \text{ implies } A(Av) = \lambda(Av).$$

We complete our discussion of eigenvalues and eigenvectors by solving the homogeneous 2×2 system of differential equations,

$$\frac{dx}{dt} = \begin{bmatrix} -1 & 2 \\ 0 & -2 \end{bmatrix} \begin{pmatrix} x_1 \\ x_2 \end{pmatrix}.$$

From the third property above, the eigenvalues are -1 and -2. If $\lambda = -1$, then $Av = \lambda v$ is the same as

$$(A - \lambda I) v = 0 \Leftrightarrow \begin{bmatrix} 0 & 2 \\ 0 & -1 \end{bmatrix} \begin{pmatrix} v_1 \\ v_2 \end{pmatrix} = \begin{pmatrix} 0 \\ 0 \end{pmatrix}.$$

Any nonzero solution is an eigenvector for $\lambda = -1$, and by inspection we select $v = (1, 0)^T$. Similarly, if $\lambda = -2$, then

$$(A - \lambda I) v = 0 \Leftrightarrow \begin{bmatrix} 1 & 2 \\ 0 & 0 \end{bmatrix} \begin{pmatrix} v_1 \\ v_2 \end{pmatrix} = \begin{pmatrix} 0 \\ 0 \end{pmatrix}.$$

In this case we select $v = (-2, 1)^T$ as the eigenvector. From the discussion above we know that any scalar multiple of a solution is again a solution, and a simple calculation further shows that the sum of two solutions is a solution. Hence,

$$x = \alpha_1 e^{-t} \begin{pmatrix} 1 \\ 0 \end{pmatrix} + \alpha_2 e^{-2t} \begin{pmatrix} -2 \\ 1 \end{pmatrix}$$

is a solution for any selection of α_1 and α_2. Moreover, since the eigenvalues are distinct, all possible solutions have this form.

A.4 Some Additional Calculus

Matrix algebra is regularly used to define functions, and gaining comfort with differentiation in this setting is helpful. Consider the real valued function of n-variables
$$f(x) = a + b^T x + x^T A x,$$
where A is symmetric. We first demonstrate that the gradient of this function is
$$\nabla f(x) = b + 2Ax,$$
a result that should be recognized as something similar to the power rule. To illustrate, let
$$a = 5, \quad b = \begin{pmatrix} 1 \\ 2 \end{pmatrix}, \quad \text{and} \quad A = \begin{bmatrix} 3 & 1 \\ 1 & 5 \end{bmatrix}.$$

Then
$$f(x_1, x_2) = 5 + (1, 2) \begin{pmatrix} x_1 \\ x_2 \end{pmatrix} + (x_1, x_2) \begin{bmatrix} 3 & 1 \\ 1 & 5 \end{bmatrix} \begin{pmatrix} x_1 \\ x_2 \end{pmatrix}$$
$$= 5 + x_1 + 2x_2 + 3x_1^2 + 2x_1 x_2 + 5x_2^2.$$

The first partials of f are
$$\frac{\partial f}{\partial x_1} = 1 + (2 \times 3x_1) + 2x_2, \quad \text{and} \quad \frac{\partial f}{\partial x_2} = 2 + 2x_2 + (2 \times 5x_2),$$
which means that the gradient is
$$\nabla f(x_1, x_2) = \begin{pmatrix} 1 + (2 \times 3x_1) + 2x_2 \\ 2 + 2x_2 + (2 \times 5x_2) \end{pmatrix} = \begin{pmatrix} 1 \\ 2 \end{pmatrix} + 2 \begin{bmatrix} 3 & 1 \\ 1 & 5 \end{bmatrix} \begin{pmatrix} x_1 \\ x_2 \end{pmatrix}.$$

The scalar multiple of 2 on the matrix is clearly the direct result of the power rule applied to the squared terms $3x_1^2$ and $5x_2^2$. However, the multiple of 2 is less clear for the cross term $2x_1 x_2$. What is important to notice is that the coefficient 2 is the sum of the off diagonal elements, i.e. $2 = 1 + 1$. When we express the gradient in terms of matrix multiplication, we only collect one of these terms, which means we have to compensate by again multiplying by 2. The symmetry of the matrix is what makes this calculation correct. Unfortunately, many alternative, non-symmetric matrices result in the same function, but they do not differentiate so easily. For example, if
$$B = \begin{bmatrix} 3 & 3 \\ -1 & 5 \end{bmatrix},$$

then f is still $a + b^T x + x^T B x$, but in this case we have

$$\nabla f(x) \neq b + 2 B x.$$

The Hessian of $f(x) = a + b^T x + x^T A x$ is found by differentiating the gradient,

$$D(b + 2 A x) = 2 A.$$

This result should again should be recognized as an extension of the power rule. In general, applying the differential operator D to any linear function Qx, where Q need not be symmetric or square, results in the constant matrix Q. In this case the derivative is the Jacobian of the function $g(x) = Qx$, which has n independent variables and m dependent variables, assuming that Q is $m \times n$.

If $f(x)$ is approximated by a second order Taylor polynomial about x_0, then for x near x_0 we have

$$f(x) \approx f(x_0) + \nabla f(x_0)^T (x - x_0) + \frac{1}{2}(x - x_0)^T D^2 f(x_0)(x - x_0).$$

Although this is an approximation of f near x_0, it is important to recognize that the approximation is, up to the second order, exact at x_0. In other words,

$$f(x_0)$$
$$= f(x_0) + \nabla f(x_0)^T (x - x_0) + \frac{1}{2}(x - x_0)^T D^2 f(x_0)(x - x_0) \Big|_{x=x_0},$$

$$\nabla f(x_0)$$
$$= D\left(f(x_0) + \nabla f(x_0)^T (x - x)_0 + \frac{1}{2}(x - x_0)^T D^2 f(x_0)(x - x_0)\right)\Big|_{x=x_0}$$
$$= \nabla f(x_0) + D^2 f(x_0)(x - x_0)\Big|_{x=x_0}, \quad \text{and}$$

$$D^2 f(x_0)$$
$$= D^2 \left(f(x_0) + \nabla f(x_0)^T (x - x)_0 + \frac{1}{2}(x - x_0)^T D^2 f(x_0)(x - x_0)\right)$$
$$= D\left(\nabla f(x_0) + D^2 f(x_0)(x - x_0)\right).$$

Index

A
acceptance probability, 165
adaptive methods, 212
affine, 7, 34, 149
Ahmdal's Law, 292
alternative hypothesis, 90
arc, 259
autocorrelation, 119

B
backward regression, 435
backward substitution, 37, 44
Belousov-Zhabotinsky reaction, 226
Bessel's correction, 78, 411
bisection, 4, 131, 140
Boltzmann constant, 419
boundary condition
   Dirichlet, 391
   Neumann, 392
bracketing method, 4
BFGS algorithm, 129
Broyden, Fletcher, Goldfarb, and Shanno algorithm, 129
BZ reaction, 226

C
central difference, 392
Central limit theorem, 80
chemical equation
   second-order, 371
   third-order, 371
Cholesky factorization, 47, 57
   inner product algorithm, 48
   outer product algorithm, 48
coefficient matrix, 69

coefficient of determination, 89
Collatz Conjecture, 284
collimator, 327
compiled language, 270
complementarity, 147
component functions, 33
concave, 149
confidence interval, 94
conjugate (vectors), 46
conjugate directions, 50
conjugate transpose, 319
constraint qualification, 146
continuity equation, 391
convergence order, 17
   linear convergence, 17
   quadratic convergence, 17
convex, 148, 420
   function, 148
   optimization problem, 149
   set, 148
convolution, 312
convolution, operator, 310
cooling schedule, 166
core, 277
correlation, 82, 410
   negative, 82
   positive, 82
   uncorrelated, 82
Coutte flow, 355
covariance, 82, 410
covariance matrix, 104, 105, 410
critical exponents, 427
cross product (of sets), 240
crossover, 171

cubic spline, 98
Curie point, 418

D
Dantzig, George, 154
DDE, 377
deflation method, 29
degrees-of-freedom, 95
delay differential equation, 377
　lag, 377
　neutral, 377
　retarded, 377
dependent variable (regression), 83
DFT, 309, 310, 315
Dirichlet boundary condition, 391
discrete Fourier transform, 309, 310, 315
discrete random variable, 76
discrete time signal, 311
dose-volume histogram, 337
doublet, 341
drift parameter, 404
dual (optimization problem), 139
dual linear program, 153
duality theorem of linear programming, 153
dummy variables, 441
Durbin-Watson test statistic, 119
DVH, 337

E
efficient frontier, 404, 411
efficient portfolio, 403
efficient solution, 410
eigenspace, 105, 462
eigenvalue, 105, 125, 461
eigenvector, 105, 125, 461
elementary matrix, 38
embedded ODE solver, 215
energy, 419
entropy, 419
equilibrium solution, 232
error
　regression, 83
　Type 1, 90
Euler's formula, 316
Euler's method, 192
　backward, 194
　forward, 192
　improved, 197
　symplectic, 235
executable (program), 270
explained variance, 89
explicit method, 194
exponential model, 219

F
F-test, 435
factorization, 36, 105
　Cholesky, 47, 57
　LU, 36
　spectral, 105, 111
　SVD, 111
fast Fourier transform, 310, 320
feasible set, 136
feature, 110
feature vector, 110, 112
Fehlberg, Erwin, 215
FFT, 310, 320
filter
　ideal high pass filter, 325
　ideal band pass filter, 325
filter, linear time invariant, 310
First order necessary condition of optimality, 124, 137
first principal component, 108
fitness, 171
fluence, 327
forward difference, 392
forward regression, 435
forward substitution, 37, 44
Fourier transform
　discrete Fourier transform (DFT), 309, 310, 315
　fast Fourier transform, 320
　fast Fourier transform (FFT), 310
Fourier, Joseph, 307
fractal, 293
　Mandelbrot set, 293
Freund, Leopold, 326

G
Galilei, Galileo, 307
Gauss, Carl Friedrich, 88
Gauss-Markov Theorem, 88
general solution (ODE), 220
genetic algorithms, 171, 238
Gibbs distribution, 419
Gibbs, John, 420
global error (ODE solvers), 204
global optimization, 164
GPU, 292
greedy neighborhood search, 166

H
Hörner's method, 16, 272
Hamiltonian, 420, 422
Healthcare Models
　insulin absorption, 365
　radiobiological transport, 329
　SIR model, 379

Index

heat capacity, 425
heat equation, 389, 391
Hermitian transpose, 319
Hessian, 125, 448
heuristic search, 164
Hilbert, David, 307
hill-climbing, 165
histogram, 116
Hubble, Edwin, 67
hypervolume, 239
hypothesis test, 90

I
ideal band pass filter, 325
ideal high pass filter, 325
ill-condition (matrix), 72
implicit method, 194
incumbent solution, 164
independent (sample), 79
independent variable (regression), 83
infimum, 124
influential point, 119
inner product, 45, 46
inner product Cholesky algorithm, 48
input (regression variables), 83
intensity modulated radiotherapy (IRMT), 326
interaction variables, 439
interior-point algorithm, 154
Intermediate Value Theorem, 4
interpolant, 7
inverse (matrix), 56
Ising model, 418
Ising, Ernst, 418
Istrail, Sorin, 418
Itō's Lemma, 406

J
Jacobian, 53, 125, 464

K
Karmarkar, Narendra, 154
KKT conditions, 147
Karush, Kuhn, Tucker conditions, 147
Khachiyan, Leonid, 154
Knuth, Donald, 123
Kutta condition, 346
Kutta, Martin, 198
Kutta-Joukowski Theorem, 347

L
lag, 377
Lagrange multipliers, 137
Lagrange polynomial, 73, 131

Lagrange, Joseph, 137
Lagrangian, 137, 420
    dual, 139
Laplace's equation, 341
Laplacian, 340, 391
Law of large numbers, 79
least squares (method), 67
least squares (ordinary), 68
least squares model, 67, 68
least squares solution, 56
left division, 55
leverage, 119
lift, 347
line search, 131
linear combination, 458
linear convergence, 17
linear interpolation, 7, 131
linear program, 151
linear regression, 74, 427
linear time invariant filters, 310
linearly ordered, 123
load balancing, 282
local truncation error, 202
logistic regression, 442
lossy (compression), 111
LU factorization, 36

M
Mandelbrot set, 293
Markov, Andrey, 88
matrix factorization, 36, 105
    Cholesky, 47, 57
    LU, 36
    spectral, 105, 111
    SVD, 111
matrix inverse, 56
mean, 77
Mean value theorem, 9
Mean value theorem of integrals, 245
metaheuristic, 165
method of least squares, 67
metric, 45, 47
mex-file, 272
minimum, 124
model (least squares), 67
model building (regression), 431
Monte Carlo integration, 239
Monte Carlo sampling, 424
Monte Carlo simulation, 239
multi-start (algorithm), 164
multicolinearity, 432
multicore processor, 276
multistart method, 299
mutation, 171

N

N-body problem, 233
natural spline, 101
negative correlation, 82
neighborhood, 166
network flow, 259
Neumann boundary condition, 392
Neumann, John von, 137, 144
neutral DDE, 377
Newton step, 15, 54, 135
Newton's Law of Cooling, 225
Newton's Method (optimization), 129
Newton's Method (single equations), 13
Newton's method (systems), 53
Newton-Cotes, 243
Newtonian fluid, 356
non-Newtonian fluid, 356
norm, 45, 46
normal equations, 69
normally distributed, 76
numerical integration, 239

O

observable, 424
offspring, 173
Onsager, Lars, 418
order (ODE solver), 206
order of convergence, 17
ordinary least squares, 68
Oregonator model, 226
orthonormal, 105
outer product Cholesky algorithm, 48
outlier, 119
output (regression variable), 83

P

p-value, 92
parallel algorithm
  load balancing, 282
  pipelined, 288
parallel programming, 277
Pareto set, 411
Pareto surface, 404
partial differential equation, 389
  central difference, 392
  Dirichlet boundary condition, 391
  explicit, 392
  forward difference, 392
  heat equation, 389
  Laplacian, 391
  Neumann boundary condition, 392
  wave equation, 397
partial pivoting, 39
partially mapped crossover, 175

partition, 241
PDE, 389
Pearson correlation, 82
pencils, 327
pendulum model
  double pendulum, 231
  linear, 227
  non-constant length, 229
  nonlinear, 227
periodic signal, 313
permutation matrix, 44, 57
permuted matrix, 43, 57
pharmacokinetics, 363
phase transition, 418
pipelined, 288
pivot, 37
portfolio
  efficient, 403
portfolio selection, 408
  weight, 408
positive correlation, 82
positive definite (matrix), 44
positive semidefinite (matrix), 148
predator-prey model, 225
prediction, 83
prediction interval, 94
primal (optimization problem), 140
primal linear program, 153
principal component, 108
principal component analysis, 103
probability density function, 76
probability function, 75
probability mass function, 76
probability plot, 118
projection, 107
proportionate selection, 173
pseudo-steady-state, 376

Q

Q-conjugate
  directions, 50
  vectors, 46
quadratic convergence, 17
quadratic program, 151
quadrature, 239
qualitative data, 437
quasi-Newton method, 129
quotient (polynomial), 16

R

R^2, 89
radiotherapy, 326
random path, 405

Index

random variable, 75
  event, 75
  outcome, 75
  sample space, 75
random walk, 405
randomized algorithm, 238
Rayleigh-Ritz quotient, 108
regression, 74
  backward, 435
  forward, 435
  linear, 74
  logistic, 442
  stepwise, 435
regressors, 83
remainder (polynomial), 16
residual, 51, 68, 84
residual analysis, 87
response (variable), 83
retarded DDE, 377
rheopectic fluid, 357
Richardson extrapolation, 212
Riemann sum, 241
Riemann, Bernhard, 241
Roentgen, Conrad, 326
rotation matrix, 348
Runge, Carl, 198
RK2, 198
RK4, 199
Runge-Kutta methods, 198, 199
  RK2, 198
  RK4, 199
Runge-Kutta-Fehlberg, 215
RKF, 215

S

saddle point, 144
sample mean, 78
sample path, 405
sample space, 75
scatter plot, 118
scripting language, 270
secant condition, 130
secants (method of), 8
Second order sufficient condition of optimality, 125, 137
second principal component, 108
sectional lift, 347
selection (of a mating cohort), 173
separation of variables, 219
serial time, 279
shear rate, 356
shear stress, 355
Sherman-Morris formula, 130
Sieve of Eratosthenes, 301

signal, 310
  discrete time, 311
  periodic, 313
signal processing, 310
similar (matrices), 47
simple linear regression, 85
simplex algorithm, 154
Simpson's rule, 239
simulated annealing, 165, 238
simulation, 238
single nucleotide polymorphism, 172
singular value decomposition, 111
SVD, 111
sink (solution of Laplace's equation), 341
SIR model, 379
slack variables, 161
Slater's interiority condition, 178
source (solution of Laplace's equation), 341
span, 458
spectral decomposition, 105, 111
spectral factorization, 105, 111
spectrum, 111
spline, 98
  cubic, 98
  natural, 101
standard deviation, 77
statistic, 77
status variable, 7
steady state, pseudo, 376
steepest ascent, 124
steepest descent, 124, 129, 132, 140
step-and-shoot, 327
stepwise regression, 435
stiff differential equations, 207
stochastic differential equation, 404
strictly convex, 125
strong duality, 140
studentized residuals, 119
surplus variables, 161
susceptibility, 425
SWIG, 273
symplectic Euler's method, 235
synthetic division, 16

T

t-distribution, 91
Taylor expansion, 19, 131
Taylor's Theorem, 19
termination criteria, 6
test statistic, 91
thixotropic fluid, 357
time delay, 377
time series, 118
total squared error, 90

total variance, 89
trade-off (risk and return), 410
traveling salesperson problem, 168
TSP, 168
tridiagonal matrix, 63, 65, 101
Type 1 error, 90

U
uncorrelated, 82
uniformly distributed, 76
unitarily similar, 47

V
variance, 77

verification index factor, 433
VIF, 433
volatility parameter, 404

W
wave equation, 397
weak duality, 140, 144
Weierstrass, Karl, 19
Weiner process, 404
wrapper file, 273

Y
yield stress, 357